CRC Series
in
MARINE SCIENCE

Editor-in-Chief
J. Robert Moore
Director, Marine Science Institute
The University of Texas
Austin, Texas

CRC HANDBOOK OF MARINE SCIENCE
Editors
F. G. Walton Smith
Dean Emeritus, School of Marine and Atmospheric Science
University of Miami
President, International Oceanographic Foundation
Miami, Florida
Frederick A. Kalbler
President, Hydrobiological Services, Inc.
Naranja, Florida

CRC HANDBOOK OF GEOPHYSICAL EXPLORATION AT SEA
Editor
Richard A. Geyer
Professor Emeritus
Department of Oceanography
Texas A & M University
College Station, Texas

CRC HANDBOOK OF MAN AND THE MARINE ENVIRONMENT
Editor
Robert A. Ragotzkie
Director, Sea Grant Institute
University of Wisconsin
Madison, Wisconsin

CRC HANDBOOK OF MARINE SCIENCE: COMPOUNDS FROM MARINE ORGANISMS
Authors
Joseph T. Baker
Vreni Murphy
Roche Research Institute of Marine Pharmacology
Sydney, Australia

CRC HANDBOOK OF COASTAL PROCESSES AND EROSION
Editor
Paul D. Komar
School of Oceanography
Oregon State University
Corvallis, Oregon

CRC HANDBOOK OF MARICULTURE
Editor
James P. McVey
Chief, National Marine Fisheries Service
Galveston Laboratory
Galveston, Texas

CRC Handbook of Geophysical Exploration at Sea

Editor

Richard A. Geyer, Ph.D.
Professor Emeritus
Department of Oceanography
Texas A & M University
College Station, Texas

Editor-in-Chief
CRC Marine Science Series
J. Robert Moore, Ph.D.
Professor
Marine Science Institute
The University of Texas
Austin, Texas

CRC Press, Inc.
Boca Raton, Florida

Library of Congress Cataloging in Publication Data

Main entry under title:

Handbook of geophysical exploration at sea.

(CRC series in marine science)
Includes bibliographies and index.
1. Ocean bottom—Handbooks, manuals, etc.
2. Prospecting—Geophysical methods—Handbooks, manuals, etc. I. Geyer, Richard A. II. Series.
GC87.H36 551.46'08'028 82-1316
ISBN 0-8493-0222-6 AACR2

This book represents information obtained from authentic and highly regarded sources. Reprinted material is quoted with permission, and sources are indicated. A wide variety of references are listed. Every reasonable effort has been made to give reliable data and information, but the author and the publisher cannot assume responsibility for the validity of all materials or for the consequences of their use.

Direct all inquiries to CRC Press, Inc., 2000 Corporate Blvd., N.W., Boca Raton, Florida, 33431.

International Standard Book Number 0-8493-0222-6

Library of Congress Card Catalog Number 82-1316
Printed in the United States

PREFACE

There is a great diversity of pertinent scientific areas comprising the broad subject of "Geophysical Exploration at Sea", but it can be classified into two major categories, namely, (a) potential field methods, as represented for example, by gravity, magnetic, electrical, and telluric current methods; and (b) those based on elastic wave theory, as it applies to the generation, transmission, reception, and display of acoustic and seismic waves, as they propagate through the sea and into unconsolidated and consolidated ocean botton sediments.

Each of these various methods depends for its success on the ability to measure certain basic physical properties of the media under investigation, such as the density for gravity, susceptibility for magnetics, resistivity/conductivity for electrical and telluric currents; and velocity and acoustic impedance properties for seismic methods. Actually, it is the differences in these properties, between the geologic target to be studied and the surrounding material that determines the degree of success in the solution of specific, pure, or applied scientific problems being studied on, in, or beneath the bottom of the sea.

A problem is never completely solved unless the location in which the survey is being conducted is known to the degree of accuracy required for the particular problem under investigation. Therefore, it is imperative that accurate navigational methods are available, especially in the open ocean where landmarks are absent. Thus it is not surprising to find that the very first chapter in the book goes into considerable detail in describing the role of satellites in providing accurate navigational information at sea.

The two basic categories of potential field and seismic-acoustic methods can be subdivided further, for purposes of discussion, into two subcategories, namely, (1) the design of specialized instrumentation to obtain data in the marine environment; and (2) the development of theoretical and applied techniques to be used successfully to interpret definitively the results of geophysical exploration at sea. These can be used to improve our understanding of various marine phenomena and processes, as they apply to the structure, composition, and geologic history of the sea, as well as exploring for its diversified resources that lie in and beneath it. It should also be emphasized that much of the information thus obtained can also be applied to help solve comparable scientific problems pertaining to the remaining 21% of the solid earth that can be studied from the land.

Participation was solicited from members of the academic, governmental and industrial communities to provide chapters describing the results of their most recent, basic and applied research and operational methods in marine geophysics. Each in its own right has important information which when shared becomes mutually advantageous. The response from these three important sectors was most gratifying and has resulted in achieving the objectives of this volume. Information in the form of case histories was also included because these provide examples of the integrated approach to the solution of special problems indigenous to methods, geographic areas, and marine geologic provinces. Seismic-acoustic methods have also been emphasized in this book because they have and will continue to dominate both pure and applied marine geophysical activities. In addition, they have in general exhibited a greater versatility and degree of resolution in solving a wide range of geophysical and geologic problems on both a regional and local scale than the potential methods.

The design of specialized instruments and instrumentation systems has also been emphasized because of the special environmental hazards encountered in the sea, as well as the broader and more efficient operational opportunities provided by conducting surveys in this medium as compared with similar ones on land. For example, continuous seismic surveying cannot be conducted on land because the seismic hydrophone cables cannot be moved continuously on land as they are at sea. The same applies to electrical and telluric current surveys, and to a lesser extent to gravity and magnetic studies. For magnetic studies, however, these can be conducted continuously from airplanes flying at low elevations over the land. Gravity surveys can be made continuously on land, but generally, must be confined to roads, if accurate results are to be obtained.

Geophysical research and operational activities are conducted at sea by citizens of many nations. Therefore, efforts were also made to obtain participation by representatives from as many different countries and parts of the world ocean as possible. Thus there are chapters written by scientists from Egypt, Canada, and the United Kingdom, as well as from the United States. Similarly, chapters are included describing research and operational methods conducted in the Atlantic and Pacific Oceans, both in shallow (continental shelf) as well as deep water, and in the Gulf of Mexico, Gulf of Alaska, and the Gulf of Suez.

The three individual sections are designed to present (a) a broad overview of the major operational geophysical methods used to obtain accurate information and to definitively interpret the results of various types of geophysical exploration at sea and (b) to provide an insight into the special needs for specific instrumentation to accurately yield the necessary data; while at the same time operating in the hostile marine environment, in terms of pressure, corrosion effects, and rapid water movement. As stated earlier, the third section, Case Histories, was included together with regional studies, as a means of presenting an integrated analysis of the solution to a particular question.

Specific chapters are provided in each section describing the results of recent diversified research and operational techniques that can be applied to markedly improve data-gathering, as well as interpretation and display procedures. These results in turn can help to solve an ever-increasingly broader spectrum of theoretical and applied problems. Thus, this leads to an improved understanding of our geological knowledge of the earth beneath the sea, which in turn can be used to search for the much needed resources contained therein.

Richard A. Geyer

EDITOR-IN-CHIEF

The Editor-in-Chief of the CRC Marine Science Series, **Professor J. Robert Moore**, received his B.S. degree (honors) from the University of Houston, in geology, his M.A. from Harvard University, also in geology, and his Ph.D. from the University of Wales (Aberystwyth), in geology and oceanography. From 1956 to 1966, he was a Senior Scientist at the Texaco Research Laboratories, Houston, Texas, engaged in research on marine sedimentation and geo-resources. From 1966 to 1977, Dr. Moore was Professor of Geology at the University of Wisconsin, Madison, where he was also Director of the Marine Research Laboratory and the Marine Mining Program. In 1977, he became Director, Institute of Marine Science, University of Alaska, and in 1979, assumed his present affiliation as Professor of Marine Studies, at the University of Texas, at Austin. Professor Moore's principal research interests are in seafloor mineral resources, marine geology, and global oceanographic processes and resources. His research has involved studies in the Bering Sea, Irish Sea, Western Pacific, Gulf of Mexico, Gulf of Alaska, Atlantic shelf and several coastal areas in the United States and abroad. He is a member of the Marine Board of the National Academy, the A.A.P.G. Marine Geology Committee, the N.A.C.O.A. Marine Mining Panel, Organizing Chairman of the CHEMRAWN IV Conference, and member of The Geochemical Society, Soc. of Economic Paleontologists and Mineralogists, Marine Technology Society, Am. Assoc. of Petroleum Geologists, Challenger Society, and a Fellow of the Geological Society of London. He is Editor of the *Marine Mining Journal*, and Exec. Secretary of the International Marine Mining Association. Since 1966, he has been an active consultant to major international corporations, chiefly on matters related to ocean resources, utilization, exploration, and development. He is currently conducting two major marine research programs, one in Palauan waters of the Western Pacific and the other in the Bering Sea off Alaska. He resides in Austin, Texas.

THE EDITOR

Dr. Richard A. Geyer has had a diversified career in geophysics and oceanography in both the industrial and academic sectors. He received his B.S. in Geology, in 1937 from New York University and his Ph.D. in Geophysics from Princeton University in 1951. Beginning at the Woods Hole Oceanographic Institute (1943-1946), he continued in oceanography as head of the Oceanographic Section of Humble Oil Co, Technical Director of Oceanography at Texas Instruments, and finally as Head of the Department of Oceanography at Texas A&M University (1966-1978). His activities in geophysics include periods when he was Manager of the Gravity and Magnetic Departments of Geophysical Service, Inc., Adjunct Professor of Marine Geology & Geophysics at the University of Houston, and is presently consulting in Exploration Geophysics.

He has received national and international recognition of his efforts in these two fields by giving invited papers; and was Vice Chairman of a presidentially appointed Committee on Marine Science, Engineering and Resources (1967-1969). He was awarded an Honorary Member of the Society of Exploration Geophysicists in 1981 and of the Dallas Geophysical Society in 1980, of which he was President in 1964-65.

Dr. Geyer has also had extensive experience as an editor of several books on marine pollution and the use of submersibles in oceanography and ocean engineering, and of the journal, *Geophysics*, together with publishing 28 papers in geophysics and oceanography. In addition, he was co-director of a NASA sponsored project (1961-1962) to determine what geological and geophysical experiments, and in what order, these were to be performed by the astronauts upon reaching the moon.

CONTRIBUTORS

A. I. Bayoumi, Ph.D.
Chairman and Head
Geology Department
Faculty of Science
Sana'a University
Yemen Arab Republic

Richard H. Bennett, Ph.D.
Senior Research Oceanographer
Principal Investigator for Marine Geotechnical Studies
Naval Ocean Research and Development Activity
Sea Floor Division
NSTL Station, Mississippi

Albert J. Berni, Ph.D.
Ph.D. Senior Research Engineer
Shell Development Company
Bellaire Research Center
Houston, Texas

Grant Blackinton
Design Engineer
Hawaii Institute of Geophysics
University of Hawaii
Honolulu, Hawaii

W. R. Bryant
Professor
Department of Oceanography
Texas A & M University
College Station, Texas

John T. Burns
Electronics Technician
Naval Ocean Research and Development Activity
Sea Floor Division
NSTL Station, Mississippi

Fred K. Duennebier
Associate Geophysicist
Hawaii Institute of Geophysics
University of Hawaii
Honolulu, Hawaii

Martin G. Fagot
Electronics Engineer
Naval Ocean Research and Development Activity
NSTL Station, Mississippi

George L. Freeland, Ph.D.
Marine Geologist
National Ocean and Atmospheric Administration
Atlantic Oceanographic and Meteorological Laboratories
Miami, Florida

Norman Gholson, Ph.D.
Electronics Engineer
Naval Ocean Research and Development Activity
NSTL Station, Mississippi

Gustave L. Hoehn
Research Associate
Mobil Research and Development Corporation
Dallas, Texas

Matthew H. Hulbert, Ph.D.
Senior Physical Chemist
Corporate Research and Development
International Minerals and Chemical Corporation
Terre Haute, Indiana

M. A. Khan, Ph.D.
Professor
Hawaii Institute of Geophysics
University of Hawaii
Honolulu, Hawaii

Douglas N. Lambert
Oceanographer
Naval Ocean Research and Development Activity
Sea Floor Division
NSTL Station, Mississippi

B. D. Loncarevic, Ph.D.
Atlanic Geoscience Centre
Geological Survey of Canada
Bedford Institute of Oceanography
Dartmouth, Nova Scotia

John L. Morack, Ph.D.
Professor
Physics Department
Geophysical Institute
University of Alaska
Fairbanks, Alaska

N. G. Pace, Ph.D.
School of Physics
University of Bath
Bath, England

L. Roemer, Ph.D.
Department of Oceanography
Texas A & M University
College Station, Texas

James C. Rogers, Ph.D.
Associate Professor
Electrical Engineering Department
Michigan Technological University
Houghton, Michigan

Carl H. Savit
Senior Vice President, Technology
Western Geophysical Company
Houston, Texas

William B. Sawyer
Oceanographer
University of Texas Institute for Geophysics
Galveston, Texas

Richard E. Sylwester, Ph.D.
Marine Geophysicist
Northern Technical Services
Seattle, Washington

Peter K. Trabant, Ph.D.
Consultant
Oceanographer/Geophysicist
Houston, Texas

Barry N. Warner
Senior Research Geophysicist
Mobil Research and Development Corporation
Dallas, Texas

TABLE OF CONTENTS

GENERAL CONSIDERATIONS

SATELLITE CONTRIBUTIONS TO GEOPHYSICAL EXPLORATION AT SEA*

M. A. Khan

One of the most interesting characteristics of our planet is that out of its 5.1×10^{14} m^2 total area, only about 1.48×10^{14} m^2 is land, the remaining 3.62×10^{14} m^2 or almost 71% of its surface being the oceans. The resource exploration, mapping and inventorying of these vast oceanic areas pose an enormous challenge. The challenge does not come only from the vastness of the areas involved but also from their expected geological and geophysical diversity and resource richness. In spite of recent efforts and advances, our knowledge of the marine potentials and prospects is based only on a small and scattered sampling. Large parts of the oceanic areas remain unexplored due to their expansiveness, the elaborate, expensive, and cumbersome technological, financial, and logistical outlays required to undertake their exploration, our limited understanding of marine environments, and curiously enough, until very recently, our limited marine technological ability.

The advent of scientific mission-oriented artificial Earth satellites in the last two decades has opened up new vistas and created new challenges in ocean sciences, as indeed in many other scientific disciplines. The satellites have certainly reduced the enormous financial outlays required for marine exploration, but what is more important, they have essentially solved the difficult problems of accessibility and position-location, reduced almost exponentially the resource requirements for data acquisition, made continuous, global scale, simultaneous mapping of diverse ocean properties possible, and contributed enormously towards the development of marine technology.

There is, of course, an important additional impact of the satellite age on scientific research, including ocean exploration. Because of the global and multifaceted nature of their scientific missions, the satellites have contributed towards ushering in a new era of global and interdisciplinary outlook in scientific research which has brought about integration of important, complementary fragments of information from the diverse segments of our highly specialized and compartmentalized world of science.

A satellite is *active* if it carries special scientific instrumentation packages which act as gauges of specific physical parameters. Otherwise it is *passive*. A passive satellite acts only as a temporal reference point in space for measurements which originate at a tracking and data acquisition system located elsewhere in space or on the ground.

Most satellite applications require an exact knowledge of satellite position in space. The satellite position can be determined by tracking it with photographic, photoelectric, laser, or other electronic tracking systems. Mostly, these tracking systems are ground-based. A satellite may be totally passive or emit an optical beacon to aid in photographic or photoelectric tracking, carry a reflector for laser tracking, and generally carry a transmitter package for other electronic tracking devices.

Satellite applications in geophysics, oceanography, and geodesy are traditionally classified as *dynamic* or *geometric*. The *dynamic* applications deal with the forces acting on a satellite in its orbit and attempt to determine these forces by observing the behavior of a satellite under their influence. The *geometric* techniques use the satellite as a temporal reference point in space for geometric positioning of specific ground points — generally the tracking stations. However, dynamic techniques also solve for the geometric quantities such as tracking station locations, and the geometric tech-

* Hawaii Institute of Geophysics Contribution No. 1178. This manuscript was prepared in May 1982.

niques must deal with dynamic quantities such as geoidal heights. In certain more recent applications, a satellite may perform *in situ mapping* or *imaging* of a certain parameter or property. Such applications may be classified as *synoptic* and may be geometric, dynamic, or both in nature. Thus, the classification into dynamic or geometric categories is not sharp. It is becoming even less distinct because of the rapid development of new satellite technologies which overlap both the categories.

This paper, because of its prescribed scope, will describe only a *few* of the many major contributions of satellites to ocean sciences. Further, it will focus only on those aspects of these contributions which are not traditionally well-known in ocean sciences.

EARTH'S GRAVITY FIELD FROM SATELLITE ORBITAL PERTURBATIONS

One of the chief dynamical contributions of satellites in geophysics and oceanography is the determination of the Earth's gravity field, particularly its long wavelength components, from the orbital motion of artificial Earth satellites. This application, of course, is not limited to the ocean areas only; it applies equally to the continental areas. But its usefulness for the oceans is especially enhanced as the marine gravimetric coverage is much more sparse than the continental gravimetric coverage. From this point of view, the satellites have made a major contribution in obtaining the long wavelength components of the gravity field over ocean areas.

The gravity vector is the gradient of the gravity potential. The gravity potential U, is the sum of the gravitational potential V, and the rotational potential. The rotational potential is a simple analytical function which can be computed at a point (r, ϕ) from the expression $\frac{1}{2}\ \omega^2 r^2 \cos^2\phi$ where ω is the angular velocity of Earth's rotation. Thus, the problem of determining the gravity field of the Earth boils down to the problem of determining V at Earth's surface. And indeed it is this gravitational potential V that is determined from the analysis of satellite orbits.

Orbital Elements

Since the position of a satellite in its orbit around the Earth depends, among other things, upon the gravitational attraction of the Earth, V, in principle, can be determined if the position of an Earth satellite is known as a function of time. The quantities which define the position of a satellite in space as a function of time are called *orbital elements.* These are illustrated in Figure 1 in which XYZ is the customary coordinate system, with its origin at the center of mass of the Earth, its Z-axis coincident with the rotation axis of the Earth, its X-Y plane coincident with the equatorial plane, and its X-axis directed towards the *vernal equinox* which is the intersection of the equatorial plane with the ecliptic. This coordinate system can be aligned with either the instantaneous or the mean directions depending upon one's objectives. The points of intersection of the orbital plane with the equatorial plane are called *nodes.* The node at which the satellite crosses the equatorial plane northwards is the *ascending node*; the node at which it crosses southwards is the *descending node.* The line joining the nodes is called the *line of nodes.* The size of the orbit is determined by its *semi-major axis a,* its shape by its *eccentricity e.* The angle between the X-axis (direction to the vernal equinox) and the line of nodes is called the *right ascension* of the ascending node and is denoted by Ω. It is measured counterclockwise and determines the position of the orbital plane in space. The angle between the equatorial plane and the orbital plane is called the *inclination i.* It determines the orientation of the orbital plane in space. The inclination lies between 0 and π. For $0 \leqslant i \leqslant \pi/2$, the orbit is called *direct*; for $\pi/2 \leqslant i \leqslant \pi$, it is called *retrograde.* The major axis of the orbit intersects the orbital

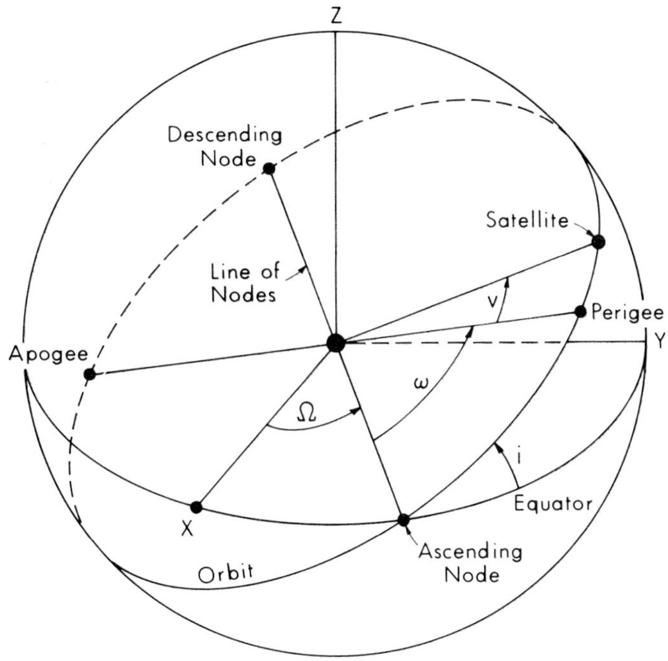

FIGURE 1. Position of a satellite in space.

ellipse at two points. The point at which the satellite is closest to the Earth is called *perigee*; the point at which it is farthest is called *apogee*. The orientation of the orbit in the orbital plane is determined by the angle between the line of nodes and the major axis of the orbit; it is called the *argument of perigee* and denoted by ω. The position of the satellite in the orbit is determined by the *epoch of perigee passage* τ. These six orbital elements define completely the position of a satellite in space at a given time.

Theory

If the Earth were a homogeneous sphere, its gravitational potential will be

$$V = \frac{GM}{r}$$

where r = radial distance, G = the gravitational constant, and M = the mass of the Earth. The orbit of a satellite around such an Earth will then be an ellipse, assuming that the motion takes place in a vacuum. The motion of a satellite in such an orbit will then be described simply by Kepler's laws of motion in a central force field. Further, the orbital ellipse will be time-invariant and the orbital elements a, e, Ω, i, ω and τ defining the orbit will remain constant with time. But the gravitational potential of the Earth is not that of a homogeneous sphere. Instead, it is of the form

$$V = \frac{GM}{r} + R \tag{1}$$

where R is the *perturbing potential* due to the departure of the Earth from spherical symmetry. It represents the irregularities of the gravity field with respect to the central

c-summation:

$$\left.\begin{matrix} p - t \leqslant m - s, o \\ p - t \geqslant m - s, p-t-m+s \end{matrix}\right] \leqslant c \leqslant \left[\begin{matrix} n-m-2t+s, p-t \geqslant n-m-2t+s \\ p-t, p-t \leqslant n-m-2t+s \end{matrix}\right.$$

t-summation:

$$o \leqslant t \leqslant \begin{matrix} p, p \leqslant k \\ k, p \geqslant k \end{matrix}$$

The function $G_{npq}(e)$ must be defined for long periodic and short periodic terms separately. For long periodic terms, it has $n-2p+q = o$, or $q = 2p-n$. Thus, for long periodic terms

$$G_{np(2p-n)}(e) = \frac{1}{(1-e^2)^{n-1/2}} \sum_{d=o}^{p'-1} \binom{n-1}{n+2d-2p'} \binom{n+2d-2p'}{d} \left(\frac{e}{2}\right)^{n+2d-2p'}$$

where

$$p' = p \text{ for } p \leqslant \frac{n}{2}$$

and

$$p' = n-p \quad \text{for} \quad p \geqslant \frac{n}{2}$$

For short-period terms, the development of G_{npq} becomes rather complicated. The resulting expressions are, however, of the form

$$G_{npq}(e) = (-1)^{\bar{q}} \left[1 + \frac{e^2}{(1+\sqrt{1-e^2})^2}\right]^n \left(\frac{e}{1+\sqrt{1-e^2}}\right)^{\bar{q}} \sum_{k=0}^{\infty} P_{npqk}$$

$$\cdot\, Q_{npqk} \left(\frac{e}{1+\sqrt{1-e^2}}\right)^{2k}$$

where

$$P_{npqk} = \sum_{r=0}^{h} \binom{2p'-2n}{h-r} \left[\frac{(n-2p'+q')(1+\sqrt{1-e^2})}{2}\right]^r \frac{(-1)^r}{r!}$$

and

$$Q_{npqk} = \sum_{r=0}^{\ell} \binom{-2p'}{\ell-r} \left[\frac{(n-2p'+q')(1+\sqrt{1-e^2})}{2}\right]^r \frac{1}{r!}$$

where

$$\bar{q} = |q|$$

$$h \begin{cases} = k+q' & \text{for } q' > 0 \\ = k & \text{for } q' < 0 \end{cases}$$

$$\ell \begin{cases} = k & \text{for } q' > 0 \\ = k-q' & \text{for } q' < 0 \end{cases}$$

$$\left.\begin{matrix} p' = p \\ q' = q \end{matrix}\right] \text{for } p \leqslant \frac{n}{2} \quad \text{and} \quad \left.\begin{matrix} p' = \ell-p \\ q' = -q \end{matrix}\right] \text{for } p > \frac{n}{2}$$

The function $Z_{nmpq}\,(\omega,M,\Omega,\theta)$ is given by

$$Z_{nmpq}(\omega,M,\Omega,\theta) = \begin{bmatrix} C_{nm} \\ S_{nm} \end{bmatrix} \begin{matrix} n-m \text{ even} \\ n-m \text{ odd} \end{matrix} \quad \cos[(n-2p)\omega + (n-2p+q)M + m(\Omega - \theta)]$$

$$+ \begin{bmatrix} S_{nm} \\ C_{nm} \end{bmatrix} \begin{matrix} n-m \text{ even} \\ n-m \text{ odd} \end{matrix} \quad \sin[(n-2p)\omega + (n-2p+q)M + m(\Omega - \theta)]$$

The expression for the perturbing potential in Equation 4 is then substituted in Equation 2, after carrying out the appropriate differentiation steps. This gives[2,3,5]

$$\frac{d\Omega_{nmpq}}{dt} = \frac{GMa_e^{\,n}\,F'_{nmp}G_{npq}Z_{nmpq}}{na^{n+3}(1-e^2)^{1/2}\sin i}$$

$$\frac{di_{nmpq}}{dt} = \frac{GMa_e^{\,n}\,F_{nmp}G_{npq}Z'_{nmpw}}{\bar{n}a^{n+3}(1-e^2)^{1/2}\sin i}\,[(n-2p\,\cos i-m]$$

$$\frac{d\omega_{nmpq}}{dt} = GMa_e^{\,n}\left[\frac{(1-e^2)^{1/2}}{e}F_{nmp}G'_{npq} - \frac{\cot i}{(1-e^2)^{1/2}}F'_{nmp}G_{npq}\right]\frac{Z_{nmpq}}{\bar{n}a^{n+3}}$$

$$\frac{da_{nmpq}}{dt} = \frac{2GMa_e^{\,n}\,F_{nmp}G_{npq}Z'_{nmpq}}{\bar{n}a^{n+2}}\,(n-2p+q)$$

$$\frac{de_{nmpq}}{dt} = \frac{GMa_e^{\,n}\,F_{nmp}G_{npq}Z'_{nmpq}}{\bar{n}a^{n+3}e}\,[(1-e^2)\,(n-2p+q) - (1-e^2)^{1/2}\,(n-2p)]$$

$$\frac{d\overline{M}_{nmpq}}{dt} = \frac{GMa_e^{\,n}\,F_{nmp}Z_{nmpq}}{\bar{n}a^{n+3}}\,[2(n+1)G_{npq} - \frac{1-e^2}{e}\,G'_{npq}] \tag{5}$$

The above equations give the time rate of change of orbital elements due to one term R_{nmpq}, of the perturbing potential R. In these equations, $\overline{M}$ denotes the perturbation of mean anomaly such that

$$\bar{M} = \int_0^t \bar{n}dt - \bar{n}(t-\tau)$$

and the functions

$$F'_{nmp}(i) = \frac{dF_{nmp}}{di}$$

$$G'_{npq}(e) = \frac{dG_{npq}}{de}$$

and $\acute{Z}_{nmpq}$ is the derivative of Z_{nmpq} with respect to its argument.

Equations 3, 4, and 5 relate the time rate of change of the orbital elements with the potential coefficients C_{nm} and S_{nm}. Since in practice, we observe these rates of change over extended periods of time, we integrate Equation 5 with respect to time to obtain the following expressions:

$$\Delta\Omega_{nmpq} = GMa_e^n \frac{F'_{nmp}G_{npq}\bar{Z}_{nmpq}}{\bar{n}a^{n+3}(1-e^2)^{1/2}\sin i[(n-2p)\dot{\omega} + (n-2p+q)\dot{M} + m(\dot{\Omega} - \dot{\theta})]}$$

$$\Delta i_{nmpq} = GMa_e^n \frac{F_{nmp}G_{npq}[(n-2p)\cos i - m]Z_{nmpq}}{\bar{n}a^{n+3}(1-e^2)^{1/2}\sin i[(n-2p)\dot{\omega} + (n-2p+q)\dot{M} + m(\dot{\Omega} - \dot{\theta})]}$$

$$\Delta\omega_{nmpq} = GMa_e^n \frac{[e^{-1}(1-e^2)^{1/2} F_{nmp}G'_{npq} - \cot i(1-e^2)^{-1/2} F'_{nmp}G_{npq}]\bar{Z}_{nmpq}}{\bar{n}a^{n+3}[(n-2p)\dot{\omega} + (n-2p+q)\dot{M} + m(\dot{\Omega} - \dot{\theta})]}$$

$$\Delta a_{nmpq} = GMa_e^n \frac{2F_{nmp}G_{npq}Z_{nmpq}(n-2p+q)}{na^{n+2}[(n-2p)\dot{\omega} + (n-2p+q)\dot{M} + m(\dot{\Omega} - \dot{\theta})]}$$

$$\Delta e_{nmpq} = GMa_e^n \frac{F_{nmp}G_{npq}Z_{nmpq}[(1-e^2)(n-2p+q) - (1-e^2)^{1/2}(n-2p)]}{na^{n+3}e[(n-2p)\dot{\omega} + (n-2p+q)\dot{M} + m(\dot{\Omega} - \dot{\theta})]}$$

$$\Delta M_{nmpq} = GMa_e^n \frac{F_{nmp}\bar{Z}_{nmpq}[2(n+1)G_{npq} - (1-e^2)e^{-1}G'_{npq}]}{\bar{n}a^{n+}[(n-2p)\dot{\omega} + (n-2p+q)\dot{M} + m(\dot{\Omega} - \dot{\theta})]} \quad (6)$$

where $\bar{Z}$ = the integral of Z with respect to its argument.

The above development was originally given by Kaula[3] and has been discussed further by Khan[5] and Mueller.[2]

Determination of Spherical Harmonic Geopotential Coefficients

Zonal Coefficients

The spelling out of Equation 5 or 6 in their literal algebraic form is quite cumbersome and tedious. But, on simplification, it is found that the largest perturbations in the orbital elements occur from the secular and the long-periodic effects of zonal harmonic coefficients. Specifically, it is found that *even* zonal harmonic coefficients produce secular changes in Ω, ω, and M, and long-periodic changes in e and *i*. The *odd* zonal harmonic coefficients produce long period changes in Ω, ω, *i*, e, and M. But in the expressions for ΔΩ and Δω, the odd zonal harmonic coefficients appear as multiples of the small factor e which mutes their long-periodic effect. Thus the predominant perturbations in Ω and ω are secular effects of the even zonal harmonic coefficients. In the expressions for Δe and Δi, the even zonal harmonic coefficients appear as multiples of the small factor e which scales down their long periodic effects on changes in e and i. Thus the predominant perturbations in e and i are the long-periodic effects of the odd zonal harmonic coefficients. Consequently, the even zonal harmonic coefficients are generally determined from the secular perturbations in Ω and ω while the odd zonal harmonic coefficients are generally determined from the variations in e and i.

The above discussion accounts for what we call *linear* perturbations. But one of the properties of the perturbing potential is that the second degree zonal geopotential coefficient C_{20}, is about 10^3 times larger than the other geopotential coefficients. Thus C^2_{20} is of the order of C_{40}. Hence the perturbations arising from the *nonlinear* terms containing factors C^2_{20}, $C_{20}C_{30}$, $C_{20}C_{40}$ etc. must also be taken into consideration. There are a number of ways to account for the nonlinear perturbations in practice. The details of such procedures are not given in this chapter. For a detailed treatment of nonlinear perturbations, see, for example, Brouwer and Clemence[6] or Kaula.[3]

Tesseral Coefficients

Tesseral harmonic coefficients produce short-periodic perturbations. A detailed examination of Equation 5 or 6 shows that short-periodic variations arise from all terms with a frequency of m $(\dot{\Omega}\text{-}\dot{\theta})$. A tesseral harmonic coefficient of order m will produce perturbations of m cycles per day. Thus, the longest period of a tesseral harmonic of order $m = 1$ is 1 day. Hence, the computational methods and the satellite arcs used in the analysis must be sensitive enough to the appropriate short-periodic perturbations.

It can be seen from Equation 5 or 6 that the geopotential coefficients depend upon i. Hence, in order to ensure mutual independence of observation equations, it is important to use data from satellites with a wide variety of inclinations. Additionally, data must have a reasonably even distribution in time.

Special Problems

Special problems can arise if inclination i, and eccentricity e are very small, in which case Ω, ω, and M may be difficult to define, or if the satellite motion is in *resonance* with a selected spherical harmonic, or when the inclination i is very close or equal to the *critical inclination* (i = 63°26′). These can be taken care of by specially designed techniques. See, for example, Reference 3 and the literature cited therein.

Sources of Other Perturbations

So far we have considered the perturbations caused by the perturbing potential R only. But there are several other factors which produce perturbations of the orbital elements. Their effects must be separated and accounted for. Chief factors in this category are *atmospheric drag, luni-solar effects, solar radiation pressure,* etc. These must be modeled or otherwise accounted for in actual computational schemes.

Remarks on Actual Computational Procedures

The above development provides an analytical basis of the relationships between the various components of the perturbing potential and perturbation of the orbit. Actual procedures require considerable numerical and statistical sophistication in data processing and preparation, and in the construction of the observational and variational equations. They range from the use of analytical theory to the numerical integration of the differential equations of motion. The latter provides the residuals used in computing the differential corrections required for refining the modeled parameters. The recovery of tesseral harmonic coefficients is often done in conjunction with tracking station coordinates and other geodetic parameters. This necessitates handling of large amounts of data resulting in considerable computer time and computer memory requirements. In these procedures, the problems requiring specific attention are the truncation of the spherical harmonic series given in Equation 3, length of orbital arcs, and origin, orientation, and scale of station coordinates. The insights gained through experience in handling such data are frequently useful in meeting these and other problems.

Combination Solutions

Because of decay of potential with altitude and progressive reduction in the size of geopotential coefficients with increasing degree, satellite orbits are not very sensitive to higher degree terms except under special conditions such as resonance. Thus, the satellite orbital data are best suited for recovering long wavelength components of the Earth's gravity field. On the other hand, surface gravity data are sensitive to high degree components. Hence they are used in conjunction with satellite orbital data to extend the recoverable frequency range. The gravity models which draw upon both the satellite and surface gravity data are called the *combination models.* More recently, the combination models are also making use of satellite altimetry data which contain useful information on high degree components.

Chief Gravity Model Development Efforts

Chief efforts in the development of spherical harmonic models of the Earth's gravity field from satellite orbits originally consisted of King-Hele[7] and Cook[8] in Europe, and Kozai,[9] Iszak,[10] Anderle,[11] O'Keefe,[12] Kaula,[13] and Guier and Newton[14] in the U.S. As the satellite data accumulated, and the growing size of gravity models demanded a sustained commitment of much greater computer time and manpower resources, the U.S. efforts have become confined principally to Smithsonian Astrophysical Observatory[15] and NASA's Goddard Space Flight Center,[16] and for defense-related applications to the Naval Surface Weapons Center. The chief focus of these efforts is to obtain a combination gravity field model that draws on the strengths of both the satellite and surface gravity data. During the last decade, there has been considerable progress in the overall accuracy and frequency range of these models. At this stage it is generally believed that the long wavelength components of the gravity field are well-determined to $n = 14$. See Reference 17 for details.

One of the recent combination models is GEM 10B, developed by Lerch et al.[18] It makes use of satellite altimetry data in addition to the satellite orbital and surface gravity data. It is complete to degree $n = 36$ and order $m = 36$. This means that the effect of features smaller than about 550 km in size (at the equator) does not show up in this gravity field representation while the larger features are reflected in it.

Table 1 lists the spherical harmonic coefficients C_{nm} and S_{nm}, of GEM 10B[18] out to 36,36. Figure 2 shows the approximate level of accuracy as a function of frequency for these coefficients. Table 2 lists the power spectra of GEM 10B coefficients and their estimated errors.[17] Note that in Figure 2 and Table 2, these estimates are given only to $n = 22$. This is because the standard against which GEM 10B has been evaluated is available to only $n = 22$. Further, the standard of comparison itself could, and probably does have some error of its own, particularly in higher degree terms. Consequently, the errors quoted here are only error estimates. Also, note that coefficients C_{10}, C_{11}, and S_{11} are set to zero. Coefficients C_{21} and S_{21} are estimated, but only to serve as a measure of error for the low degree coefficients.

Table 1
NORMALIZED SPHERICAL HARMONIC COEFFICIENTS
OF EARTH'S GRAVITY MODEL GEM 10B[a]

n	m	C_{nm}	S_{nm}
2	0	-0.48416551D-03	0.0
3	0	0.95867438D-06	0.0
4	0	0.54111657D-06	0.0
5	0	0.68167394D-07	0.0
6	0	-0.15045905D-06	0.0
7	0	0.93775913D-07	0.0
8	0	0.50380844D-07	0.0
9	0	0.26875950D-07	0.0
10	0	0.52713709D-07	0.0
11	0	-0.47737609D-07	0.0
12	0	0.39140038D-07	0.0
13	0	0.42947460D-07	0.0
14	0	-0.23215176D-07	0.0
15	0	0.25416043D-08	0.0
16	0	-0.62210036D-08	0.0
17	0	0.15687650D-07	0.0
18	0	0.10402509D-07	0.0
19	0	0.17376701D-08	0.0
20	0	0.24493175D-07	0.0
21	0	-0.88998884D-09	0.0
22	0	-0.39052695D-08	0.0
23	0	-0.20440900D-07	0.0
24	0	-0.13466914D-08	0.0
25	0	0.43540233D-09	0.0
26	0	0.18624153D-08	0.0
27	0	0.64896289D-08	0.0
28	0	-0.14464851D-07	0.0
29	0	-0.66105453D-08	0.0
30	0	-0.16060110D-08	0.0
31	0	-0.70988293D-09	0.0
32	0	-0.71650235D-08	0.0
33	0	-0.37622261D-08	0.0
34	0	-0.74545822D-08	0.0
35	0	0.92526737D-08	0.0
36	0	-0.35266714D-09	0.0
2	1	0.97058783D-09	-0.15398455D-10
3	1	0.20310057D-05	0.25346868D-06
4	1	-0.53315357D-06	-0.47316396D-06
5	1	-0.54424306D-07	-0.95999204D-07
6	1	-0.76961200D-07	0.26640402D-07
7	1	0.27402457D-06	0.98011475D-07
8	1	0.27704665D-07	0.50795815D-07
9	1	0.15230509D-06	0.15903089D-07
10	1	0.79394909D-07	-0.12952320D-06
11	1	0.58660808D-08	-0.15778261D-07
12	1	-0.56504103D-07	-0.49494914D-07
13	1	-0.46147243D-07	0.34909745D-07
14	1	-0.17307032D-07	0.32504658D-07
15	1	0.69557499D-08	0.36183436D-08
16	1	0.30058165D-07	0.13989032D-07
17	1	-0.32606451D-07	-0.27289235D-07
18	1	-0.23406836D-08	-0.21494255D-07
19	1	-0.28411719D-07	0.40981065D-08
20	1	0.13019021D-08	-0.12800117D-07
21	1	-0.35608928D-07	0.31656204D-07

Table 1 (continued)
NORMALIZED SPHERICAL HARMONIC COEFFICIENTS OF EARTH'S GRAVITY MODEL GEM 10B[a]

n	m	C_{nm}	S_{nm}
22	1	0.97439603D-08	0.93221537D-08
23	1	-0.43205584D-08	0.23503480D-07
24	1	-0.98690243D-08	0.11645526D-08
25	1	-0.43280198D-08	0.40150504D-08
26	1	0.52806282D-09	-0.73517675D-08
27	1	-0.60435712D-09	-0.14076459D-08
28	1	-0.22793223D-08	0.42318759D-11
29	1	0.24022903D-08	0.41691787D-08
30	1	-0.23385736D-08	-0.36299329D-08
31	1	0.57066357D-08	-0.10557668D-07
32	1	-0.79952253D-09	0.27577329D-08
33	1	0.17484773D-08	-0.598966160-08
34	1	0.10074620D-09	-0.67810014D-08
35	1	0.68787513D-08	-0.19744786D-07
36	1	0.51208442D-08	0.10237473D-08
2	2	0.24326520D-05	-0.13971495D-05
3	2	0.89403337D-06	-0.62129824D-06
4	2	0.35205129D-06	0.66983366D-06
5	2	0.65231516D-06	-0.32511907D-06
6	2	0.46362672D-07	-0.35952862D-06
7	2	0.32549003D-06	0.10486604D-06
8	2	0.72004324D-07	0.71494023D-07
9	2	0.23789137D-07	-0.27903723D-07
10	2	-0.88584538D-07	-0.30323416D-07
11	2	0.26390130D-07	-0.89394193D-07
12	2	0.46999861D-08	0.28086733D-07
13	2	0.43902995D-07	-0.55435860D-07
14	2	-0.30748523D-07	0.14670745D-07
15	2	-0.87355691D-08	-0.18301147D-07
16	2	-0.23365479D-07	0.35543265D-07
17	2	-0.12418449D-07	0.15467526D-07
18	2	0.95431035D-08	0.19216671D-07
19	2	0.36407754D-07	-0.18850588D-08
20	2	0.48503779D-08	0.81131321D-08
21	2	0.13160957D-08	-0.13395656D-08
22	2	-0.12210801D-07	-0.99389366D-08
23	2	0.50485201D-08	-0.10307560D-07
24	2	0.24098256D-08	0.14432305D-07
25	2	0.22896865D-07	0.65406521D-10
26	2	-0.10893905D-07	0.88623138D-08
27	2	0.70133337D-08	-0.11389358D-07
28	2	-0.23205945D-07	-0.12066973D-07
29	2	0.72259777D-08	-0.15958979D-07
30	2	-0.19068227D-07	0.25061737D-08
31	2	0.96494509D-08	-0.50543580D-08
32	2	0.77215218D-09	-0.38040473D-08
33	2	-0.20520226D-08	-0.54850121D-08
34	2	-0.30613573D-08	-0.23819473D-08
35	2	-0.15522803D-07	0.26120876D-08
36	2	-0.84148513D-08	0.86961266D-08
3	3	0.71319370D-06	0.14185773D-05
4	3	0.99068551D-06	-0.20496630D-06
5	3	-0.46158302D-06	-0.21616595D-06
6	3	0.60918941D-07	0.44827727D-08
7	3	0.24675398D-06	-0.21336858D-06
8	3	-0.17581190D-07	-0.93438034D-07

Table 1 (continued)
NORMALIZED SPHERICAL HARMONIC COEFFICIENTS
OF EARTH'S GRAVITY MODEL GEM 10B[a]

n	m	C_{nm}	S_{nm}
9	3	-0.16844187D-06	-0.87673062D-07
10	3	-0.98687352D-08	-0.15740506D-06
11	3	-0.41250529D-07	-0.13482629D-06
12	3	0.47215746D-07	0.17066914D-07
13	3	-0.29555737D-07	0.89826533D-07
14	3	0.38102637D-07	0.53678724D-08
15	3	0.42823209D-07	0.14892226D-07
16	3	-0.19403084D-07	-0.29629186D-07
17	3	0.49892739D-08	0.95587205D-08
18	3	-0.14611291D-07	-0.35886996D-08
19	3	-0.10917347D-07	-0.35066058D-08
20	3	-0.16203684D-07	0.27348006D-07
21	3	0.20410234D-07	0.15207240D-07
22	3	-0.97481405D-08	0.16353264D-07
23	3	0.38895908D-08	-0.15515231D-07
24	3	-0.13414594D-07	0.36011670D-09
25	3	-0.35808013D-08	-0.18552418D-08
26	3	-0.45655005D-10	0.75650683D-08
27	3	0.11158020D-07	0.12923847D-07
28	3	-0.15752799D-08	0.36173311D-08
29	3	0.77067891D-08	-0.51997583D-08
30	3	0.12879665D-08	-0.14372718D-07
31	3	0.28111852D-08	-0.43851414D-08
32	3	-0.37505661D-08	0.88782120D-08
33	3	-0.67134828D-09	0.15218434D-07
34	3	0.20691683D-07	0.14386205D-07
35	3	0.14296740D-07	0.12552227D-08
36	3	0.68659111D-08	-0.69433259D-08
4	4	-0.19340572D-06	0.30698508D-06
5	4	-0.30153230D-06	0.46296116D-07
6	4	-0.97257177D-07	-0.46387733D-06
7	4	-0.27903638D-06	-0.12537619D-06
8	4	-0.24723182D-06	0.75663782D-07
9	4	-0.19666966D-07	0.10878612D-07
10	4	-0.92448997D-07	-0.70934666D-07
11	4	-0.45862600D-07	-0.62514960D-07
12	4	-0.72535296D-07	-0.82546126D-09
13	4	-0.96030075D-08	-0.17309515D-07
14	4	-0.59497608D-08	-0.11769494D-07
15	4	-0.55218277D-07	-0.14631414D-08
16	4	0.39553952D-07	0.63476292D-07
17	4	-0.90739887D-08	0.11445128D-07
18	4	0.52761654D-07	0.90843040D-08
19	4	0.65614227D-08	-0.29300567D-07
20	4	0.17771273D-08	-0.24359208D-07
21	4	-0.83939214D-08	-0.59336896D-08
22	4	0.96011611D-09	0.30726347D-07
23	4	-0.13502581D-08	0.11854308D-07
24	4	0.10515240D-08	0.10696914D 07
25	4	0.13572946D-07	0.64233643D-08
26	4	-0.39808335D-09	-0.12289860D-07
27	4	-0.23504097D-08	0.70892344D-08
28	4	-0.11868878D-07	0.87872130D-08
29	4	-0.20127538D-07	-0.89729831D-09
30	4	-0.51969445D-08	0.34107112D-08
31	4	0.11802715D-07	-0.10477557D-07

Table 1 (continued)
NORMALIZED SPHERICAL HARMONIC COEFFICIENTS OF EARTH'S GRAVITY MODEL GEM 10B[a]

n	m	C_{nm}	S_{nm}
32	4	0.33918369D-08	-0.21864887D-09
33	4	-0.19562311D-08	-0.10038846D-11
34	4	-0.67736243D-08	0.43672515D-08
35	4	-0.53726583D-08	-0.36987733D-08
36	4	-0.16998044D-08	0.10826589D-08
5	5	0.15720986D-06	-0.66483158D-06
6	5	-0.26409780D-06	-0.53906753D-06
7	5	0.60535215D-08	0.26713783D-07
8	5	-0.19634025D-07	0.85116660D-07
9	5	-0.18486818D-07	-0.52761800D-07
10	5	-0.55683330D-07	-0.42472030D-07
11	5	0.42816886D-07	0.57377614D-07
12	5	0.42422502D-07	0.11276993D-07
13	5	0.64016925D-07	0.65300857D-07
14	5	0.33257638D-07	-0.23211005D-07
15	5	0.58513460D-08	-0.10553374D-07
16	5	-0.38819206D-08	-0.42407377D-08
17	5	-0.13948633D-07	0.14389126D-08
18	5	0.43863245D-08	0.32166139D-07
19	5	0.11951053D-07	0.10188599D-07
20	5	-0.68149066D-08	0.48563579D-08
21	5	0.24394981D-08	-0.15697773D-07
22	5	-0.69311638D-08	0.13301924D-07
23	5	0.12892110D-08	-0.28207454D-08
24	5	-0.10738683D-07	0.72045122D-08
25	5	-0.16016134D-07	-0.65905883D-08
26	5	0.35705097D-09	0.16832611D-07
27	5	0.16208816D-07	-0.10892070D-08
28	5	-0.31259728D-08	-0.50628145D-08
29	5	-0.97567396D-08	-0.18910988D-08
30	5	0.45098276D-08	-0.11744013D-07
31	5	-0.28252152D-08	0.42844595D-08
32	5	0.37958875D-08	0.18940566D-08
33	5	-0.46892349D-08	0.13677796D-07
34	5	-0.10402572D-08	0.49741846D-08
35	5	-0.13335806D-07	-0.45379225D-09
36	5	-0.94745043D-08	0.20323387D-08
6	6	0.51526292D-08	-0.23594647D-06
7	6	-0.36276579D-06	0.13658664D-06
8	6	-0.67333446D-07	0.31235200D-06
9	6	0.55161469D-07	0.21740580D-06
10	6	-0.39828354D-07	-0.77971268D-07
11	6	-0.10035017D-07	0.34848428D-07
12	6	0.86825088D-08	0.38736657D-07
13	6	-0.30096357D-07	-0.56235197D-08
14	6	-0.16656195D-07	0.69377585D-08
15	6	0.28771063D-07	-0.37634494D-07
16	6	0.14174584D-07	-0.35519938D-07
17	6	-0.34193117D-08	-0.45033247D-07
18	6	0.17989069D-07	-0.18065960D-07
19	6	-0.46761425D-08	0.22004835D-07
20	6	0.91366337D-08	0.11105890D-07
21	6	-0.22440340D-10	-0.20736256D-08
22	6	0.17324814D-07	0.83040564D-08
23	6	-0.82910238D-08	0.15855445D-07
24	6	-0.16620239D-08	0.72110425D-08

Table 1 (continued)
NORMALIZED SPHERICAL HARMONIC COEFFICIENTS
OF EARTH'S GRAVITY MODEL GEM 10B[a]

n	m	C_{nm}	S_{nm}
25	6	0.87041702D-08	-0.15220274D-08
26	6	0.96291331D-08	0.24484925D-08
27	6	0.70891813D-08	0.42845501D-09
28	6	0.10140807D-07	0.40506470D-08
29	6	0.13569664D-07	0.65076021D-08
30	6	0.80392817D-08	0.56633991D-08
31	6	-0.54301986D-08	0.20714859D-08
32	6	0.34535051D-08	-0.11348798D-07
33	6	0.16600421D-08	-0.15435255D-08
34	6	0.19117676D-07	-0.14443542D-08
35	6	-0.22807818D-08	0.30679132D-10
36	6	0.96894592D-08	-0.68426457D-08
7	7	-0.47486560D-08	0.22379513D-07
8	7	0.66508224D-07	0.72124164D-07
9	7	-0.10953669D-06	-0.89911321D-07
10	7	0.83826892D-08	0.59443013D-08
11	7	0.88969360D-08	-0.93851231D-07
12	7	-0.17911267D-07	0.36443544D-07
13	7	0.75987200D-10	-0.12088519D-07
14	7	0.25476928D-07	-0.12728759D-07
15	7	0.57593335D-07	0.54554059D-08
16	7	-0.15783899D-07	-0.98771517D-08
17	7	0.33007111D-07	-0.45942223D-08
18	7	0.64090288D-08	-0.29061754D-08
19	7	0.59877134D-08	-0.10876629D-07
20	7	-0.27769673D-07	-0.37487331D-08
21	7	-0.53015818D-08	0.11263708D-07
22	7	-0.83083485D-08	0.47813880D-08
23	7	0.13012940D-08	0.88496733D-08
24	7	-0.16308086D-07	-0.92698777D-08
25	7	0.16152107D-07	-0.77714901D-09
26	7	-0.11242777D-07	-0.49840952D-08
27	7	0.31104917D-08	0.10836965D-07
28	7	-0.36276323D-08	0.49020836D-08
29	7	0.87460578D-09	-0.96844684D-09
30	7	-0.88205417D-09	0.33564057D-08
31	7	-0.22513006D-08	0.17386931D-08
32	7	-0.18866683D-08	-0.34791254D-09
33	7	-0.13238775D-07	0.87625687D-09
34	7	0.36654897D-08	0.73257522D-09
35	7	-0.29934548D-08	0.72108067D-08
36	7	0.59617319D-09	0.77831549D-08
8	8	-0.12536046D-06	0.12140074D-06
9	8	0.19239752D-06	-0.80289674D-08
10	8	0.40856095D-08	-0.82080665D-07
11	8	-0.35845869D-08	0.23439910D-07
12	8	-0.27591269D-07	0.21293719D-07
13	8	-0.11662625D-07	-0.50906532D-08
14	8	-0.35343180D-07	-0.46522954D-08
15	8	-0.23978948D-07	0.38541121D-07
16	8	-0.24399102D-07	0.80989238D-08
17	8	0.28518580D-07	0.28903079D-08
18	8	0.24631140D-07	-0.30866250D-08
19	8	0.28470171D-07	-0.19000414D-07
20	8	-0.50454511D-09	0.12590983D-09
21	8	-0.76541753D-08	0.10792806D-07

Table 1 (continued)
NORMALIZED SPHERICAL HARMONIC COEFFICIENTS OF EARTH'S GRAVITY MODEL GEM 10B[a]

n	m	C_{nm}	S_{nm}
22	8	-0.28974260D-07	0.17272196D-07
23	8	-0.17856639D-08	-0.56515611D-08
24	8	0.13291030D-07	-0.14079345D-07
25	8	0.15253240D-07	-0.71721560D-08
26	8	0.85533816D-08	0.25848623D-09
27	8	0.20548522D-08	-0.16699134D-07
28	8	0.10650875D-07	0.23934522D-09
29	8	-0.11108237D-07	0.11422505D-07
30	8	0.20797217D-08	0.14180939D-07
31	8	0.95251997D-09	-0.14892164D-08
32	8	0.95561098D-08	0.61351684D-08
33	8	0.18529437D-08	0.55709511D-08
34	8	-0.19044622D-08	0.50814524D-08
35	8	0.45087028D-08	0.36653091D-08
36	8	-0.43283388D-10	-0.31995333D-08
9	9	-0.48967179D-07	0.96795683D-07
10	9	0.12639265D-06	-0.46329986D-07
11	9	-0.28132731D-07	0.35484935D-07
12	9	0.38835956D-07	0.14756298D-07
13	9	0.24170114D-07	0.43878927D-07
14	9	0.30210846D-07	0.17607414D-07
15	9	0.12450846D-07	0.37037160D-07
16	9	-0.23010414D-07	-0.38029086D-07
17	9	-0.51350396D-08	-0.29293264D-07
18	9	-0.22265684D-07	0.37974740D-07
19	9	0.74692429D-08	0.89463169D-08
20	9	0.20503986D-07	-0.91712246D-09
21	9	0.22681472D-07	0.81966240D-08
22	9	0.39881324D-08	0.98456604D-08
23	9	-0.27289048D-08	-0.69438220D-08
24	9	-0.22241841D-07	-0.13226307D-07
25	9	-0.32561459D-07	0.65743534D-08
26	9	-0.14391724D-07	0.38544263D-08
27	9	0.67861628D-08	0.49488644D-08
28	9	0.12434999D-07	0.37869019D-08
29	9	-0.66374419D-09	0.29423068D-08
30	9	-0.80862015D-08	0.52966501D-08
31	9	-0.69135018D-08	0.34294502D-08
32	9	0.65910959D-08	-0.12966002D-08
33	9	-0.37207499D-08	-0.18897829D-08
34	9	0.66600234D-09	-0.95978417D-09
35	9	-0.31944807D-08	-0.19271812D-08
36	9	0.32701154D-08	-0.35442877D-08
10	10	0.10173156D-06	-0.29930950D-07
11	10	-0.54683476D-07	-0.13490481D-07
12	10	-0.17045449D-08	0.29185369D-07
13	10	0.49055928D-07	-0.37309121D-07
14	10	0.50658411D-07	0.57832748D-08
15	10	0.19765668D-07	0.85400008D-08
16	10	0.27789066D-08	0.11181416D-07
17	10	0.94296655D-08	0.15782651D-07
18	10	0.16758363D-08	-0.12332424D-07
19	10	-0.29219534D-07	-0.18673284D-07
20	10	-0.26854527D-07	-0.10924909D-07
21	10	-0.35596903D-08	0.50331880D-08
22	10	0.10081700D-08	0.27955414D-07

Table 1 (continued)
NORMALIZED SPHERICAL HARMONIC COEFFICIENTS OF EARTH'S GRAVITY MODEL GEM 10B[a]

n	m	C_{nm}	S_{nm}
23	10	0.40021749D-08	0.64982978D-10
24	10	0.78558169D-08	0.19931590D-07
25	10	0.56789463D-08	-0.28382994D-08
26	10	-0.13947235D-07	-0.11235883D-07
27	10	-0.51054542D-08	-0.10289966D-07
28	10	-0.10160883D-08	0.12713018D-07
29	10	0.77463119D-08	0.33655374D-09
30	10	-0.53213638D-08	0.33119023D-08
31	10	-0.10817641D-07	-0.69287181D-08
32	10	0.14716654D-08	-0.56901835D-08
33	10	-0.50394481D-08	-0.52434766D-08
34	10	-0.34467713D-09	0.24340411D-08
35	10	-0.22752233D-08	0.34146808D-08
36	10	0.55930472D-08	-0.17411687D-09
11	11	0.48398690D-07	-0.58800050D-07
12	11	0.13758010D-07	-0.31481607D-08
13	11	-0.41031672D-07	0.31536657D-09
14	11	0.21808671D-07	-0.37303530D-07
15	11	0.80168083D-08	0.19461395D-07
16	11	0.21204344D-07	0.21009630D-08
17	11	-0.22567833D-07	0.12245764D-07
18	11	-0.11916001D-07	0.10361784D-07
19	11	0.68533465D-08	0.22666237D-07
20	11	0.19495408D-07	-0.14396429D-07
21	11	0.21455409D-08	-0.26217155D-07
22	11	0.19157435D-10	-0.27502751D-07
23	11	0.65475757D-08	0.97432039D-08
24	11	0.13105870D-07	0.19587435D-07
25	11	-0.49089877D-08	0.15564987D-08
26	11	-0.29324885D-08	-0.75955939D-08
27	11	0.29595129D-08	-0.74515308D-08
28	11	0.84661278D-08	0.89219661D-09
29	11	0.20352561D-08	0.11667116D-07
30	11	-0.96563729D-08	0.16887521D-08
31	11	-0.33813211D-08	0.12698065D-07
32	11	-0.40995538D-08	0.12381851D-08
33	11	-0.36089835D-08	-0.79371335D-08
34	11	0.15950159D-08	0.35206096D-10
35	11	-0.21814922D-08	0.69233322D-10
36	11	0.26913556D-08	0.55231163D-08
12	12	-0.68580102D-08	-0.10328716D-07
13	12	-0.32248271D-07	0.93092571D-07
14	12	0.85899052D-08	-0.29082469D-07
15	12	-0.34239874D-07	0.17910251D-07
16	12	0.17062256D-07	0.11851650D-07
17	12	0.29912157D-07	0.15844535D-07
18	12	-0.34138155D-07	-0.19786313D-07
19	12	-0.12491839D-07	-0.31196950D-08
20	12	-0.64459113D-08	0.26274379D-07
21	12	-0.57590374D-09	0.12882662D-07
22	12	-0.11599113D-07	-0.13467665D-07
23	12	0.10872785D-07	-0.14830201D-07
24	12	0.11970725D-07	-0.69547400D-08
25	12	-0.89828671D-08	0.72377591D-08
26	12	-0.18224311D-07	-0.39129547D-09
27	12	-0.37730105D-08	-0.28504969D-08

Table 1 (continued)
NORMALIZED SPHERICAL HARMONIC COEFFICIENTS OF EARTH'S GRAVITY MODEL GEM 10B[a]

n	m	C_{nm}	S_{nm}
30	17	-0.10606807D-07	-0.28880705D-08
31	17	-0.33103757D-08	0.88490160D-08
32	17	0.25264827D-08	0.90968026D-08
33	17	0.49443963D-09	0.89946404D-08
34	17	-0.56690044D-09	-0.18849902D-08
35	17	-0.56477835D-09	-0.13694238D-07
36	17	0.67930465D-08	-0.44794481D-08
18	18	0.52037601D-08	-0.29178848D-07
19	18	0.39148071D-07	-0.25941339D-07
20	18	0.41193895D-08	0.76970611D-08
21	18	0.19984044D-07	-0.13325109D-07
22	18	0.19994435D-07	-0.15620705D-07
23	18	0.78321154D-08	-0.90103992D-08
24	18	-0.17449874D-08	-0.97550394D-08
25	18	-0.62555009D-08	-0.26158611D-08
26	18	-0.18291913D-07	0.90645597D-08
27	18	-0.84223502D-08	0.88139399D-08
28	18	0.10871233D-07	-0.64142127D-08
29	18	0.51132314D-08	0.26362715D-09
30	18	-0.75585461D-08	-0.87272485D-08
31	18	-0.67335927D-08	0.41209920D-09
32	18	0.31457833D-08	-0.24816949D-08
33	18	-0.94481591D-08	-0.40676847D-08
34	18	-0.90441066D-08	-0.11577938D-08
35	18	-0.32153260D-08	-0.15127897D-08
36	18	0.36528024D-08	0.98877220D-08
19	19	-0.68537934D-08	0.81594160D-08
20	19	0.54833209D-08	0.23351964D-08
21	19	-0.12835855D-07	0.69449131D-08
22	19	0.50717678D-08	-0.13391765D-07
23	19	-0.17832989D-09	-0.55314985D-08
24	19	-0.11883153D-07	-0.84616281D-08
25	19	-0.28450461D-08	0.44103069D-08
26	19	-0.35822400D-08	-0.48600804D-09
27	19	-0.90366950D-08	-0.66282688D-08
28	19	0.10947230D-07	0.20709337D-07
29	19	-0.53393666D-08	0.14147462D-07
30	19	-0.75055017D-08	0.48742345D-08
31	19	0.55874096D-08	0.19008512D-08
32	19	-0.14137706D-08	0.13281584D-08
33	19	0.64117426D-08	-0.45476849D-08
34	19	-0.33015071D-08	0.22394070D-09
35	19	0.13875039D-08	-0.79342387D-08
36	19	-0.41815492D-08	-0.18665438D-08
20	20	0.22428401D-08	-0.94120039D-08
21	20	-0.38797504D-07	0.94653091D-08
22	20	-0.36384591D-10	0.20539817D-07
23	20	0.12437140D-07	-0.56371968D-08
24	20	0.70934596D-08	0.64452408D-08
25	20	-0.34035721D-09	-0.52895389D-08
26	20	-0.74902946D-09	-0.15726618D-07
27	20	0.23944083D-08	-0.91929951D-08
28	20	-0.47346023D-09	0.11430807D-09
29	20	-0.51001609D-08	0.16498255D-10
30	20	0.88197419D-09	0.98724901D-08
31	20	-0.70941823D-08	0.82092000D-08

Table 1 (continued)
NORMALIZED SPHERICAL HARMONIC COEFFICIENTS OF EARTH'S GRAVITY MODEL GEM 10B[a]

n	m	C_{nm}	S_{nm}
32	20	0.48672999D-08	0.30391239D-09
33	20	0.22115703D-09	-0.87581829D-08
34	20	0.44495142D-08	-0.36002538D-08
35	20	0.36065026D-08	0.10361750D-08
36	20	-0.29160866D-08	0.26955308D-08
21	21	0.37589144D-09	-0.10478099D-07
22	21	-0.27205978D-07	0.94656913D-08
23	21	0.20597207D-07	0.16294229D-07
24	21	0.65714052D-08	0.18100355D-07
25	21	0.48822133D-08	0.92473052D-08
26	21	0.11892530D-08	0.49623678D-08
27	21	0.91893865D-08	-0.47555246D-08
28	21	0.59315740D-08	0.50784976D-08
29	21	-0.25247656D-08	0.16755961D-08
30	21	-0.22416547D-08	0.18094019D-09
31	21	-0.48451211D-08	-0.20808116D-08
32	21	0.22382429D-08	0.33149910D-08
33	21	-0.23686877D-08	0.29587557D-08
34	21	-0.14803489D-09	-0.68719340D-08
35	21	0.35178192D-08	0.18353294D-08
36	21	-0.17300073D-08	-0.33589644D-08
22	22	0.31050443D-08	0.67625259D-08
23	22	-0.14550592D-07	-0.14749367D-08
24	22	-0.61306062D-09	-0.49214995D-08
25	22	-0.11396125D-07	0.10359640D-08
26	22	0.12384743D-07	0.56582561D-08
27	22	-0.49093056D-08	0.54072854D-08
28	22	-0.67467325D-08	-0.56945581D-08
29	22	0.86321338D-08	-0.79879995D-09
30	22	-0.23265344D-08	0.22184597D-08
31	22	-0.97149595D-08	-0.87287371D-08
32	22	-0.62878364D-08	0.28837254D-08
33	22	-0.14343741D-08	-0.10896039D-07
34	22	0.66002442D-08	-0.40224937D-08
35	22	0.86519458D-09	0.30841177D-08
36	22	-0.35741120D-08	-0.27931745D-08
23	23	0.43714923D-08	-0.10971668D-07
24	23	0.19640273D-09	-0.28212740D-08
25	23	0.16306136D-07	-0.69686684D-08
26	23	0.22031379D-08	0.13718756D-07
27	23	-0.88610100D-08	-0.19819974D-08
28	23	0.61384946D-08	-0.11617278D-07
29	23	-0.32809625D-09	-0.96784402D-08
30	23	0.26563561D-08	-0.77653432D-08
31	23	0.80504197D-08	-0.51574863D-09
32	23	0.26694826D-08	-0.43253306D-08
33	23	-0.64718731D-08	-0.78937361D-08
34	23	0.67662008D-09	-0.55152308D-08
35	23	-0.98411428D-09	0.59614030D-08
36	23	0.27391859D-08	-0.32831001D-08
24	24	-0.25464680D-08	-0.69397490D-08
25	24	-0.44779351D-09	-0.37047730D-08
26	24	-0.57883898D-08	0.21929532D-07
27	24	0.54452621D-08	0.13520631D-08
28	24	0.64926688D-08	-0.80693760D-08
29	24	0.43026515D-08	0.29281702D-08

Table 1 (continued)
NORMALIZED SPHERICAL HARMONIC COEFFICIENTS OF EARTH'S GRAVITY MODEL GEM 10B[a]

n	m	C_{nm}	S_{nm}
30	24	0.75149552D-08	-0.55102442D-08
31	24	0.51623267D-08	0.29596853D-08
32	24	0.21602603D-09	0.83745939D-08
33	24	0.77477734D-08	-0.10529860D-08
34	24	0.37082368D-08	-0.11428986D-08
35	24	0.20463872D-08	0.58307268D-09
36	24	0.15497107D-08	0.28574417D-09
25	25	0.16743265D-07	0.32817397D-08
26	25	0.19716450D-08	-0.39652609D-08
27	25	-0.53149552D-08	-0.30098382D-08
28	25	-0.79933164D-08	-0.22947255D-07
29	25	-0.26472101D-08	0.11035364D-07
30	25	-0.46512458D-09	-0.79902282D-08
31	25	-0.54828228D-08	0.44150872D-08
32	25	-0.11370033D-07	0.84750846D-08
33	25	0.12591879D-08	0.15323197D-08
34	25	0.36872034D-08	0.25671595D-08
35	25	0.33906291D-08	0.60152930D-08
36	25	0.59977767D-08	0.93403359D-08
26	26	-0.90761028D-08	0.12290155D-07
27	26	-0.36431909D-08	-0.64004772D-08
28	26	0.82987191D-08	-0.23483813D-08
29	26	-0.16227995D-08	-0.15594271D-07
30	26	-0.91119095D-08	0.18225034D-08
31	26	-0.70121308D-08	0.19969380D-08
32	26	0.12551808D-08	-0.88651630D-09
33	26	0.34591609D-08	0.14291411D-08
34	26	-0.35284851D-08	-0.71264253D-08
35	26	-0.91854900D-08	0.21183528D-08
36	26	-0.91285002D-09	0.28477507D-08
27	27	0.58046233D-09	-0.78504793D-09
28	27	-0.12974341D-07	0.13276551D-09
29	27	-0.25723722D-09	-0.35900129D-08
30	27	-0.76889869D-08	0.11770360D-07
31	27	-0.69922887D-08	0.12576083D-08
32	27	-0.95913863D-08	-0.45511753D-08
33	27	-0.39711679D-08	0.15600057D-08
34	27	0.95499272D-08	-0.20766601D-08
35	27	0.59342270D-08	-0.81986942D-08
36	27	-0.53307148D-08	0.33015899D-08
28	28	0.93876328D-08	0.85863422D-08
29	28	-0.31036363D-08	-0.99569211D-08
30	28	-0.17425188D-07	-0.34115637D-07
31	28	0.13232742D-07	-0.66395515D-09
32	28	0.21449650D-08	-0.27967431D-08
33	28	0.72318444D-08	-0.41164910D-08
34	28	-0.72544163D-09	-0.14288820D-07
35	28	0.21963398D-08	-0.81547912D-08
36	28	-0.13983505D-08	-0.86793443D-09
29	29	0.39604134D-08	-0.80856804D-08
30	29	0.29173331D-08	0.19867383D-08
31	29	-0.61380258D-08	0.11010648D-08
32	29	0.74023566D-08	0.16334769D-08
33	29	-0.12994623D-07	-0.23943729D-08
34	29	0.31160551D-08	-0.35230308D-08
35	29	0.66258439D-08	-0.33527843D-08

Table 1 (continued)
NORMALIZED SPHERICAL HARMONIC COEFFICIENTS
OF EARTH'S GRAVITY MODEL GEM 10B[a]

n	m	C_{nm}	S_{nm}
36	29	0.24294498D-08	-0.45908186D-08
30	30	-0.52123030D-08	0.11144587D-07
31	30	-0.13666349D-08	-0.56676035D-08
32	30	-0.58674772D-09	-0.16519921D-09
33	30	0.18986217D-08	-0.15788174D-07
34	30	-0.11871936D-07	0.11759477D-08
35	30	-0.22846320D-09	-0.66218480D-09
36	30	-0.39167848D-08	-0.91160455D-09
31	31	-0.83296446D-08	-0.39074208D-08
32	31	-0.37904170D-08	-0.15099211D-08
33	31	0.37746411D-08	-0.33637716D-08
34	31	-0.29887289D-09	0.91088585D-08
35	31	0.11917417D-07	0.81466324D-08
36	31	-0.33838417D-08	0.63204829D-08
32	32	0.12523843D-07	-0.13283820D-09
33	32	0.62566218D-08	-0.38480071D-08
34	32	0.69893350D-09	0.29327947D-08
35	32	-0.29799762D-09	-0.71317923D-08
36	32	0.10480769D-07	0.22054074D-08
33	33	0.91088555D-09	0.10384000D-07
34	33	0.88743139D-08	0.79451510D-08
35	33	0.12106222D-07	-0.41256160D-08
36	33	0.88900660D-08	0.40135876D-09
34	34	-0.48656484D-08	-0.34327377D-08
35	34	-0.49226736D-08	0.82902976D-08
36	34	-0.75176290D-08	0.66207880D-08
35	35	-0.79154536D-08	-0.73547895D-08
36	35	0.35990157D-08	-0.73197024D-08
36	36	0.24936818D-08	-0.88664947D-09

[a] The symbol D-xx denotes units of 10^{-xx}, i.e., D-07 = 10^{-7}.

Geoidal Undulations and Gravity Anomalies

Once the coefficients C_{nm} and S_{nm} have been determined, the geoidal undulations and gravity anomalies can be computed readily.[19]

The gravity anomaly g is given by

$$g = \left(\frac{GM}{a_e^2}\right) \sum_{n=2} \sum_{m=0}^{n} \left(\frac{a_e}{r}\right)^{n+2} (n-1)\,(\delta C_{nm} \cos m\lambda + \delta S_{nm} \sin m\lambda) P_{nm}(\sin \phi) \tag{7}$$

and the geoidal undulation N is given by

$$N = \left(\frac{GM}{a_e \gamma}\right) \cdot \sum_{n=2} \sum_{m-0}^{n} \left(\frac{a_e}{r}\right)^{n+1} (\delta C_{nm} \cos m\lambda + \delta S_{nm} \sin m\lambda) P_{nm}(\sin \phi) \tag{8}$$

where

$$\delta C_{nm} = \text{observed } C_{nm} - \text{reference } C_{nm}$$

$$\delta S_{nm} = \text{observed } S_{nm} - \text{reference } S_{nm}$$

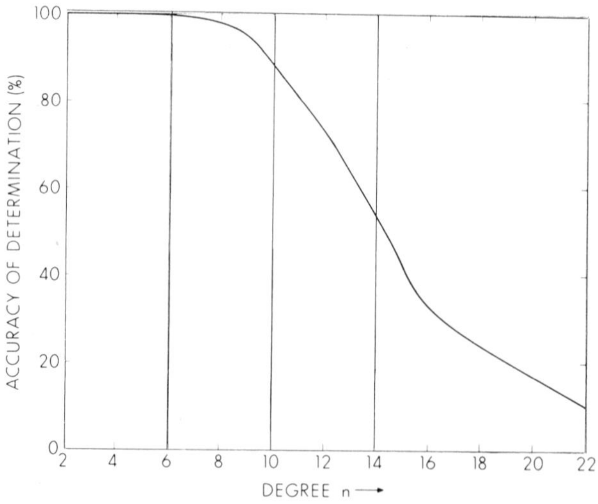

FIGURE 2. Estimated accuracy of gravity field model coefficients.

Table 2
POWER SPECTRA AND ERROR ESTIMATES FOR EARTH'S GRAVITY MODEL GEM 10B

Degree n	Power spectra ($mgal^2$)	Power spectra of estimated errors ($mgal^2$)
2	8.4	0.0
3	33.9	0.0
4	21.9	0.0
5	21.0	0.1
6	19.3	0.0
7	19.4	0.3
8	11.4	0.2
9	11.3	1.1
10	9.6	0.5
11	6.2	1.4
12	2.9	1.1
13	7.6	1.8
14	3.4	1.5
15	3.6	1.7
16	4.2	2.8
17	3.2	2.4
18	4.3	3.0
19	4.3	4.3
20	3.1	2.5
21	3.5	3.3
22	4.2	4.2

In these equations, the reference C_{nm} and S_{nm} define the reference figure. A customary reference figure is an ellipsoid of revolution. Because of axial or rotational symmetry of such an ellipsoid, all its $C_{nm} = S_{nm} = 0$ for $m \neq 0$. Also, because of its equatorial symmetry, all its odd zonal harmonics are zero. Thus only the even zonal harmonics are left. Of these, the coefficients of sixth degree or higher are negligibly small in a second order theory and are ignored. Thus, in practice, the reference ellipsoid is defined by second and fourth degree zonal harmonics. Consequently, all δC_{nm} and δS_{nm} in Equations 7 and 8 are equal to their corresponding observed coefficients except for δC_{20} and δC_{40} which are the differences of the appropriate observed and referenced values.

Note that the *perturbing potential* R, of satellite orbital theory is different from the *disturbing potential* of geodesy and geophysics. In orbital theory, R is the irregularity with respect to the central term of spherical attraction. In geodetic and geophysical applications, the disturbing potential represents the terms which are additional to the potential of an ellipsoid of revolution or the reference ellipsoid.

Geoidal undulation and gravity anomalies based on the coefficients given in Table 1, with C_{20} and C_{40} adjusted with respect to the Geodetic Reference System 1980 (GRS 80), are given in Figures 3 and 4, respectively. They give a global picture of the long wavelength geoidal and gravity variations over the entire oceanic areas (as well as the continental areas). Since the marine gravimetric surveys are generally confined to areas of tectonic interest, the satellite-derived gravity data gave, for the first time, a complete gravity representation over the oceanic areas, albeit in a relatively low frequency range.

Geoidal Undulations from Combination of Satellite and Surface Gravity

Since the satellite-determined geopotential yields the long wavelength components of the gravity field, and in the geoidal computations, the distant areas make a predominantly long wavelength contribution, the satellite-determined gravity can be combined with surface gravity data in the neighborhood of the computation point to obtain total geoidal undulation. If the gravity anomaly Δg is known on a uniform grid of reasonable size on the Earth, the geoidal undulation N at point P (ϕ,λ) is given by Stokes' formula[4]

$$N = \frac{a_o}{4\pi\gamma} \int_0^{2\pi} \int_{-\pi/2}^{\pi/2} \Delta g(\phi',\lambda')\, S(\Psi) \cos\phi'\, d\phi'\, d\lambda' \qquad (9)$$

where

ϕ,λ = latitude and longitude of point P
(ϕ',λ') = latitude and longitude of the variable point P′
a_o = mean radius of the Earth reference ellipsoid

The function S (Ψ) is called the Stokes' function and is given by

$$S(\Psi) = \csc\left(\frac{\Psi}{2}\right) - 6 \sin \frac{\Psi}{2} + 1 - 5 \cos \Psi - 5 \cos \Psi$$
$$- 3 \cos \Psi \ln\left(\sin \frac{\Psi}{2} + \sin^2 \frac{\Psi}{2}\right)$$

The quantity Ψ is the spherical distance between P and P′ and is

$$\Psi = \cos^{-1} \left[\sin\phi \sin\phi' + \cos\phi \cos\phi' \cos(\lambda' - \lambda)\right]$$

Note that as $\Psi \rightarrow 0$, $S(\Psi) \rightarrow \infty$. Hence, in the immediate neighborhood around P, the computation of N requires special treatment which changes Equation 9 to

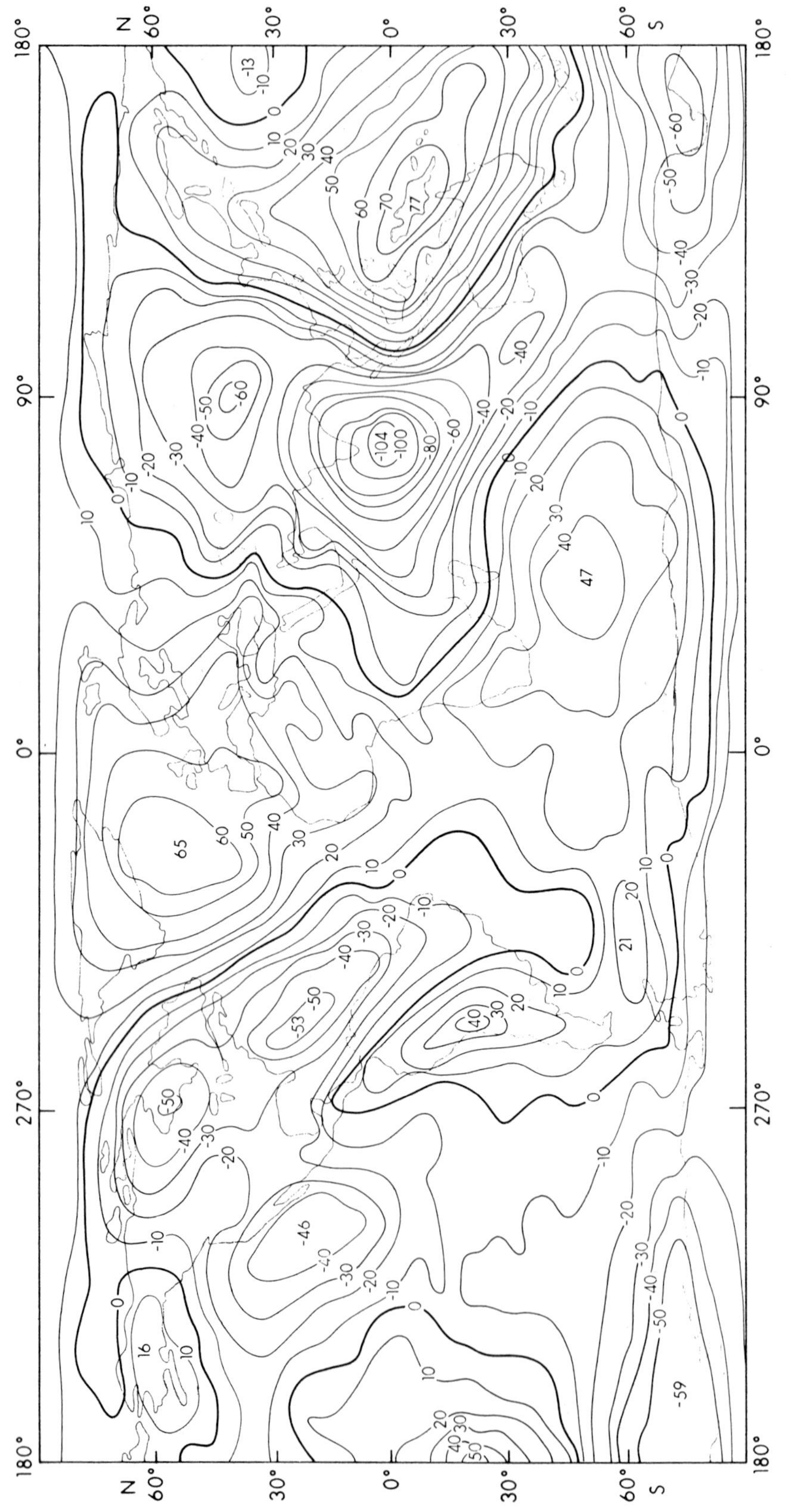

FIGURE 3. Geoidal undulations based on gravity model GEM 10B with reference to Geodetic Reference System 1980. Contour interval: 10 m.

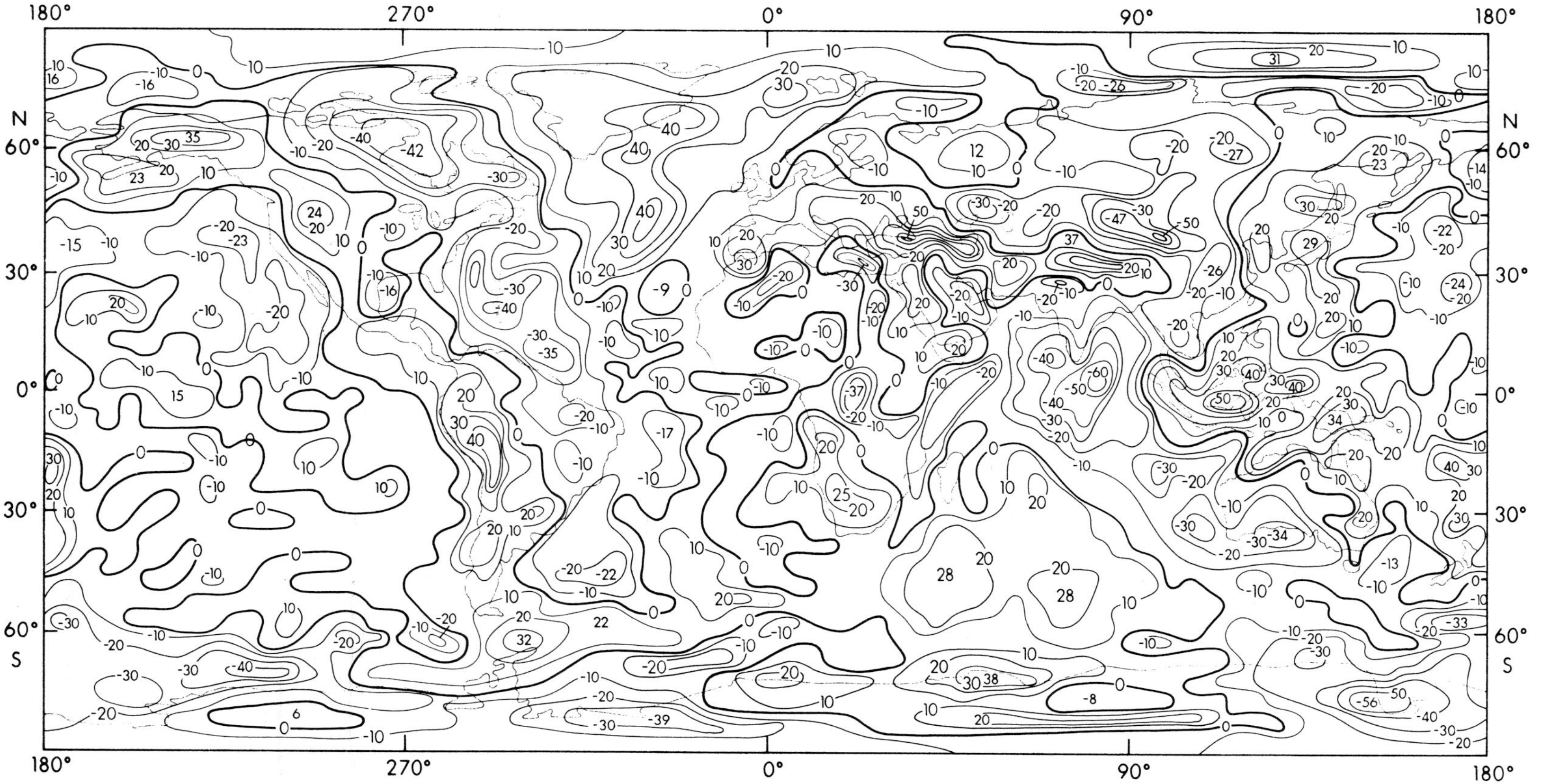

FIGURE 4. Gravity anomalies based on gravity model GEM 10B with reference to Geodetic Reference System 1980. Contour interval: 10 mgal.

$$N_p = \frac{S_o}{\gamma} \Delta g_p \tag{10}$$

where S_o is the radius of a small circular zone around P, and N_P and Δg_p are the geoidal undulation and gravity anomaly, respectively, at the point P.

The computation of N, however, can be designed so that the long wavelength components of the geoid are computed from satellite-derived gravity data while the Stokes' formula, applied to a limited area in the neighborhood of the computation point P, yields the local component of the geoid. Additionally, the effect of the innermost zone around P in which S(Ψ) goes wild, can also be computed separately. To achieve this, let the total gravity anomaly Δg be

$$\Delta g = \Delta g_s + \delta g$$

where

Δg_s = satellite determined gravity anomaly
δg = the difference between the total anomaly and Δg_s

The total geoidal undulation N can also be written as

$$N = N_p + N_s + \delta N$$

where

N_p = geoidal height component due to the innermost zone around P
N_s = geoidal height component due to Δg_s
δN = geoidal height component due to δg

The component N_p is computed from Equation 10 by substituting δg_p for Δg_p N_s is computed from Equation 9 by substituting Δg_s for Δg or directly from Equation 8. The component δN is computed from a modified form of Equation 9 so that at point (ϕ,λ)

$$\delta N = \frac{a_o}{4\pi\gamma} \int_{\lambda-\delta\lambda}^{\lambda+\delta\lambda} \int_{\phi-\delta\phi}^{\phi+\delta\phi} \delta g(\phi',\lambda') S(\Psi) \cos\phi' \, d\phi' \, d\lambda' \tag{11}$$

where the integration is clearly carried only over the area bounded by longitudes $\lambda-\delta\lambda$, $\lambda+\delta\lambda$ and latitudes $\phi-\delta\phi$, $\phi+\delta\phi$. With $\delta\lambda = 5°$, $\delta\phi = 5°$, the integration area for Equation 11 will be 10° × 10°. A 10° × 10° or 20° × 20° area around P is generally regarded large enough to account for the local contribution of surface gravity data adequately. Note that the innermost zone S_o, should be skipped in carrying out the integration in Equation 11. In general, this is taken care of by the choice of appropriate integration interval. The extent of the integration area in Equation 11 is generally determined by the amount of surface gravity data available, geoidal accuracy desired, and the computer resources available.

This technique of utilizing the strengths of both the surface and satellite gravity data in the computation of the geoid is originally due to Strange and Khan,[20,21] and has been applied subsequently by a number of investigators.[22] In applying this technique, it must be ensured that both the satellite-derived and surface gravity data are referred to the same datum. The geoids computed by this method are generally called "detailed gravimetric geoids".[22] Such geoidal heights are important as they provide an uninterrupted picture of the geoidal surface over ocean areas and have been used as an initial calibration standard for radar altimeters mounted on satellites which map the ocean surface in a synoptic fashion. The accuracy of these geoids is controlled, in addition to the satellite gravity models, by the quality, quantity, and distribution of surface gravity data and therefore varies from area to area.

Potentials and Limitations

Analysis of satellite orbits via perturbation theory techniques, and their combination with existing surface gravimetric data, have yielded gravity field models to degree $n = 36$ and order $m = 36$. This corresponds to a spatial resolution of about 550 km. However, there is some question about the reliability and individual geophysical significance of terms higher than $n = 14$ or 15, corresponding to a spatial resolution of about 1550 to 1650 km, though these higher terms are predominantly controlled by contributions from surface gravimetric data. These satellite techniques provided, for the first time, a global picture of large scale gravity variations over oceans which marine gravimetric data had failed to provide because of their sparse coverage. This was a major addition to our knowledge of the marine environments. Since then, this knowledge has been used to study the structure of the oceanic crust and mantle,[23] their relationships with bathymetry and topography, and their implications in terms of isostasy,[24-26] and numerous other geophysical problems.[27,28] The geophysical applications of these satellite results will not be discussed here as they have become a subject in their own right and cannot be condensed in a brief review of this type. It is clear from the foregoing, however, that these satellite techniques have made a major contribution to the exploration of the oceans.

Most of the satellite orbits available for such analysis are relatively high. At high altitude, the resolution of the gravity field is limited by several factors. For example, the small perturbing effects of the short wavelength components of gravity may not be within the measuring capability of satellite tracking systems. Even with improved precision laser ranging technology, the attenuation of short wavelength components with altitude makes their effect difficult to detect except for cases of resonance which, however, enable determination of only a few selected harmonic coefficients. For low satellite orbits, the atmospheric drag becomes a predominant factor. This creates the problem of selecting orbital altitudes as low as possible while ensuring at the same time that the atmospheric drag effects are low enough to permit both accurate orbital determination and reasonable orbital lifetime. For orbits much deeper in the atmosphere, satellite hardware must include drag-free systems. Although drag-free technology has been tried on some navigation satellites, it is still in an experimental stage. But even if the satellite orbits could be lowered to, say, about 200 km, their nominal spatial resolution will be no better than 100 km, due to, among other things, the attenuation of short wavelength components with altitude. This limitation is generally handled by combining surface gravity and satellite tracking data to strengthen the short wavelength information but surface gravity coverage is relatively sparse and is concentrated selectively. In addition, a spatial resolution of 100 km corresponds to a 180×180 ($n = 180$, $m = 180$) gravity field model as compared to the currently available 36×36 models. Continual updating and refinement of a gravity model with such a large array of coefficients and associated parameters, and the handling of the large data base required for their refinement requires enormous computer resources. For all these reasons, such techniques are not ideal for recovering the short wavelength components of the Earth's gravity field.

However, features with a horizontal scale of a few tens or hundreds of kms are among the most intriguing in marine geophysical studies. For example, in the compression belts of the plate tectonic system which are widely regarded as sources of continental crust creation, the ocean trenches are generally associated with very narrow negative gravity anomalies while the adjacent island arcs and contiguous areas have broad positive gravity anomalies. To resolve such fine detail, the short wavelength components of the order of a few tens of kms must be determined. Such requirements have stimulated interest in other space techniques which show greater promise and potential of mapping the finer structure of the Earth's gravity field. One such technique is satellite altimetry which is discussed in the next section of this chapter.

SATELLITE ALTIMETRY

Satellite altimetry is one of the relatively new space techniques which has proven its ability to yield the finer structure of the Earth's gravity field, not directly in its classical form of gravity anomalies, but in terms of sea level variations that are a close approximation to the familiar geoidal variations. While our poor knowledge of sea level variations caused by nongravitational oceanographic factors limits us from converting sea level variation into geoidal variations, and there is the subsequent problem of transforming an integral function, i.e., geoid, into its differential components, i.e., gravity anomalies, with its attendant loss of resolution, satellite altimetry has already made significant contributions in filling the void left by the satellite orbital analysis techniques.

Traditionally, our knowledge of *open ocean surface* has been extremely limited, clearly due to the virtual impossibility of frequent and extensive sampling through conventional means. Satellite altimetry obtains synoptic mapping of the instantaneous sea level almost directly. Unlike the satellite orbital analysis techniques which do not discriminate between oceanic and continental areas, satellite altimetry applications are limited, at least as of now, to ocean areas only.

The basic idea behind satellite altimetry is simple. A satellite is used as a moving platform mount for a microwave sensor which makes range measurements to the ocean surface by logging the round trip travel time of microwave transmissions. The microwave sensor is called *radar altimeter.* It is capable of transmitting very short pulses of radar frequency range with an antenna directed towards the ocean surface and, after their reflection and scatter from the ocean surface, picking them up on return to process them into altitude and average pulse shape data with the help of sophisticated techniques and electronics. The microwave frequency of the altimeters used so far is 13.9 GHz, which corresponds to an electromagnetic wavelength of 2.16 cm. Their spatial resolution is determined, *in part,* by the altimeter *footprint* which is a circular area *illuminated* by the incident microwave beam and which depends on the transmitted pulse width, beam pattern, and return pulse detection process. The satellite altimeters used so far have employed limited pulsewidth leading edge detection systems.

The first experimental radar altimeter was flown aboard a SKYLAB mission. An improved version was flown on GEOS-3 satellite mission which lasted much beyond its expected lifetime and yielded large quantities of useful data. A more advanced and improved version was flown on SEASAT-A mission. This mission ended prematurely but did yield considerable amounts of high quality data. During these missions, the altimeter measurement precision advanced from about ±1 m rms to ±20 cm rms for altitude measurements, and from 25% of wave height range of 1 to 10 m to ±1 m or

Table 3
SATELLITE ORBIT AND ALTIMETER CHARACTERISTICS OF GEOS-3 AND SEASAT-A MISSIONS

	GEOS-3	SEASAT-A
	Satellite Orbit	
Period	101.8 min	100.75 min
Altitude	843 km	800 km
Inclination	114.98°	108°
Orbits/day	14.1	14.3
Launch date	April 9, 1975	June 26, 1978
Orbital lifetime	≃ 3½ years	≃ 5 months
Residual orbit determination error	1—5 m	0.7—2.5 m
	Altimeter	
Antenna beamwidth	2.6°	1.5°
Altitude control	1.0°	0.5°
Frequency	13.9 GHz	13.9 GHz
Peak power	2.5 kW	2.5 kW
Pulsewidth	12.5 nsec	3 nsec
Footprint size	3.6—14.2 km	1.6—12 km
Measurement precision		
Altitude	±60 cm	±10 cm
Significant wave height ($H_{1/3}$)	25% of $H_{1/3}$ for $2 \leqslant H_{1/3} \leqslant 10$ m	larger of 1 m or 10% of $H_{1/3}$ for $1 \leqslant H_{1/3} \leqslant 25$ m

10% (whichever is larger) of waveheight range of 1 to 25 m. The spatial resolution has improved from 3.6 to 14.3 km for GEOS-3 to 1.6 to 12 km for SEASAT-A. Note that actual spatial resolution depends upon data acquisition rate and actual averaging interval in addition to footprint size. For SEASAT-A altimeter, the spatial resolution is not determined by beam width but by compressed pulse width and sea state conditions. Table 3 gives the basic data for GEOS-3 and SEASAT-A missions.

Altimeter Measurement Geometry

The principal behind obtaining sea level heights from satellite altimetry data is simple. Let $\overline{H}(\phi,\lambda)$ be the altimeter observation vector, $\varrho(\phi,\lambda)$, the geocentric radius vector of the satellite and $\overline{r}(\phi,\lambda)$, the geocentric radius vector of the subsatellite point P on the *reference ellipsoid* — an agreed upon reference figure best describing the shape of the Earth (see Figure 5). Then as a first approximation, the sea level height (relative to the reference ellipsoid) $h(\phi,\lambda)$, is[29]

$$h(\phi,\lambda) = \bar{\rho}(\phi,\lambda) - \overline{H}(\phi,\lambda) - \bar{r}(\phi,\lambda) - \epsilon \qquad (12)$$

where (ϕ,λ) are the latitude and longitude of the satellite and ε denotes the resultant error from all sources not considered so far.

The actual sea surface which intercepts and reflects the altimeter pulse is disturbed by such factors as sea swell, wind waves, tides, the complicated interactions of meteorological parameters and ocean perturbations, normal seasonal variations, changes in currents, etc. It is called *instantaneous sea surface.* It is the height $h(\phi,\lambda)$, of this surface

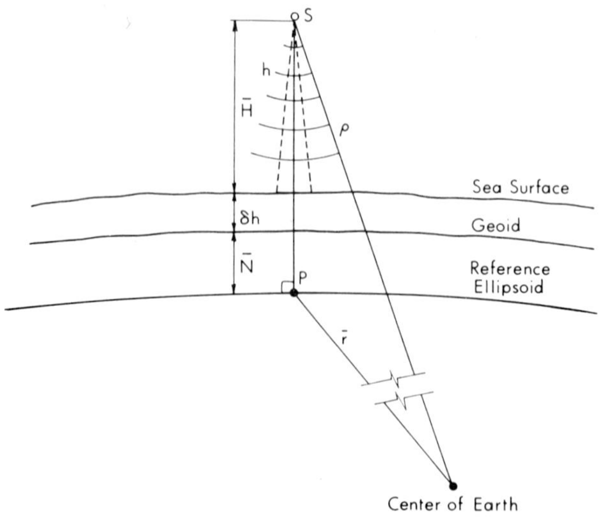

FIGURE 5. Principle of altimeter measurements.

which is mapped by the altimeter at a certain instant. It can be averaged for specific temporal variations to yield *mean sea surface* also denoted here by $h(\phi,\lambda)$. The exact definition of mean sea surface may vary depending upon the specific scientific objectives and even one's professional socialization and taste. In general, however, this mean surface can be considered coincident with the well known *geoid* so that geoidal height $N(\phi,\lambda) \simeq h(\phi,\lambda)$. Bear in mind, however, that mean sea level is *not always* coincident with the geoid and that one of the chief aims of satellite altimetry is to study the differences $\delta h = N(\phi,\lambda) - h(\phi,\lambda)$ which are known as *sea surface topography* and which have great significance in dynamical oceanography. Depending upon the choice of definition of mean sea level, sea surface topography may contain certain relatively long-term time dependent sea surface deviations from the geoid, such as Gulf Stream rings.[30] If $h(\phi,\lambda)$ is taken as instantaneous sea surface height or some variation of it, the difference $\delta h = N(\phi,\lambda) - h(\phi,\lambda)$ will also contain information on time dependent sea surface deviations.

Although the principle stated in Equation 12 is simple, the actual computations must contend with a number of observational data errors and instrument limitations that require elaborate data processing and analysis procedures.

Principal Sources of Error

Satellite Orbit Errors

It is clear from Equation 12 that errors in the radial or vertical components of satellite position will be transferred directly to sea level height determinations. Horizontal errors in satellite position present less serious problems but they do affect horizontal location of the satellite and its associated ground measurement points. Orbital errors arise from the following sources:

1. Gravity field model errors, caused by the uncertainties[17] or unmodeled effects in our knowledge of the gravity field and can be of the order of 10 m or more though recently developed of mission-specific geopotential models[31] can be used to reduce the orbital residuals of a specific satellite considerably.
2. Tracking station position errors, which arise from errors in the assumed positions

of tracking stations and can be of the order of 10 m or so. These mislocation errors are somewhat smaller in recent solutions but still are of the order of several meters.

3. Tracking system noise and bias errors, which include a wide array of errors associated with various tracking systems and can consist of calibration error, timing error, phase angle error, scale error, dynamic time lag error, etc. depending upon the system or systems contributing to observational data. The amplitude of these errors can range from a few centimeters to several meters.[34]
4. Errors due to unmodeled atmospheric drag which vary greatly in amplitude depending upon atmospheric variations, satellite configuration, satellite altitude and orbital arc length.
5. Errors due to solar radiation pressure variations. These again vary greatly in amplitude depending upon satellite shape and altitude, solar activity, relative position of the sun, etc.
6. Atmospheric propagation errors affecting the optical and electronic tracking data. For example, in the case of optical tracking systems, the propagation of an optical ray will be affected by atmospheric refraction, aberration due to the relative velocity of the satellite with respect to the observer, shimmer due to atmospheric turbulence, etc. In case of electronic tracking systems, the transmission of a radiowave will be affected by electronic refraction consisting of ionospheric refraction determined by variations of electron density, frequency of transmission, etc. and tropospheric refraction determined by variations of pressure, temperature, vapor content, etc.[2] These propagation errors vary from a few centimeters to a few meters.

The orbital computation programs handle these errors through a variety of modeling, differential improvement, and other statistical techniques during data preprocessing and data analysis stages. The *residual* errors which linger after these correction factors have been applied, determine the overall accuracy of the computed orbit. The residual errors are due to the unmodeled or incorrectly modeled factors and are caused by inadequate environmental data or knowledge, which itself is being improved with increasing orbital data. Their magnitudes are generally much smaller than those cited above except for categories (1) and (2). Thus, the overall residual orbital uncertainty can range from a few decimeters to several meters[31,32] depending upon the type, precision, and spatial and temporal distribution of tracking data, type of satellite and its altitude, length of orbital arc in question, the available error models, etc., and can vary from satellite to satellite. Current realistic estimates of orbital uncertainty for altimeter-carrying satellites are in the range of several meters, despite claims to the contrary. The most predominant and persistent sources contributing to residual orbital error are those in categories (1) and (2). Recently, mission-specific gravity models have been used to reduce the radial orbital error considerably.[31]

Altimeter Measurement Errors

Any errors in the altimeter measurement $\overline{H}(\phi,\lambda)$ in Equation 12 will affect the sea level height $h(\phi,\lambda)$. Major sources of altitude measurement errors by a satellite altimeter are as follows.

Propagation Errors

These are caused by the effect of the atmosphere on microwave transmissions and largely dependent on refractive index variation. These consist of *ionospheric effects* which depend upon electron density along the ray path and microwave frequency (the principal determinants of refractive index in the ionosphere), and *tropospheric effects* which depend upon pressure, temperature and relative humidity (which determine the refractive index in the lower atmosphere). These effects can introduce a radar measurement error of about 3 to 5 m[36] at altimeter frequency of 13.9 GHz, with most of the error caused by tropospheric effects. Most of this error can be removed if these effects are known.

The tropospheric refraction correction for GEOS-3 altimetry data[37] was based on an average or standard atmosphere which incorporates the *latitudinal* and *seasonal* (monthly) variations of the appropriate meteorological parameters but not their *diurnal* and *random* variations to reflect their *in situ* values. This leaves a residual error of a few decimeters from tropospheric causes[38] which can be reduced to just below a decimeter level if meteorological measurements are known in the vicinity of the altimeter measurement. The ionospheric correction, which is of the order of a decimeter or so, was not applied to GEOS-3 altimetry data. The residual tropospheric error and ionospheric error must be considered in any dynamic oceanographic applications and can be scaled down to a decimeter level for SEASAT-A altimetry data with the help of the accompanying microwave radiometer measurements.

Instrumentation Errors

These can be random or bias errors and are generally of three types.

1. Those caused by electronics limitations and include *closed loop tracker noise* — a timing jitter caused by thermal noise, *timing clock drift, altimeter drift, dynamic time lag* caused by the time lag between transmission and return of the altimeter pulse and resultant change of satellite altitude, etc. Their effects are of the order of a few decimeters, most of which can either be compensated electronically or modeled.
2. Errors caused by ocean surface reflecting characteristics. These can be random or bias errors and can be compensated electronically and by calibration. The bias caused by the shift away from mean sea level toward the wave troughs in proportion to surface roughness which was not compensated in GEOS-3 altimeter, was compensated in SEASAT-A altimeter.
3. Errors caused by antenna beam geometry. These consist of errors caused by *antenna offset* which can shift the reflection point away from the surface point nearest to the satellite or cause an asymmetric signal distribution around the measurement point, *spacecraft libration* causing similar effects as antenna offset, *beam spreading* which causes difference in path lengths between the central and extreme rays in the beam, etc. These errors can be removed by modeling techniques or hardware constraints.

A more detailed discussion of the expected errors and specific corrections for GEOS-3 altimeter data is given in Reference 37.

Geodetic Factors

These affect the derivation of sea level height h, from satellite altimetry measurements and are illustrated in Figure 6. An altimeter, employing a leading edge detection system, measures the height of the satellite in the direction of first signal return. This is along the local vertical to the instantaneous sea surface which passes through the satellite. Thus, the altimeter measurement $\overline{H}$ is along SM. The geocentric radius $\overline{\varrho}$ of the satellite is computed along the direction OS. The ellipsoidal radius *r* is determined along the direction $O\acute{S}$ as ϕ,λ are geodetic coordinates of the satellite. Sea level height

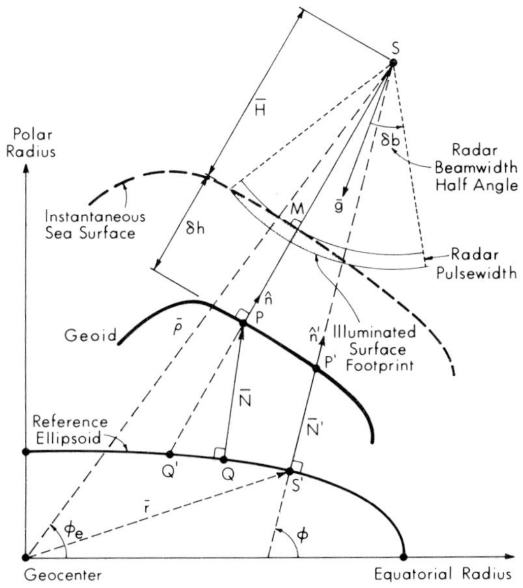

FIGURE 6. Detailed altimeter measurement geometry.

h = N + δh where δh is the deviation of sea surface with respect to the geoid. Note that N and δh are not necessarily colinear. Further complications result if sea level and geoidal surfaces are not parallel. The geodetic factors can introduce errors of several centimeters, but most of these can be removed by certain geodetic transformations.

Sea Level Heights and Geoidal Undulations

Major thrust of satellite altimetry data analysis techniques for recovering sea level heights is to correct for the radial orbital error, and often the altimeter drift error, as most of the other errors discussed above are generally removed during data preprocessing[37] or can be otherwise modeled. Satellite orbit can be determined either from the tracking data alone or in combination with altimetry data which are used as satellite range data in the adjustment. In the latter, certain sea surface wavelengths may alias into the computed orbit thus robbing the altimeter-derived sea surface signal of certain frequencies. The orbits can be computed in short segments (usually less than a revolution) or long arcs (several days). In a least square sense, short arc methods generally provide better fit to data but suffer from scale errors. On the other hand, in a long arc, the residuals can build up considerably. Figure 7 shows a typical radial orbital error[33] associated with a GEOS-3 altimeter pass. Note that the wavelength of the orbital error is much greater than the length of the altimeter pass.[32,33] This facilitates the filtering of the error by relatively simple manipulations such as a low degree polynomial filter.[29,33] Such filtering, however, can also deprive the altimetry signal of wavelengths comparable to the orbital error wavelength. More recently, mission-specific models[31] have been developed that reduce the radial orbital error of a specific satellite to less than 1 m.

A variety of techniques can be employed to filter out the orbital and altimeter drift bias from recovered sea level heights. One such method is to proceed by recovering h in Equation 12 by ignoring the residual error model ε initially. Define[29,33]

$$h = h^1 + h^2 + \epsilon^1 + \epsilon^2 \qquad (13)$$

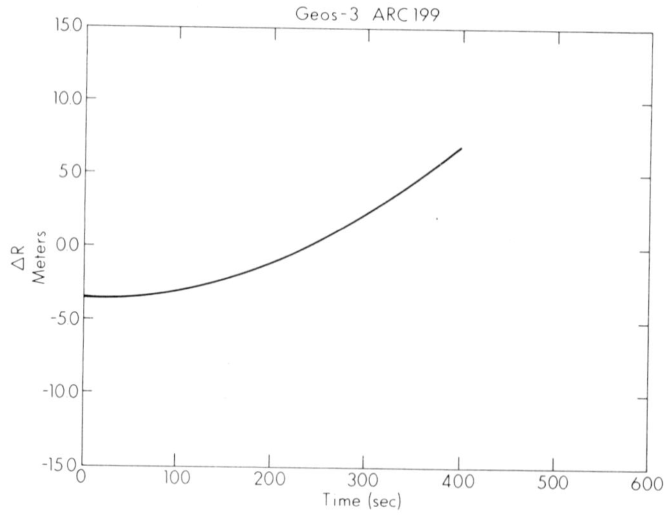

FIGURE 7. Typical radial orbital error associated with a GEOS-3 satellite altimeter profile. (From Khan, M. A., *E.O.S.*, 57, 901, 1976. With permission.)

where

$$h^1 = \sum_{i}^{n} h_i \quad , \quad h^2 = \sum_{i=n+1}^{n} h_i$$

with ε^1 and ε^2 defined correspondingly. Now *suppose* h^1 is known from, say, the satellite-determined geopotential model from techniques described earlier. The quantity ε^1 contains frequencies with $i \leqslant N$. Let us pass the signal $h - h^1$ through a filter of threshold frequency $i = N$. The output quantity will be $h^2 + \varepsilon^2$. By a careful choice of the threshold frequency, ε^2 can theoretically be made independent of the orbital and drift errors. If ε^2 can be *eliminated* or *identified* by statistical adjustment techniques, h will simply be

$$h = h^1 + h^2 \tag{14}$$

In this procedure, quantity h^1 is the long wavelength component of geoidal undulation N and is computed from Equation 8 by truncating n at a suitable frequency that is controlled by the accuracy level of the gravity model as well as the nature of expected orbital error. These procedures can be expanded to estimate sea surface topography and tidal effects by special treatment of residual error budget ε (which will further constrain the choice of the degree of truncation). The value of h can be refined further by comparing h at *crossover points* and adjusting the differences in a least squares process in a minimum residuals sense (much in the same way as marine gravity readings are adjusted on the intersections of ship tracks).

Note that in this procedure any errors in the long wavelength components of the reference geopotential model will spill into the sea surface heights. Also, any correctional information about those long wavelength components contained in the original altimetry signal will be eliminated. Other limitations of this method are discussed in References 29 and 33.

A variation of the above procedure is to attempt to absorb the radial orbital error directly in the adjustment process. In this case we can write

$$h_{rec} = h + x + y(t - t_o) + \epsilon \qquad (15)$$

for each data point in an altimetry pass. Here, t_o is the starting time of the altimeter data pass, t is the time of specific observation H from which h was derived, x and y are the bias and tilt corrections,respectively,due to the orbital and drift errors, and ε the residuals. From a network of altimetry data passes, one can set up difference equations at the crossover points and solve for the above parameters. Note that x and y remain the same for a single pass while h and ε change for each data point.

Another procedure[39] attempts to remove the radial orbital and drift errors as a part of an error model by the following adjustment scheme:

$$h + B\Delta = N + v \qquad (16)$$

where

- Δ = parameters of error model
- B = coefficients of parameters in Δ
- v = residual after adjustment
- h = altimetry-derived sea level height estimate, such as recovered initially from Equation 12
- N = geoidal height based on a geopotential model, such as from Equation 8

The recovered geoidal undulation N_{rec}, then is[39]

$$N_{rec} = h + B\Delta$$

Several other procedures exist which cannot all be summarized here. For details, refer to Reference 40 which describes some results of the GEOS-3 project.

The procedures developed for GEOS-3 satellite altimetry data analysis are now being honed further to avail of the opportunities created by the improved accuracy of SEASAT-A altimetry data and much more accurate (in a relative sense) orbital data provided by the relatively recent mission-specific gravity models which seem to have improved the radial orbital error from several meters to less than 1 m.[35] Because of the much better precision of the SEASAT-A altimeter, it is now reasonable to distinguish between altimeter-recovered sea surface and the geoid, and to estimate sea surface topography δh. Such distinctions were useful for GEOS-3 altimeter data only for the experimental development and refinement of analysis procedures. In general, the sea level heights recovered from GEOS-3 altimetry data can be regarded as geoidal undulations to an rms of $\pm$ 2m.

Figure 8 shows a few selected geoidal profiles[29,33] recovered from GEOS-3 altimetry data and compares them with geoidal information available from other sources.

Figure 9 shows geoidal undulations derived from GEOS-3 satellite altimetry data[39] over the Tonga-Kermadec Trench and the Patton seamounts areas.

Careful analyses of satellite altimetry data and treatment of error sources, combined with the reduction of orbital error with the help of mission-specific gravity models, seem to have achieved a *submeter level accuracy* ($\simeq \pm$ 50 cm) in sea surface height determinations for $1° \times 1°$ blocks on a global scale. This accuracy is expected to improve even further by further analyses of SEASAT altimetry data.

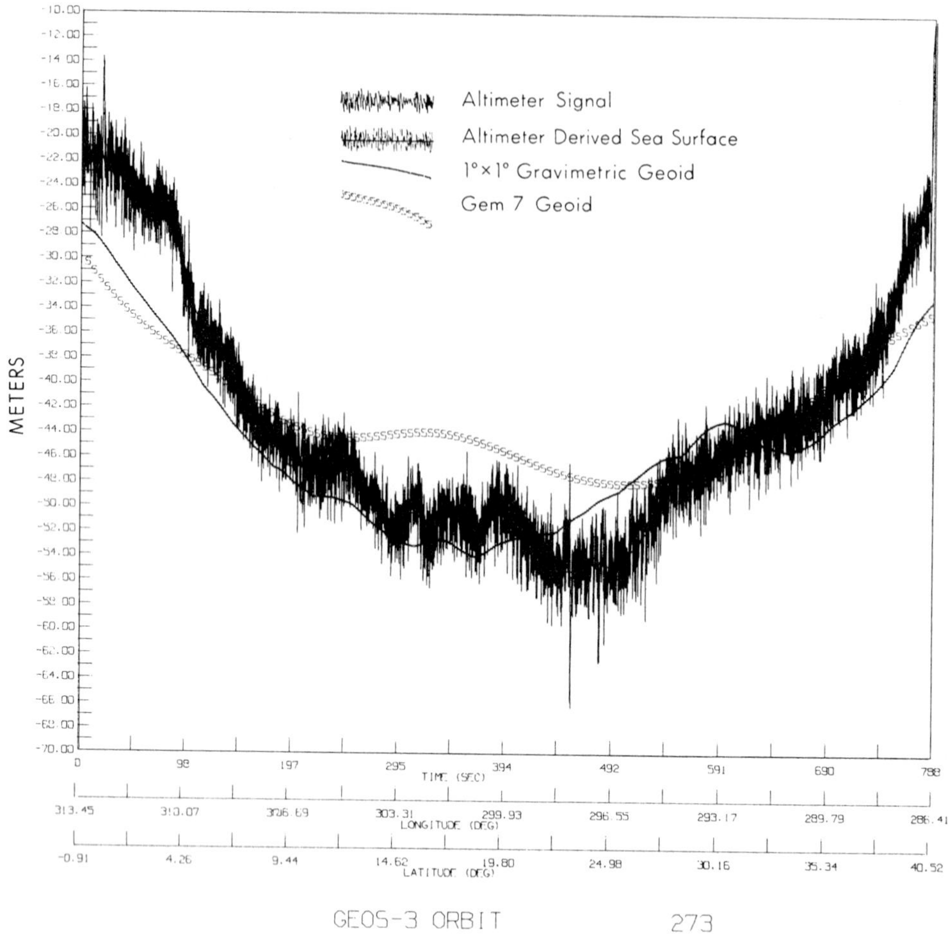

FIGURE 8. Selected geoidal profiles recovered from GEOS-3 satellite altimetry data. (From Khan, M. A., *E.O.S.*, 57, 901, 1976. With permission.)

Gravity Anomalies

The problem of deriving gravity anomalies from geoidal undulations suffers from the traditional theoretical limitation of estimating differential components from an integral quantity as the high frequency information is generally averaged and lost in the integration process. A number of techniques to derive gravity anomalies from geoidal information have been suggested. These fall in two principal categories.

The first category consists of procedures which depend essentially on an inversion of the process outlined in Equations 9 through 11. The quantity δN in this case is the difference between the altimetry-derived geoidal height and the geoidal height based on a reference geopotential model such as in Equation 8. In principle, then, the observation equations can be set up for a limited area (for ease of computation) by essentially an inversion of Equation 11, and δg values can be estimated by a weighted least squares estimation procedure. The recovered δg values can be finally added to the corresponding gravity anomaly component derived from the above reference geopotential model via Equation 7 to obtain the total gravity anomaly. It is clear that any contribution from the satellite altimetry data comes via δg. Several statistical variations of these procedures are available. See Reference 40 and the literature cited therein. It

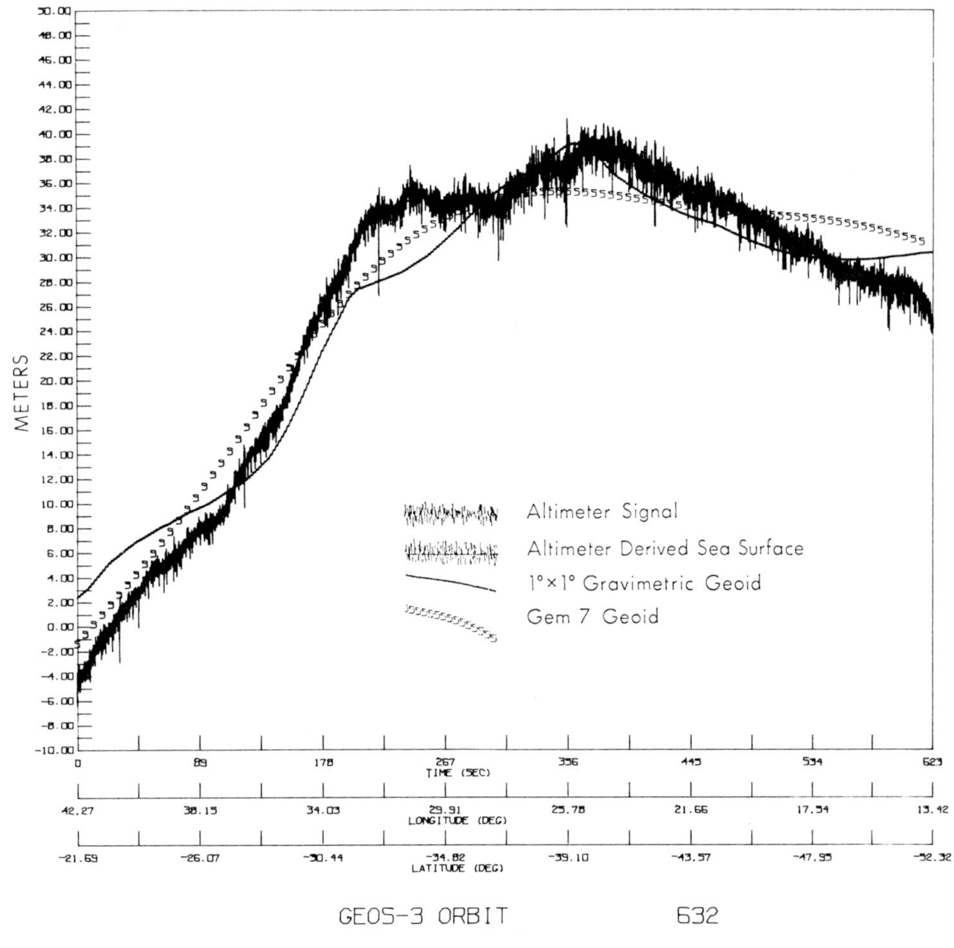

FIGURE 8. (continued)

has been suggested that the method of least square collocation provides more flexibility in the gravity anomaly estimation process.[39] These methods provide gravity anomaly information in areas with good altimetry data coverage. The accuracy of the computed gravity depends upon such factors as the density of altimetry data coverage, the size of the unit block area for which anomaly is sought, etc. For example, point anomaly computation seem to be constrained to an accuracy of a few tens of mgals while for $1° \times 1°$ blocks, accuracies ranging from ±2 mgal[42] to ±7 mgal[39] have been claimed. Other accuracy estimates range from ±3 mgal to ±8 mgal for $5° \times 5°$ blocks.[41] Analyses of SEASAT-A altimetry data is likely to improve these accuracies. Figure 10 compares the altimetry-derived gravity anomalies[39] with surface gravity data[43] over the Patton seamounts.

The second category of methods consists of deriving gravity anomalies from geoidal undulations by determining the geopotential coefficients. These methods are essentially an inversion of the process outlined in Equation 8. The altimetry-derived geoidal undulations are used in Equation 8 or some variation of it, to set up observational equations. Each data point yields an observation equation. Given enough geoidal data points — certainly the case in satellite altimetry, a large enough system of observation equations can be set up so that the total number of observation equations is much

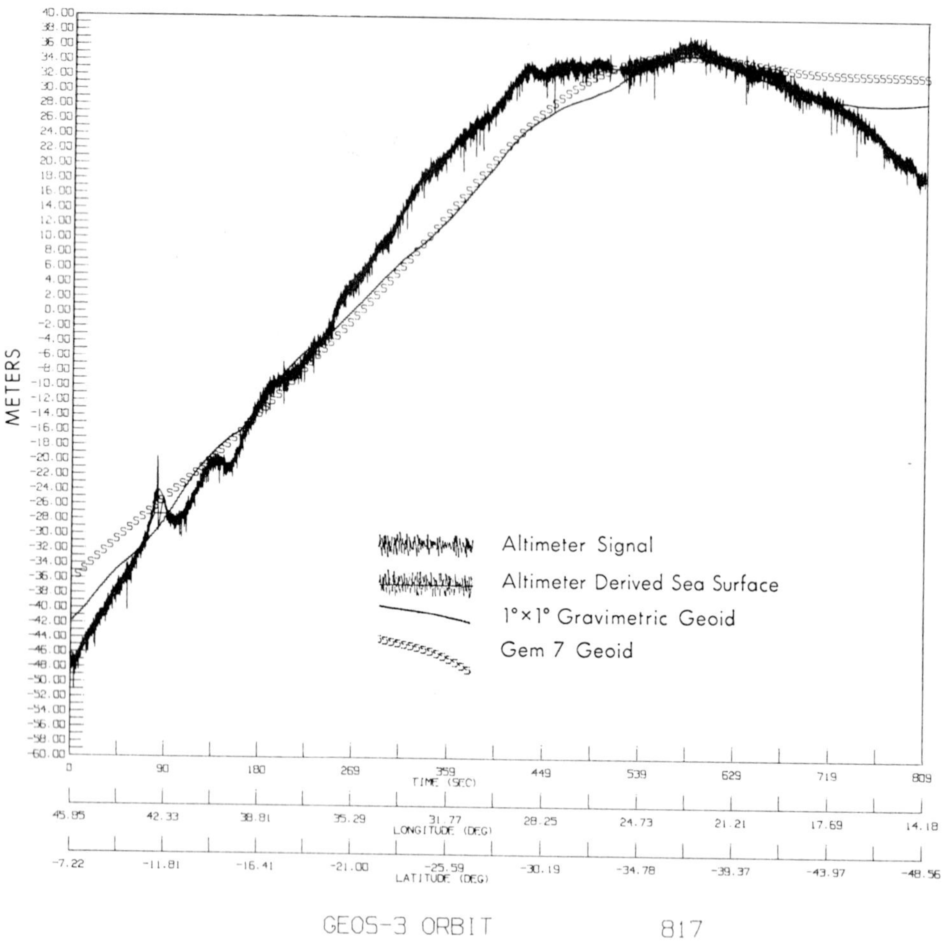

FIGURE 8. (continued)

greater than the total number of parameters to be determined, and thus a least squares estimation process can be used to compute a set of geopotential coefficients out to degree n. In general, however, the altimetry data are not used alone but in combination with other surface gravity and satellite tracking data to develop combination geopotential models. Once the geopotential coefficients are determined, gravity anomalies can be readily obtained from Equation 7.

Sea Surface Topography

Sea surface topography is the deviation of the mean sea surface from the geoid, though what exactly should constitute the geoid and mean sea level surface has been a matter of debate, caused at least in part by a lack of interdisciplinary communication between geodesists and oceanographers (though this has improved somewhat, lately). If $h(\phi,\lambda)$ of Equation 12 is taken as the instantaneous sea surface height, the differences $N(\phi,\lambda) - h(\phi,\lambda)$ will contain information ranging from temporal oceanographic effects such as eddies, storm surges, swells, tidal effects, etc., to dynamic oceanographic factors and can be processed to retrieve these factors. As an example, Figure 11 illustrates the tidal surface, along a subsatellite track over Patagonian shelf, recovered from SEASAT-A data[44] and compares it with the tide estimated from coastal data.

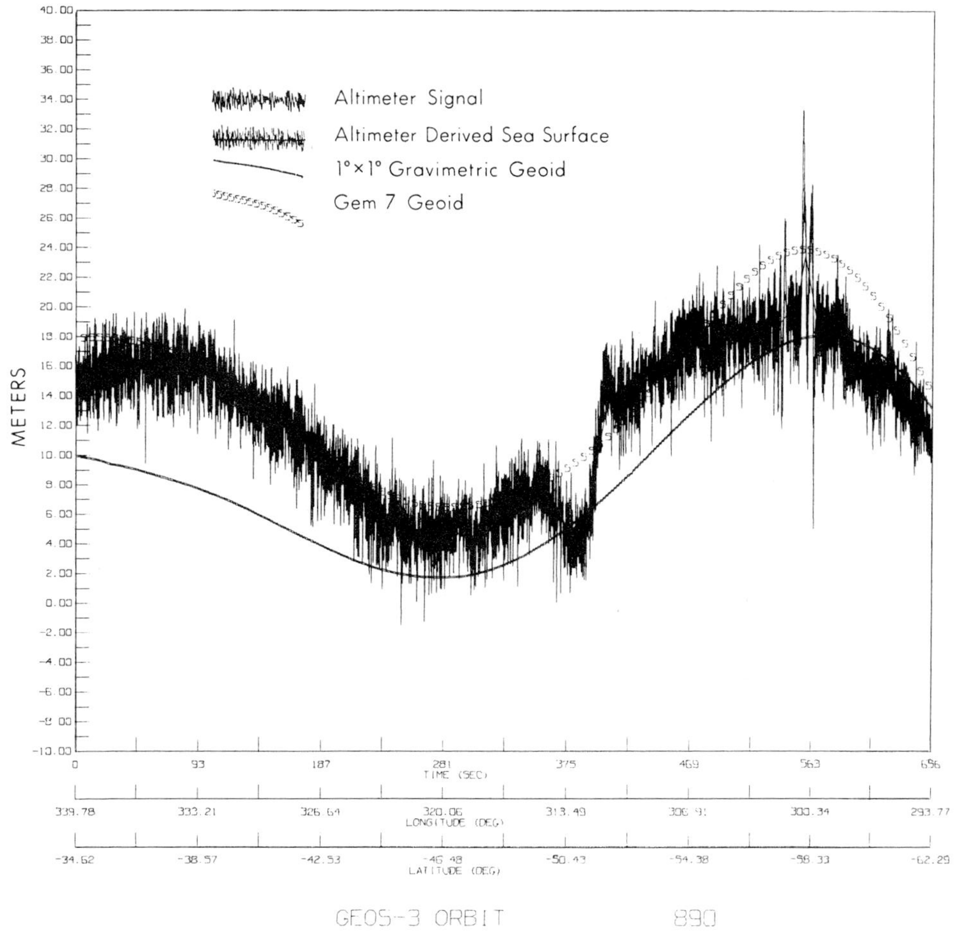

FIGURE 8. (continued)

Such time-variant phenomena can be periodic or have variable time scales determined by their causing agents. If these time-dependent factors are averaged out or otherwise extracted from $h(\phi,\lambda)$, the differences $N(\phi,\lambda) - h(\phi,\lambda)$ yield information on certain ocean dynamics phenomena. Such differences, however, should be determined to better than ± 10 cm to make a significant impact in such dynamic oceanographic applications as ocean circulation studies, etc. Table 4 summarizes the accuracy requirements in oceanographic applications of satellite altimetry.

It is clear that the determination of sea surface topography requires determination of both the geoid and mean sea level as well as a universal adoption of their exact definitions to make the results easily interpretable and comparable. Satellite altimetry has provided the means to determine mean sea surface to a submeter level accuracey and it is only a matter of time before much better accuracies, of decimeter or even several centimeter level, are achieved. The determination of the geoid to a comparable precision of ± 10 cm, however, imposes prohibitive constraints in terms of the required density or accuracy of gravity or astrogeodetic data, though the geoidal accuracy goal is easier to achieve for local rather than global geoids. Developments in satellite-to-satellite tracking and space gradiometry show promise in achieving the geoidal accuracy goals at sometime in the future. In spite of the restraints imposed by the limited

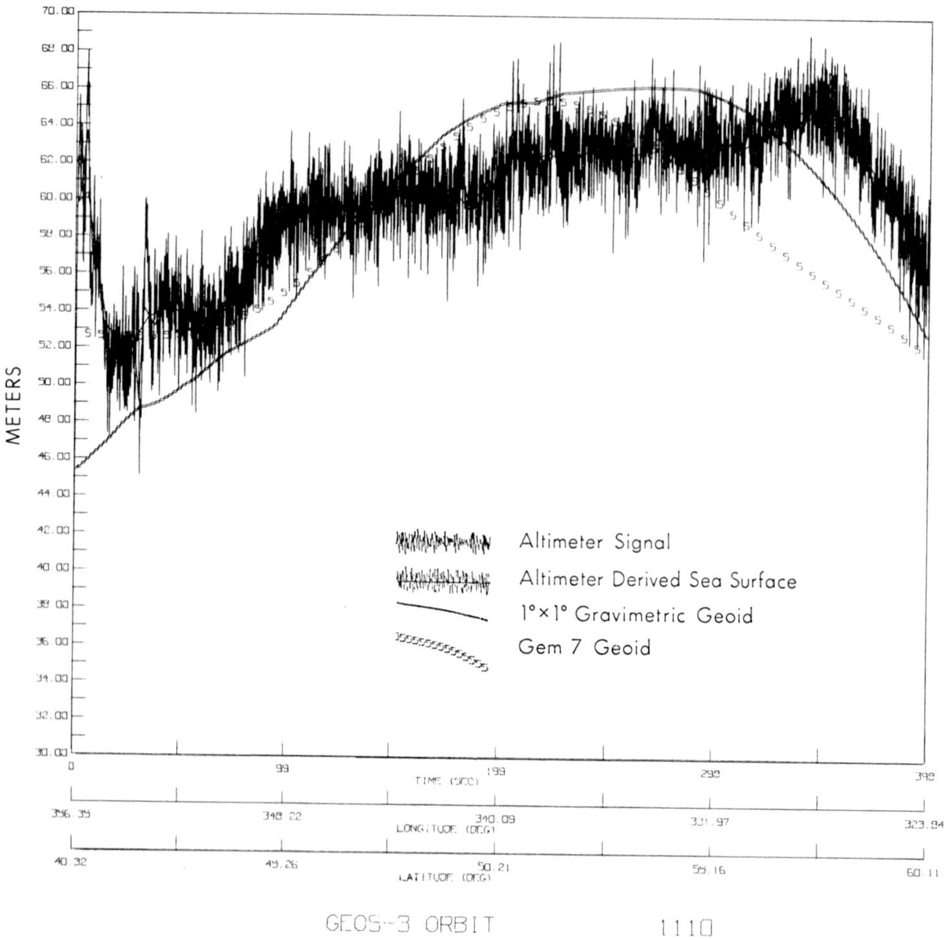

FIGURE 8. (continued)

geoidal accuracy presently achievable, satellite altimetry has already made significant contributions to oceanography. GEOS-3 altimetry data have been successfully used, even though only on feasibility study basis, to delineate ocean surface features such as Gulf Stream boundaries[45] and its associated rings.[30] Attempts have also been made to estimate meterological determinants associated with sea state, such as surface wind speed, from GEOS-3 data.[46,47] More accurate SEASAT-A altimetry data are now being used much more extensively in oceanographic studies such as a two-dimensional variability of the surface ocean circulation,[48] geostrophic current velocity determination,[49] mesoscale temporal changes in sea surface topography,[50] etc. A complete review of these contributions must await the final analyses results of SEASAT-A altimetry data.

Sea State Mapping

A major contribution of satellite altimetry to ocean sciences, perhaps not so well-known outside the circle of specialists, is the mapping of sea state or wave height measurements by determining the *significant wave height* (SWH) parameter $H_{1/3}$ which is defined as the average of the heights of the one-third highest waves observed at a point. These wave height measurements are obtained by studying the shape of the altimeter pulse, particularly the stretching of the leading edge of the return pulse due

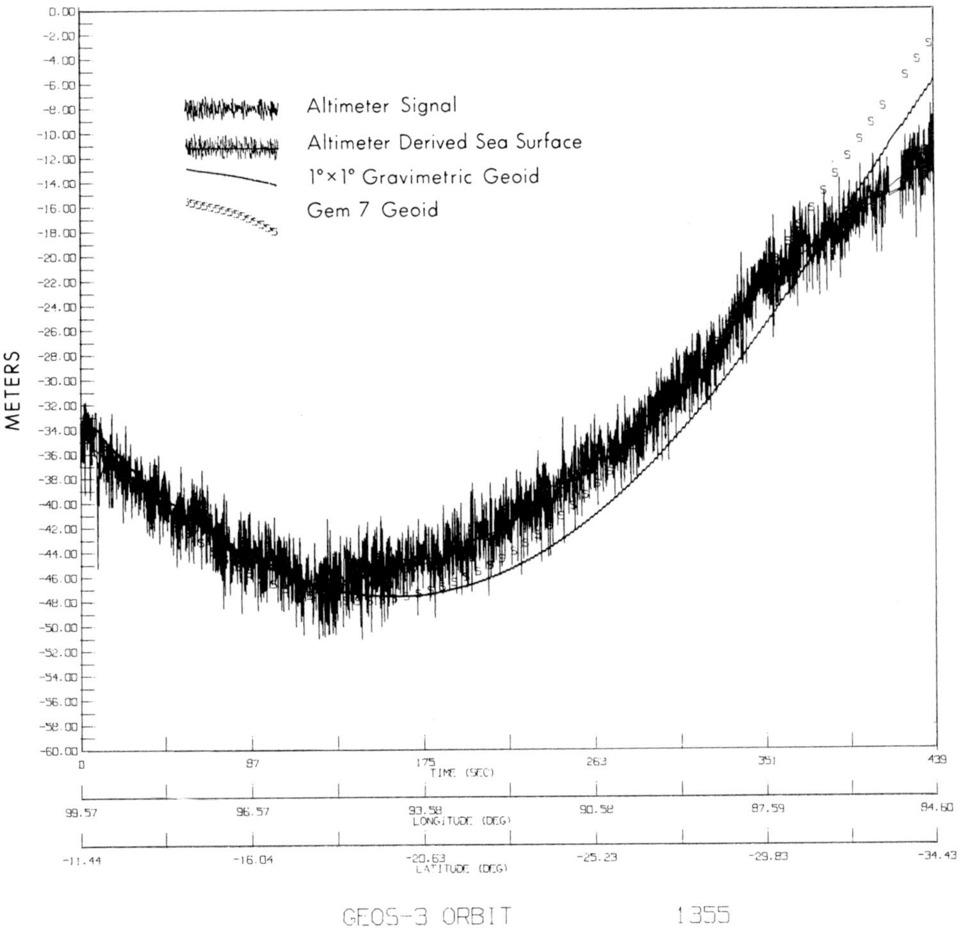

FIGURE 8. (continued)

to changes in ocean surface roughness. A radar wave emitted by the satellite altimeter is intercepted first by the crests of ocean waves directly below the satellite and is backscattered by their properly oriented facets. As the wave proceeds, the backscattered energy reflected from the rough ocean surface returns to the receiver from differencing depths, thus creating a random stretching of the leading edge of the return pulse. The determination of this stretching by appropriate statistical techniques provides a measure of $H_{1/3}$. To illustrate the principle,[51] let t_p be the rise time (stretching) of the leading edge of the compressed pulse of half-power width τ due to the waves of rms height σ, then

$$H_{1/3} = 4\sigma = \left[\left(\frac{2ct_p}{2^{1/2}}\right)^2 - \left(\frac{c\tau}{(2 \ln 2)^{1/2}}\right)\right]^{2\frac{1}{2}} \qquad (17)$$

where

c = speed of light in nsec

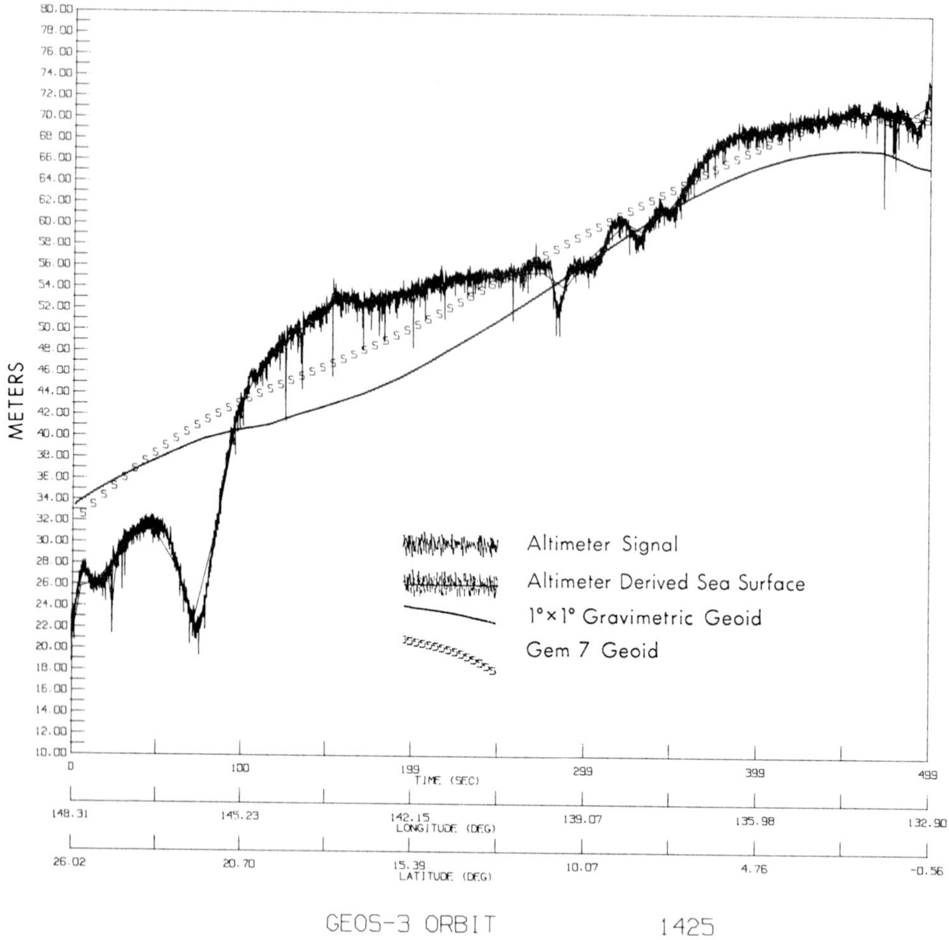

FIGURE 8. (continued)

and

$$t_p = (2/c)\left[\left(\frac{c\tau}{4(\ln 2)^{1/2}}\right)^2 + 2\sigma^2\right]^{1/2}$$

A number of techniques and algorithms have been developed to extract $H_{1/3}$ information from satellite altimetry data. All yield essentially the same results, which compare very well with surface truth data. Figure 12 shows, as an example, significant wave height measurements recovered from GEOS-3 altimetry data and compares them with shipboard wave height observations made by Ocean Weather Station vessels located in the North Atlantic.[52] Although there are some bias problems, the two types of measurements show promising agreement. The sea state derived from satellite altimetry has also been interpolated for comparisons with the forecasts of spectral ocean wave model (SOWM) of the Navy Fleet Numerical Weather Central (FNWC), the forecasts of Spaceflight Meteorology Group (SMG), and of National Meteorological Center (NMC). The results of such comparisons[52] were found very promising. The satellite altimetry results of significant wave height measurements are regarded so good that they have been already used to refine certain features of the SOWM.

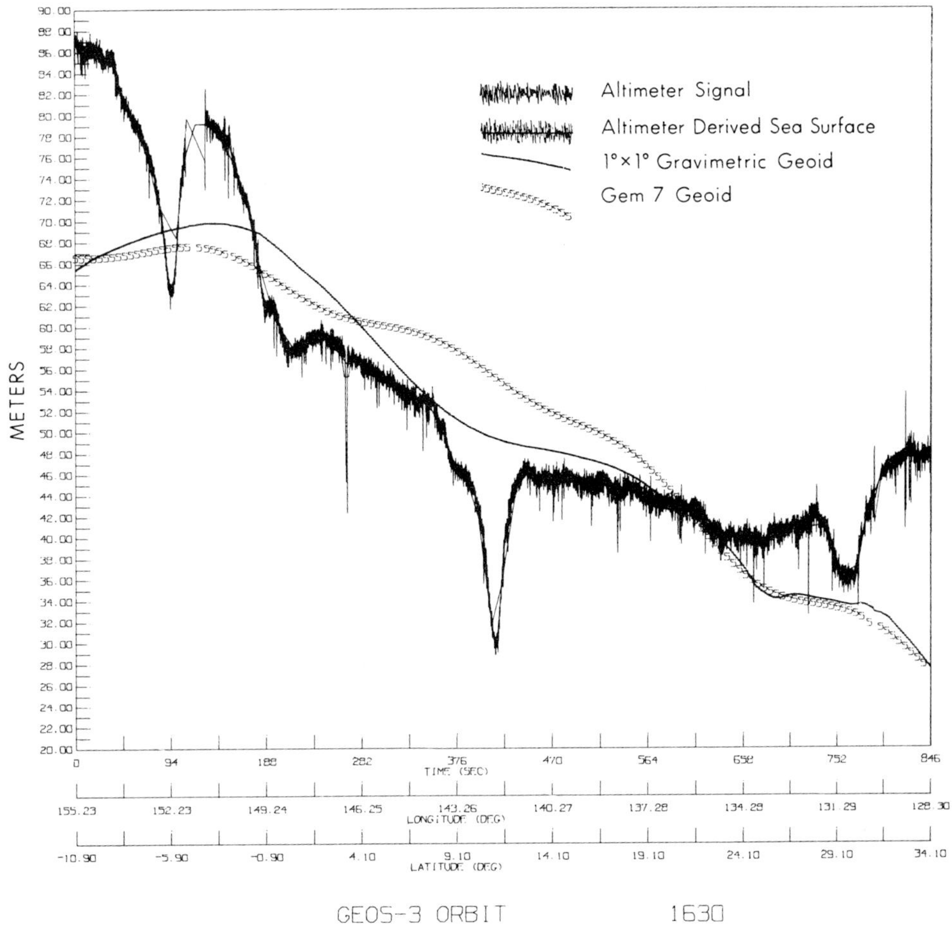

FIGURE 8. (continued)

POSITION LOCATION AT SEA BY SATELLITES

One of the major problems in ocean exploration is to tie the geophysical and oceanographic measurement sites to their geographical locations. These measurements will be clearly useless if their locations are not known correctly. The methods of determining such site locations fall into two broad categories: the traditional *Earth-based radiolocation methods* and the relatively recent methods of *positioning by satellites.* In the Earth-based methods, the position of a point is determined by nearly simultaneous measurements of radio signals from several fixed transmitters from a well-distributed network of ground station transmitters. These methods are very useful in good transmission area but their coverage is relatively small on a global basis. They also suffer from constraints of weather, signal interference, grid distortion, limited range, and the limited choice of ground station locations imposed by political, economic, geographic, and security-related considerations. Satellites, on the other hand, are *generally* free from such limitations and are a natural alternative for an accurate global navigation system.

The principle of position location by a satellite is quite simple. The satellite is regarded as a reference point in space whose position is known as a function of time in

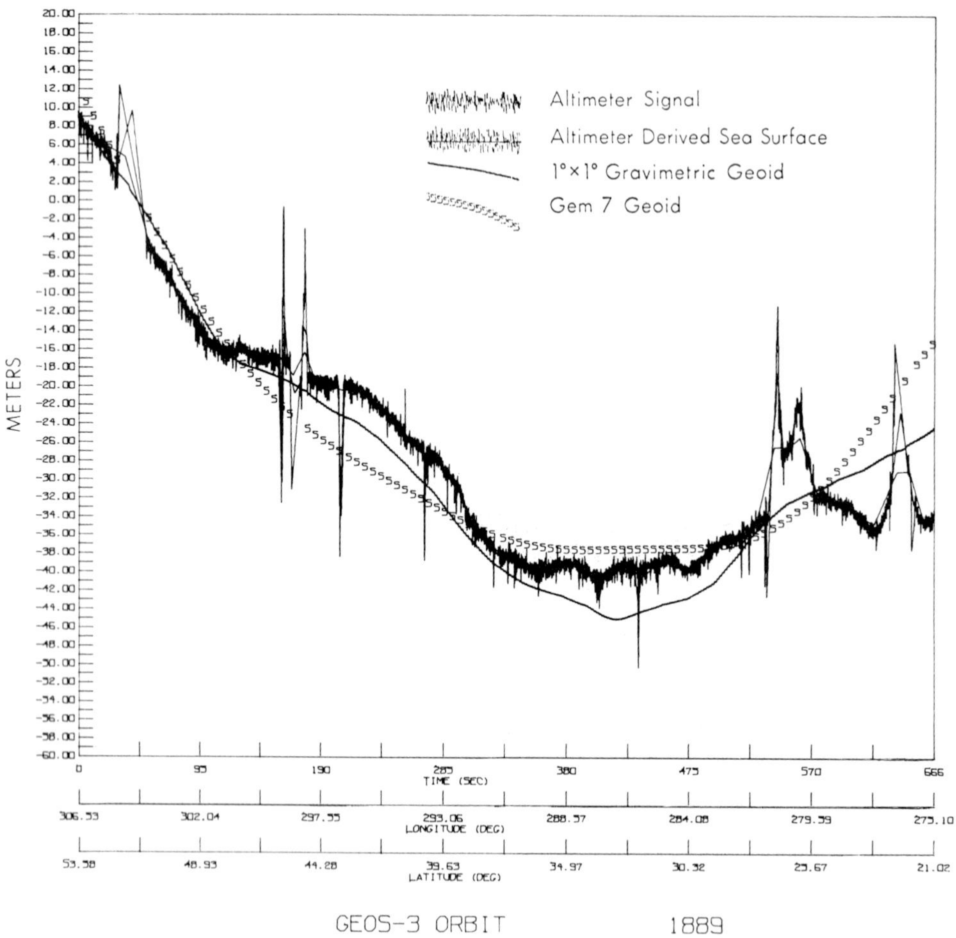

FIGURE 8. (continued)

a given reference coordinate system. The position of an observer relative to the satellite can be determined by monitoring, with specially designed equipment, the time of travel or frequency shift of electromagnetic signals emitted from the satellite. From this information, the position of the observer can be located in the reference coordinate system in quite a straightforward manner.

Positioning by satellites is used in a wide variety of applications, mainly of operational nature. The most visible of these is to locate the position of a vessel on the open ocean. In general, this entails position location from an individual satellite pass or from simultaneous sighting of several satellites. More recently, however, satellites are also being used for precise *geodetic positioning.* In the geodetic applications, either a single fixed station is occupied for several days to obtain a large number of satellite passes to recover the precise position of the fixed station by the *point positioning technique,* or two or more fixed stations are occupied to track a common set of satellite passes (mainly to identify orbital error) to recover the differential or absolute locations of the tracking stations by the *translocation technique.* Clearly, the geodetic positioning will be far more accurate than the positioning of a moving vessel from a single satellite pass. However, geodetic positioning by satellites is not within the purview of this chapter and will not be discussed in detail.

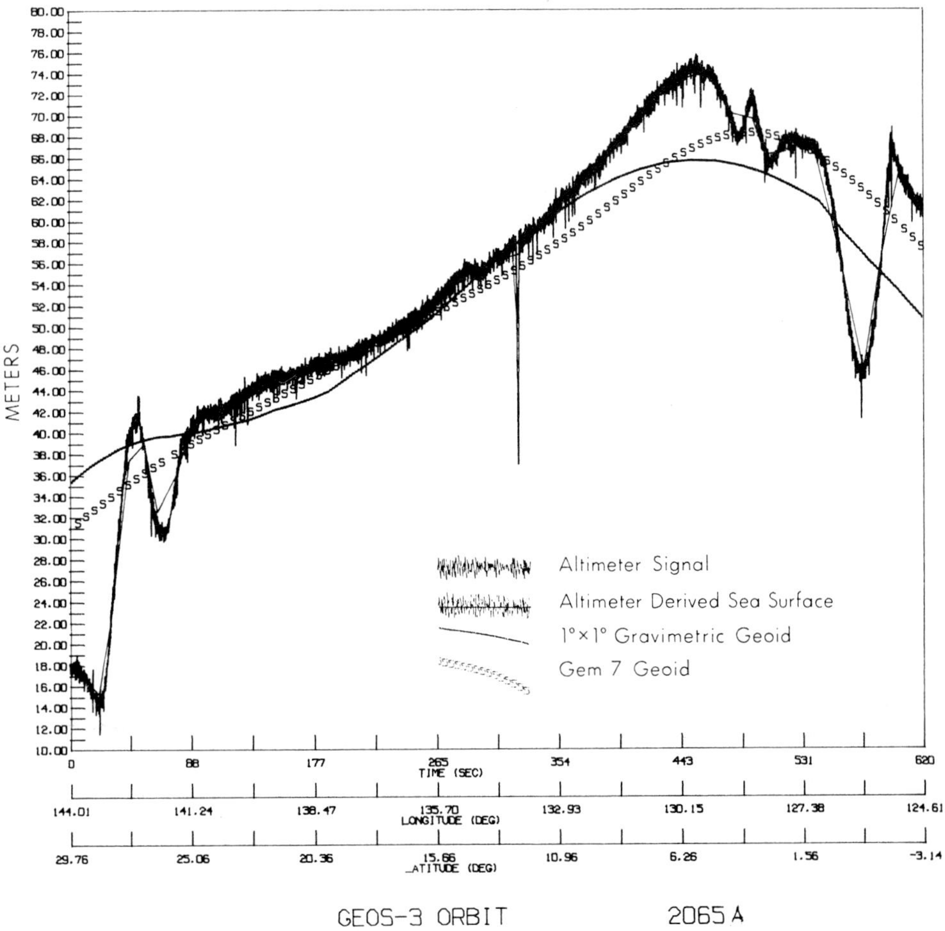

FIGURE 8. (continued)

At present, we have two major satellite systems for navigation purposes: the Navy Navigation Satellite System (NNSS), also called the *Transit*, and the Global Positioning System (GPS), also called the *Navstar*. The former has been fully operational since 1964; the latter is in the initial stages of implementation.

Navy Navigation Satellite System or Transit System

The system was conceived as a response to the urgent need for an accurate global navigation system for Polaris submarines. It became operational in 1964 for defense uses only, but was opened up for civilian use in 1967 and has provided reliable, uninterrupted, global position determination service since then.

The system consists of five operational satellites (previously six but Satellite 30180 turned *rogue* and started transmitting meaningless messages) in circular, polar orbits, circling the Earth every 107 min at a mean altitude of about 1075 km. This is illustrated in Figure 13, adapted from Reference 53. A *position fix* of a navigation receiver can be obtained whenever a satellite is above the horizon. During a satellite pass, which lasts typically 10 to 18 min, traversing an orbital arc of 4400 to 7900 km, sequential measurements of range differences to satellite position are taken via Doppler shift while the satellite transmits its orbital position information as a function of time to

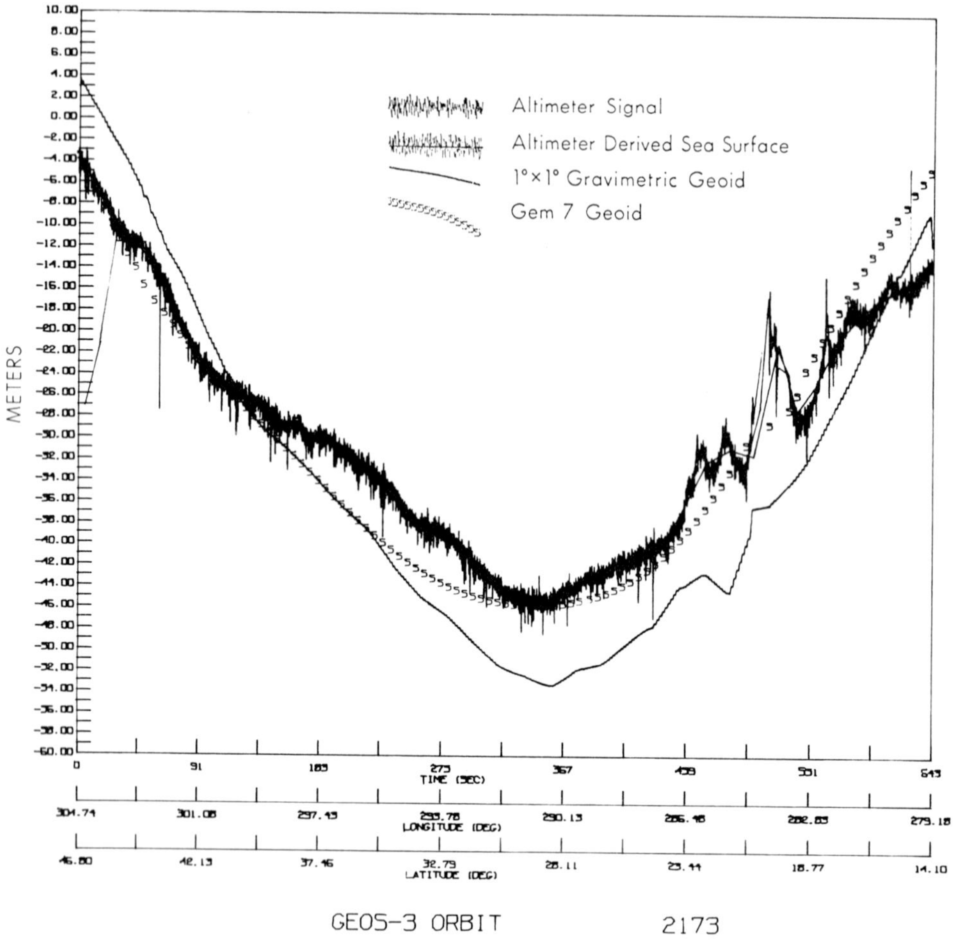

FIGURE 8. (continued)

the receiver. The quality of receiver equipment (single channel or double channel) plays a crucial role in the accuracy of position fix. Since these measurements are *sequential* and *not instantaneous*, the motion of the receiver (vessel velocity) must also be taken into account.

Position Determination for Geophysical and Oceanographic Surveys

A position fix can be for a *stationary observer* (receiver) or for a *mobile observer.* A stationary position fix requires (1) satellite position, (2) observer position relative to the satellite, and (3) an initial estimate of observer position. A mobile position fix also requires knowledge of observer's motion.

Satellite Orbital Position

Information about satellite orbital position is available in two codes: *precise ephemeris* and *broadcast ephemeris.* Precise ephemeris is available mostly for defense-related and geodetic applications with high precision requirements. It has an accuracy of ±2 m in components of position.[54] It is derived from tracking data gathered over periods of 48 hr from a global tracking network of about 13 *Tranet* stations, four mobile stations, and the current ground support network of four tracking stations for Transit

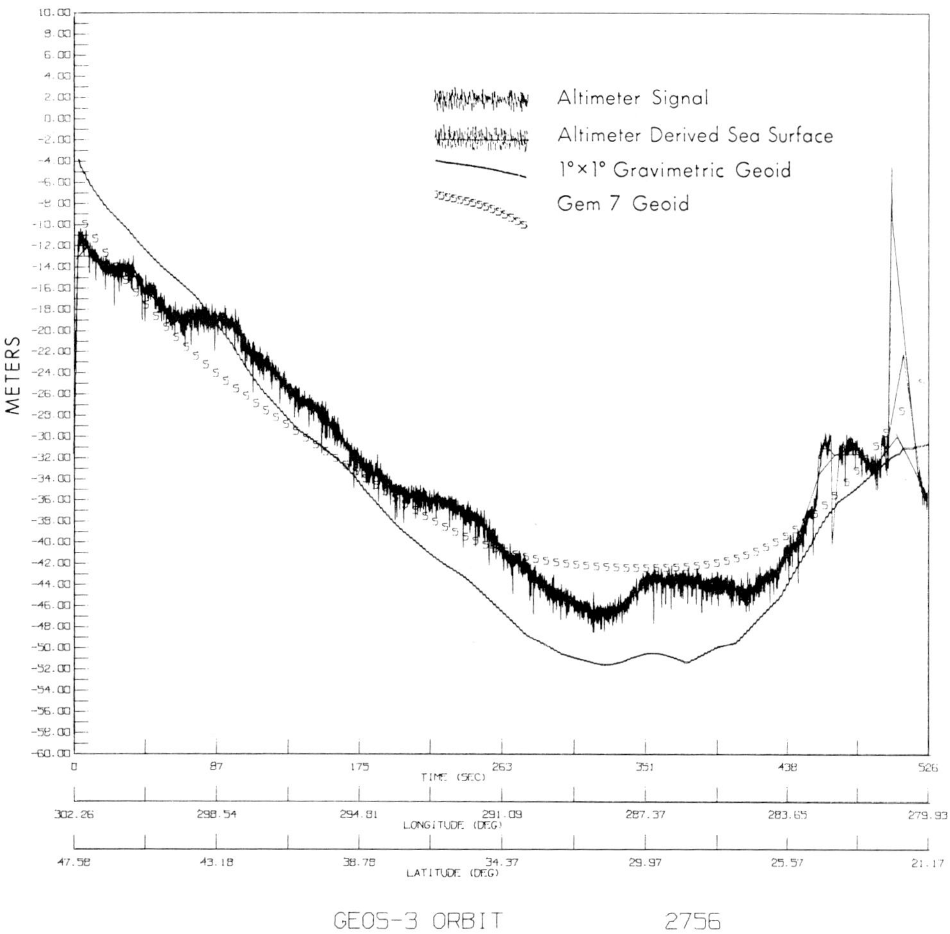

FIGURE 8. (continued)

satellites. Broadcast ephemeris is available to all users. It is based on the latest satellite tracking data obtained by the current ground support networks of four tracking stations for Transit satellites. It has a degraded accuracy of 17 m, 16 m, and 8 m in along-track, cross-track, and radial components of satellite position.[55]

The predicted orbital ephemeris (precise or broadcast) is stored in the satellite and is updated every 12 hr. The satellite transmits this orbital information to the ground receiver in a *navigation message* with highly precise timing marks. From the information supplied in the navigation message, the computer obtains satellite position as a function of time.

Position of the Observer Relative to Satellite

This is determined by measuring the Doppler shift due to the variation of the distance between the satellite and the observer during the observation interval. Let f_G be the *ground reference frequency* of the navigation receiver oscillator, f_T the frequency transmitted by the satellite, and f_R the frequency received from the satellite at the receiver. The receiver from the frequency offset $f_G - f_R$, the integral of which gives the Doppler count,[56] i.e.,

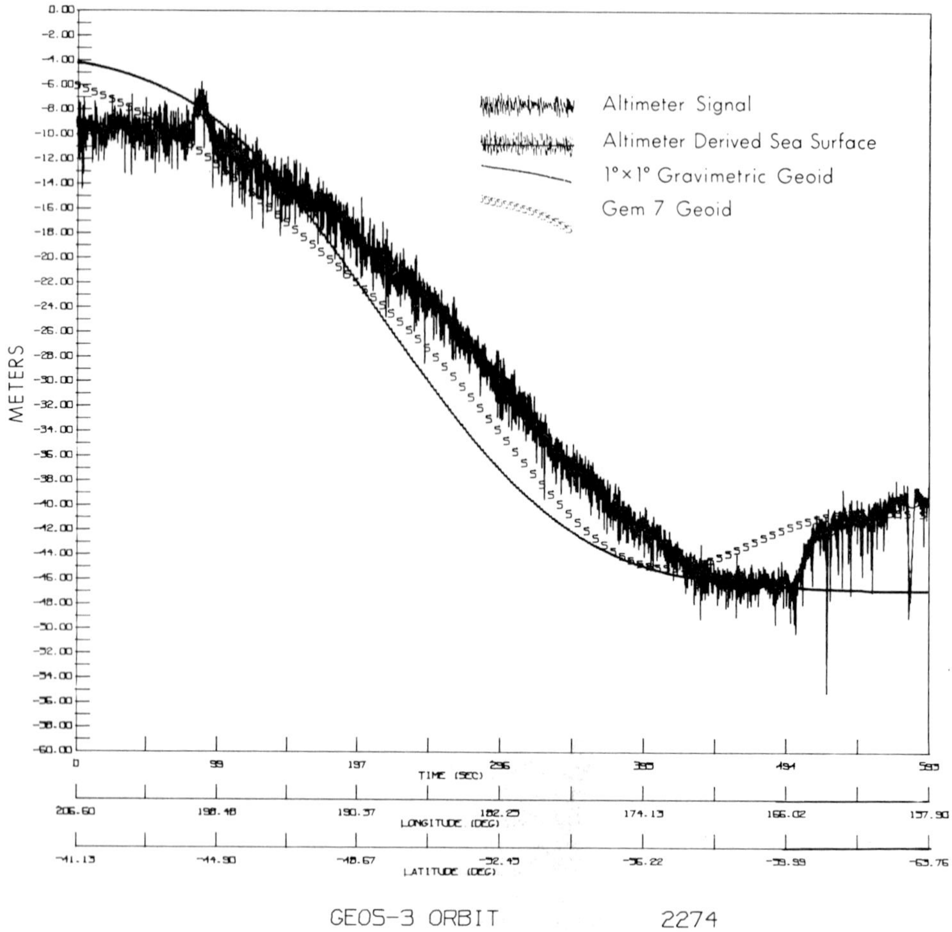

FIGURE 8. (continued)

$$N_1 = \int_{t_1 + r_1/c}^{t_2 + r_2/c} (f_G - f_R)dt$$

where r_1/c is the time taken by the satellite timing mark transmitted at t_1 to travel the distance r_1, from satellite to receiver, c being the velocity of light.

This can be integrated, under certain assumptions, to yield

$$N_1 = (f_G - f_\Upsilon)(t_2 - t_1) + \frac{f_G}{c}(r_2 - r_1) \qquad (18)$$

Doppler count intervals generally depend on the navigation message format. During a typical satellite pass of 10 to 18 min duration, which will form an orbital arc baseline of about 4400 km to 7900 km, one can obtain about 20 to 40 good counts.

Motion of the Observer

The motion of the observer/receiver during the satellite pass can be taken from the

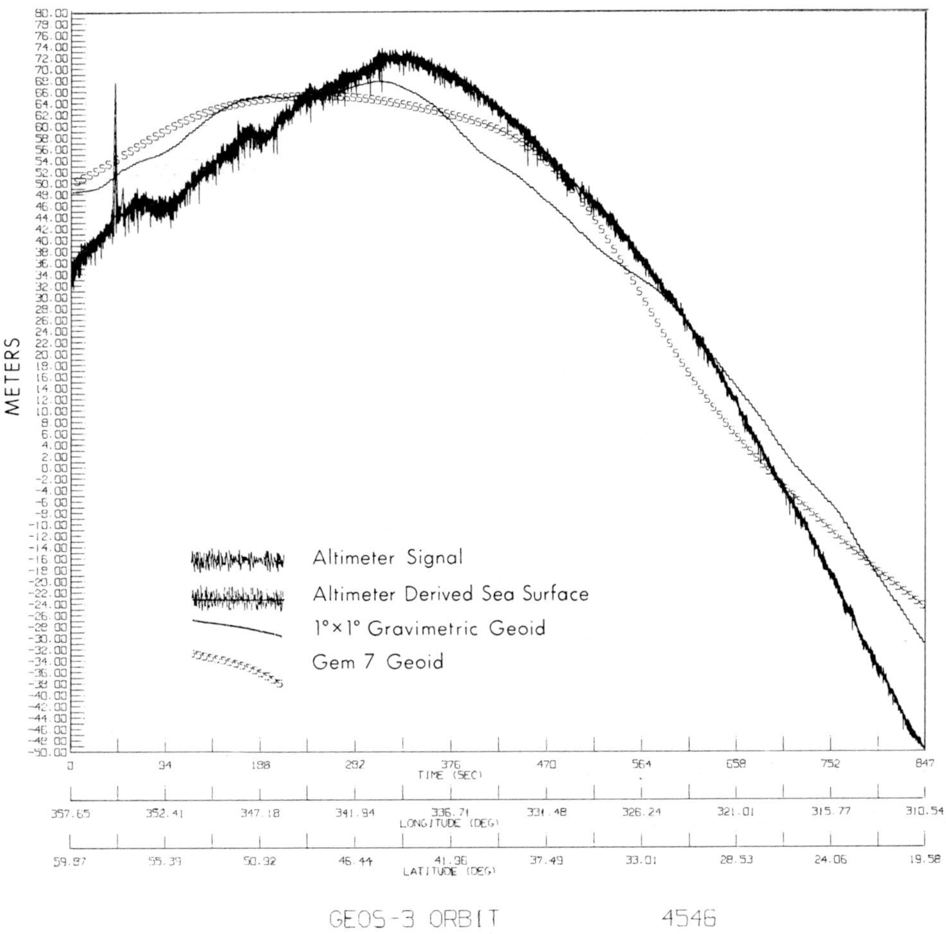

FIGURE 8. (continued)

speed and heading record of the vessel and can be made available to the onboard computer, frequently as an automatic input. From this information and from an initial estimate of observer's position, the computer can prepare a table of the estimated positions of the observer as a function of time.

Position Fix Computation

From the information contained in the navigation message, the onboard computer computes the satellite position at the beginning and end of each Doppler count. It also has access to estimated position of the observer at the beginning and end of each Doppler count. From this information, it computes the *slant range* to each satellite position and compares this *estimated* slant range change to the *observed* slant range change given by the Doppler count to form *residuals* between the two quantities. These residuals are then *minimized* in a least squares estimation process to derive the best estimate of the receiver position. The results of the least square adjustment are generally in the form of latitude correction $\Delta\phi$, and longitude correction $\Delta\lambda$ (the correction to the frequency offset is also computed) to the initial position of the receiver with corresponding adjustment for subsequent positions.

The onboard computer is generally preprogrammed for the entire run of computations to yield the adjusted positions.

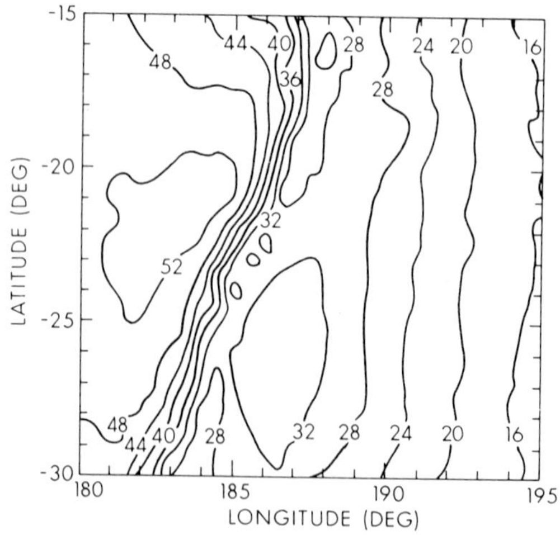

Altimeter geoid in Tonga-Kermadec Trench area, contour interval: 4 m.

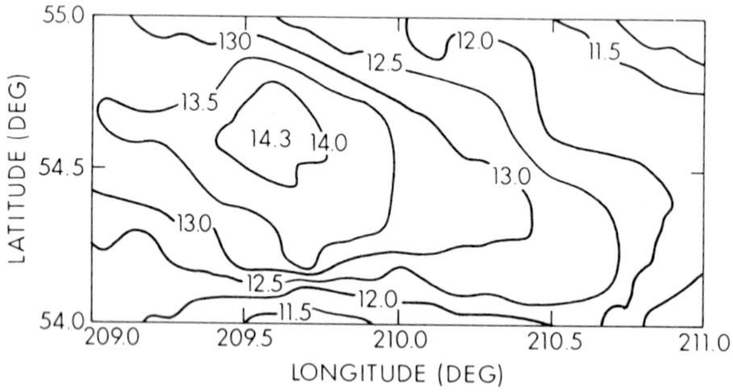

Altimeter geoid over Patton seamounts, contour interval: 0.5 m.

FIGURE 9. Geoidal undulations derived from satellite altimetry data over typical bathymetric features. (From Rapp, R. H., *J. Geophys. Res.*, 84, 3784, 1979. With permission.)

Sources of Error

The *residual errors* arising from the *unmodeled effects* of error sources affecting the positioning accuracy of a ship from an *individual satellite pass* are given below.

Satellite ephemeris errors — These consist of *gravity model uncertainties* ≃ 10 to 20 m, unmodeled and incorrectly modeled effects of *atmospheric drag* and *solar radiation pressure* ≃ 10 to 25 m, and *ephemeris rounding* error ≃ 5 m.

The residual error in satellite position from these sources is about 15 to 33 m (rms). This compares with an rms error of about 25 m estimated for the broadcast ephemeris[55] (excluding the rounding error). The error in the precise ephemeris is much smaller and is about 2 m in components of satellite position.[54]

Propagation errors — These consist of *ionospheric refraction effect* which is a frequency shift in the received signal due to the rate of change of signal wavelength stretching caused by dispersion in the ionosphere and is inversely proportional to the square of the transmitted frequency, and *tropospheric refraction effect* caused by the slowdown of propagation speed due to signal wavelength compression in the troposphere and is directly proportional to the transmitted frequency. The ionospheric refraction correction can be obtained by combining Doppler measurements at two distinct frequencies which, in fact, is the reason why *Transit* satellites transmit two

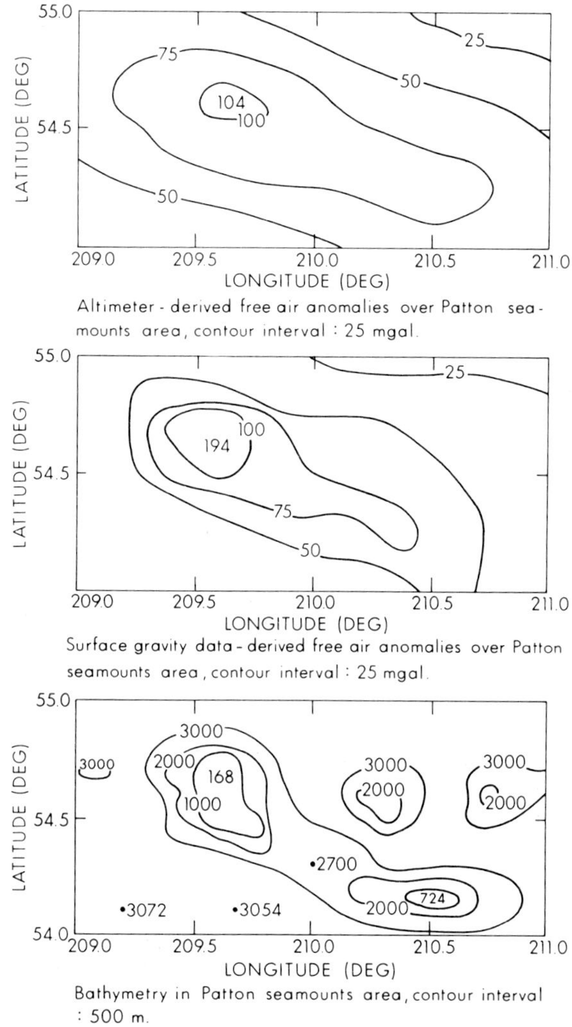

FIGURE 10. Comparison of gravity anomalies based on altimetry data and surface gravity data over the Patton seamounts area. (From Rapp, R. H., *J. Geophys. Res.*, 84, 3784, 198; Watts, A., Rep. MC-12, Geological Society of America, Boulder, Colo., 1976. With permission.)

frequencies (150 MHz and 400 MHz) and why *dual channel receivers* are used for monitoring both frequencies. This is also the reason why *single channel receivers* show position fix errors which are several times larger than those given by dual channel receivers. The *tropospheric refraction* effect is generally corrected for by either *estimating* the frequency shift due to a model troposphere or by *deleting* the Doppler data close to the horizon. The *residual error* due to the ummodeled effect of propagation errors can range from 1 to 5m.

Receiver altitude error — As noted before, the position fix requires an estimate of initial receiver position (ϕ,λ,h) to compute estimated slant range change and extrapolate receiver position as a function of time. Here ϕ is latitude, λ is longitude and h is altitude or height above the reference ellipsoid of the receiver antenna. In practice,

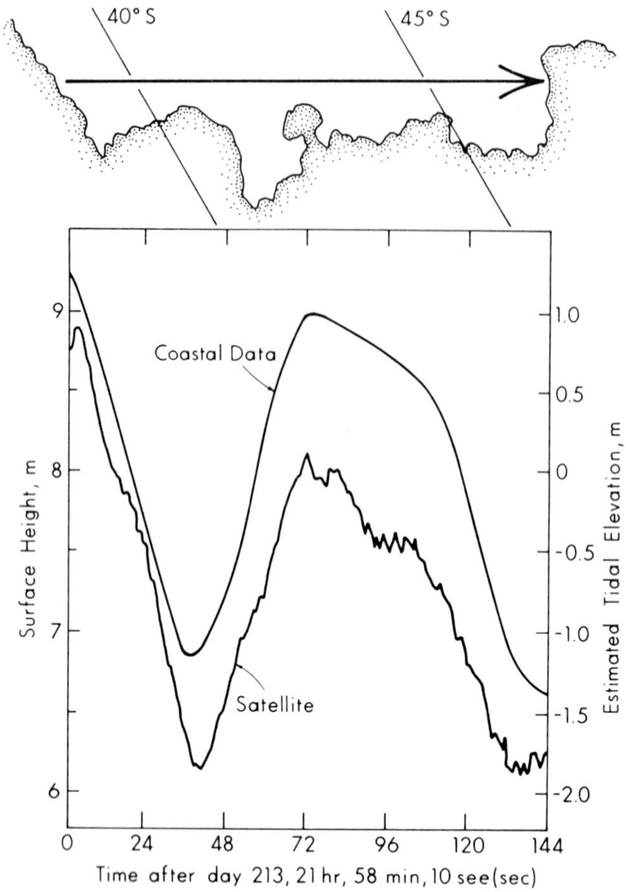

FIGURE 11. Subsatellite tide over Patagonian shelf. (From Hendershott, M. C., *Applications of Marine Geodesy in Support of National Objectives in Ocean Science, Engineering, and Operations,* Saxena, N., Ed., University of Hawaii, Honolulu, 1980. With permission.)

Table 4
SEA LEVEL AND GEOIDAL ACCURACY REQUIREMENTS FOR SOME DYNAMICAL OCEANOGRAPHIC APPLICATIONS

Phenomenon	Required geoidal accuracy	Required sea level accuracy
General basinwide Circulation	< + 10 cm	< ± 10 cm
Currents	< + 10 cm	< ± 10 cm
Other sea surface topography	< + 50 cm useful but frequently < ± 5 cm	< ± 50 cm useful but frequently ⩽ ± 5 cm

this is taken as the geoidal height plus the height of vessel antenna above sea level (the instantaneous sea surface topography is generally ignored). The sensitivity of satellite position fix to any error in h is shown in Figure 14. It depends upon the error in the estimated altitude and the satellite elevation angle. For example, an error of 10 m in

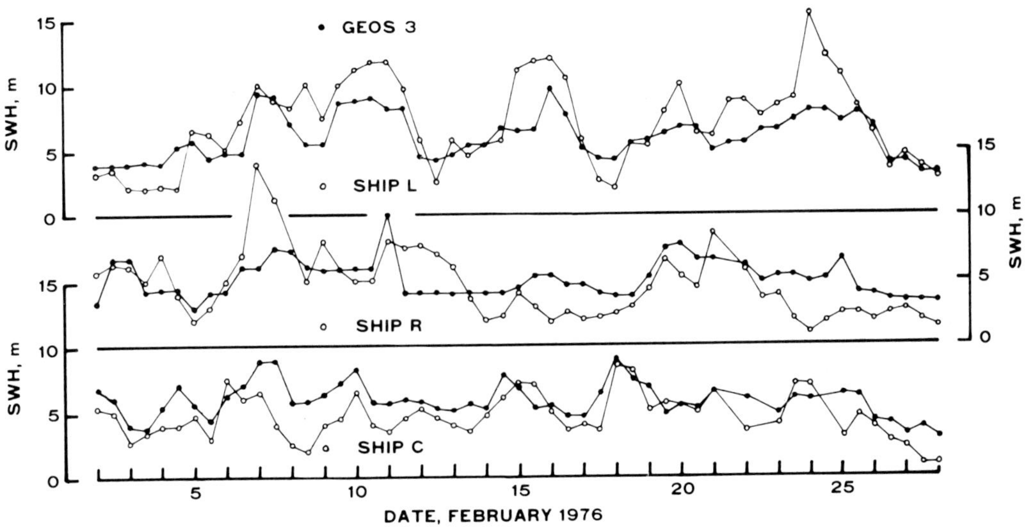

FIGURE 12. Comparisons of significant wave heights derived from GEOS-3 altimeter data with shipboard wave height measurements. (From Parson, C. L., *J. Geophys. Res.*, 84, 4011, 1979. With permission.)

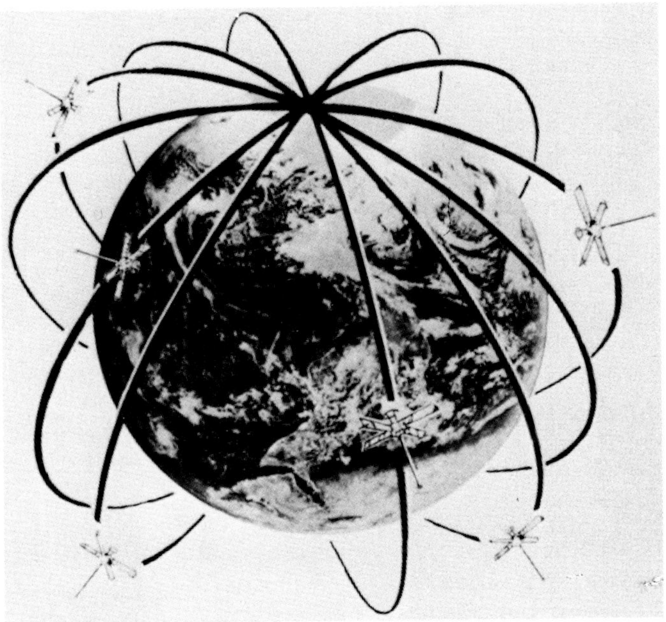

FIGURE 13. Schematic configuration of Transit satellites in polar orbits around the Earth. (From Stansell, T. A., *Navigation,* 25, 55, 1978. With permission.)

estimated altitude can translate to about 30 m in longitude for a satellite elevation angle of about 70° and only to about 6 m for a satellite elevation angle of 15°.

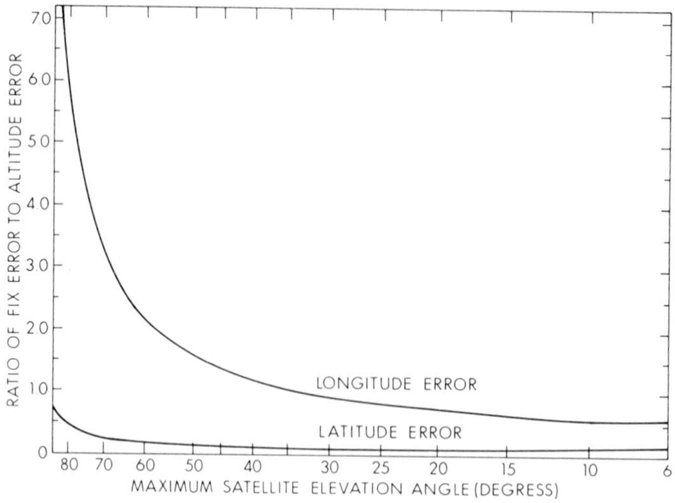

FIGURE 14. Sensitivity of position fix error to altitude estimate error. (From Stansell, T. A., *Electronic Surveying and Navigation,* Laurila, S. M., Ed., John Wiley & Sons, New York, 1976. With permission.)

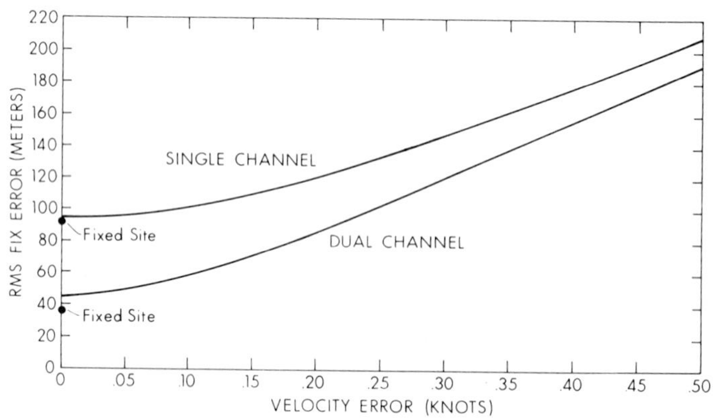

FIGURE 15. Position fix error as a function of vessel velocity error. (From Stansell, T. A., *Navigation,* 25, 55, 1978. With permission.)

Receiver motion error — As noted before, if the receiver/observer is in motion, the motion must be recorded accurately. Otherwise small errors in vessel speed can produce very large errors in position fix. This is illustrated in Figure 15. Notice that with a dual channel receiver equipment, a 0.5 kn error in velocity will cause a position fix error of about 190 m as opposed to about 40 m error for the stationary case. It is relatively more complex to correct for the velocity error than correcting for other error types. The velocity can be included in the least squares adjustment process as a parameter to be improved but that generally degrades the accuracy of other recovered parameters. Consequently, velocity is included as an adjustable parameter only if the velocity errors are suspected to be high (> 0.4 kn).

Hardware and Measurement Noise Errors — These are due to the receiver and satellite oscillator phase jitter, the receiver time clock error, etc., and can range from 3 to 6 m.

Overall Positioning Accuracy

The *overall accuracy* of a position fix for a stationary observer from a single pass is generally in the range of 27 to 37 m rms under field conditions.[53] The position fix error for a survey/research vessel in the open ocean can amount to *several hundred meters.*

The *overall accuracy ofgeodetic positioning* by satellite is much higher, principally because the observations can be continued over longer periods for several satellite passes, there is generally no observer velocity error, initial estimates of position are generally better, and precise ephemeris is generally accessible. Also, two and more stations can participate in tracking a common set of satellite passes. Further, because of the amount and span of data, more sophisticated analysis procedures can be employed. By a combination of these factors, it is possible to achieve an overall accuracy of 1 m or better, in geodetic positioning.[57]

Age Jitters of Transit System

Three of the Transit System satellites are over 14 years old. Orbital precession and degradation with age has caused the orbits to shift so that the spacing among orbits is no longer uniform with certain orbits crowding together and others moving apart. Orbit crowding leads to simultaneous satellite sightings causing signal interference. Wide spacing in orbits leads to long delays in between satellite sightings and thus long *outages* in between position fixes. These factors degrade the quality of navigational positioning. Additionally, as pointed out earlier, satellite transmission system malfunctioning causes problems.[53]

Improvements in the Transit System

A number of the above problems are being taken care of. A new generation of Transit satellites for Transit Improvement Program (TIP) have been developed. The TIP satellites have many improved features[53] such as the Disturbance Compensation System which eliminates the effects of atmospheric drag and thus preserves the orbit integrity better and longer, making orbital predictions more stable and suitable for longer use, an onboard computer which can receive ground commands and has the memory capability to store orbital ephemeris of 10 days without updating, stronger signal, very precise clock control with frequency resolution of 1 in 10^{12}. Additionally, the use of part-time navigational satellites such as TRANSAT can solve some of the orbital spacing problems. The arrangements for improvements in the Transit System are being made only as a contingency against unforeseen delays in the completion of the new Global Positioning System discussed next.

NAVSTAR Global Positioning System

NAVSTAR Global Positioning System (GPS) is a space-based global navigation system which has the projected capability of providing global, continuous, three dimensional navigation to a large array of suitably equipped users. The intended objective of the completed system is to provide an accurate navigation solution consisting of three dimensional position, velocity, and time (seven dimensional information) *all the time on any location* on or near the Earth, by insuring direct, line of sight measurements continuously from at least four satellites to any user anywhere.

Originally, a fully operational GPS was supposed to consist of 24 satellites located in three uniformly spaced (120° apart) orbital planes, with 8 satellite uniformly distributed (45° apart) in each of the three 12-hr circular orbits, at a mean altitude of 20, 183km. However, due to funding constraints, GPS satellite constellation is scheduled to be pared down to 18 satellites.[58] The system is slated to be fully operational in 1987. Figure 16 gives a schematic illustration of the original 24-satellite constellation NAV-

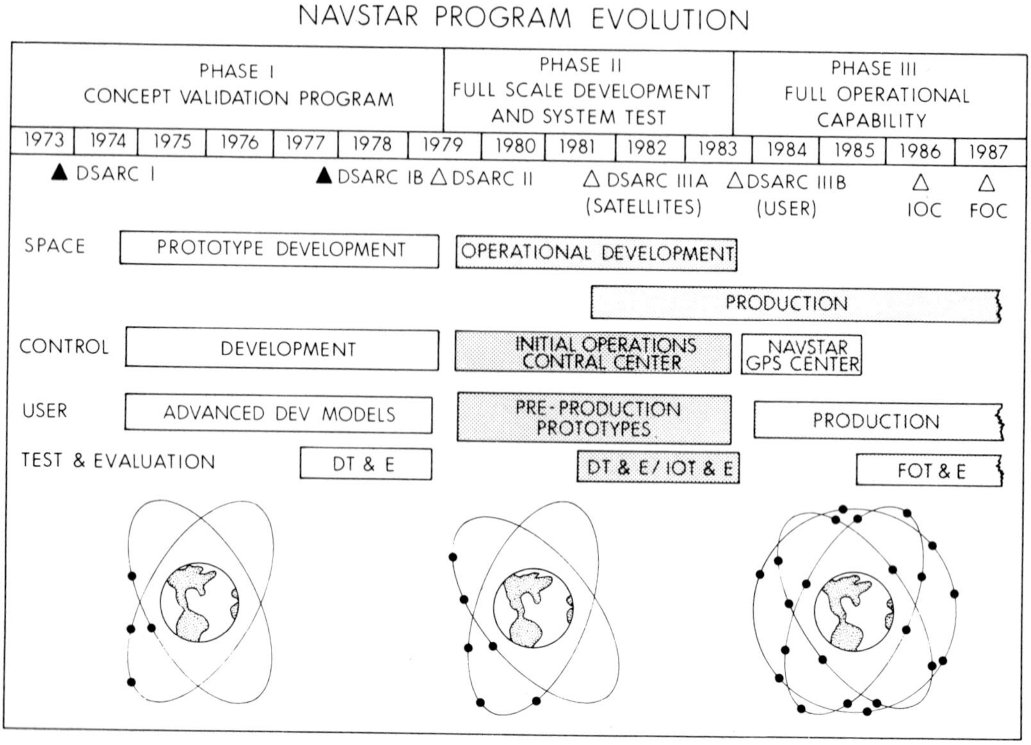

FIGURE 16. NAVSTAR program evolution of the original 24-constellation system. (From Henderson, D. W. and Coriat, H., *Navigation,* 27, 54, 1980. With permission.)

STAR program evolution.[59] The program has successfully completed Phase 1, which consisted of program concept validation, and is already in Phase 2 of full scale development and system test.

NAVSTAR System Components

The system consists of the following components:

Space Segment

The fully operational system will consist of 18 satellites, with probably some adjustments in satellite spacing, and perhaps orbital spacing and number of orbital planes, to accommodate the reduction in satellite constellation size. The reduction will probably cause occasional outages or blind spots for some users but the purpose of an optimal configuration of the satellite constellation is to insure the minimum possible impact on the program objective of providing continuous, global coverage.

At present, the system consists of five satellites in two planes inclined at 63° to the equator. This allows about 3 hr of daily testing with three to five satellites in view in selected areas, principally at Yuma test site which is the primary test area for GPS evaluation.

The satellites transmit navigation messages containing satellite ephemeris, atmospheric propagation correction, and satellite clock bias information. The satellite navigation signal is a dual composite code — one for *precise navigation* (P-code) and the other for *acquisition and coarse navigation* (C/A code for *clear access or coarse/acquisition).* The P-code information will be restricted to qualified users with high precision requirements. The C/A code information will be available to all civilian and commercial users. GPS satellites will transmit at two frequencies — at 1575. 4 MHz called L_1 and 1227.6 MHz called L_2. The P-code will be transmitted both at L_1 and L_2 while the C/A code will be transmitted at L_1 only. The dual L-band frequency transmission will permit the derivation of ionospheric correction.

Control Segment

A ground control network will track the satellites to determine the orbit and the atomic frequency standard (clock error) of each satellite precisely, extrapolate them, and store them in the memory system of each satellite on a daily basis, thus equipping it for transmission of the navigation message to the user.[60]

At present, the master control station, currently at Vanderberg Air Force Base in California, collects ranging data from four ground-based monitor stations located in Alaska, Guam, Hawaii, and Vanderberg Air Force Base, computes the orbital and clock information, formulates the navigation message, and loads it in each satellite daily.[59] The primary function of the control segment is to generate precise navigation data for the user.

User Segment

This segment consists of a receiver and navigation processer, capable of tracking four satellite signals, either simultaneously or sequentially, depending upon the dynamics of the user. It also has the capability of selecting the appropriate satellites providing the best geometry to secure optimal accuracy in the position fix.[61] The primary purpose of the user segment is to receive the navigation message, demodulate it, and use it in computing the position fix and clock offset. Signals from at least four satellites are required to solve continuously for user's three-dimensional position and clock offset.

Navigation Solution

The navigation solution requires measurement of the time of arrival of navigation signals from four satellites for the nominal determination of four independent variables which are generally three components of user position and user clock offset. From these time measurements, the *pseudorange* R_j to the jth satellite is computed as

$$R_j = c(t^u - t_j^s) - c\,\Delta t_j^A \qquad j = 1 \text{ to } n \geq 4 \qquad (19)$$

where

- c = the speed of light
- t^u = the signal reception time as logged by the user clock and assumed simultaneous from all satellites
- t_j^s = the signal transmission time as logged by j^{th} satellite's atomic clock
- Δt_j^A = propagation delay for signal from the jth satellite

Since there is no requirement for GPS users to have their clocks synchronized with GPS time, the user clock generally has an offset of Δt^u with respect to GPS time so that the actual reception time $\bar{t}^u$ in GPS time reference is

$$t^u = \bar{t}^u + \Delta t^u$$

The satellite clock may also be offset with respect to GPS time by Δt^s so that the GPS-referred transmission time $\bar{t}^s$ is

$$t^s = \bar{t}^s + \Delta t^s$$

Hence

$$R_j = c(\bar{t}^u - \bar{t}_j^s) + c(\Delta t^u - \Delta t_j^s) - c\,\Delta t_j^A \tag{20}$$

Information to determine Δt^s and Δt^A for each satellite is contained in the navigation message. Thus the only unknown is the user clock offset Δt^u which must be determined as one of the four unknowns in the navigation solution. Consequently, Equation 20 can be written as

$$R_j = \bar{R}_j + c\,\Delta t^u = \bar{R}_j + B \tag{21}$$

where R_j is the *exact* slant range to the jth satellite and B is the user clock range bias. Note that R_j is the actual pseudorange measurement to the jth satellite.

Now let x_i (i = 1 to 3) be the exact position of the user (unknown) and x_{ij} (i = 1, . . . 3, j = 1, . . . n $\geqslant$ 4) be the positions of j satellites (all known) in an Earth-centered inertial coordinate system X_i(i = 1,2,3) which at zero time has its X_1-axis oriented towards the intersection of the equator and prime meridian, X_3-axis passing through the North Pole, with X_2-axis completing the right-handed coordinate system. Let x'_i and B′ be the *nominal estimates* of the user position and user clock range bias. Then[58,62]

$$x_i = x_i + \Delta x_i$$

$$B = B' + \Delta B$$

Also

$$R_j = R'_j + \Delta R_j$$

where R'_j is the nominal estimate of slant range to the jth satellite and is given by

$$R'_j = \sqrt{\sum_i (x'_i - x_{ij})^2} + B'$$

Also,

$$\bar{R}_j = \sqrt{\sum_i (x_i - x_{ij})^2}$$

Thus Equation 21 can now be written as

$$R_j = \sqrt{\sum_i (x_i - x_{ij})^2} + B$$

or

$$R'_j + \Delta R_j = \sqrt{\sum_i (x'_i + \Delta x_i - x_{ij})^2} + B' + \Delta B$$

which, to the first order, simplifies to

$$\Delta R_j = \sum_i \frac{x_i - x_{ij}}{R' - B'} \cdot \Delta x_i + \Delta B \tag{22}$$

The corrections Δx_i and ΔB in Equation 22 are the desired corrections to be computed. All other quantities are known. The coefficients of Δx_i are the direction cosines of the user to satellite range. The coefficients of ΔB are unity. The linearized relationships in Equation 22 can be expressed in the matrix notation as

$$\alpha x = r, \quad \text{or}$$

$$x = \alpha^{-1} r \tag{23}$$

where

$$\alpha = \begin{vmatrix} \alpha_{11} & \alpha_{12} & \alpha_{13} & 1 \\ \alpha_{21} & \alpha_{22} & \alpha_{23} & 1 \\ \alpha_{31} & \alpha_{32} & \alpha_{33} & 1 \\ \alpha_{41} & \alpha_{42} & \alpha_{43} & 1 \end{vmatrix}$$

$$x = \left| \Delta x_1 \ \Delta x_2 \ \Delta x_3 \ \Delta B \right|^T$$

$$r = \left| \Delta R_1 \ \Delta R_2 \ \Delta R_3 \ \Delta R_4 \right|^T$$

Equation 23 gives the desired user position correction Δx_i and user clock range bias ΔB, which are added to the nominal estimates of user position x'_i and user clock range bias B' to obtain computed user position x_i and user clock range bias B. In the above, we have assumed measurements from four satellites only. If measurements from more than four satellites are available, a general least squares solution can be used to compute r in Equation 23.

Effect of Satellite Geometry on Navigation Solution

Since the relationship in Equation 23 is linear, it can also be used to express the relationship between the expected errors in range measurements and user quantities. The covariance matrices of these expected errors in x and r are related as

$$\text{COV}(x) = [\alpha^T \text{COV}(r) \alpha]^{-1} \tag{24}$$

Note that this relationship is a function of matrix α only, which itself consists of direction cosines of line of sight measurements to the satellites. Thus the *error relationships are a function of satellite geometry only.* The strength of the geometry of the system is generally measured by a quantity called *geometric dilution of precision* (GDOP). There are other associated criteria for evaluating the geometry of the system such as

- **Position Dilution of Precision (PDOP)**, which is the square root of the sum of the squares of the three components of position error.
- **Horizontal Dilution of Precision (HDOP)**, which is the square root of the sum of the squares of the horizontal components of the position error.
- **Vertical Dilution of Precision (VDOP)**, which is the altitude error.
- **Time Dilution of Precision (TDOP)**, which is the user clock range bias.

A good approximation of the effect of satellite geometry on user errors is obtained by assuming that COV(r) equals the identity matrix. This implies that each pseudorange measurement has an *error of one* with *zero expected mean*, and that the *correlation of errors between satellites is zero.* Thus Equation 24 becomes

$$COV(r) = [\alpha^T\alpha]^{-1} = \left|\begin{matrix} V_{x_1x_1} & V_{x_1x_2} & V_{x_1x_3} & V_{x_1t} \\ V_{x_2x_1} & V_{x_2x_2} & V_{x_2x_3} & V_{x_2t} \\ V_{x_3x_1} & V_{x_3x_2} & V_{x_3x_3} & V_{x_3t} \\ V_{tx_1} & V_{tx_2} & V_{tx_3} & V_{tt} \end{matrix}\right|$$

where the diagonal terms are the *variance* or squares of 1σ errors in user position and time while the off-diagonal terms are the *covariance* between the errors in user quantities and reflect the correlations of these errors. The above defined dilution of precision quantities can be expressed in terms of these errors as following:

$$GDOP = (V_{x_1x_1} + V_{x_2x_2} + V_{x_3x_3} + V_{tt})^{1/2}$$

$$= (PDOP^2 + TDOP^2)^{1/2}$$

$$PDOP = (V_{x_1x_1} + V_{x_2x_2} + V_{x_3x_3})^{1/2}$$

$$= (HDOP^2 + VDOP^2)^{1/2}$$

$$HDOP = (V_{x_1x_1} + V_{x_2x_2})^{1/2}$$

$$VDOP = (V_{x_3x_3})^{1/2}$$

$$TDOP = (V_{tt})^{1/2}$$

Since COV(r) was assumed to be an identity matrix, the product of the above defined dilution of precision factors and the errors in range measurements gives the corresponding errors in user position and time. Thus, the best satellite geometry corresponds to the lowest values of the above parameters.

Sources of Error

Errors in pseudorange measurements, which form the basis of the navigation solution, arise from the following sources.[62]

Satellite clock errors — Individual satellite clocks, though highly stable, may have a time offset relative to GPS reference time. Information to apply the clock correction is contained in the navigation message. The residual error from this source is of the order of a few decimeters, and is generally lumped with satellite orbital position error.

Propagation errors — These consist of ionospheric and tropospheric errors. The ionospheric refraction error, which causes reduction of speed and bending of the ray, can be removed by the information provided by transmission of satellite signal at two frequencies — L_1 and L_2. Information about propagation delays is also available in the navigation message for those who cannot compute the correction from dual frequency transmission. Tropospheric errors, which are independent of frequency, can be modeled quite simply. The residual error from these sources is from 2.4 to 5.2 m.

Satellite ephemeris errors — These errors have been discussed in detail in the section on Satellite Altimetry and Transit Satellites. In spite of a good distribution of tracking stations and continual orbital refinement computations to improve model parameters for better orbit prediction precision, the satellite ephemeris errors are still of the order of *several meters*. In this particular case, only the satellite position errors which affect range errors are of primary importance. The part of the range error, caused by satellite ephemeris uncertainties, which is common to the four satellites will appear in the solution as user clock range bias and will get corrected. The residual error can still be of the order of several meters and as discussed earlier in this section, depends upon whether the user has access to P-code or C/A code.

Group delay errors — These are caused by the uncertainty in the processing of the signal by satellite equipment and their effect is generally included in satellite clock offset. The residual error from this source is about 1 m.

Multipath and interference effects — These result from a combination of signals from more than one path and depend upon the nature and location of reflective surfaces peculiar to a user location. These range between 1.2 to 2.7 m.

Receiver vehicle dynamics — These are caused by motion of the user vehicle and will contribute significantly to the total ranging error which will eventually be transferred to uncertainties in the user position fix.

System Accuracy

The magnitude of the ranging errors combined with the geometry of the measurement system will determine the *overall accuracy* of the user position fix. It is anticipated that with a fully operational GPS and P-code access, a *mobile* user, with nominal receiver vehicle dynamics, who selects the best four satellites from those that are visible 5° or more above the horizon (having the lowest value of GDOP) will be able to attain an overall accuracy of 9.4 to 16.4 m in his position fix.[60,62] A *stationary* user will be able to improve the overall accuracy of his position determination to about 1 m or better, by occupying the location for several days under the above stated conditions. The C/A code will degrade the overall accuracy considerably, to about 100 m or so for users with nominal dynamics. High dynamics will degrade accuracy even further.

Integrated Positioning Systems

The increasingly sophisticated navigation and other positioning requirements demand increasingly higher levels of positioning coverage continuity and accuracy. The new geophysical techniques for offshore oil exploration impose high positional accuracy requirements.[61,63] The new three-dimensional seismic surveys require continuous navigation of about 5 m or better. Precise positioning of offshore drill rigs requires continuous and highly accurate position determinations which reportedly have achieved accuracies of about 1 m. Offshore pipe laying must be done in precisely surveyed corridors for a variety of reasons. Hydrographic charting requires accuracies of 5 to 8 m. To meet these needs of continuous and accurate positioning, the trend has been the development of integrated systems which capitalize on the advantages of each component, to achieve high accuracy and continuous positioning information. There have been several advances in the integration of relative positions obtained from underwater acoustic devices with satellite positioning systems. The development of shipboard systems independent of acoustic transponders offers new possibilities of integration with satellite systems. The combination of GPS with underwater acoustic capabilities, and perhaps shipboard systems, should offer unlimited applications, particularly for those requiring submeter level accuracy in nonreal time. The combination of inertial navigation systems with GPS offers excellent possibilities for high dynamics environments.[64]

Applications of Positioning by Satellites

The positioning by satellites has a wide spectrum of applications that is fast expanding. Some of these are

- Military surface and submarine navigation
- Commercial navigation
- Navigation and positioning for scientific research vessels
- Navigation and positioning for resource exploration vessels
- Relocation of offshore oil exploration grids and prospecting sites
- Positioning of offshore platforms and drilling rigs
- Aircraft traffic monitoring
- Spacecraft navigation
- Coastal zone monitoring
- Geodetic positioning for multiple applications

SUMMARY AND CLOSING REMARKS

This chapter discussed three major applications of satellites to ocean exploration: gravity field determination from satellite orbital perturbations, satellite altimetry, and positioning at sea by satellites. An attempt was made to give details of only those aspects which are not traditionally familiar in ocean sciences. The aspects that are traditionally well known in ocean sciences are only briefly covered. There are *several other satellite contributions* which have not been covered in this chapter because of its prescribed scope. Some of the areas *not* covered are such potentially major satellite applications as satellite gradiometry, gravity field from satellite-to-satellite tracking, satellite-mounted microwave sensor mapping and imaging of ocean surface wind velocity (scatterometer data), sea surface temperatures (radiometer data), multispectral scanning and thermal data, etc.

It is not possible to cite all the relevant literature in a paper of this type. Thus, the bibliographic references cited in this chapter do not claim to cover all the relevant literature; instead they are intended as a guide to further literature on the subject.

REFERENCES

1. **Smart, W. M.,** *Spherical Astronomy*, Cambridge University Press, London, 1965.
2. **Mueller, I. I.,** *Introduction to Satellite Geodesy*, Frederick Ungar Publishing Co., New York, 1964.
3. **Kaula, W. M.,** *Theory of Satellite Geodesy*, Blaisdell Publishing Co., Waltham, Mass., 1966.

4. **Heiskenan, W. A. and Mortiz, H.,** *Physical Geodesy,* W. H. Freeman and Co., San Francisco, 1967.
5. **Khan, M. A.,** A Review of the Perturbation Theory as Applied to the Determination of Geopotential, Rep. No. HIG-68-1, Hawaii Institute of Geophysics, Honolulu, 1968.
6. **Brouwer, D. and Clemence, G. M.,** *Methods of Celestial Mechanics,* Academic Press, New York, 1961.
7. **King-Hele, D. G., Brookes, C. J., and Cook, G. E.,** Odd zonal harmonics in geopotential from analysis of 28 satellite orbits, *Geophys. J. R. Astron. Soc.,* 64, 3, 1981.
8. **Cook, A. H.,** Resonant orbits of artificial satellites and longitude terms in Earth's external potential, *Geophys. J. R. Astron. Soc.,* 4, 53, 1961.
9. **Kozai, Y.,** New determination of zonal harmonic coefficients in the Earth's gravitational potential, *Astron. Soc. Jpn.,* 16, 263, 1964.
10. **Izsak, I. G.,** A new determination of nonzonal harmonics by satellites, in *Trajectories of Artificial Celestial Bodies as Determined from Observations,* Kovalevsky, J., Ed., Springer-Verlag, Berlin, 1966.
11. **Anderle, R. J.,** Geodetic parameter set NWL-SE-6 based on Doppler satellite observations, in *2nd Int. Symp. Geodetic Uses of Satellites,* Veis, G., Ed., National Technical University, Athens, Greece, 1966.
12. **O'Keefe, J. A., Eckels, A., and Squires, R. K.,** The gravitational field of the Earth, *Astron. J.,* 64, 245, 1959.
13. **Kaula, W. M.,** Analysis of Gravitational and Geometric Aspects of Satellites, Tech. Note D-572, National Aeronautics and Space Administration, Washington, D.C., 1961.
14. **Guier, W. H. and Newton, R. R.,** The Earth's gravity field deduced from the Doppler tracking of five satellites, *J. Geophys. Res.,* 70, 4613, 1965.
15. **Gaposchkin, E. M. and Mendes, G.,** Geopotential model to degree and order 30 from satellite and terrestrial data, *E. O. S. Trans. Am. Geophys. U.,* 58(6), 371, 1977.
16. **Lerch, F. J., Klosko, S. M., Laubscher, R. E., and Wagner, C. A.,** Gravity model improvement using GEOS-3 (GEM 9 and 10), *J. Geophys. Res.,* 84, 3897, 1979.
17. **Khan, M. A.,** Accuracy of Earth's gravity field models, in preparation.
18. **Lerch, F. J.,** private communication.
19. **Khan, M. A.,** Some geophysical and geodetic contributions of satellite-determined gravity results, *Geophys. Surveys,* 2, 469, 1976.
20. **Strange, W. M. and Khan, M. A.,** The prediction of gravity in unsurveyed areas using a combination of satellite and surface gravity data, *Trans. Am. Geophys. U.,* 46, 3, 1965.
21. **Khan, M. A. and Strange, W. M.,** Investigations on the shape of geoid in North America, *Trans. Am. Geophys. U.,* 47, 1, 1966.
22. **Marsh, J. G. and Vincent, S.,** Global detailed geoid computation and model analysis, *Geophys. Surveys,* 1, 481, 1974.
23. **Moberly, R. and Khan, M. A.,** Interpretation of the sources of the satellite-determined gravity field, *Nature (London),* 223, 263, 1969.
24. **Kaula, W. M.,** Comparisons and combination of satellite with other results for geodetic parameters, in *The Use of Artificial Satellites for Geodesy, II,* Veis, G., Ed., National Technical University, Athens, Greece, 1967.
25. **Khan, M. A.,** Nature of satellite-determined gravity anomalies in *The Use of Artificial Satellite for Geodesy,* Henriksen, W., Mancini, A., and Chovitz, B. H., Eds., Geophys. Monogr. No. 15, American Geophysical Union, Washington, D.C., 1972.
26. **Woollard, G. P. and Khan, M. A.,** Prediction of Gravity in Ocean Areas, Rep. No. HIG-72-11, Hawaii Institute of Geophysics, Honolulu, 1972.
27. **Hide, R. and Malin, S. R. C.,** Novel correlations between global features of Earth's gravitational and magnetic fields, *Nature,* 225, 605, 1970.
28. **Khan, M. A.,** Some geophysical implications of the satellite-determined geogravity field, *Geophys. J., R. Astron. Soc. London,* 23, 15, 1972.
29. **Khan, M. A.,** Sea level heights from GEOS-3 altimetry data, *E.O.S.,* 57, 901, 1976.
30. **Douglas, B. C. and Gaborski, P. C.,** Observation of sea surface topography with GEOS-3 data, *J. Geophys. Res.,* 84, 3893, 1979.
31. **Lerch, F. J., Marsh, J. G., Klosko, S. M., and Williamson, R. G.,** Gravity model improvement for SEASAT, *E.O.S.,* 62, 257, 1981.
32. **Anderle, R. and Hoskin, R.,** Correlated errors in satellite altimetry geoids, *Geophys. Res. Lett.,* 4, 421, 1977.
33. **Khan, M. A.,** Gravity model improvement using GEOS-3 data, presented to GEOS-3 Principal Investigators Meet., NASA Wallops Flight Center, Wallops Island, Va., 1975.
34. **Brown, R. D. and Fury, R. J.,** Determination of the Geoid from Satellite Altimeter Data, NASA Goddard Space Flight Center, Rep. No. X-550-72-268, National Aeronautics and Space Administration, Greenbelt, Md., 1972.

35. **Lerch, F. J., Klosko, S. M., Laubscher, R. E., and Wagner, C. A.**, Gravity model improvement using GEOS-3 (GEM 9 and 10), *J. Geophys. Res.*, 84, 3897, 1979.
36. **Greene, A. H.**, Accuracy of satellite radar altimeter measurements, in *The Use of Artificial Satellites for Geodesy*, Proc. 3rd Int. Symp., American Geophysical Union, Washington, D.C., 1972.
37. **Stanely, H. R. and Dwyer, R. E.**, NASA Wallops Flight Center GEOS-3 altimeter data processing report, NASA Ref. Publ. No. 1066, National Aeronautics and Space Administration, Washington, D.C., 1980.
38. **Frey, E. J., Harrington, J. V., and von Arx, W. S.**, A study of satellite altimetry for geophysical and oceanographic measurements, in Proc. 16th Int. Astronaut. Cong., Athens, Greece, 1965.
39. **Rapp, R. H.**, GEOS-3 data processing for the recovery of geoid undulations and gravity anomalies, *J. Geophys. Res.*, 84, 3784, 1979.
40. **Anon.**, Scientific Results of GEOS-3 Mission, *J. Geophys. Res.*, 84, 3779-4079, 1979.
41. **Kahn, W. D., Siry, J. W., Brown, R. D., and Wells, W. T.**, Ocean gravity and geoid determination, *J. Geophys. Res.*, 84, 3872, 1979.
42. **Gopalapillai, G. S. and Mourad, A. G.**, Detailed gravity anomalies from GEOS-3 satellite altimetry data, *J. Geophys. Res.*, 84, 6213, 1979.
43. **Watts, A.**, Gravity field of the northwest Pacific ocean basin and its margin: Phillipine Sea, Rep. MC-12, Geological Society of America, Boulder, Colo., 1976.
44. **Park, M. E., quoted by Hendershott, M. C.**, Marine geodesy and physical oceanography, in *Applications of Marine Geodesy in Support of National Objectives in Ocean Science, Engineering, and Operations*, Saxena, N., Ed., Rep. No. CEGS-TR-80-1, University of Hawaii, Honolulu, 1980.
45. **Leitao, C. D., Huang, N. E., and Parra, C. G.**, A note on the comparison of radar altimetry with IR and *in situ* data for the detection of Gulf Stream surface boundaries, *J. Geophys. Res.*, 84, 3969, 1979.
46. **Brown, G. S.**, Estimation of surface wind speeds using satellite-borne radar measurements at normal incidence, *J. Geophys. Res.*, 84, 3974, 1979.
47. **Mognard, N. and Lago, B.**, The computation of wind speed and wave heights from GEOS-3 data, *J. Geophys. Res.*, 84, 3979, 1979.
48. **Parks, M. E. and Chelton, D. B.**, Error analysis of SEASAT altimeter sea surface height, *E.O.S.*, 62, 298, 1981.
49. **Lazarewicz, A. R., White, J. V., and Sailor, R. V.**, Geostrophic current velocity determination from satellite altimetry, *E.O.S.*, 62, 298, 1981.
50. **Thompson, J. D., Born, G. H., and Maul, G. F.**, SEASAT altimetry in the Gulf of Mexico, *E.O.S.*, 62, 298, 1981.
51. **Fedor, L. S., Godbey, T. W., Gower, J. F. R., Guptill, R., Hayne, G. S., Rufenach, C. L., and Walsh, E. J.**, Satellite altimeter measurement of sea state — an algorithm comparison, *J. Geophys. Res.*, 84, 3991, 1979.
52. **Parson, C. L.**, GEOS-3 wave height measurements: an assessment during high sea state conditions in North Atlantic, *J. Geophys. Res.*, 84, 4011, 1979.
53. **Stansell, T. A.**, The many faces of Transit, *Navigation*, 25, 55, 1978.
54. **Anderle, R. J.**, Error model for geodetic positions derived from Doppler satellite observations, *Bull. Geod.*, 50, 47, 1976.
55. **Arur, M. G.**, Experiments for improved positioning by means of integrated Doppler satellite observations and the NNSS broadcast ephemeris, Dept. of Geodetic Sci., Rep. No. 258, Ohio State University, Columbus, Ohio, 1977.
56. **Stansell, T. A.**, Positioning by satellites, in *Electronic Surveying and Navigation*, Laurila, S. M., Ed., John Wiley & Sons, New York, 1976.
57. **Brown, D. C.**, Positioning by satellites, *Rev. Geophys. Space Phys.*, 17, 1424, 1979.
58. **Jorgensen, P. S.**, NAVSTAR Global Position System 18-satellite constellation, *Navigation*, 27, 89, 1980.
59. **Henderson, D. W. and Coriat, H.**, Status report — Global Positioning System, *Navigation*, 27, 54, 1980.
60. **Russell, S. S. and Schaibly, S. S.**, Control segment and user performance, *Navigation*, 25, 166, 1978.
61. **Stansell, T. A.**, Civil marine applications of the Global Positioning System, *Navigation*, 25, 224, 1978.
62. **Milliken, R. J. and Zoller, C. J.**, Principle of operation of NAVSTAR and system characteristics, *Navigation*, 25, 95, 1978.
63. **Mourad, G. A.**, Marine geodesy (1975-79) — Rep. to the Int. Assoc. Geodesy, Sec. 1, *Mar. Geodesy*, 4, 293, 1980.
64. **Cox, D. B.**, Integration of GPS with inertial navigation systems, *Navigation*, 25, 236, 1978.

ACQUISITION OF SEISMIC DATA AT SEA

Carl H. Savit

INTRODUCTION

While the objective is the same in both land and marine surveys, actual operations are generally easier and results are better at sea. Water is an excellent conductor of seismic signals. In addition, many of the logistic and acoustic difficulties that plague land surveys are absent or minimized at sea.

LOGISTIC ADVANTAGES OF THE WATER MEDIUM

Seismic exploration at sea is conducted with the detector arrays in continuous motion. Thus, operations are faster and more uniform, and the cost per kilometer of survey line is much lower than it is on land.

Both seismic source and seismic receivers can be kept at an optimum depth below the surface without stopping and without the need to drill holes.

The absence of culture, irregular terrain, and property lines makes it possible for marine survey lines to maintain a continuous, uniform coverage without the gaps or intervals of reduced coverage so common on land.

Space is available on an ocean-going vessel for sophisticated recording and processing systems, for storage of supplies and data, and to provide crew quarters for use during off-duty periods. Consequently, the most sophisticated systems can be used at sea and a 24-hr operation is possible.

Because both the source of seismic signals and the receivers are under water, and because current radar and navigation systems do not require visibility, operations can be conducted on a 24-hr basis even in foggy, rainy, or moderately bad weather.

ACOUSTIC ADVANTAGES AND DISADVANTAGES OF THE WATER MEDIUM

For all practical purposes, water is homogeneous and isotropic. That is, on any one prospect we can expect the seismic characteristics of the water to be uniform both laterally and vertically. The variations in water properties noted in sonar operations are too small to be of consequence because seismic reflection sound paths are nearly vertical, and travel through the water represents a small fraction of the total travel path.

Transverse waves (shear waves) cannot be transmitted by water and hence do not produce the interference effects that they do on land. Surface waves such as Rayleigh waves are also effectively absent.

The upper surface of the water is, for all practical purposes, a horizontal plane and the velocity of sound in the water is substantially constant and well known; thus, first layer or ''weathering'' effects can be readily corrected out.

Because the velocity of sound in water (about 1.5 km/sec) is more than four times higher than the velocity of sound in air (0.36 km/sec), the reflection coefficient at the water-air interface can be 90% or more. Because the sound is being reflected from an interface with a lower-velocity medium, the reflected signal is opposite in phase to the incident signal. Consequently the system of source and surface reflector acts as a notch filter, rejecting wavelengths equal to twice the depth of the source. Thus, for example, a source at a depth of 6 m will be deficient in energy at 125 Hz (sound velocity of 1.5

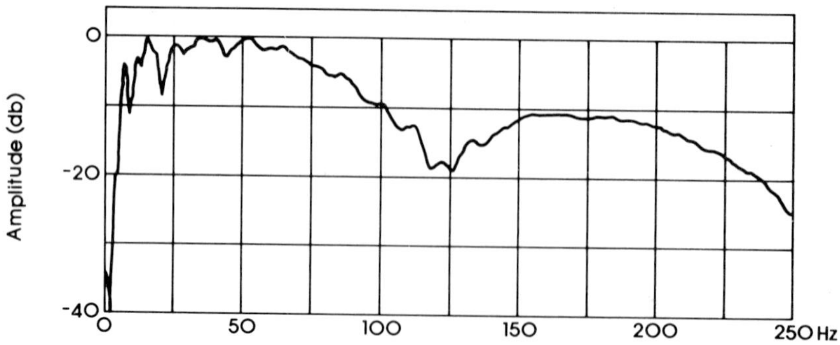

FIGURE 1. Amplitude spectrum for a tuned array of 10 high-pressure air guns at a depth of 6 m. The "notch" at about 125 Hz is the result of the surface reflection.

km/sec divided by a wavelength of 12 m). A similar relationship exists with respect to the receiver and the surface.

To select the proper depth for source and receivers balance must be struck between competing factors. Sources are more efficient at greater depths, while deeper towing of receivers reduces background noise. Keeping both source and receiver at shallower depths increases the high-frequency content and bandwidth of the signal and hence improves resolution. Recently, operating depths have been reduced from an older standard of about 13 m to about 6 or 7 m and the resulting poorer signal-to-noise ratio has been compensated by increasing the redundancy of data gathering (CDP multiplicity).

Water is an excellent conductor of seismic energy and acts as an ideal medium to transmit seismic waves from the marine source into the geologic formations underneath. As the water depth increases, the energy transfer from the water to the formation takes place over larger areas, thereby reducing the flux density of acoustic energy transfer per unit area and increasing the seismic energy transmitted into the formation. Averaging over minor irregularities in the bottom may also contribute to improved results.

The reflection coefficient for seismic waves at the sea bottom is large so that reverberations, ghosts, point-source diffractions, and multiple reflections often interfere seriously with valid data. Before the advent of digital data processing, long spreads, and CDP shooting, virtually all marine seismic records and sections were effectively saturated with invalid data that tended to obscure reflections from geologic interfaces below the sea bottom.

In water deeper than a few hundred meters, the shallow subbottom reflections are repeatedly imaged as a strong sequence of multiple reflections. Contemporary data processing methods do not provide routine procedures to cope adequately with this problem. Effective suppression of multiples in water more than a few hundred and less than about 2500 to 3000 m deep usually requires specialized computer programs and the exercise of judgment by an experienced, skilled geophysicist. When the water is deeper than about 2.5 to 3 km, the useful information arrives before the first interfering multiple so special processing to remove multiples is not required.

MARINE SEISMIC SOURCES

The Bubble Problem

The history of the development of marine seismic sources has been dominated by a search for solutions to the bubble problem.

When seismic operations first moved to sea in the late 1940s, the source of seismic energy was the familiar charge of explosive used on land. When such charges were detonated at an adequate depth to provide the confinement or tamping necessary to generate the maximum seismic energy, geophysicists did not get the single, sharp pulse they were accustomed to getting on land, but, instead, got a machine-gun-like succession of pulses. What had happened was that the expanding gases from the explosion produced a bubble which, because of the inertia of the water mass, got so big that the pressure inside was below that of the surrounding water. At that point the bubble collapsed rapidly, compressing the gases inside and making a second signal. Successive expansions and collapses of the bubble produced the machine gun effect and with it, a hopelessly confused seismic record.

To get around the bubble problem, geophysicists before about 1966 detonated their charges near the surface and allowed the expanding bubble, as well as most of the energy of the explosion, to escape into the air. From 1966 until the present, a large number of explosive and nonexplosive sources have been designed and tested in attempts to resolve the dilemma and at the same time to produce adequate signal characteristics.

Three Bubble Problem Solutions

While at least 14 different sources have been or are being used commercially in deep seismic surveying (an additional half dozen or so have been used in engineering and reconnaissance work), the three most commonly used sources exemplify the three basic solutions to the bubble problem:

1. Distributing the bubbles in time by using arrays of sources
2. Confining the bubble in a container and venting it to the surface
3. Letting the bubbles be produced but mathematically combining their signals into the equivalent of the record from a simple pulse

With air guns, the typical solution is to deploy the guns in arrays of different size guns so that the bubbles are distributed and hence less troublesome.

The air gun is a device that holds a quantity of compressed air at high pressure in an underwater metal chamber. Two types of air guns are presently used: one operates with larger volumes of lower pressure air (2000 lb/in.^{2} or 140 bar, nominal), and the other with smaller volumes of higher pressure air (5000 lb/in.^{2} or 350 bar, nominal). On command from the recording system the chamber suddenly opens and releases the air into the water.

When several air guns of different sizes are fired simultaneously, the initial shock waves (or seismic pulses) combine together to make a single, strong pulse. Because the time between successive bubble pulses varies according to the amount of gas in the bubble, the secondary and subsequent bubble pulses from the guns of different size do not occur at the same time. The result is a big initial pulse followed by a weaker train of irregular small pulses, i.e., the bubble pulse train is stretched out and subdued.

Much effort is expended in providing control systems to assure that all air guns in an array are fired simultaneously. Systems commonly in use today electronically track gradual variations in the firing times of the guns and correct for those variations by delaying firing signals to those guns that respond more quickly than others. The best available systems generally fire 10 to 20 or more guns within a 1 msec interval.

The quality of an air gun array is often specified in terms of the strength of the initial pulse in bar-meters (the acoustic pressure in bars at a distance of one meter from the source in meters) together with the ratio of the strength of the initial pulse to the strength of the maximum bubble peak. To maximize both of these quantities, "tuned" arrays are used in which one or more pairs of guns are placed at carefully calculated locations, so that the individual guns of the pair are far enough apart that the initial pulses do not materially interfere with each other but close enough together to allow certain of their bubbles to interact.

Array performance figures are highly dependent on the filter response of the recording instruments. Within the common seismic band of about 10 to 50 Hz a peak-to-bubble ratio less than about 3 will significantly affect the ability to distinguish a weak reflection closely following a strong one. Ratios much greater than 3 are not required if deconvolution is applied to the data during the processing sequence.

One class of marine seismic sources, the confined-gas gun produces acoustic energy by detonating a mixture of gases in a flexible-walled container. Because the detonation products are vented to the atmosphere, the bubbles do not form.

The confined-gas gun is a device which holds a combustible mixture, usually propane and oxygen, inside an expandable underwater chamber. On command from the recording system, the mixture is ignited by a spark plug, causing the contents of the chamber and the chamber walls to expand suddenly and impart a shock wave to the water. The chamber is prevented from collapsing completely by its internal structure and the gaseous products of combustion are exhausted or vented to the air. As a result, no bubble is produced in the water and the secondary pulse is relatively weak.

Some sources are designed to use the bubble train rather than reject it. The bubbles are allowed to form and then used to contribute to the useful seismic signal.

A small (225 g, or about ½ lb) explosive charge is injected into the water at a depth of about 13 m (40 ft). The charge detonates after a delay of about 1 sec. (The delay allows the delivery mechanism to get out of the way before the explosion.) The expected sequence of bubble collapses and expansions is allowed to proceed unhindered.

A detector on the delivery device picks up the whole sequence of pulses and transmits them to the seismic recorder on board. Later the data reduction computer sorts out the received signals and compares them with the recorded pulse sequence. The computer output is equivalent to a record that would have been produced by combining all the bubble pulses into a single, simple pulse.

Even with some of the nonexplosive sources and source arrays, a detector is sometimes used in an attempt to detect and record the exact shape of the original pulse and the train of subsequent bubble events. Computer processing is used to sharpen the received seismic reflection data on the basis of the detected pulse train. It is rarely possible, however, to position the detector at an equal distance from all sources in an array. Consequently, the recorded pulse does not properly represent the shape of the pulse that actually enters and travels through the earth. The pulse sharpening process, therefore, is much less effective in the case of an array than it is with the single charge.

Choice of Source

Nearly all of the commonly used sources today are satisfactory in most prospects.

The air gun array represents an acceptable compromise source for regular use everywhere. The higher pressure guns are effective in obtaining more power over a broader band. Confined-gas guns are able to produce superior detail in at least the upper part of the section. Small explosive charges are among the most powerful sources in use today, and are preferred by many where penetration is a problem; detail is at least as good as can be obtained from any other source.

Source selection is generally determined by availability, cost, company preference, and suitability for data requirements of the prospect.

In worldwide operations, one may occasionally encounter one of the less commonly used sources. These include:

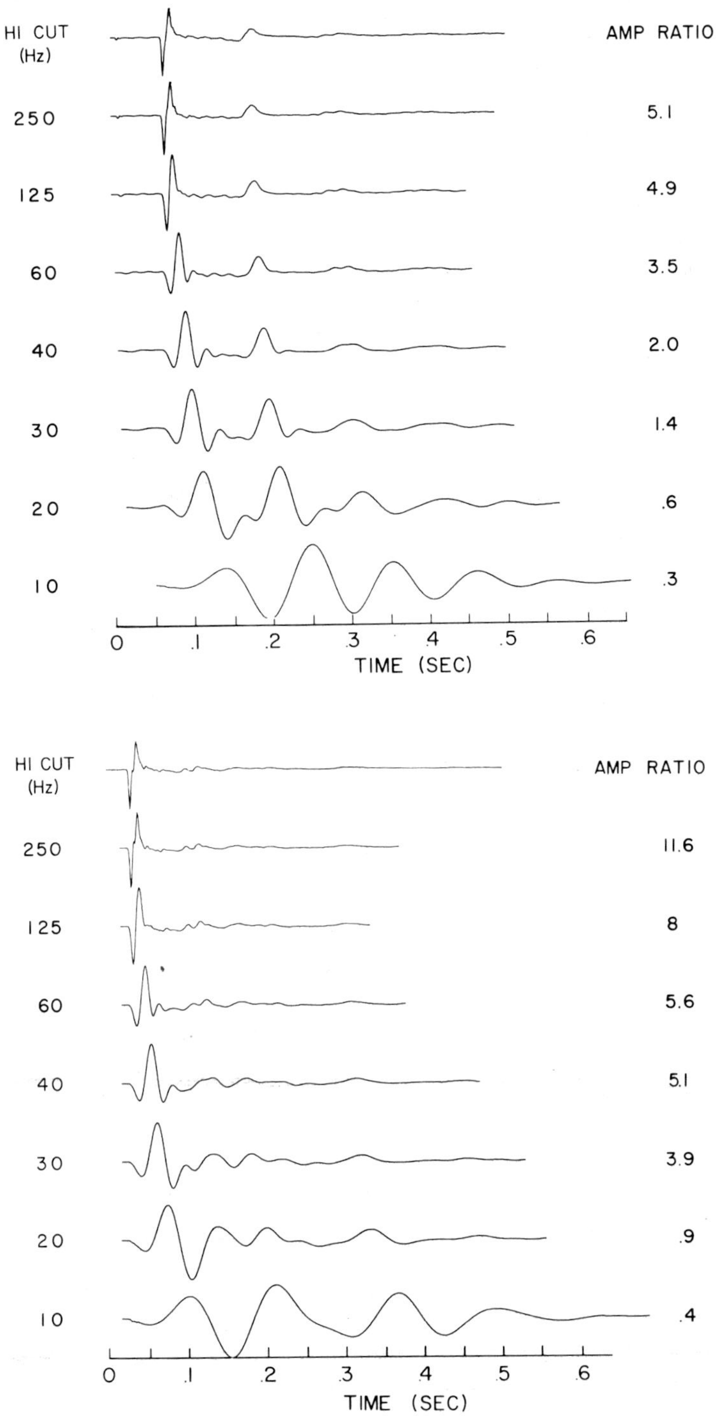

FIGURE 2. Broadband peak-to-bubble ratios quoted in typical specifications have little relevance in the critical part of the seismic passband which is generally below 30 Hz. Tuning markedly improves the peak-to-bubble ratio at all frequencies.

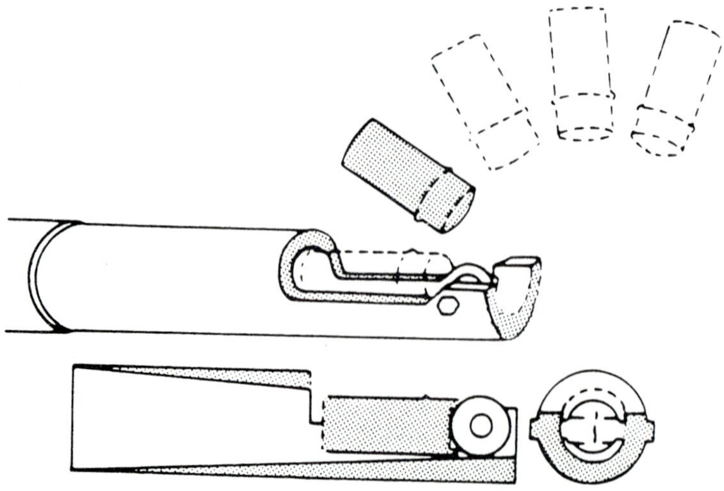

FIGURE 3. MAXIPULSE gun and charge. Top — perspective view of the gun with the charge shown successively striking the firing wheel and being ejected. Bottom — cross-sectional and end views of the charge in the gun at instant of striking the firing wheel.

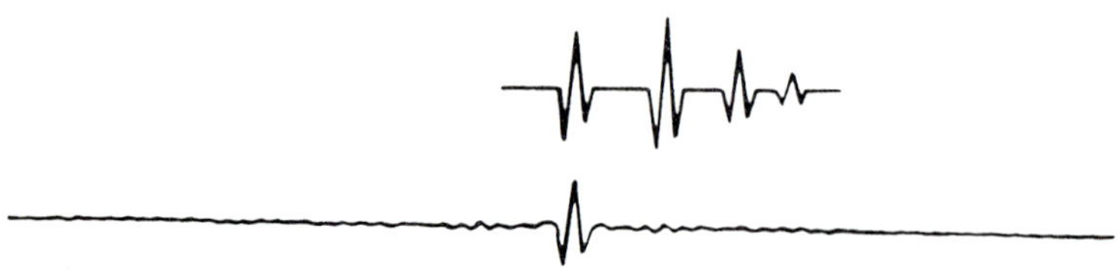

FIGURE 4. Computer printout illustrating the bubble pulse collapse technique. (A) Original bubble pulse sequence showing the original shot and three bubble pulses. (B) Bubble pulse signature collapsed into an initial pulse with a 401 point operator. Additional side-lobe reduction can be achieved with longer operators.

High-pressure or high-volume air guns used singly — A detector near the single source obtains a reasonably good record of pulse shape so that computer processing can be used to collapse the bubble pulses.

Explosive cords and small explosives inside perforated spheres — Large bubbles are replaced by many small ones that produce reduced interference.

Added-air guns and steam guns — The bubble effect is not as great as with other sources, and those that remain are further reduced by routine deconvolution procedures.

Imploders — (devices that abruptly create a cavity in the water). These are in a sense "ideal" sources in that they produce no bubbles and produce a sharp pulse suitable for fine detail work. They are not widely used because of the massive, expensive machinery required.

Sparkers — (devices that form an electric spark under water). Because simple sparkers produce pulses that are relatively weak in the lower frequency seismic range, they are used primarily for detailed shallow surveys with a penetration not exceeding about 1 km below the sea bottom.

Marine Seismic Detector Systems

Historically, deep-water streamer cables have developed from velocity-sensitive land geophones on or in stationary floats, to the modern pressure-sensitive phones in a neutral-buoyant cable in continuous motion.

The basic marine sensor today is a pressure-sensitive piezoelectric crystal in an acceleration-canceling configuration. Acceleration is usually canceled by electrically connecting two "bender" crystals so that if they move together, their outputs cancel, while if they move towards or away from each other, their outputs add. Cylindrical crystals can also be configured to reduce acceleration sensitivity.

Sensors are connected together in groups. Group lengths on older cables vary from about 30 to about 100 m. Newer cables with more groups will have group lengths of 3 to 15 m. Phones in each group are connected in parallel, and through a stepdown transformer to conductors in a cable. Some recent cables have been built without transformers and employ charge-coupled amplifiers instead. Phones in a group are either uniformly spaced or are distributed in a noise-canceling pattern in which phones are more concentrated near the center of the group.

Detector cables consist of 48, 96, 200 or more groups usually arrayed over lengths of 2400 or 3200 m. Sometimes the groups are overlapped.

Systems are now in use in which the hydrophone groups are connected to amplifiers and digitizers in the section couplers of the cable and all data are sent on to the ship through a coaxial line in the cable. Such construction is generally used for configurations requiring more than about 200 groups. At least one 500 group cable is in regular use.

Streamers are designed to be approximately neutrally buoyant so that they tend to remain at a constant depth. The streamer cables are oil-filled plastic tubes, usually 60 to 80 mm in diameter, containing phones, conductors, transformers, and strain members. They normally come in interchangeable sections 50 to 75 m long.

"Live" sections contain hydrophones. "Dead" sections are identical in all respects but contain no phones and are sometimes used as spacers. "Stretch" sections are sections in which resilient ropes replace the usual steel-braid strain members. They are used as shock absorbers between the towing ship and the live sections.

Cables are kept at a uniform, optimum depth (usually between 13 and 6 m, 40 and 20 ft) by a dynamic system. A compact, heavy-cable lead-in from the boat serves to depress the instrument cable. The speed of the boat and the length and weight of the lead-in control the depth of the stretch sections from which the main cable is towed. At intervals on the cable, pressure-sensitive depth-control "birds" automatically compensate for irregularities of cable depth by rotating their wings to give more or less lift.

Streamer cables are normally operated under tow at a speed of 5 or 6 km. At these speeds, the combination of noise-suppressing systems reduces the noise in the seismic frequency band to the order of 1 μbar. That is, there is about as much equivalent ambient pressure fluctuation as would be produced by vertical movement of about 0.01 mm.

MARINE POSITIONING AND NAVIGATION

Shore-based systems depend on radio ranging from two or more onshore stations.

Range-range systems, such as shoran, have accuracies of the order of 10 to 20 m, but a limited range of the order of 80 to 250 km.

Phase comparison systems such as Raydist, Lorac, Decca®, Loran C, etc. have accuracies of the order of 25 to 100 m and ranges of the order of 300 km, except for Loran C which has a range of 2000 to 3000 km.

Some commercial maritime navigation systems such as Loran A and Omega have been generally available, but are not sufficiently accurate for seismic work.

So-called "self-contained" systems do not require special shore-based transmitting stations. All rely for absolute position fixes on U.S. "Transit" satellites which permit location accuracies of 30 to 45 m or so, when they pass overhead. Although, on the average, a satellite fix is obtainable about every hour at higher latitudes and about every 2½ hr at the equator, frequently there are longer intervals. Gaps of 16 hr between usable fixes are not unusual.

Between fixes, most systems in use today rely on Doppler sonar to obtain their position information. Such systems depend on bouncing sonar beams off the bottom, and are ineffective in water much deeper than a few hundred meters. However, accuracies of the order of 100 m can be attained with such systems.

The only systems that are presently operational in deep water, and far from a hospitable shore, are those that rely on inertial guidance. Inertial systems are highly sophisticated gyroscope-based instruments that can calculate positions between fixes to within about 200 to 300 m.

The accuracy quoted for the self-contained systems depends on the accuracy with which the location of the ship is known at the time the determination is made. When all of the location data for a prospect are brought to shore, computer programs are applied to refine and adjust the real-time positions. The so-called "post plot" positions are often subject to less than half the quoted error.

In performing three-dimensional surveys required for precise delineation of oil fields or oil-field structures, it is necessary to know the configuration of the streamer cable in detail and to record the location of each hydrophone group within, at most, a few tens of meters. Elaborate systems exist for determining the cable configuration relative to the ship. At this time, the techniques in use are under intensive development. Most techniques use remote reading compasses at several locations in the cable. Elaborate curve fitting techniques are used to translate these compass positions, the known length of cable segments, and radio or radar fix positions of a trailing tail buoy into an accurate estimate of the cable configuration and orientation.

All shore-based systems rely on accurate determination of onshore base-station positions for absolute accuracy. Frequently, benchmarks and grids on different bodies of land, or from different surveys on the same land mass, are inconsistent by several kilometers.

When base-station positions are unknown, or suspected of being inaccurate, signals from those stations may nevertheless be used to determine velocity of the ship in order to interpolate between satellite fixes. Satellite positions are geocentric and consistent worldwide. They may not agree with local surveys.

SINGLE-CHANNEL, HIGH-RESOLUTION, SEISMIC-REFLECTION PROFILING: A REVIEW OF THE FUNDAMENTALS AND INSTRUMENTATION

Richard E. Sylwester

INTRODUCTION

The development of continuous seismic-reflection profiling equipment[1,2,3] has made possible a means for studying the structure of rocks and sediments beneath the floor of water-covered areas. These areas include the world oceans, estuaries, lakes, and rivers.

The technique requires a ship with either a hull-mounted or towed device that emits an acoustic pulse at regular intervals as the ship travels along a selected course or survey track (Figure 1). These acoustic pulses, when reflected from the sea floor and underlying geology, are intercepted by a hydrophone that converts the acoustic pressure waves into electrical signals. Onboard the ship these electrical signals are processed and displayed on a seismic-reflection recorder. This display, the seismic-reflection record, is an acoustical image of the sea floor and subbottom structure along the survey track (Figure 2).

In recent years the advent of computers and digital processing techniques, together with advances in instrumentation and multichannel data acquisition, have greatly improved the capabilities of the basic single-channel reflection system.[4] These advances have been necessitated by the requirements for deep structural and stratigraphic information as well as for data concerning the lithology, porosity, and seismic velocities of such buried rock units.

In spite of these sophisticated and powerful data gathering and processing techniques, there is still a need for the basic single-channel seismic-reflection system, since it is capable of providing high-resolution information with moderate penetration. This, together with the relative simplicity and low cost of operation, makes this system particularly attractive and completely adequate for many investigations.

This chapter reviews the fundamental principles of acoustic theory and certain aspects of the reflection process, as well as the basic instrumental components of the reflection profiling system. This order of presentation provides the necessary background for understanding the importance of the acoustic characteristics of the various instruments and profiling systems.

Theory of Acoustic Reflection

The reflection of an acoustic signal at the air-water, water-sediment, or sediment-sediment interface results from the change in acoustical impedances at these interfaces. The acoustical impedance, which is an intrinsic property of all substances, is the product of the velocity of a compressional wave through the material and the bulk density of the material, i.e.,

$$Z = \varrho C$$

where

Z = the acoustical impedance (g/cm²/sec)
ϱ = the density of the medium (g/cm³)
C = the compressional wave velocity of the medium (m/sec)

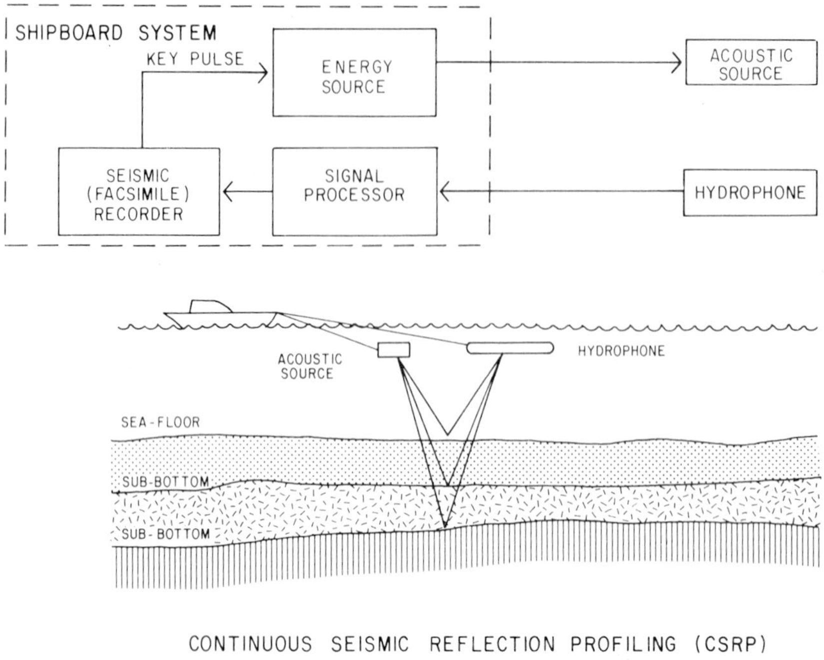

FIGURE 1. The basic components of the continuous seismic-reflection profiling system are shown in the upper illustration. The lower illustration depicts the continuous seismic profiling technique.

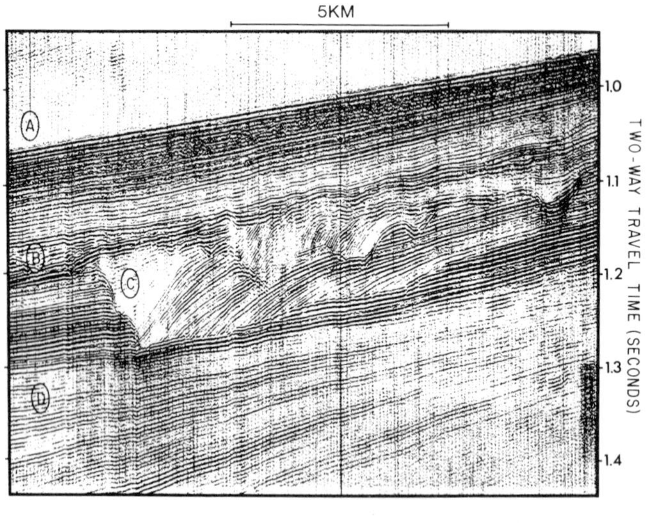

FIGURE 2. A seismic-reflection record showing the sea floor (A) and several easily identified subbottom horizons (B), (C), (D).

It has been shown[5] that for unconsolidated marine sediments the acoustical impedance is mainly a function of the compressional wave velocity in the sediments since there is only a small variation in their bulk density. The compressional wave velocity is given by the equation:

$$C = \left[\frac{K + 4/3\,G}{\rho}\right]^{1/2}$$

where

C = the compressional velocity of the medium (km/sec),
K = compressibility of the medium $(dyn/cm^2)^{-1}$,
G = shear modulus or rigidity of the medium (dyn/cm^2), and
ϱ = bulk density of the medium (g/cm^3)

Because the bulk density of these unconsolidated sediments changes, at most, by a factor of 2,[5] most of the variation in the compressional wave velocity in unconsolidated sediment is caused by (1) differences in the compressibility of mineral grains, and (2) factors that alter the rigidity of the system such as stronger interparticle bonds of the mineral grains.

For consolidated or semiconsolidated sediments, the bulk density becomes an important factor contributing to the variation in the acoustic impedance. This consolidation or lithification of unconsolidated sediments results from (1) overburden pressures due to water, ice, or other sediments, (2) desiccation of sediments during low water levels, (3) cementation by silica or calcium carbonates found in pore solutions, and (4) diagenetic processes whereby chemically dissimilar grains produce new minerals which increase the sediment coherence.[6]

The ratio of the amplitude of the reflected wave to the amplitude of the incident wave for a plane wave incident on a boundary between two media having different acoustical impedances is expressed by the Rayleigh reflection coefficient. At normal incidence the Rayleigh reflection coefficient, R, is given by

$$R = \frac{A_r}{A_i} = \frac{Z_2 - Z_1}{Z_2 + Z_1}$$

where

A_r, A_i = amplitude of reflected signal and incident signal, respectively
Z_2, Z_1 = acoustic impedance of the media on either side of the interface

For a large contrast in acoustic impedance such as at the air-water interface (density of air is approximately 0.001 that of water; velocity of sound through air is approximately 0.25 the velocity through water) nearly all of the normally incident energy is reflected. On the other hand, the reflected energy at a sediment-sediment interface will be quite small where the acoustical impedances are nearly equal, and if they are equal the signal will be transmitted through the interface with no signal reflection occurring (Figure 3).

It must be kept in mind that the coefficient of reflectivity only takes into consideration the acoustical impedance for normal incidence. Other factors that have a profound effect on the amplitude of the reflected signal are (1) the angle of incidence, (2) attenuation of the acoustic signal by the sediments, (3) loss of acoustical energy due to spherical spreading, and (4) loss of signal due to scattering by surface and subsurface irregularities.[7]

Physical Aspects of the Reflection Process

In the previous section the process of reflection was discussed by considering only a

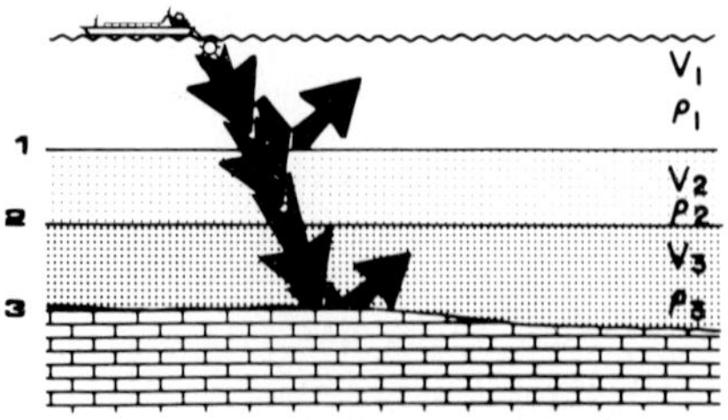

ACOUSTICAL IMPEDANCE $(Z) = \rho v$

RELECTION COEFFICIENT $= \frac{Z_2 - Z_1}{Z_2 + Z_1}$

TYPICAL REFLECTION COEFFICIENT VALUES

WATER→LIMESTONE	.5	MUD → CLAY/SILT	.1
" →SAND	.3-.4	CLAY→SAND	.1
" →CLAY/SILT	.1-.2	SAND→LIMESTONE	.2
" →MUD	.05-.1	CLAY→LIMESTONE	.3
→AIR	-1	SAND→GRANITE	.4

FIGURE 3. (Upper) A contrast in the acoustical impedance results in a reflection at interface 1 and 3. Interface 2 shows no reflection indicating a lack of acoustical impedance contrast across this boundary. (Lower) Reflection coefficient values for some typical geologic boundaries.

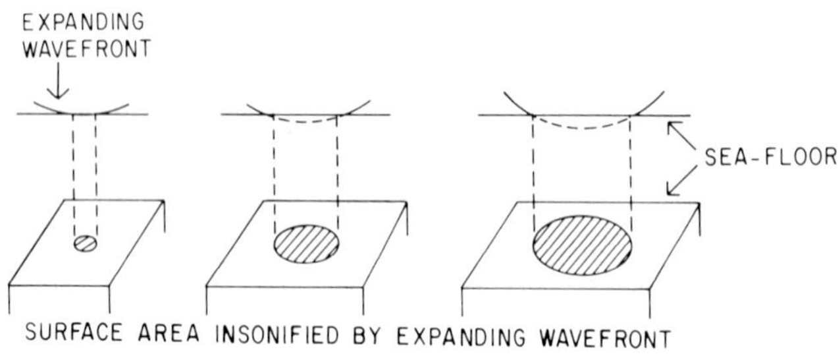

FIGURE 4. This illustration (left to right) shows how an area of the sea floor is isonified by an acoustic wavefront. This area increases with time due to the spherical spreading of the wave front.

ray of acoustic energy. In actuality, a wavefront formed at a sound source spreads spherically as it moves through the water column. When this wavefront contacts the sea floor, a large area of the sea floor is insonified with acoustic energy (Figure 4). However, only a relatively small area of the irradiated surface (erroneously referred to as the reflection point) produces the signal on the seismic record. The size of this area, known as the first Fresnel zone from optics, depends upon the wavelength of the acoustic energy and the water depth. Specifically, it is acoustic energy that is received from the reflecting surface within a half-cycle.[8] Therefore the energy comes from an area that lies within a circular region whose center and outer boundary are determined by one-quarter wavelength. As shown in Figure 5, a high frequency (short wavelength) acoustic signal has a smaller Fresnel zone than a low frequency (long wavelength) signal. The graph in Figure 6 illustrates the relationship between the Fresnel zone radius, the frequency of the acoustic signal, and the water depth.

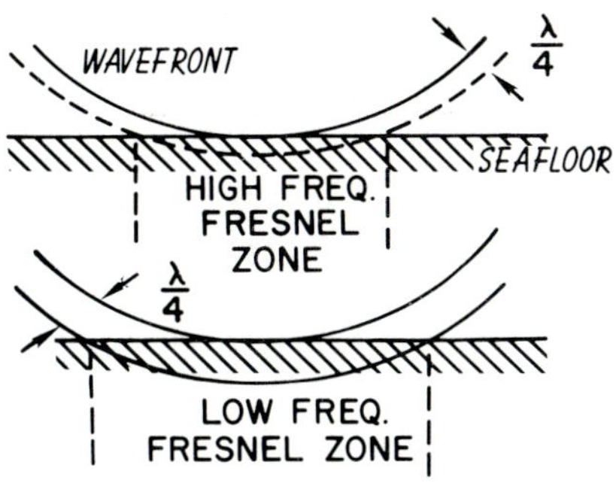

FIGURE 5. The first Fresnel zone, which is frequency dependent, defines the area involved in the reflection process. The high frequency acoustic signal has a smaller Fresnel zone than the low frequency signal.

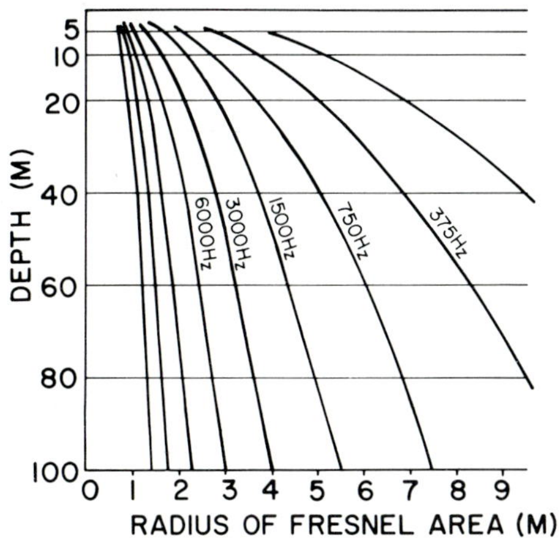

FIGURE 6. Graph showing the radius of the Fresnel zone as a function of the frequency of the acoustic source and the depth to a reflecting horizon.

SEISMIC REFLECTION RECORD

Seismic Record and the Geologic Section

The seismic-reflection record is an acoustic cross section of the geologic section. The degree to which this record truly represents the geologic section is a function of the instrumentation and the complexity of the geology from which the acoustic signals are reflected.

The most obvious departure between the two is that the vertical axis on the seismic record is scaled in units of time, whereas the geologic section is scaled in units of length (depth in meters). Furthermore, the seismic record is seldom displayed at a 1:1 scale between the vertical and horizontal axes. Additional complications result from the fact that the normal incident reflections often do not originate from features directly beneath the sound source and, furthermore, can arrive from surfaces outside of a vertical plane extending beneath the source. Instrumental limitations reduce the ability to both resolve closely spaced reflectors and locate reflectors that are at great depth. Finally, discontinuities such as faults and other features that are small in scale relative to the acoustic wavelength, can produce coherent arrivals by diffraction that greatly complicate the seismic record.

In this section the causes and identifying characteristics of some of these distortions on the seismic record will be discussed.

Vertical Exaggeration

Slopes on the seismic-reflection record are exaggerated due to the difference in the horizontal and vertical scales (Figure 7). The horizonal axis represents the distance covered in a given amount of time and is dependent upon the ship speed and the rate at which the paper displaying the data is moved out of the seismic recorder. The vertical axis represents the time for the sound wave to travel from the sound source to the hydrophone following reflection from the sea floor and subbottom reflectors. This scale is selected on the basis of water depth, firing interval, and expected subbottom penetration.

The resulting vertical exaggeration generally ranges from 5:1 to 20:1. It is possible to correct this distortion later in the laboratory, particularly for the sea floor, by knowing the velocity of sound through the water column. However, corrections for subbottom dips require a knowledge of the velocity structure of the overlying strata. Measurements of the velocity variations are difficult to obtain.

Record Distortion Resulting from Dip and Curvature

The inclination of slopes on reflection records is not a true representation of the inclination of the sea floor. As the ship profiles over a slope, the acoustic energy received by the hydrophones is reflected at normal incidence to the slope (Figure 8). Therefore, the vertically plotted data on the seismic record can represent the sea floor in front of or behind the ship. The resulting effect is that the apparent slope is not as steep as the true slope. The trigonometric relationship for this difference in slope is

$$\sin\Theta = \tan\Phi$$

where Θ = angle of the true slope and Φ = the angle of the apparent slope. However, the $\tan\Theta$ (true slope of sea floor) and $\tan\Phi$ (apparent slope of the sea floor) are nearly equal for angles less than 15°, for which the maximum error would be 3%.[9]

Folds such as synclines and anticlines also exhibit this type of dip distortion. That is, a reflection that is normal to the curved surface of the structure will be displayed on the seismic record as though arriving from a surface directly beneath the survey vessel (Figure 9). The degree to which curved features are distorted depends on the center of curvature and depth of the fold, and the velocity structure in the overlying strata.[10]

Side Echoes

The omnidirectional acoustic signal produced by the sound source (certain narrow-beam echo sounders are an exception), often results in the arrival of reflections from structures on or beneath the sea floor that are not in the vertical plane of the survey track. When these side echoes or sideswipes appear on the reflection record, they can easily be interpreted as subbottom structure. This is particularly true for a side echo

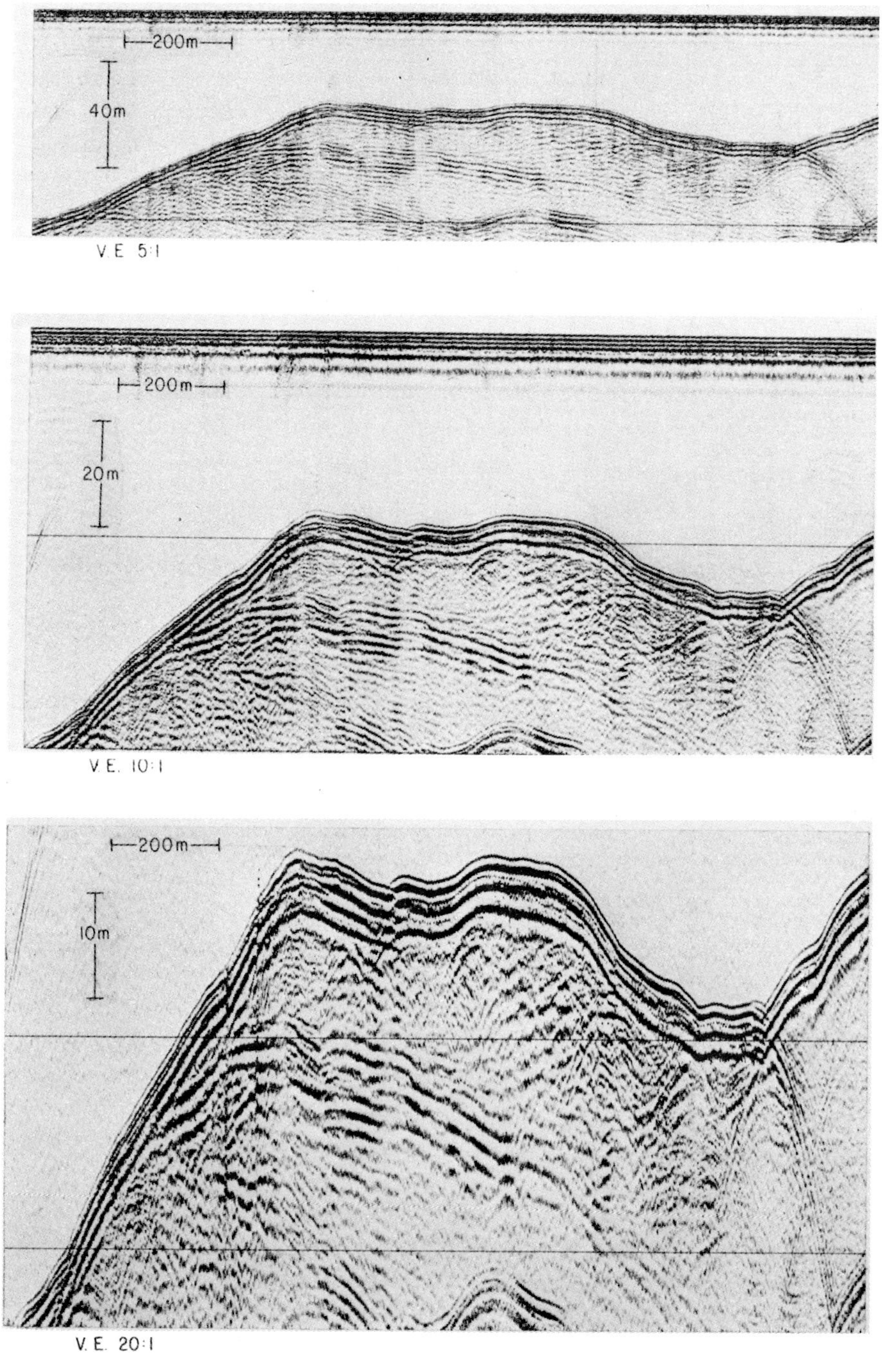

FIGURE 7. Vertical exaggeration on the seismic-reflection record.

produced from a feature that parallels the survey track for some distance, such as a steep slope. Figure 10 illustrates how side echoes can be generated; Figure 11 shows a seismic-reflection record with a prominent side echo.

Because acoustic signals transmitted through rock or sediments are selectively fil-

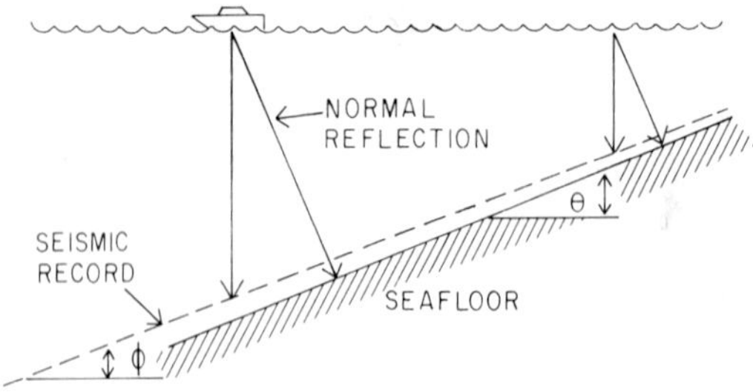

FIGURE 8. Reflections from normal incidence plotted vertically below the source result in a decrease in the sea floor dip displayed on the seismic-reflection record.

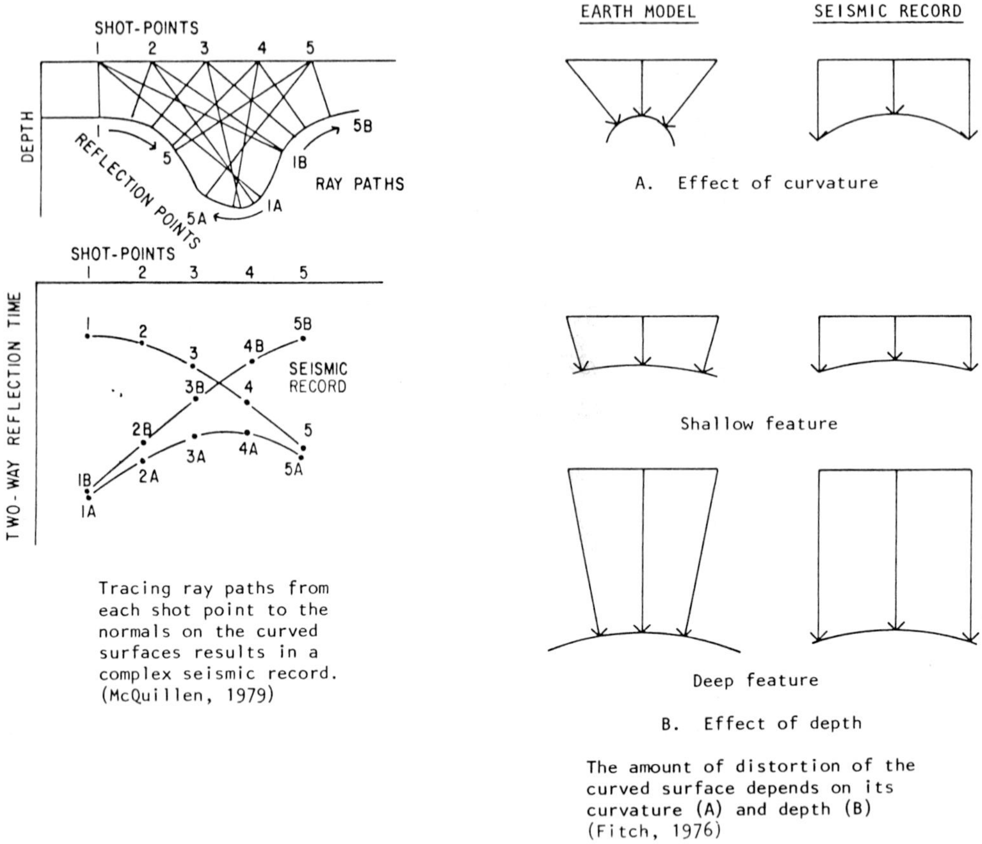

FIGURE 9. Normal incidence reflections from curved surfaces and their appearance on the seismic-reflection record.

tered (high frequencies are attenuated more rapidly than the low frequencies), it is possible, in certain cases, to determine if a signal on the record is a real feature or a side echo.[11] This technique is possible only with a transducer that produces a signal having broad frequency spectrum or high frequency components. The procedure calls for the passing of the received signal through a band-pass filter set for a high pass of 1 to 2 kHz. Under these circumstances all subbottom reflectors should be greatly atten-

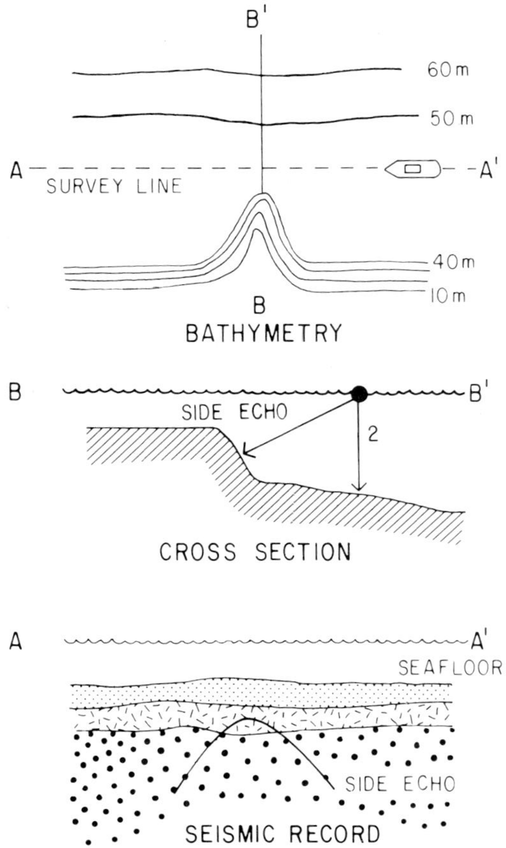

FIGURE 10. Side echoes are produced by reflections from surfaces not in the vertical plane of the survey transverse.

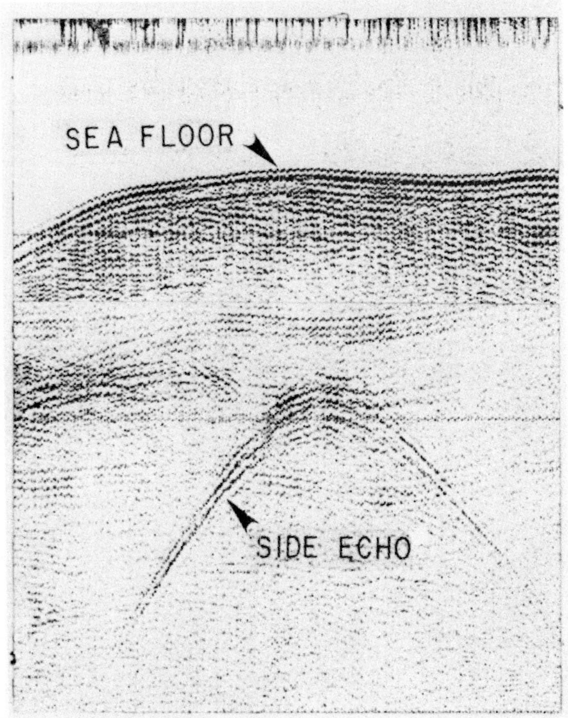

FIGURE 11. Example of a side echo on a reflection record.

uated because energy in this frequency band is strongly absorbed by the sediments.[12] Because most side echoes are produced by reflections from surface features, they will still show up on the seismic record as relatively strong signals. High amplitude late arrivals that appear on the seismic record at these frequencies are almost certain to be side echoes.

It is necessary to determine travel times to the sea floor beneath the ship and to surrounding features in order to interpret surface structures on the record that are suspected of being side echoes. A good contour chart of the bathymetry in the survey area is a valuable aid in carrying out these calculations.

Multiple Reflections

The acoustic signal reflected from the sea floor is often again reflected at the air-water interface after it has been received by the hydrophone. This interface is nearly a perfect reflector, having a reflectivity coefficient close to -1. The negative sign indicates that the reflected signal is 180° out of phase with the incident signal, and a value of 1 indicates that nearly all of the acoustic signal is reflected.

If enough energy remains in this rereflected signal (as is often the case in shallow water where the bottom sediments have a low coefficient of attenuation and spreading loss is negligible), it can travel a number of times between the air-water interface and the sea floor. Each multiple reflection from the sea floor will produce an output on the seismic record until the signal level is too low to be detected by the hydrophones (Figure 12).

In shallow water these multiples will be closely spaced and will tend to obscure shallow subbottom reflections. It has been shown that these multiples can often have a greater signal strength than the primary reflection.[13] In this situation, it would be nearly impossible to detect the weaker subbottom reflectors on the record beneath the multiple reflection.

Where the multiple reflections are weak, it is possible that they can be interpreted as subbottom reflectors. This is particularly true in those areas where the bottom and subbottom reflectors are horizontal. In survey areas where the bottom slopes or changes in depth, the multiple reflection will accentuate these changes (first multiple will double the slopes, and so on) which is helpful in identifying and eliminating these multiples from real subbottom reflectors.

There is very little that can be done to remove multiple reflections from single-channel seismic-reflection records. Because it is one of the principal difficulties encountered in shallow water surveys (water depths in the 1 to 100 m range), meticulous care is required in the analysis of the reflection records in order to preclude erroneous interpretations.

Diffractions

Acoustic energy returned from features having dimensions that are comparable to the wavelength of the acoustic signal or from structures that terminate abruptly, produce arcuate or hyperbolic features on the seismic-reflection record. These features, produced by the diffraction process, can be generated by undulating surface features such as sand waves, small pinnacles, large erratics, ice scour marks, erosional surfaces, or faults. (If the features are much smaller than a wavelength, they scatter the energy and if they are much larger, they reflect the energy.[14])

To demonstrate the concept of diffraction, a model in Figure 13 has been constructed showing a small object beneath the sea floor. As the ship travels from shotpoint 1 to 7, the hydrophone receives acoustic energy from the object that acts as a point source for reflected acoustic energy (Figure 13, upper); hence the often heard

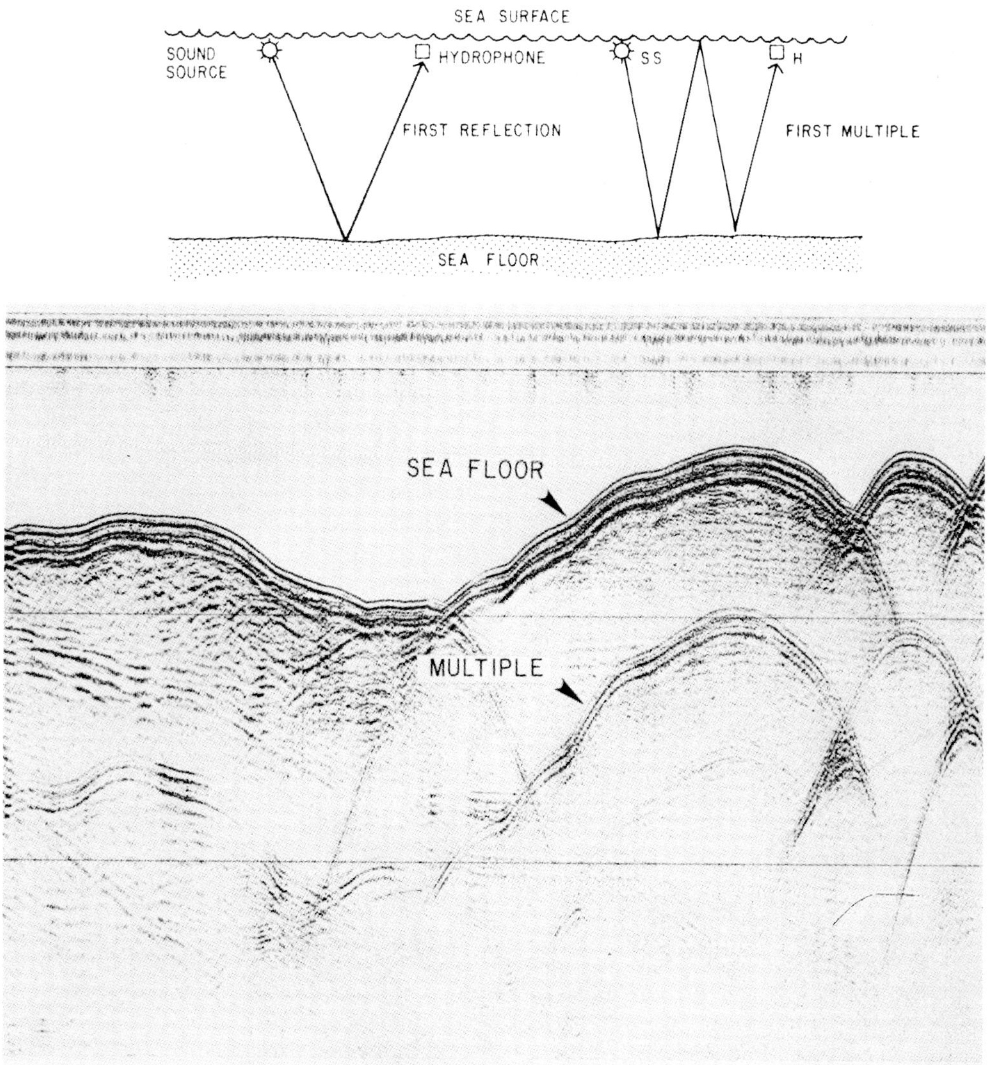

FIGURE 12. Reverberation of the acoustic signal between the sea floor and the air-water interface produces a multiple echo of the sea floor. Note the increase in the curvature of the multiple relative to the first sea floor reflection.

phrase that diffractions are the result of point sources. At shotpoint 1, acoustic energy is received from the point source and plotted on the seismic record as an event beneath the shot point (Figure 13, lower). All subsequent shotpoints (2 to 7) also show a return from the point source. The cumulative effect on the record, then, is a hyperbola which would be very continuous for shotpoints spaced closely together. Figure 14 shows, rather clearly, this effect from an irregular sea-floor surface and from fish in the water column.

A sudden structural change, such as a fault or a dip in the sea floor, will produce diffractions from the area around the discontinuity.[15] These diffractions have a smoothing effect that will show the termination as a curved surface. In Figure 15 an example has been plotted for both cases, Generally, it is not possible on the seismic record to observe the limb of the hyperbola beneath the flat portion of the sea floor

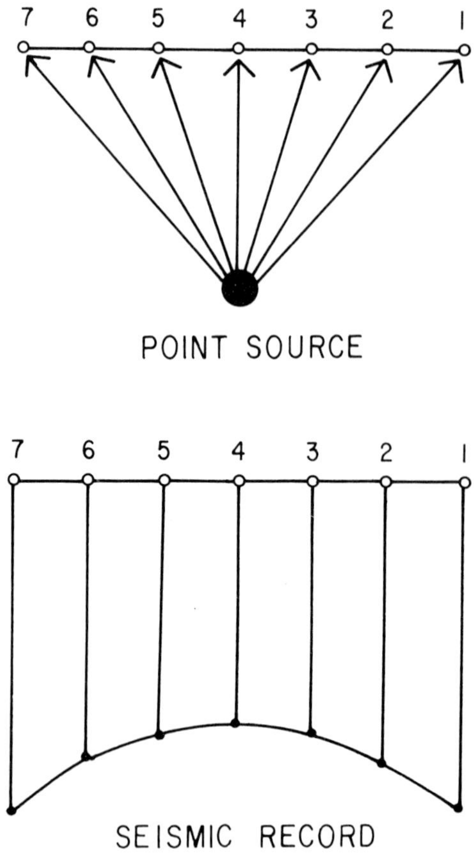

FIGURE 13. (Upper) Diffractions occur when a small bottom or subbottom feature acts as a point source of acoustic energy. (Lower) The seismic-reflection record plots the acoustic signal from the point source as an arrival coming from vertically beneath the seismic profiler resulting in a crescent-shaped or hyperbolic feature. Only shotpoint 4 shows the true location of the object.

depicted on the fault model. Furthermore, if the survey track crosses the feature producing the diffraction at an angle or off to one side, the hyperbolas will not exhibit the symmetry shown in the fault model.[16]

Vertical Resolution and Depth of Penetration

The ability to detect and display on the reflection record stratigraphic horizons that are very closely spaced is dependent on the vertical resolution of the seismic-reflection system. The resolution will vary depending on the sweep speed of the recorder, the signal-to-noise ratio, and the acoustic source characteristics. The source characteristics, particularly the pulse-length and frequency component, are generally the parameters considered when determining the vertical resolution.

An approximation of the resolution for an acoustic signal can be made in the following manner:[17]

$$T = \frac{0.5V}{2f}, \text{ or } 1/4\,\lambda$$

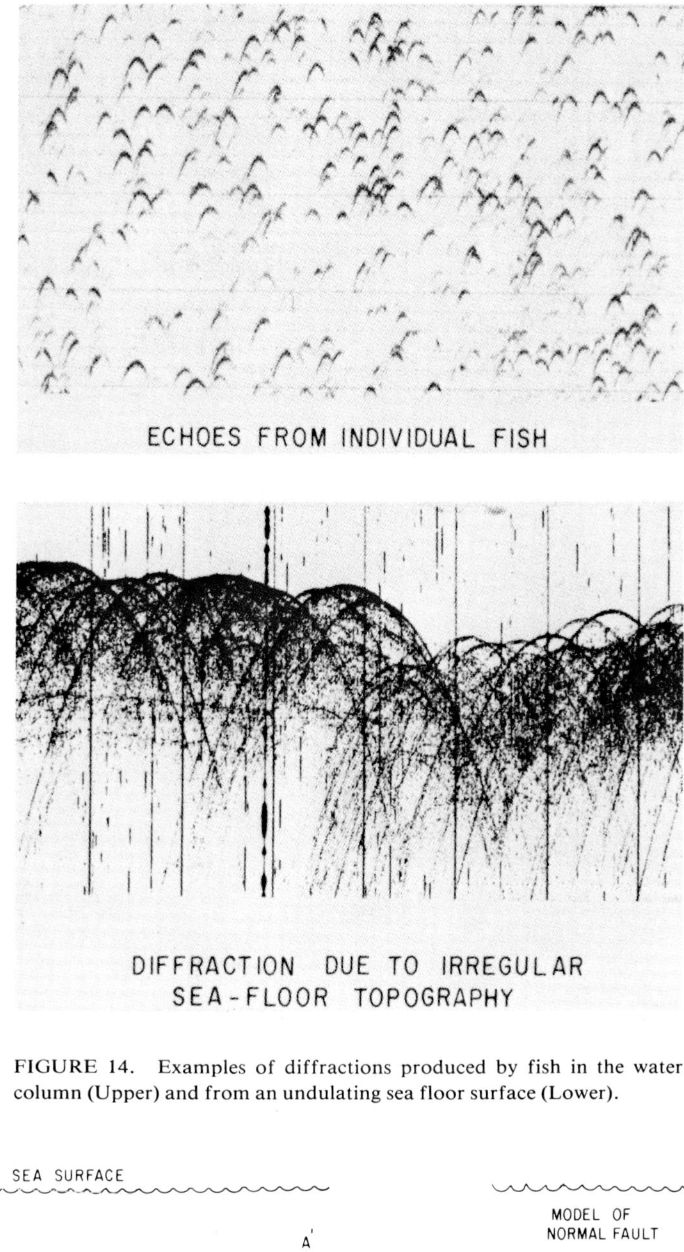

FIGURE 14. Examples of diffractions produced by fish in the water column (Upper) and from an undulating sea floor surface (Lower).

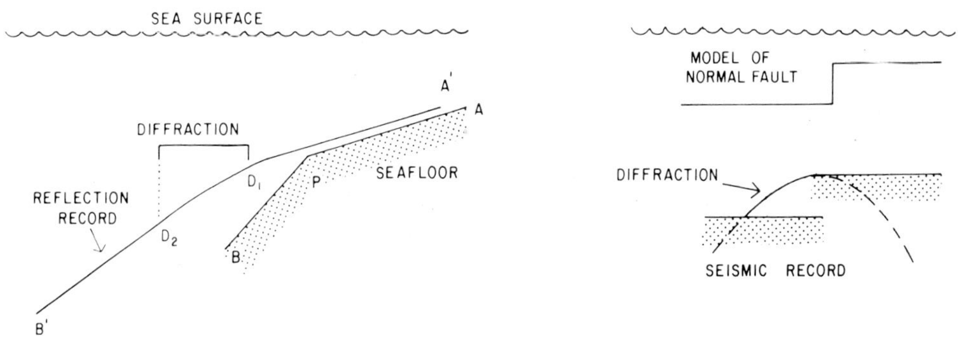

FIGURE 15. The surfaces A-P and P-B appear on the reflection record as A^1-D_1 and D_2-B^1. The section D_1-D_2 results from diffracted energy produced by the break in slope at P. The fault model illustrates the diffractions generated at the fault edge. Both examples show the smoothing effect that diffractions have on abrupt structural changes.

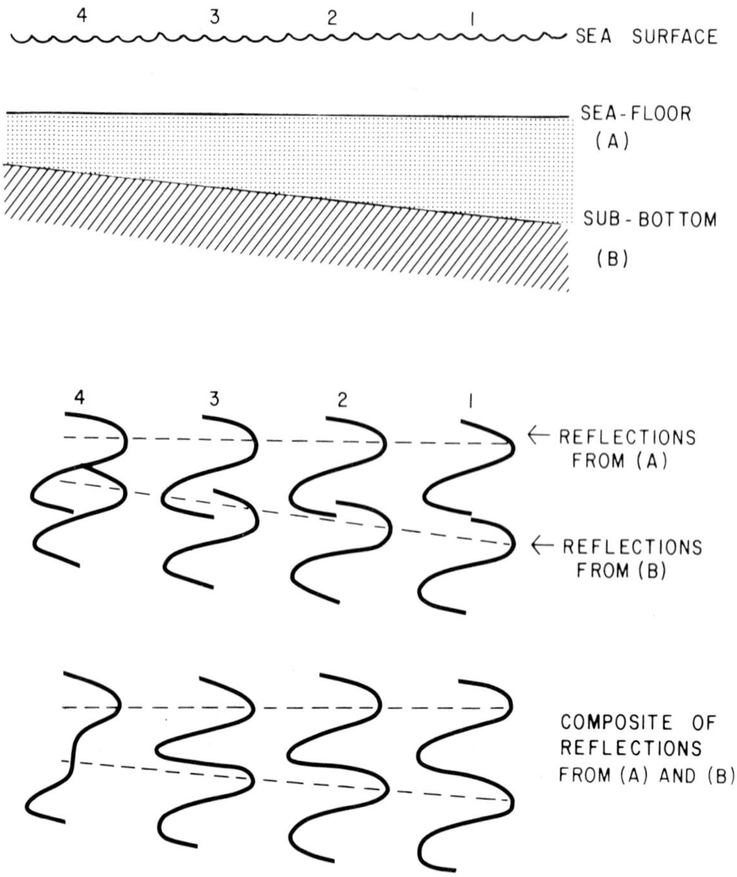

FIGURE 16. The model for a thinning geologic unit (upper) and the reflection from the two interfaces (middle) produce the reflection record (lower) shown. At shotpoint 4 (¼ wavelength), the two reflectors are no longer distinct features.

where

T = the thickness of the geologic unit that can be resolved
V = the compressional wave velocity of the acoustic medium
f = dominant frequency of the acoustic pulse
λ = the wavelength of the dominant frequency

For example, if the dominant frequency of the acoustic source is 100 Hz and the compressional wave velocity of the sediments is 1800 m/sec then the minimum resolvable layer thickness is

$$T = \frac{(0.5)\,(1800)}{2(100)} = 4.5 \text{ m}$$

Therefore to increase the resolution requires a higher frequency (shorter wavelength) acoustic source.

In Figure 16 a geologic feature is shown to pinch out or decrease in thickness along the survey track.[18] At location 1 there are two distinct reflections, one from the top and one from the bottom of the unit. Between locations 1 and 2 the thickness of this unit is equal to the wavelength of the acoustic signal, and from here to the left the two separate reflections overlap. When the thickness of the unit is less than ¼ wavelength, the upper and lower boundaries can no longer be identified as individual reflectors. Vertical resolution is important not only to detect thinning of beds laterally, but also for mapping small vertical offsets due to faulting and for mapping thin stratigraphic units.

Concurrent with maintaining good vertical resolution is the need for receiving reflection information from deep subbottom reflectors. The acoustic signal is attenuated by various processes (spherical spreading, scattering, viscous losses, etc.) as it passes through the earth. The cumulative effect of these processes is a loss of energy in the acoustic signal which is frequency dependent.[19] The higher the frequency the greater the attenuation; therefore, deep penetration requires an acoustic signal containing low frequency components. Because resolution decreases with decreasing frequency, the increase in depth of penetration is achieved at the expense of a reduction in resolution.

It is necessary to know the objectives of a given seismic-reflection survey, in terms of resolution and penetration, to select the appropriate seismic-reflection system. If both high resolution and deep penetration are important then at least two reflection profiling systems, one for high resolution and one for deep penetration, are required.

Horizontal Resolution

In addition to resolving vertical relationships of the geologic section, horizontal or spatial resolution is needed. As discussed earlier, the reflected acoustic signal is not a narrow beam or ray reflected from a single point on the reflecting surface. The "reflection point" is an area defined by the first Fresnel zone. That is, the acoustic energy from all reflecting surfaces within this area add together in a constructive manner to provide a composite signal. This composite signal appears on the seismic-reflection record as a single small mark.

A feature having dimensions less than those of the Fresnel zone will not appear on the seismic-reflection record as a discrete identifiable object. Any reflection information regarding its position and shape will be added with the acoustic information from the remaining portion of the Fresnel zone. Accordingly, the horizontal resolution of an acoustic signal is limited by the dimensions of the first Fresnel zone.

This problem can be overcome, particularly when surveying with high repetition rate profiling systems.[20] If the speed of the survey vessel is slow enough and the pulse repetition rate is high enough, then the Fresnel zones will overlap along the survey track and the multiplicity of data received contain significant information about the dimensions of the object.

Information arriving from outside of the Fresnel zone, which does not contribute to the actual reflecting wavefront, tends to smear the record. An improvement in the sharpness of the reflection record can be achieved by eliminating the reception of these late arrivals as well as reducing the size of the Fresnel zone. Comparison of Figures 17 and 18 will illustrate this point. Figure 17, a seismic-reflection record obtained with the acoustic source and hydrophone at the sea surface, contains obvious geometric distortions and appears out of focus. When the hydrophone is towed near the sea floor (Figure 18), the record is not only better focused, but is also a truer representation of the actual sea-floor geometry.[21]

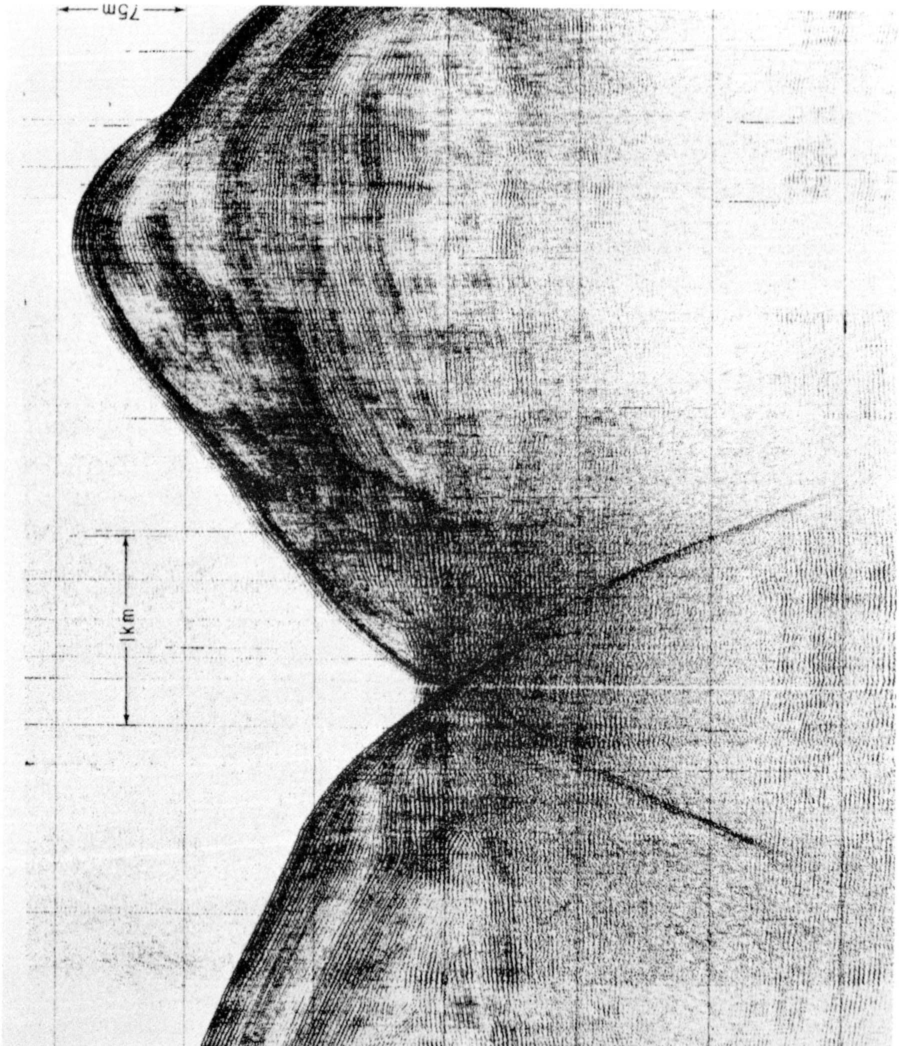

FIGURE 17. This reflection record obtained with acoustic source and hydrophone at the surface exhibits geometric distortion and poor horizontal resolution.

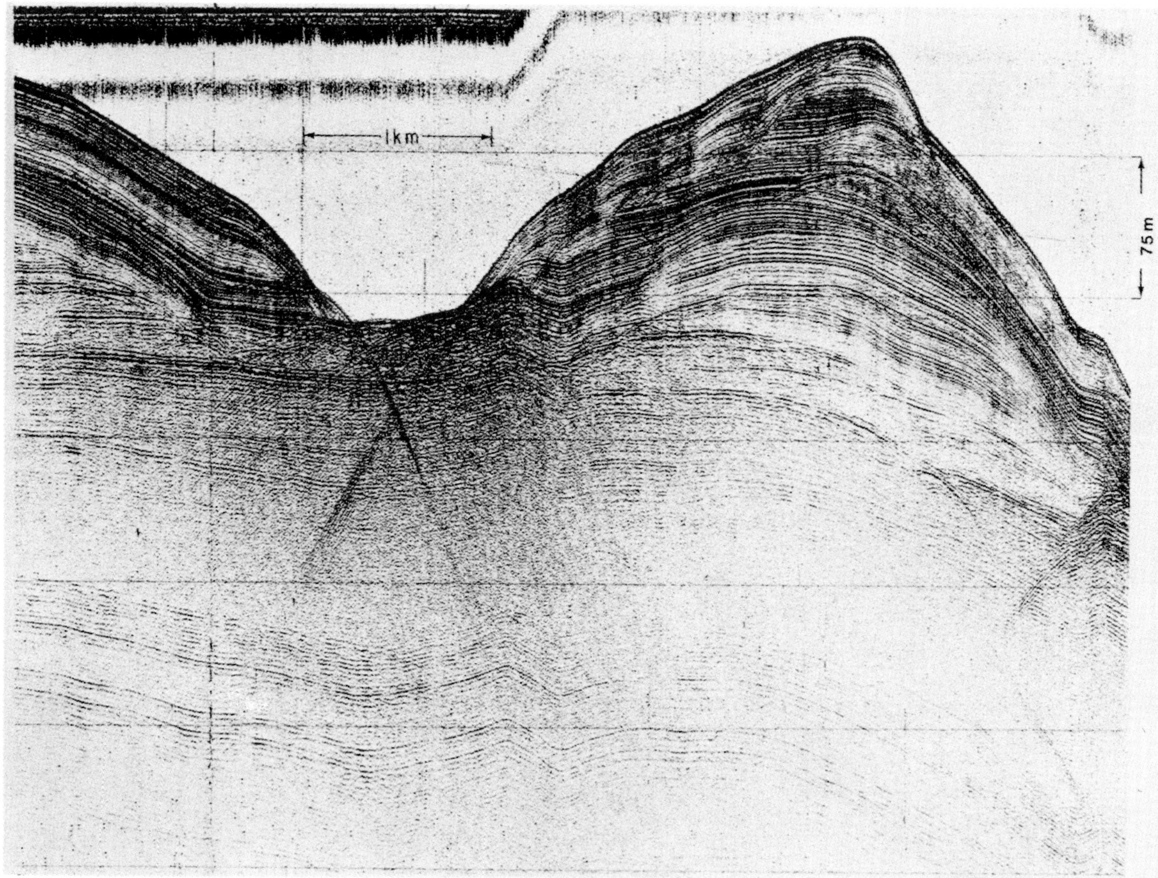

FIGURE 18. Towing a hydrophone near the sea floor results in a considerable improvement in the data shown in Figure 17.

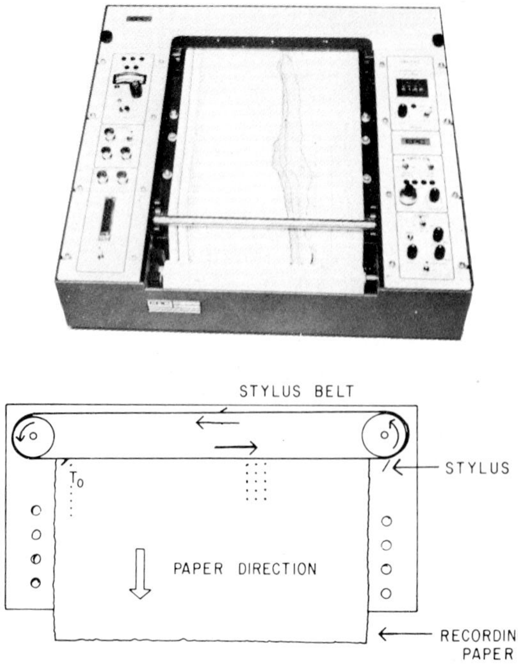

FIGURE 19. (A) Seismic-reflection recorder (EPC Labs Inc., Danvers, Mass.) (B) The reflection record is produced by a burning process at the stylus as it travels across the recording paper. The burn marks represent the time at which an acoustic signal was received by the hydrophone. After one sweep the recording paper is advanced, the sound source keyed again at T_o, and another trace of data is burned on the paper. The alignment of these burned dots produces the continuous reflection record.

INSTRUMENTATION

Introduction

This section will discuss the basic components comprising the seismic-reflection profiling systems. Advantages and disadvantages of each profiling system will be given as they relate to the effectiveness of the system for the study of the sea-floor bottom and subbottom structures. Only four reflection-profiling systems are covered in this review, out of the many that are available.[22] These four (tuned transducer, Uniboom®,* sparker, and air gun) are most commonly used and are readily available. Furthermore, they illustrate the range of resolution and penetration that is typical of subbottom profiling systems. Several ancillary instruments that improve the quality of the seismic-reflection record or assist in the survey work are discussed at the close of the section.

Seismic Recorder

The seismic recorder (line recorder, facsimile recorder, graphic recorder) keys or triggers the acoustic source and also displays the acoustic information received by the hydrophone (Figure 19). The process of keying and printing occurs in the following manner. The stylus, which is on a continuously moving belt, keys the acoustic source

* Any use of trade names is for descriptive purposes only and does not imply endorsement by the USGS.

at the left edge of the paper at the start of the sweep. It receives electrical signals from the hydrophone which results in small marks being left on the paper by a burning process. These marks, which are the reflection data, vary in intensity depending on the amplitude of the reflected signal. Once the sweep of the stylus is completed, the recording paper is advanced slightly and the process begins again. After several sweeps, the data marks will appear as a continuous line forming the seismic reflection-record.

The reflection data on the record represent the time for the acoustic signal to travel from the sound source to the reflecting surface and back to the hydrophone. Thus it is called the two-way travel time. For example, if the sweep speed of the recorder is 1 sec and the sea floor appears halfway down the record, then it has a two-way travel time of ½ sec. If, as is often the case, it is necessary to express this depth in units of length (meters, feet, fathoms), the following formula is used:

$$D = \frac{V}{2} \times t$$

where

D = depth to the reflector
V = velocity of sound through the medium; approximately 1500 m/sec for water
t = the two-way travel time

In our example, then

$$D = \frac{1500 \text{ m/sec}}{2} \times 0.5$$

$$= 375 \text{ m}$$

Accurate determination of depth to subbottom reflectors requires that the velocity of sound through the overlying sediments be known. For shallow subbottom reflectors (<0.1 sec) in unconsolidated sediments, a velocity of 1500 to 1600 m/sec is commonly used.

The sweep speed is variable on the recorder providing a means to select the most appropriate vertical scale. For instance in shallow water, a sweep speed of 1/10 sec representing a depth scale of 75 m might be selected for a very high resolution, shallow penetration reflection survey. When deep penetration is expected, a sweep speed of 1 sec (750 m), 2 sec (1500 m), or 4 sec (6000 m) might be more desirable.

Acoustic Source

The selection of the acoustic source is often based upon the requirements for resolution and depth of penetration. However, equally important considerations are the cost, size, ease of operation, maintenance, and availability of the energy source and acoustic source.

The acoustic sources that will be discussed, listed in order of decreasing resolution and increasing penetration, are: the tuned transducer, the Uniboom®, the sparker, and the air gun. Other acoustic sources, not discussed here, should also be considered when planning a survey.

The acoustic source requires some form of primary energy (electrical, pneumatic, or hydraulic) which is developed by systems on board the survey vessel. And each employs a device in the water that converts stored energy to an acoustic pulse.

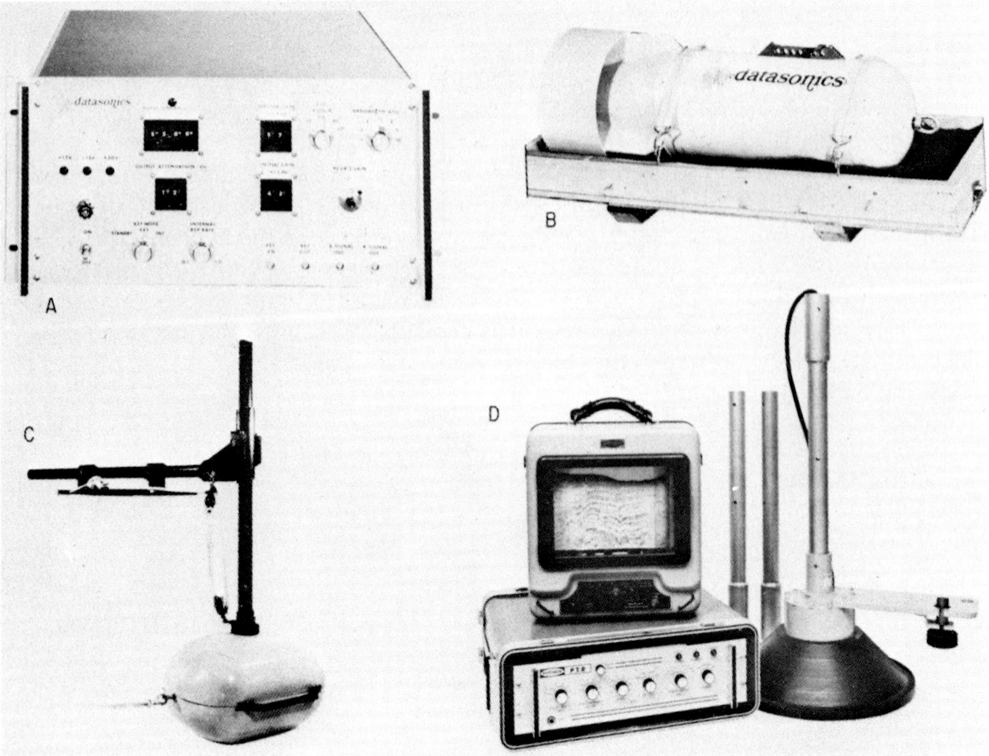

FIGURE 20. (A) Transceiver with 3.5 kHz transducer; (B) In-tow vehicle (Data Sonics, Falmouth, Mass.); (C) Four transducer array mounted in over-the-side housing (O.R.E., Inc., Falmouth, Mass.); (D) Multiple frequency 3.5 kHz/7 kHz transducer system (Raytheon, Inc., East Providence, R.I.).

Tuned Transducer Systems

Piezoelectric crystalline materials undergo deformation when subjected to an electrical signal. Conversely, if this same material is deformed, for example by an acoustic pressure wave, it will generate an electrical signal.

A tuned transducer is constructed by forming these crystalline materials into various shapes and then placing them in a mechanically rugged housing. This device is capable of emitting and receiving acoustic signals at the frequency for which it was designed and constructed. Additional important characteristics that depend on the design are the conversion efficiency, the power handling capability, and the beam width of the transducers.

The complete system consists of the transducer and a transceiver (Figure 20). The transceiver contains a transmitter that controls the pulse length and provides the electrical power at the proper frequency for driving the transducer. The receiver section of the transceiver amplifies the electrical signals coming from the transducer for display on the seismic recorder.

Tuned transducer systems for subbottom profiling are selected on the basis of their frequency (generally 1.5 kHz to 7.5 kHz) and power available from the transceiver (500 W to 12 kW). The high frequency systems provide good resolution but are limited in penetration (Figure 21). To improve penetration, low frequency transducers are used with increasing amounts of electrical power in order to provide a high acoustic source level. If the additional power exceeds the capacity of a single transducer, then an array of several transducers is used. This distributes the power among the individual transducers and increases the acoustic source level.

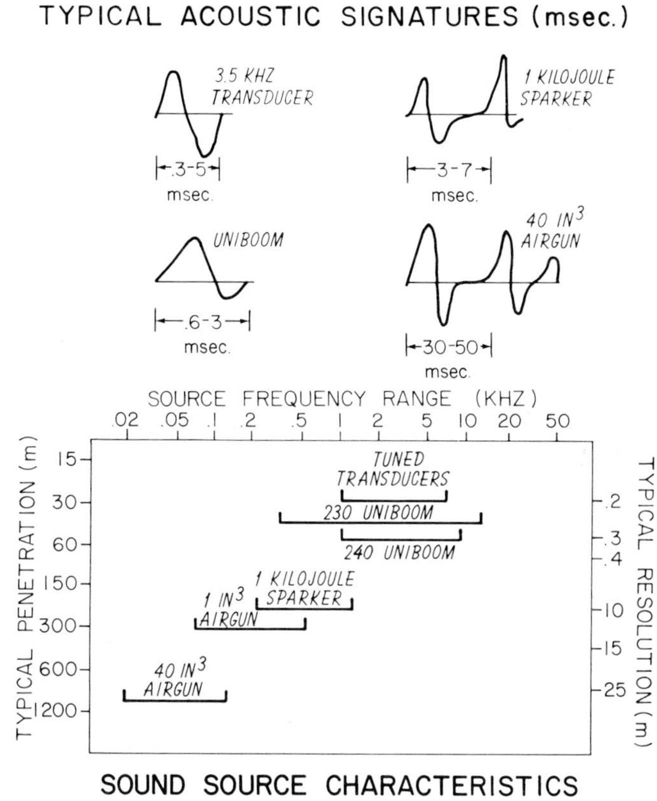

FIGURE 21. Acoustic signature for various sound sources (Upper) and a relative comparison of their depth of penetration and resolution capabilities (Lower).

Tuned transducer systems commonly are mounted in three ways. If they are to be used permanently with a given survey vessel, they are placed in a water-filled compartment near the hull of the ship. The acoustic signal is coupled through the hull to the seawater and back again for the returning signal. Considerable attenuation of the acoustic signal as it passes through the hull plate may require cutting a small window in the plates and covering it with thinner material.

The transducer or transducer array can also be installed in a towed vehicle that is suspended from a davit or frame on the survey vessel. The towed vehicles (or "fish") are designed to maintain a constant depth at survey speeds, assuming the ship motion due to adverse sea conditions is not excessive. Because these vehicles are often quite heavy (200 to 600 lb), a small hand-powered or electrically driven winch is required for deployment and retrieval.

It is also possible to mount the transducers on a frame that is fastened to the rail of the survey vessel. These systems are generally smaller and, as such, are quite portable. Although they can be mounted on any size of vessel, they are ideal for shallow, protected waters (harbors, lakes, rivers) which can be surveyed with smaller boats (10 to 30 ft in length).

Tuned transducer systems are primarily used for subbottom profiling studies that require high-resolution information. Subbottom penetration, although limited, can be expected if the geologic conditions are favorable. Several examples illustrating the capability of these systems are shown in Figures 22 and 23.

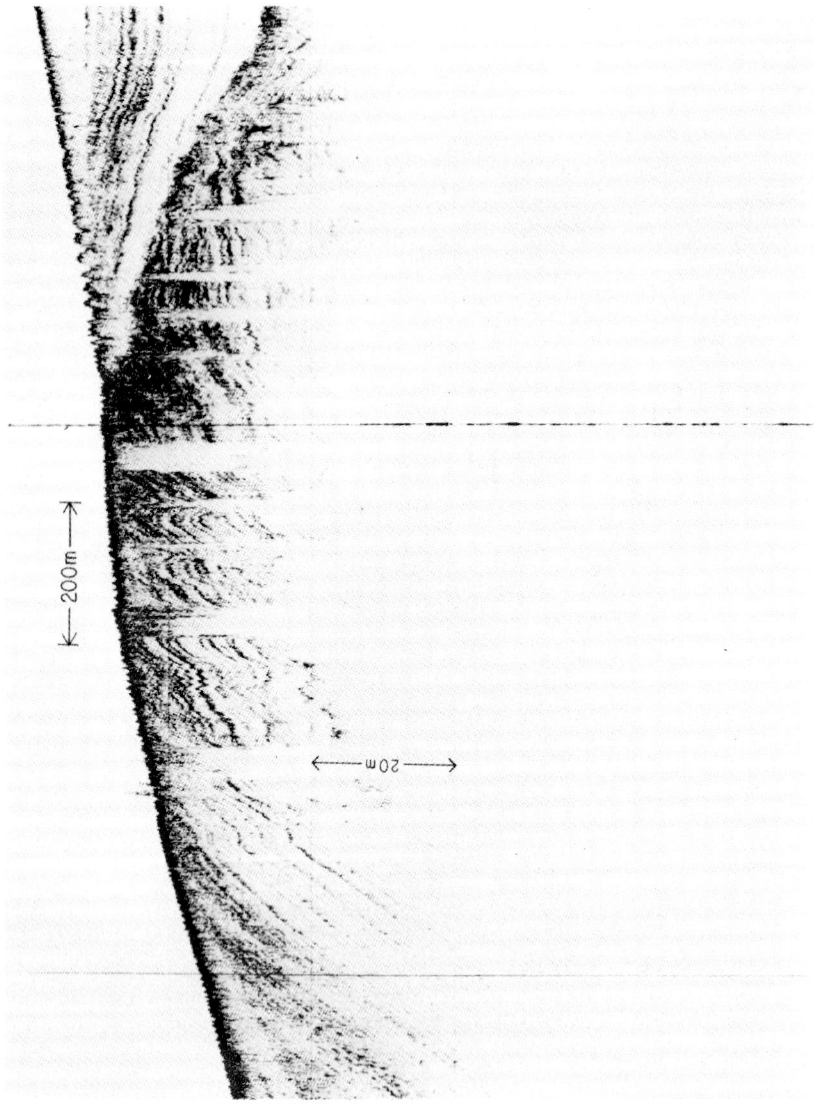

FIGURE 22. Seismic-reflection data from a 3.5 kHz transducer system.

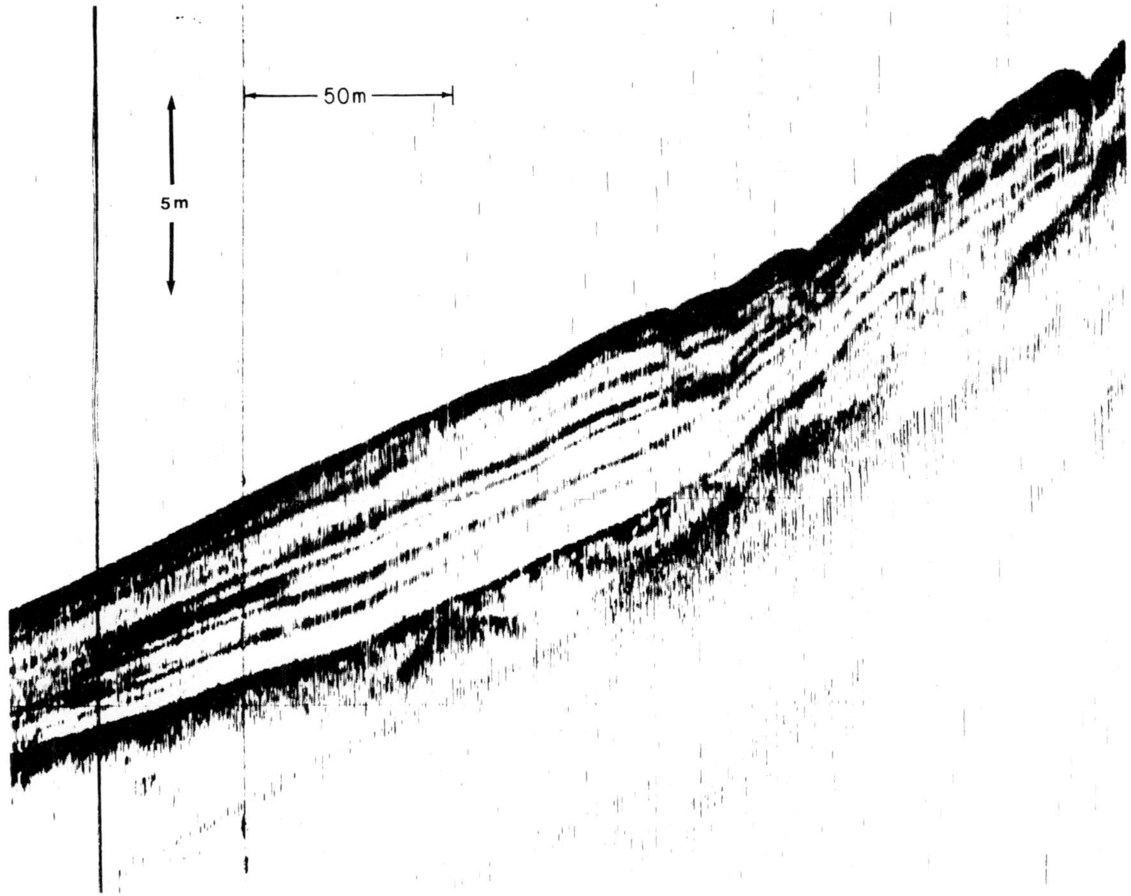

FIGURE 23. Seismic-reflection data from a 7 kHz transducer system.

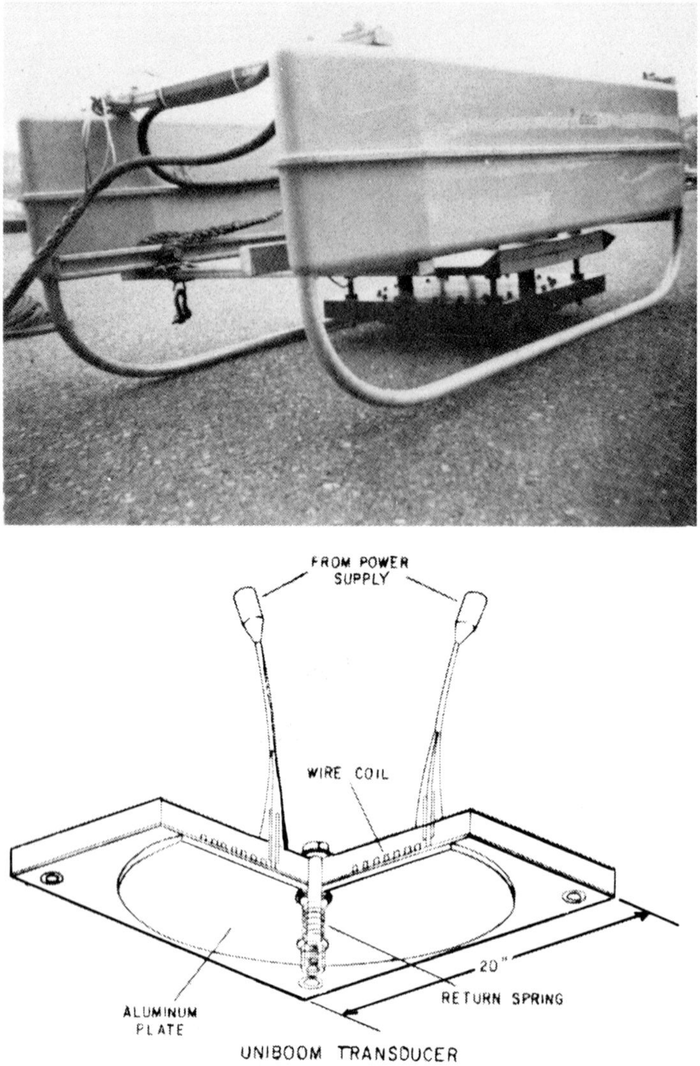

FIGURE 24. The Uniboom® electrical-mechanical transducer and catamaran (EG&G, Waltham, Mass.).

Uniboom®

The Uniboom® is often classed as a displacement type device because the acoustic signal is produced by the rapid movement of a flat, round plate against the water (Figure 24). The motion of the plate is produced in the following manner. When electrical energy stored in the power supply is rapidly discharged through the coil (initiated by the key pulse from the recorder) located in the transducer, a magnetic field is developed in the coil. This magnetic field opposes the induced magnetic field generated in the aluminum plate, resulting in a rapid repulsion action of the plate. The fields decay very rapidly and a return spring forces the plate back to the initial position. The acoustic signature generated is of short duration and relatively high amplitude (Figure 21), having a characteristic band width of 400 to 4000 Hz.

The Uniboom® transducer is usually mounted on a catamaran that is towed astern of or alongside the survey vessel. This towing arrangement, due to the instability of

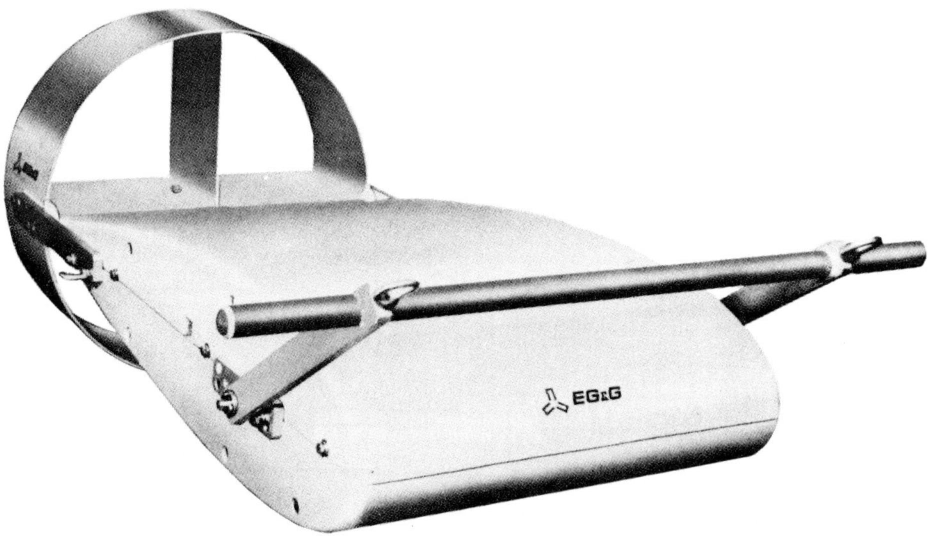

FIGURE 25. The Uniboom® Sub-Tow vehicle containing the electrical mechanical transducer and the hydrophone (EG&G, Waltham, Mass.).

the catamaran, places limits on the speed of the survey vessel and also subjects the device to wave motion. These wave motions impart a wave-like distortion to the bottom and subbottom reflectors that make data interpretation difficult. A recent design change, whereby the unit is towed beneath the surface, attempts to overcome these limitations (Figure 25). Unibooms® have also been installed in seachests inside the hull with considerable success.

The amplitude and frequency characteristics of the generated acoustic signal results in good resolution (0.5 to 1 m) and penetration (25 to 50 m). In geologic environments where the bottom and subbottom materials are acoustically opaque to a tuned transducer system, for instance sands and gravels or glacial deposits, good subbottom records can be obtained (Figure 26).

Sparker

The sparker is probably the simplest device for generating an acoustic pulse, both with regards to its power supply and its transducer design. When the power supply is keyed by the seismic recorder, the stored electrical energy is discharged to the sparker transducer (Figure 27). The electrical discharge travels from the firing electrode through the seawater to the ground return of the transducer. Seawater subjected to this high voltage, high current energy is immediately vaporized into steam and ionized particles. This rapid vaporization produces the initial acoustic pulse. Unfortunately, the expanding steam bubble cools and eventually collapses, at which point another intense pressure wave is emitted. Therefore, each discharge of the energy source produces two pulses (the second one is called the bubble pulse), which means that each reflector will be represented at least twice on the reflection record.

The sparker has much versatility that results in a certain degree of control over its penetration and resolution capabilities. Such control can be accomplished by using various power supply and/or transducer configurations.

The power supply (Figure 28) consists of a high voltage transformer, storage capac-

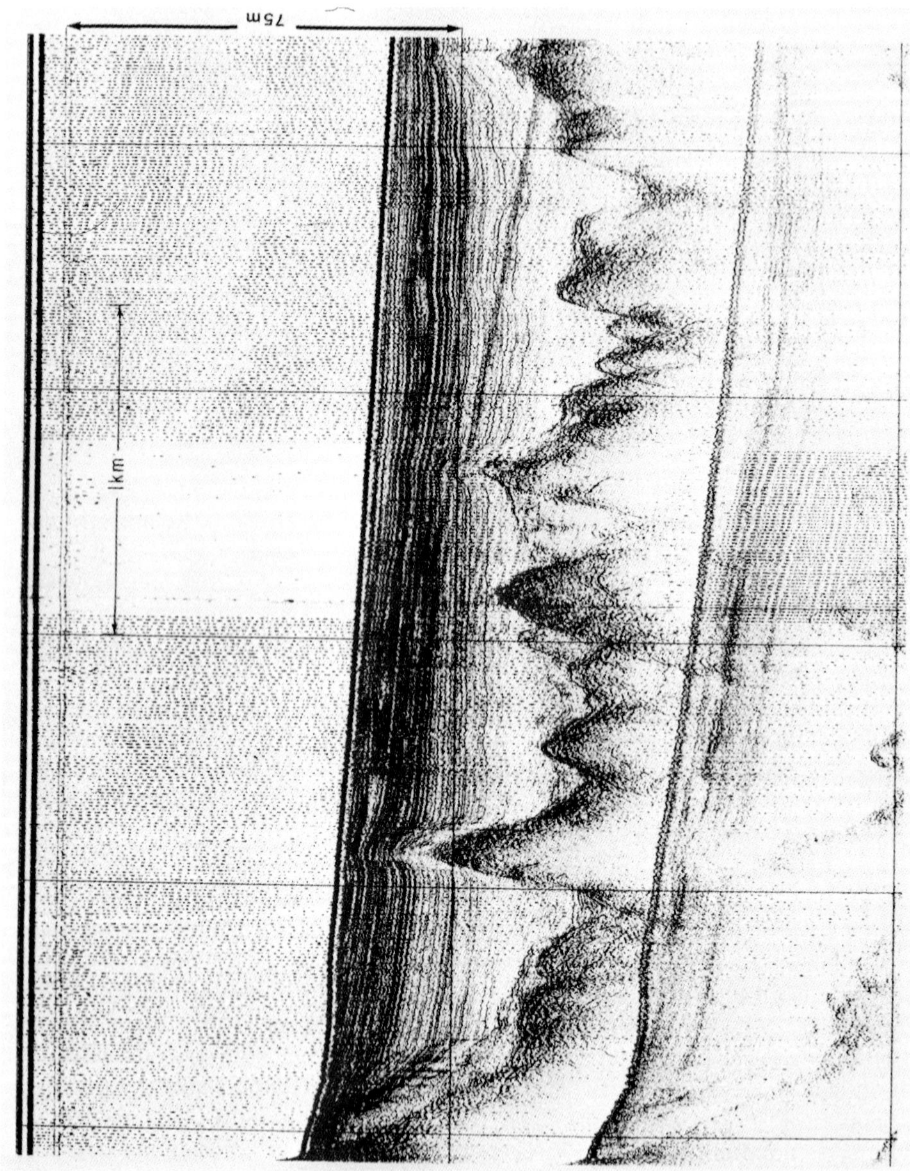

FIGURE 26. Seismic-reflection data from Uniboom® system.

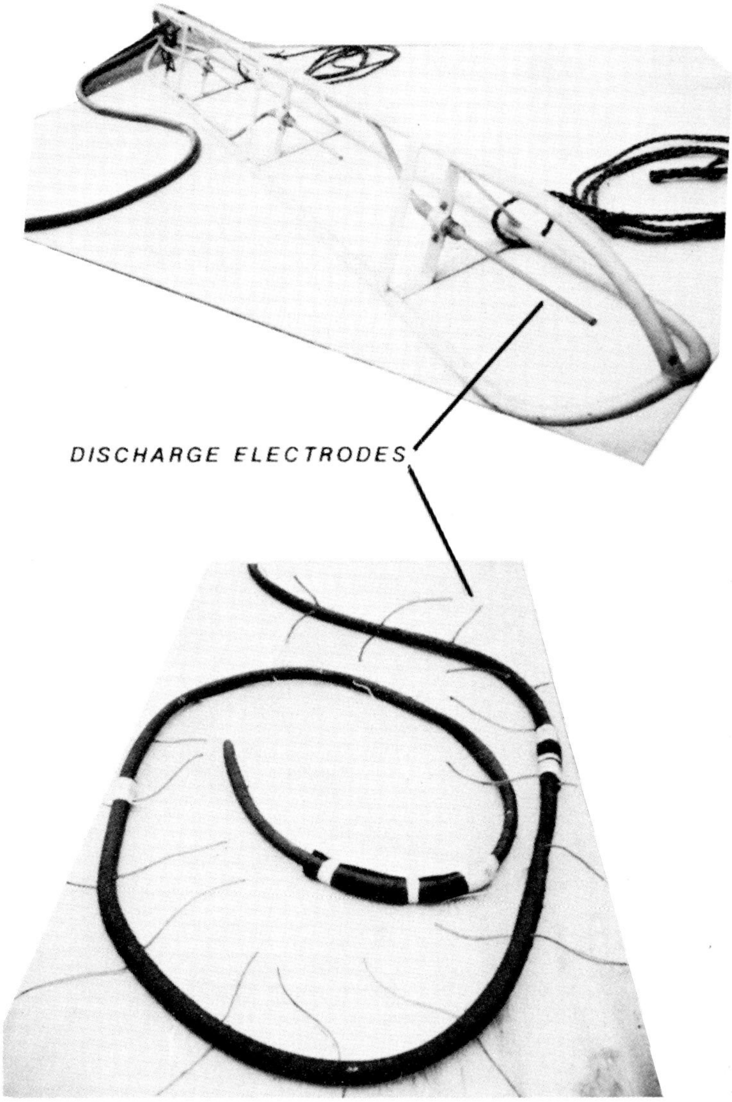

FIGURE 27. (Upper) Low frequency Sparkarray (EG&G), and (Lower) multitip sparker array used with medium (300 to 1000 J energy power supplies (Teledyne Exploration, Houston, Tex.).

itors, and some type of high voltage switch that transfers the energy stored in the capacitors to the sparker transducer. The energy capacity of the power supply (in joules) is determined in the following way.

$$E = 1/2\ CV^2$$

where

E = energy in joules
C = capacitance in microfarads (μF)
V = voltage in volts

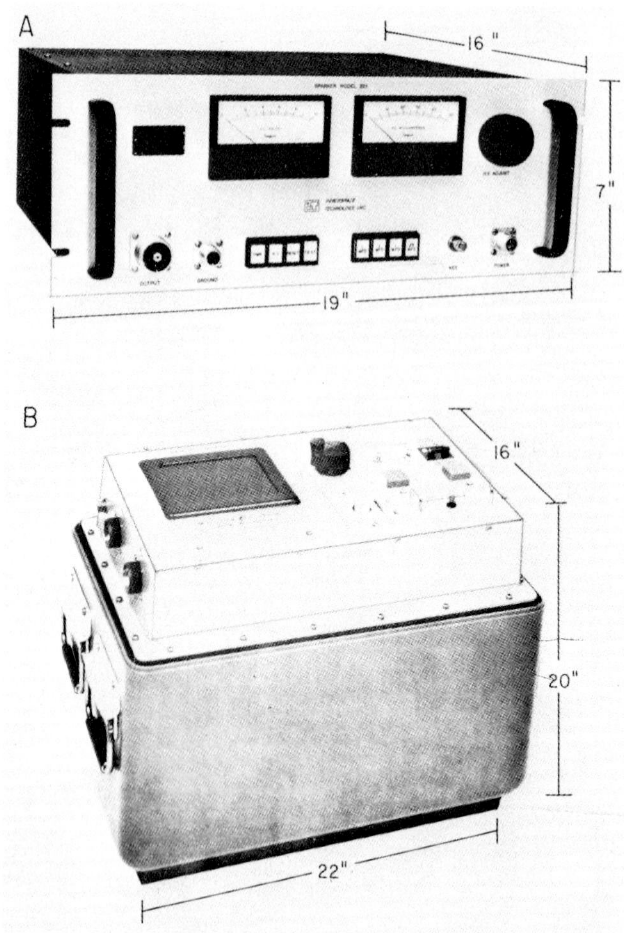

FIGURE 28. Two examples of sparker power supplies (A) 25 J energy source (Innerspace Technology, Inc., Waldwick, N.J.); (B) 800 J energy source (Teledyne Exploration, Houston, Tex.).

For example, the stored energy in a power supply operating at 10 kV, having 20 μF of capacitance would be

$$E = 1/2\,(20 \times 10^{-6})\,(10 \times 10^{3})^{2}$$

$$= 100 \text{ J}$$

By doubling the capacitance to 40 μF, the stored energy would double and the amplitude of the outgoing signal would increase. This added capacitance also increases the low frequency characteristics of the acoustic signal. Together, the increased amplitude and lower frequency increase the penetration capabilities of the system. Correspondingly, a decrease in the capacitance reduces the low frequency characteristics as well as the amplitude of the outgoing signal. To offset this latter effect, an increase in the high voltage produces an increase in the peak pressure of the discharge.

In summary, by varying the capacitance and/or the high voltage, a sparker system can be designed for good resolution and minimum penetration or low resolution and

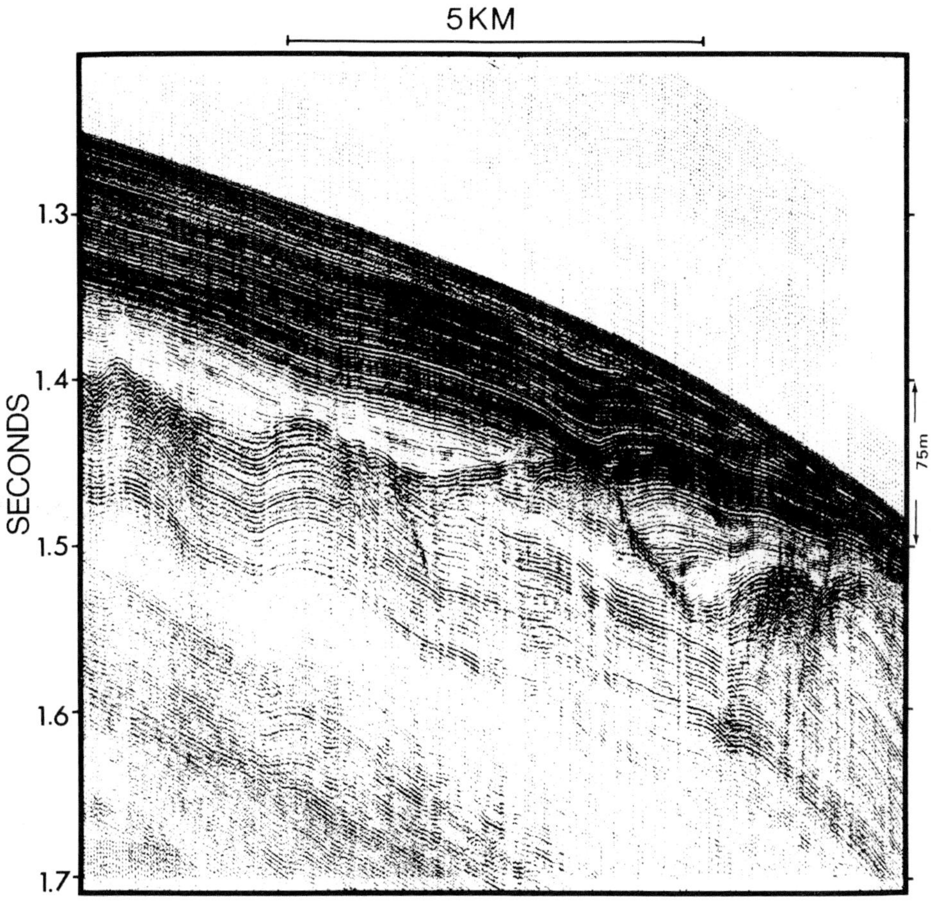

FIGURE 29. Seismic-reflection data obtained with an 800 J sparker system.

substantial penetration. The various combinations presently available provide a range of 10J to 160 kJ of energy.

There are numerous ways to design sparker transducers (Figure 27). The principal design criteria are for a device that efficiently converts the electrical energy to a high amplitude, wide bandwidth acoustic signal. These characteristics depend on several parameters such as the number, size, and spacing of electrodes, and energy discharge per electrode. Hence there are numerous types of sparker transducer devices available, of differing optimum signal characteristics. Additional attention is given to the reliability, ease of maintenance, and longevity of the electrodes.

The sparker system, relative to acoustic sources having equivalent energy, is capable of fair resolution (5 to 10 m) and good penetration (200 to 300 m). They are particularly suited to continental shelf regions where good penetration in hard sands and semiconsolidated material is required. In deeper offshore slope regions (Figures 17, 18, and 29), good penetration results from their relatively high amplitude, low frequency acoustic output.

Air Gun

The air gun is a pneumatic acoustic sound source. The system (Figure 30) consists of an onboard air compressor and storage bottom (analogous to the sparker power

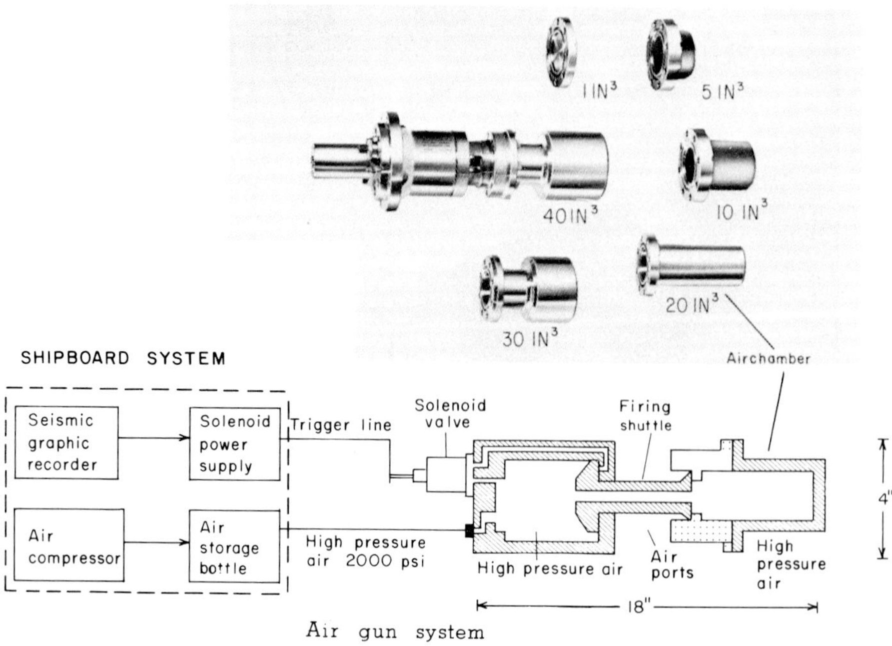

FIGURE 30. (Lower) Basic components of the air gun system. (Upper) Air gun with various interchangeable air chambers used to vary the characteristics of the acoustic signature (Bolt Associates, Inc., Norwalk, Conn.).

supply), a shipboard electrical firing circuit controlled by the seismic recorder, and one or more air guns towed astern of the survey vessel.

On command from the seismic recorder, the air gun releases a specified volume of high pressure air (1000 to 4000 lb/in²) into the water. This explosive release of air produces a steep-fronted shock wave followed by several oscillations resulting from the repeated collapse and expansion of the air bubble (Figure 21). This oscillation is similar to the bubble pulse associated with the collapse of the steam bubble produced by the sparker discharge. For the air gun, however, the oscillation continues for a number of cycles.

The acoustic signature of the air gun can be modified in a number of ways to adjust both the resolution and the depth of penetration. An increase in the air chamber size produces an acoustic signal having both a higher amplitude and lower frequency components (Figure 31). The resulting improvement in penetration can be significant, as is the accompanying reduction in resolution.

A reduction in the size and number of oscillations is an important factor in improving the resolution capabilities of the air gun. One technique for accomplishing this is by placing a small insert (Wave Shape Kit, WSK®) in the air chamber. This device releases air into the initial bubble at a rate that dampens the oscillations. It is also a common practice, particularly in offshore petroleum exploration surveys, to use arrays of air guns of different sizes (air chamber volume), each size characterized by a particular oscillation frequency. The proper selection of various sizes and geometrics for the array may result in cancellation of the oscillations produced by the individual air guns. An added benefit of this procedure is that an increase in the initial pressure pulse accompanies the decreased oscillations.

A reduction of the oscillation period can be achieved by towing the air gun near the surface. If it is sufficiently shallow, the bubble vents to the atmosphere and once

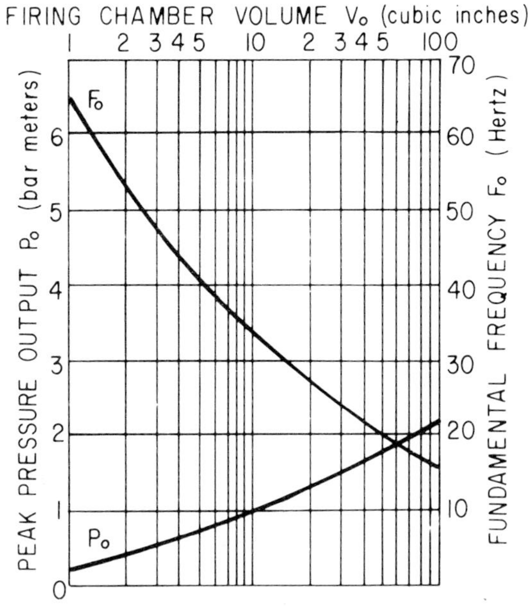

FIGURE 31. Graph shows the relationship between the volume of the air gun (Vo), the source level (Po), and the fundamental frequency (Fo) of the acoustic signal.

vented, no bubble pulse can form. Unfortunately this also reduces the amplitude of the initial pressure wave. This loss of energy can be overcome by towing the air gun at a relatively deep depth (a function of the size of the air gun), which will reduce the bubble pulse period due to the loading effect of the hydrostatic pressure.

The air gun, relative to the previously discussed acoustic sources, has poor resolution (15 to 30 m). However, it is capable of considerable penetration (200 to 2000 m), particularly when large air guns or air gun arrays are used. Figure 32 is an example of data obtained with a small air gun. The good penetration is particularly evident.

Hydrophone Streamer

All acoustic sources, with the exception of tuned transducers, require a separate receiving device referred to as the hydrophone streamer or hydrophone array. Its two principal functions are to receive the reflected acoustic pressure wave and to convert this pressure wave to an electrical signal which is processed and displayed on the seismic recorder.

The hydrophone streamer is constructed of a number of transducers, most commonly piezoelectric crystalline elements (Figure 33), that are electrically wired together to act as one receiving system. This string of elements is placed in a flexible sleeve or tube that is filled with oil and then sealed. The oil provides a means for coupling the pressure wave arriving at the hydrophone through the sleeve and to the hydrophone elements, which convert the pressure wave to an electrical signal.

Various design and construction techniques are used in the production of hydrophone streamers in an attempt to optimize certain important characteristics. For example, the array should exhibit good sensitivity to reflected acoustic information arriving from a direction normal to the towing axis.[23] This reduces the contribution of noise arriving from the propulsion system of the ship along the streamer axis. A stretch section at the head of the streamer decouples the strumming action produced by the

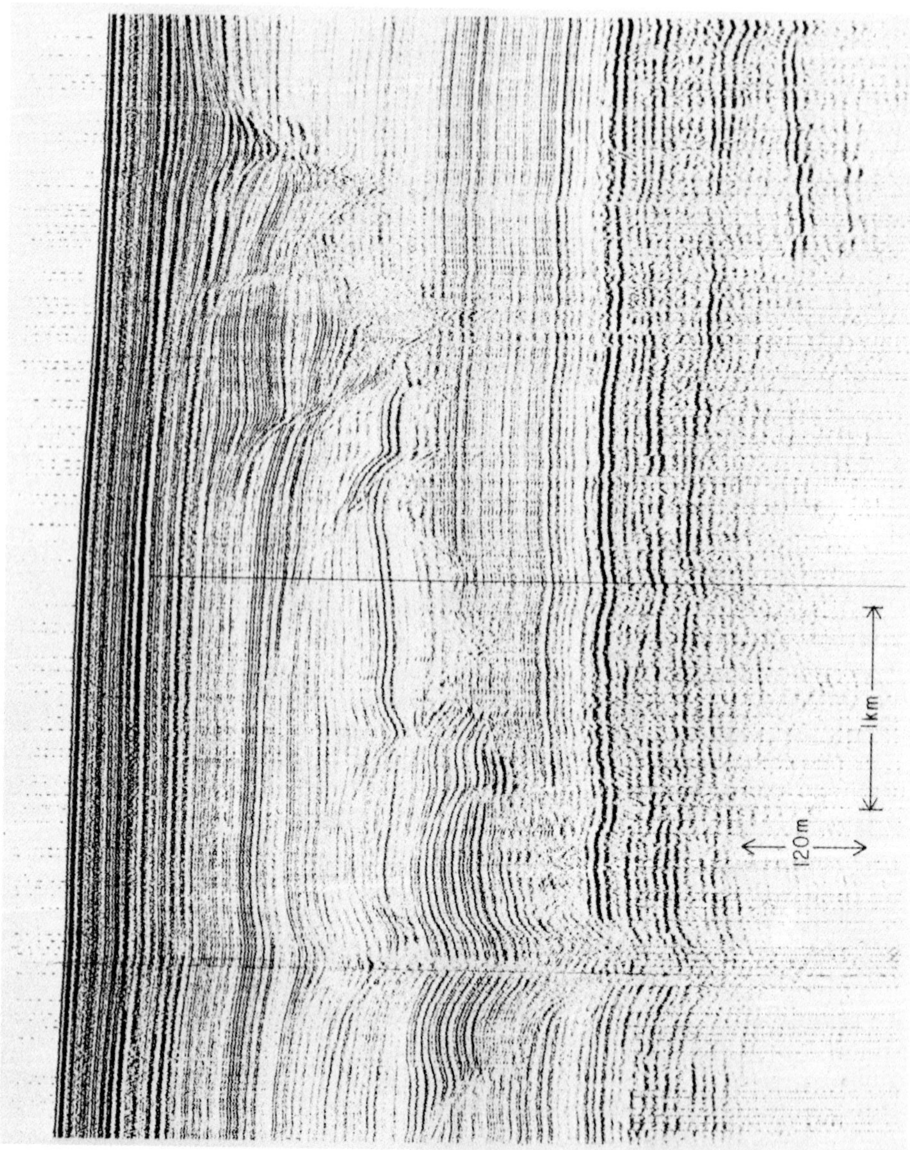

FIGURE 32. Seismic-reflection data obtained with a 5 in^3 air gun.

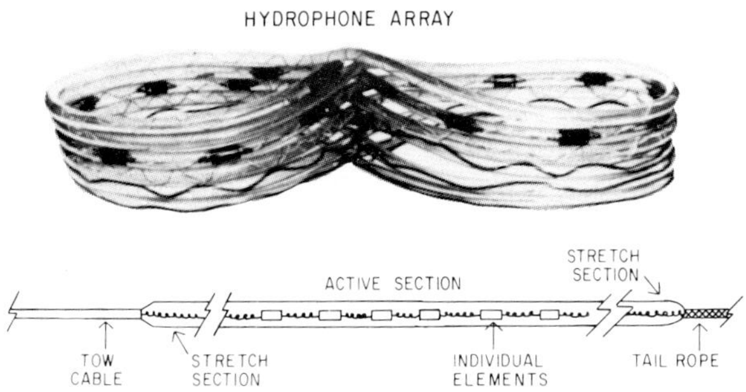

FIGURE 33. A high-resolution hydrophone streamer (Upper) and its basic construction (Lower).

tow cable. An identical stretch section, with a rope tail, attached to the end of the streamer reduces the whipping action that would normally act on the hydrophone active section. Finally, the number of elements used, their spacing, and the total length are calculated in order to optimize the hydrophone array receiving characteristics for a given acoustic spectrum and direction of reception. The low frequency directional characteristics and sensitivity for a linear streamer depend on the total acoustical length (a summation of the number of elements and their spacing); and the high-frequency directionality depends on the spacing of the individual elements.[24]

Seismic Signal Processor

The reflected acoustic signal received by the hydrophone is usually not of sufficient amplitude to be printed on the seismic recorder without additional amplification. This is particularly true for reflections from subbottom reflectors that are greatly attenuated as they pass through the earth. In addition, the hydrophone is also sensitive to pressure waves, other than the desired reflected information, that can degrade the reflection record. This ambient noise results from both mechanical and acoustical causes (e.g., acoustic noise from the ship propulsion system, mechanical noise produced by strumming of the tow cable, and noise created by turbulence of the water flowing past the hydrophone being towed[25]).

The purpose of the signal processor, then, is to increase the amplitude of the electrical signal from the hydrophone and to discriminate against or filter out unwanted acoustic information.

Amplifiers can increase the signal in several ways. A straight gain or linear amplifier produces the same increase in all reflected signals, depending on the position of the gain control. That is, the reflected signal from the sea floor and all subbottom reflectors will be amplified equally. If the sea-floor reflection is particularly strong, the amplifier gain will have to be reduced in order to prevent excessive burning of the recording paper. If the recorder gain is reduced, low amplitude signals from subbottom reflectors will not be displayed because they do not exceed the threshold printing level of the seismic recorder.

To overcome this problem, amplifiers with automatic gain controls (AGC) are used (Figure 34A). In these amplifiers, the gain is automatically adjusted as the signal level varies. The gain would be reduced for the high amplitude signal returning from the sea floor and would be increased for the weaker subbottom returns. Separate controls are used to select the rate at which the gain changes (attack rate) and then returns to the normal setting (release rate).

A third technique, used particularly with tuned transducer systems, is called time

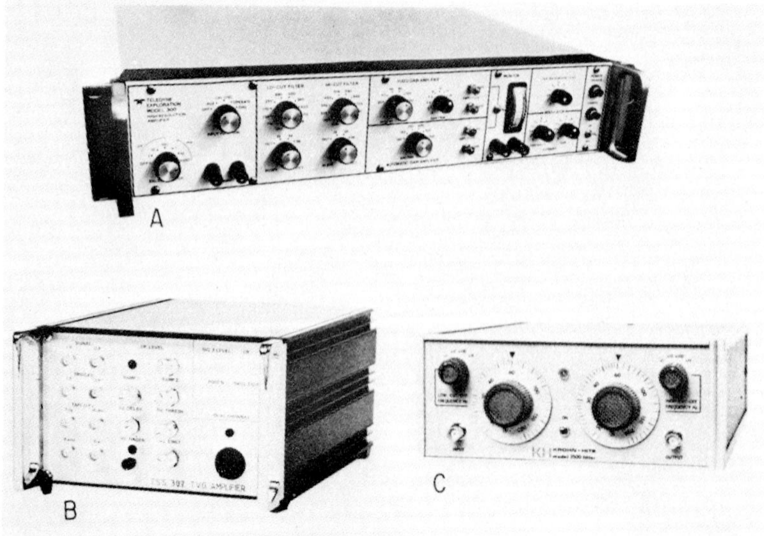

FIGURE 34. Examples of seismic processing equipment. A. AGC amplifier and band-pass filter integrated into one unit (Teledyne Exploration, Houston, Tex.). B. TVG amplifier (TSS Ltd., Oxon, England). C. Band-pass filter (Krohn-Hite Corp., Avon, Mass.)

variable gain amplification (TVG). In these amplifiers (Figure 34B), the gain automatically increases with time as the stylus moves across the recording paper. The time at which the increase begins and the rate of increase can be individually selected.

Electronic band-pass filters are used to attenuate or remove unwanted acoustic signals (Figure 34C). The band-pass is the range of frequencies allowed to pass through the filter to the amplifier. It is selected by separate controls: one for the low cut-off frequency and one for the high cut-off frequency. This allows a great deal of flexibility in selecting both the band pass frequency range (100 to 200 Hz, 90 to 220 Hz) and the location of the center frequency within the acoustic spectrum (100 to 200 Hz; 1000 to 2000 Hz).

The selection of the band-pass range and center frequency for a survey is dependent on the frequency characteristics of the sound source and the frequency of the ambient noise. For example, a band-pass of 400 to 800 Hz, typical for a small sparker system, would reduce or eliminate noise due to the previously mentioned causes, outside of this range.

A possible improvement in resolution can be achieved by using the band-pass filter to pass only the high frequencies. On the other hand, by changing the center frequency to a lower end of the spectrum, it is possible to improve on the depth of penetration as observed on the reflection record. With the high frequency information now filtered out, the amplifier gains can be increased. This results in enhancing the low frequency information reflected from deeper in the geologic section. The seismic processor consisting of the amplifier and the band-pass filter, can be two separate components or integrated as a single unit.

Ancillary Instrumentation

A continual effort is made to improve on the shortcomings and limitations of the basic components of the seismic-reflection system. Concurrent with these efforts are the use of instruments that further enhance the reflection record by solving some par-

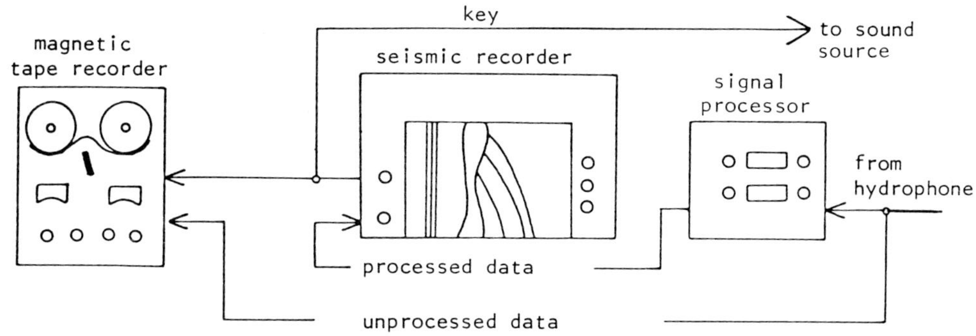

During the seismic reflection survey the key pulse from the seismic recorder triggers the sound source and is stored on magnetic tape. The seismic signal from the hydrophone is also stored on magnetic tape in addition to being processed for immediate display on the seismic recorder.

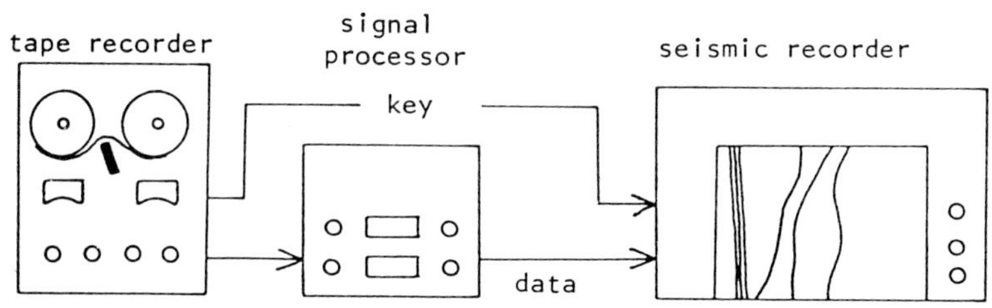

For data replay the key signal from the magnetic tape is used to initiate the display on the seismic recorder. The seismic data (which was recorded unprocessed) is amplified and filtered and then displayed on the seismic recorder.

FIGURE 35. Procedure for tape recording and replaying seismic-reflction data.

ticular problem. In this section several of these instruments, some of which show ever-increasing use as part of the basic reflection system, will be discussed.

Magnetic Tape Recorder

Analog tape recording of seismic-reflection data, although not a recent development,[26] is surprisingly seldom used. It is not difficult to justify taping as a routine practice for all seismic-reflection surveys. The taping and replay procedures are simple and the costs are minimal relative to the costs of the survey vessel, seismic-reflection equipment, and support personnel. The potential for improving the reflection record by various postcruise processing and display techniques can be significant.

The block diagram in Figure 35 shows the minimal requirements and electrical connections needed for taping and replaying the seismic-reflection data. As shown, a two channel analog tape recorder records the key signal and the unprocessed data during the actual survey. When the data are replayed, the key signal on tape is used to synchronize the recorder sweep to the seismic-reflection data. The recorded data are processed by the band-pass filter and displayed on the seismic-reflection recorder. By trying several combinations of band-pass filter setting and display ranges, it is possible to improve on the original field record.

If the tape recorder speed can be increased during playback (1 7/8 in/sec recording speed, 7 1/2 in/sec playback speed), then the playback time can be reduced relative to the time required for obtaining the actual field data (a factor of 4 in this case). Maintaining a vertical scale on the replayed data that corresponds to the original reflection record requires a decreased sweep speed of the seismic-reflection recorder. That is, increasing the tape recorder speed during playback by a factor of 4 requires reducing the sweep speed of the seismic recorder by the same factor. The band-pass filter must also be adjusted to compensate for this increase in speed. If the desired band-pass of the replayed data is 400 to 800 Hz, then it is necessary to use a range of 1600 to 3200 Hz — a fourfold increase in the filter setting. Figures 36-38 illustrate data replayed and reprocessed from magnetic tape.

The availability of a tape recorder with additional channels provides a means of recording other types of information which can be beneficial when the data are replayed. For instance, voice annotation indicating information regarding location, date, time, and equipment parameters can be useful. Recording time information from digital electronic clocks, which then decode and display this information during playback, provides a means to accurately correlate the field data and the playback data. Finally, recording two or more reflection systems used for the survey can be accomplished if the tape reconformation from digital electronic clocks which then decode and display this information during playback provides a means to accurately correlate the field data and the playback data. Finally, recording two or more reflection systems used for the survey can be accomplished if the tape recorder has the necessary additional channels (Figure 41).

Motion Compensation

The quality of the reflection data is often degraded as a result of the vertical motion (heave) of the tuned transducer, or sound source and receiver in heavy seas. This motion causes the reflectors to have a sawtooth appearance and produces a loss in continuity that often destroys the resolution of subbottom horizons. Several instruments have been designed that reduce or eliminate adverse effect of motion on the reflection record. One such device, called a heave compensator, uses an accelerometer (Figure 40) to sense the motion of the tuned transducer. This information is then used to apply the appropriate delay to the transceiver key pulse, thereby compensating for the vertical motion (Figure 41). An accelerometer can be used to compensate for motion when the source and receiver are attached to the boat or to the same towed vehicle. When the source and receiver are towed separately, and have independent motion, other techniques must be used to eliminate ocean wave and swell effects. The Swell Filter® is a shipboard unit (Figure 42) that solves the problem of independent source and receiver motion by filtering the periodic wave action or swell. This filtering technique compensates for the apparent vertical movement of the sea floor and subbottom reflectors (Figure 43). Filtering can be done as part of the field operation or during postcruise processing of magnetically recorded data.

Digital Stacking

The coherency or continuity of the reflection record can be enhanced by improving the signal-to-noise ratio. One method of doing this is by storing the reflection information for two or more shotpoints in an electronic buffer and then displaying the summed or "stacked" output on the seismic recorder. The reflection information adds constructively providing a signal increase, and the noise, being random, is reduced by the summation process. This processing can be done in real time during the survey or as part of postprocessing the data from analog tape recordings.

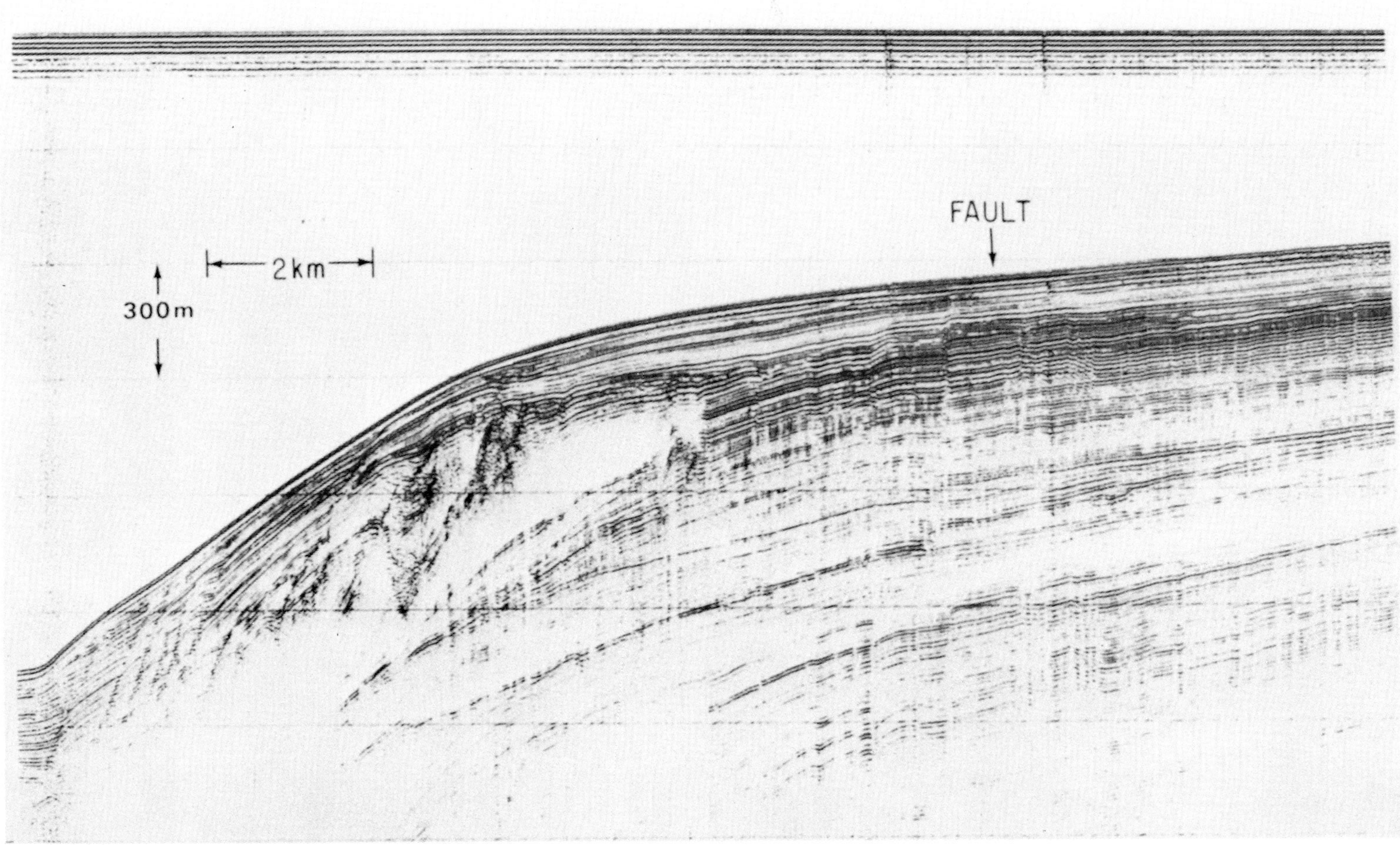

FIGURE 36. Replay of seismic-reflection data at original scale. Compare the fault zone (F) on this figure and Figure 37.

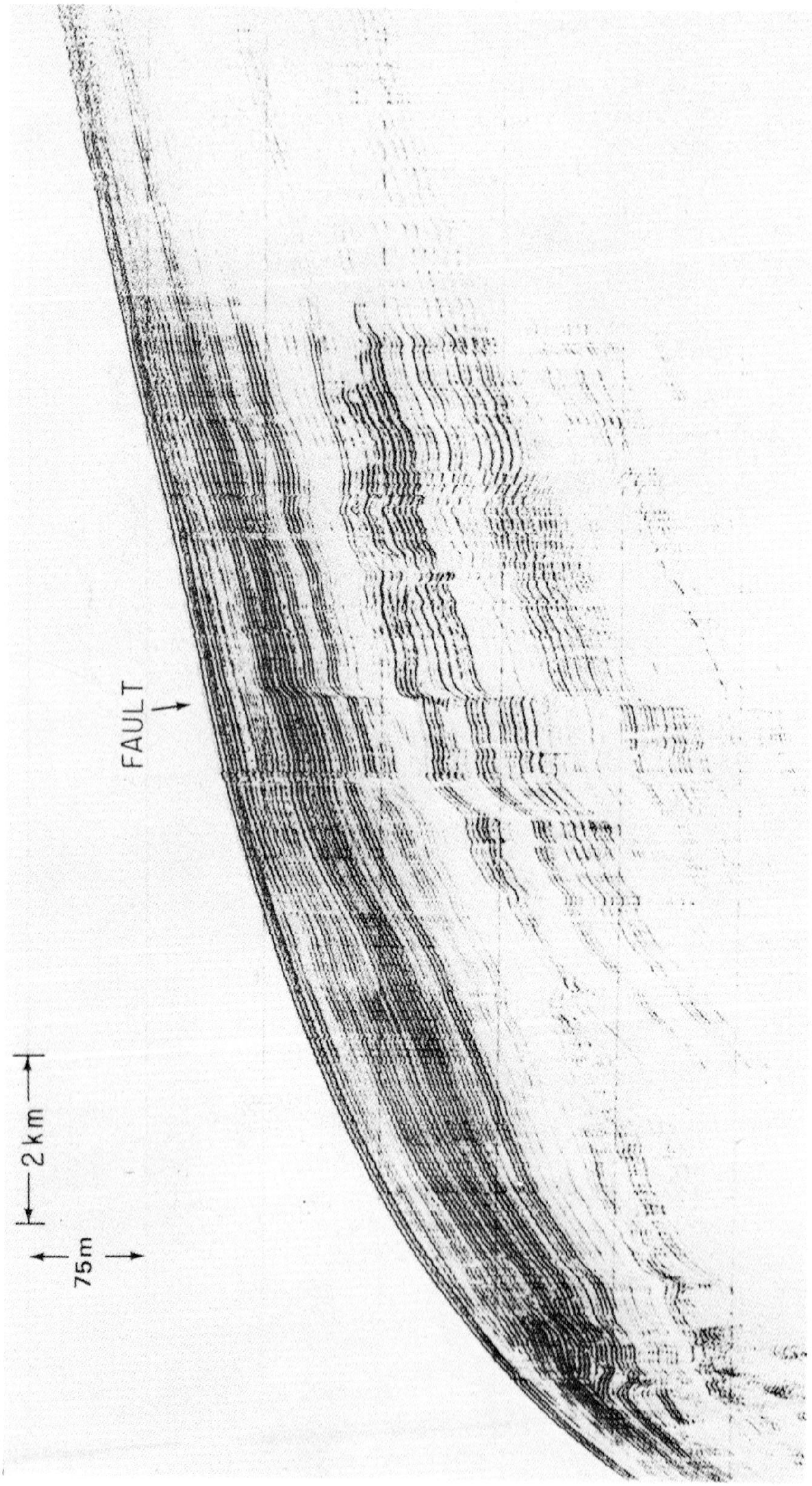

FIGURE 37. Replay of seismic-reflection data shown in Figure 36, but at an expanded vertical scale. The fault zone is quite easily seen at this scale.

FIGURE 38. Replay of seismic-reflection data using a split trace mode to display the same data on two separate channels on the seismic recorder. Upper channel filtered at 40/80 Hz and lower channel at 250/500 Hz. Note the difference in the vertical scale.

Record Annotation

Annotation of the seismic-reflection record during the survey is necessary for postcruise data analysis. The required information, in particular the date, time of day, and track line number, is periodically written on the edge of the seismic record. During the analysis of the data this information provides a means of correlating the data with the plotted positions on the cruise track.

Electronic record annotators that interface with the seismic recorder and perform this task automatically are available (Figure 44). These instruments are programmed by means of a keyboard to provide the desired information for annotation. The time between annotation and the scale or dimensions of the annotation can also be varied to correspond with the sweep speed.

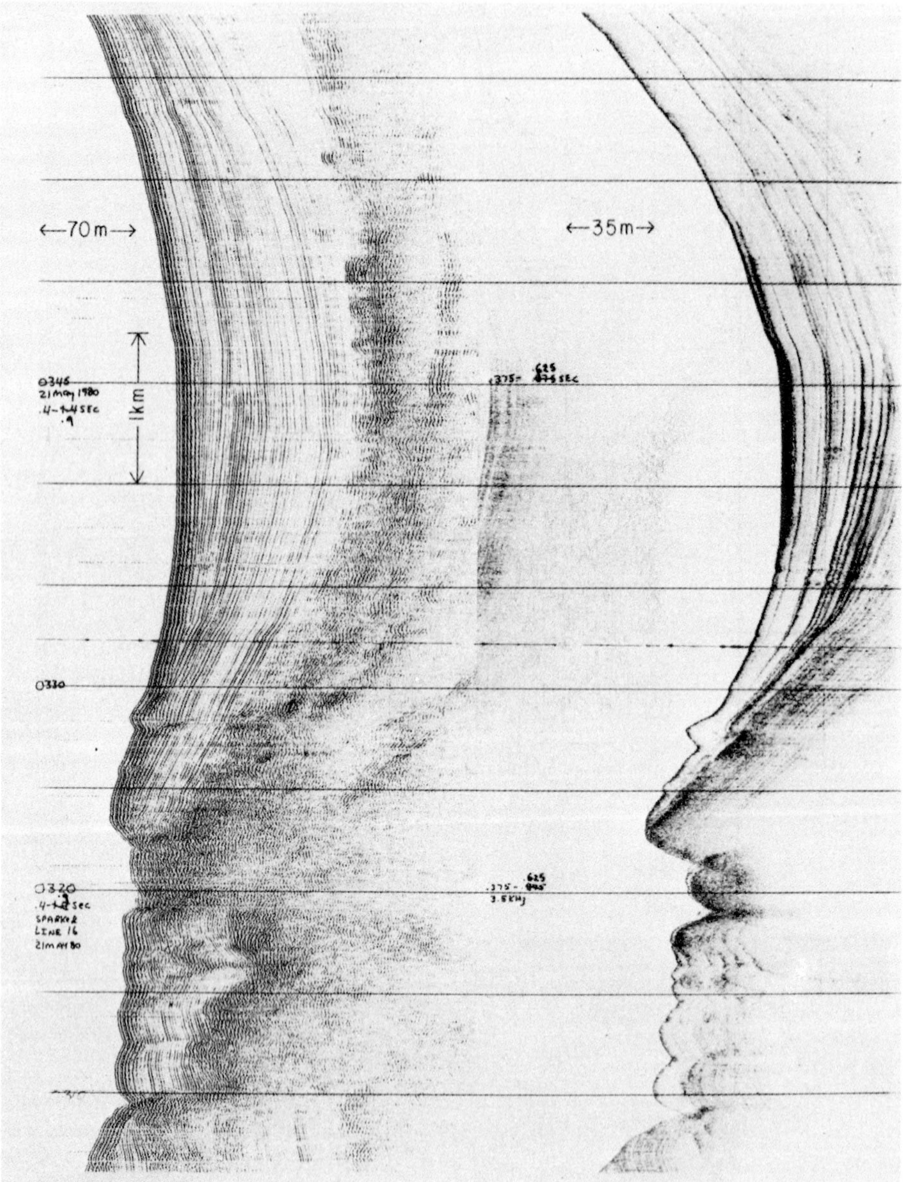

FIGURE 39. Replay of seismic-reflection data from two different profiling systems. Upper channel is data from an 80 J sparker and lower channel is a 3.5 kHz hull-mounted tuned transducer array.

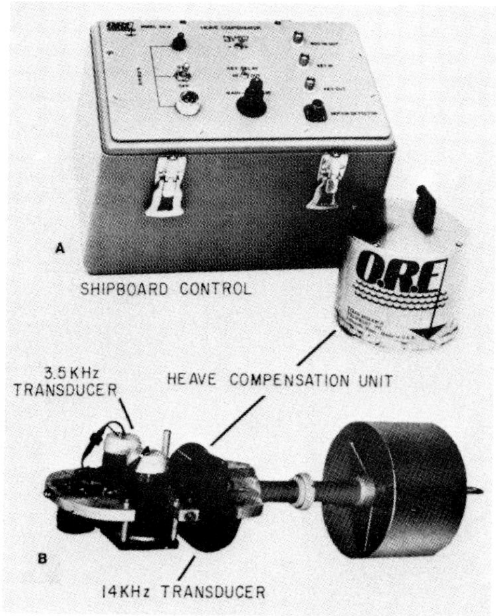

FIGURE 40. (A) Heave compensator deck unit and heave compensator. (B) Tow vehicle with multitransducer system and heave compensator (O.R.E., Inc., Falmouth, Mass.).

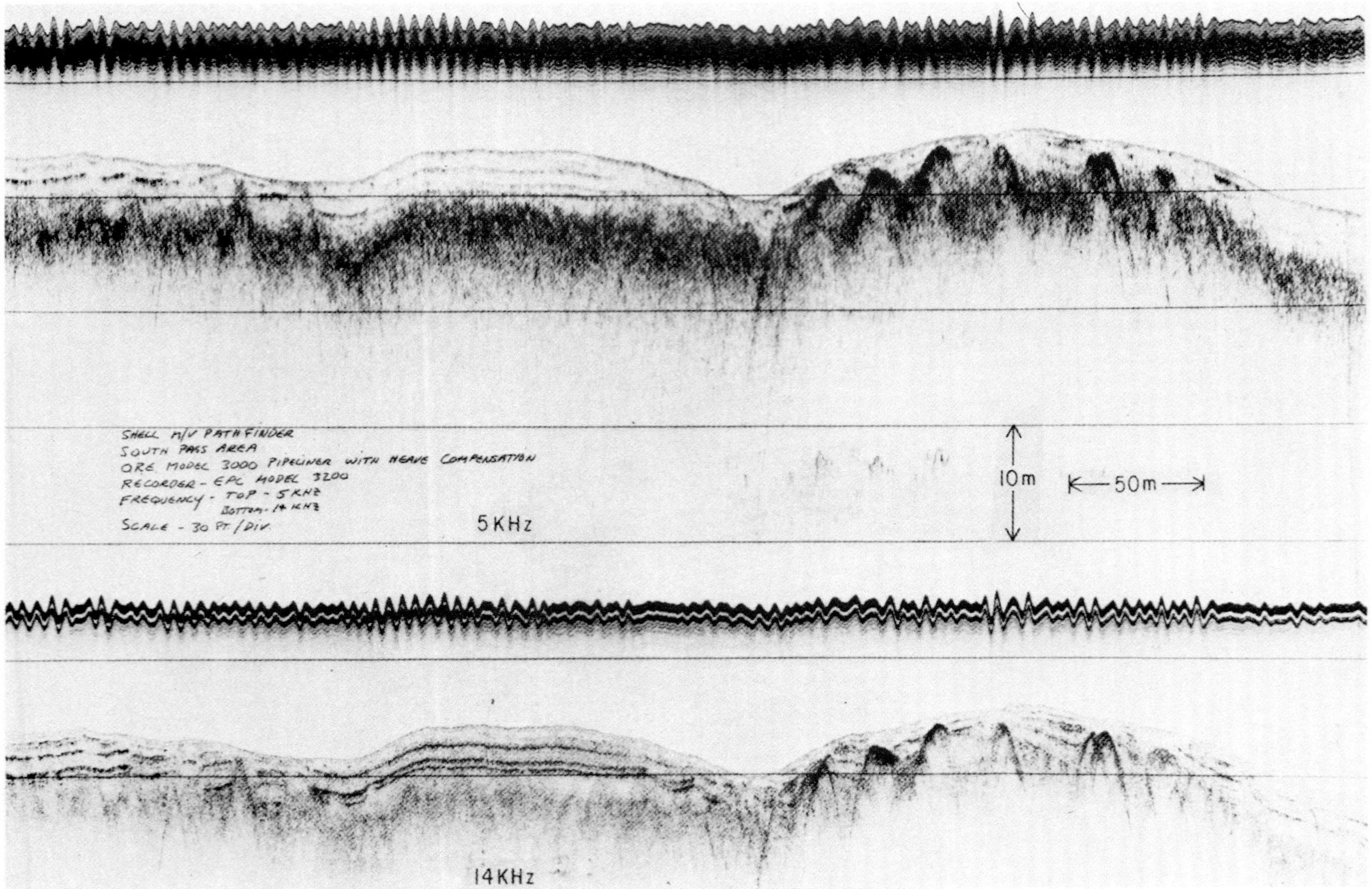

FIGURE 41. Seismic-reflection data that have been corrected for heave motion. The oscillating signal above the records indicates the amount of vertical motion removed by the system. Data is from multitransducer system shown in Figure 40.

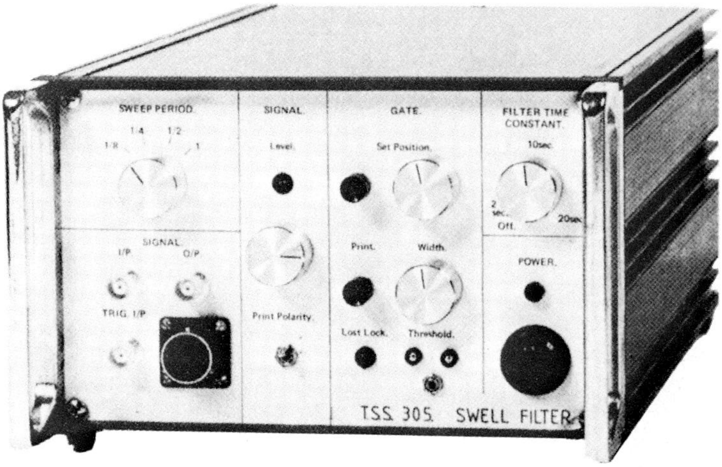

FIGURE 42. Swell filter for removing source and receiver motion effects on the reflection record. (TSS Ltd., Oxon, England).

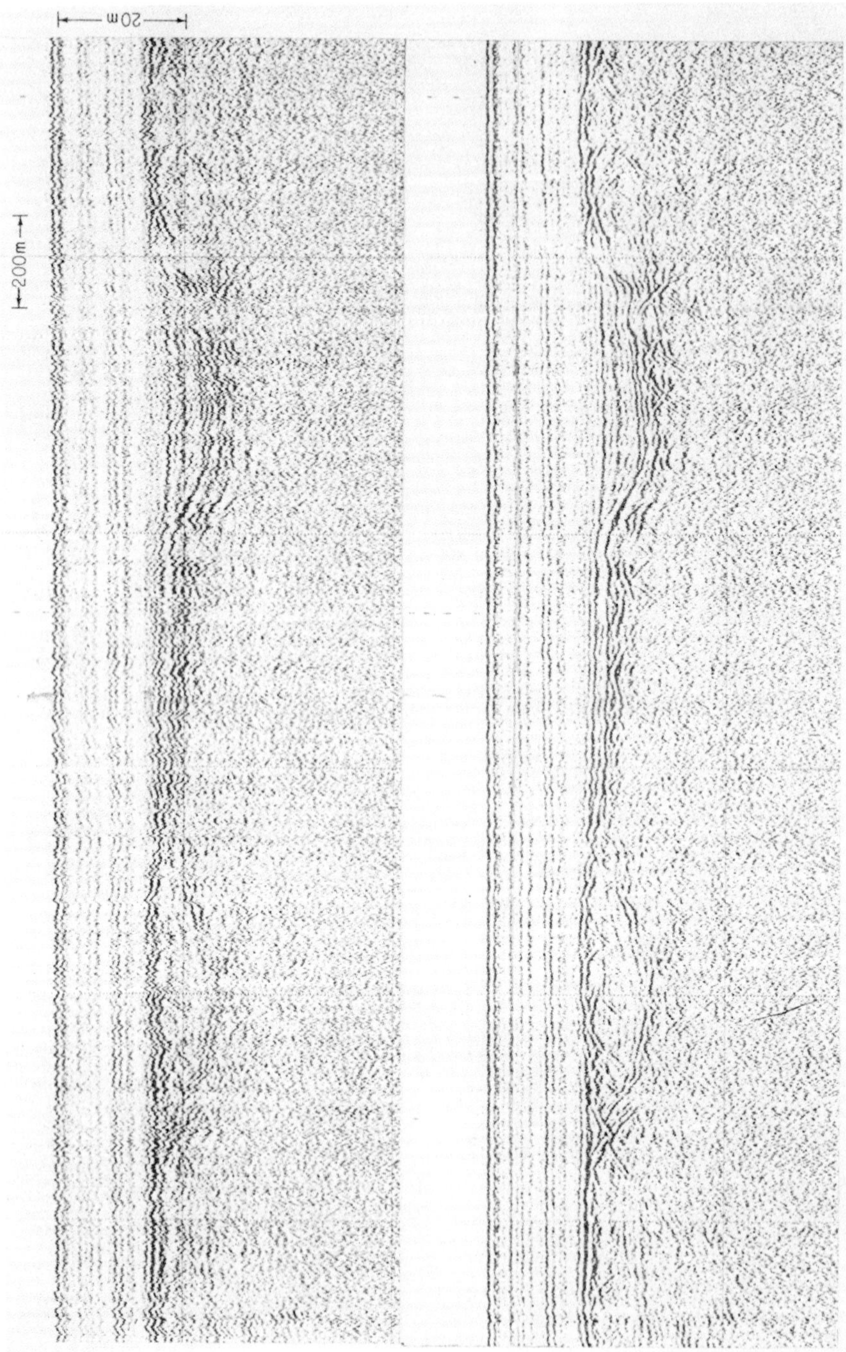

FIGURE 43. Seismic-reflection data before (Upper) and after (Lower) processing by a swell filter. (TSS Ltd., Oxon, England).

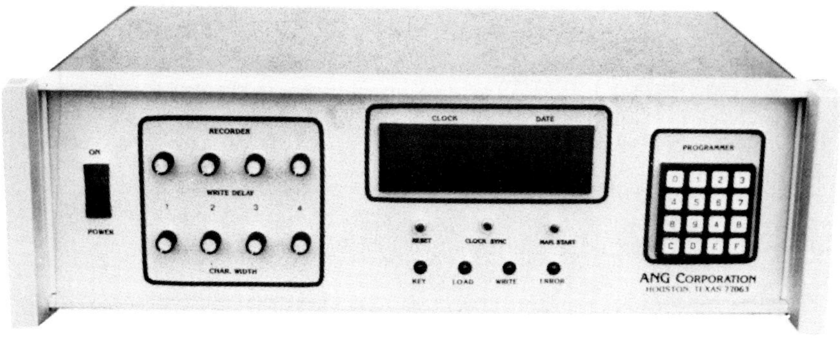

FIGURE 44. Automatic seismic-reflection record annotator (ANG Corp., Houston, Tex.).

REFERENCES

1. **Edgerton, H. E. and Hayward, G. G.,** The Boomer sonar source for seismic profiling, *J. Geophys. Res.,* 64, 3033, 1964.
2. **Ewing, J. I. and Tirey, G. B.,** Seismic profiler, *J. Geophys. Res.,* 66, 2917, 1961.
3. **Shor, G. G., Jr., Moore, D. G., and Huckabay, W. B.,** Deep-sea test of a new nonexplosive reflection profiler, *J. Geophys. Res.,* 68, 1567, 1963.
4. **Anstey, N. A.,** *The New Seismic Interpreter,* International Human Resources Development Corporation, Boston, 1976.
5. **Nafe, J. E. and Drake, C. L.,** Variations with depth in shallow and deep water sediments of porosity, density, and the velocities of compressional and shear waves, *Geophysics,* 22, 523, 1957.
6. **Hamilton, E. L.,** Elastic properties of marine sediments, *J. Geophys. Res.,* 76, 579, 1971.
7. **Sheriff, R. E.,** Factors affecting seismic amplitudes, *Geophys. Prospecting,* 23, 125, 1975.
8. **Sheriff, R. E.,** Limits on resolution of seismic reflections and geologic details derivable from them, in *Seismic Stratigraphy — Applications to Hydrocarbon Exploration,* Mem. 26, Payton, C. E., Ed., American Association of Petroleum Geologist, Tulsa, 1977, 3.
9. **Krause, D. C.,** Interpretation of echo sounding profiles, *Int. Hyd. Rev.,* 39, 65, 1962.
10. **Fitch, A. A.,** The geometry of seismic sections, in *Seismic Reflection Interpretation,* Fitch, A. A., Ed., Gerbrüder Borntrager, Berlin, 1976, 16.
11. **Bennett, L. C. and Savin, S. M.,** The natural history of the Hardanger Fjord, *Sarsia,* 14, 79, 1963.
12. **Shumway, G.,** Sound speed and absorption studies of marine sediments by a resonance method, *Geophysics,* 25, 451, 1960.
13. **Backus, M. M.,** Water reverberations — their nature and elimination, *Geophysics,* 24, 233, 1959.
14. **Woods, J. P.,** A seismic model using sound waves in air, *Geophysics,* 40, 593, 1975.
15. **Trorey, A. W.,** A simple theory for seismic diffractions, *Geophysics,* 35, 762, 1970.
16. **Sylwester, R. E., Dillon, W. P., and Grow, J. A.,** Active growth fault on the seaward edge of the Blake Plateau, in *Geomathematical and Petrophysical Studies in Sedimentology,* Gill, D. and Merriam, D. F., Eds., Pergamon Press, New York, 1979, 197.
17. **Ricker, N.,** Wavelet contraction, wavelet expansion, and the control of seismic resolution, *Geophysics,* 18, 769, 1953.
18. **Sheriff, R. E.,** Inferring stratigraphy from seismic data, *Am. Assoc. Pet. Geol. Bull.,* 60, 528, 1976.
19. **Hamilton, E. L.,** Compressional wave attenuation in marine sediments, *Geophysics,* 19, 430, 1972.
20. **Van Overeem, A. J. A.,** Shallow penetration, high resolution subbottom profiling, *Mar. Geotechnol.,* 3, 61, 1978.
21. **Robb, J. M., Sylwester, R. E., and Penton, R.,** A simplified method for deep tow seismic profiling, *Geo-Mar. Lett.,* 6, 1, 1981.

22. **Bunce, E. T.,** Assessment of Seabed Seismic Systems, Rep. WHOI-80-24, Woods Hole Oceanographic Institution, Woods Hole, Mass., 1980.
23. **Nowalk, R. T.,** A comparison between the response of a continuous line array and a discrete linear array, Rep. WHOI 69-12, Woods Hole Oceanographic Institution, Woods Hole, Mass., 1969.
24. **Knott, S. T.,** Construction and performance of the ceramic cylinder array in use for seismic profiling at Woods Hole — 1968 to 1973, Rep. WHOI-73-64, Woods Hole Oceanographic Institution, Woods Hole, Mass., 1973.
25. **Schoenberger, M. and Mifsud, J. F.,** Hydrophone streamer noise, *Geophysics,* 39, 781, 1974.
26. **Knott, S. T.,** Operational technique for continuous seismic profiling and related studies, Rep. WHOI 70-43, Woods Hole Oceanographic Institution, Woods Hole, Mass., 1970.

STRUCTURE OF THE CONTINENTAL SHELF AND SLOPE OF THE NORTHERN GULF OF MEXICO AND ITS GEOHAZARDS AND ENGINEERING CONSTRAINTS

William R. Bryant and Lamar B. Roemer

INTRODUCTION

The continental margins of the Gulf of Mexico and the U.S. East Coast are classified as passive margins. However, the amount and type of geological activity that takes place in the continental shelf and slopes of these two areas are anything but passive. The continental shelves of these areas are fairly flat, and on a large-scale chart appear to be featureless, while the continental slopes are steeply dipping areas and have very rugged topography.

Close examination of the topography of the continental shelf of the northern Gulf of Mexico will reveal that numerous irregularities dot the surface of the shelf. Examination of the internal structure of the shelf reveals a very complex system of faults, intrusive structures, and sedimentary sections.

The subbottom or internal structure of the continental slope of the northern Gulf of Mexico consists of numerous intrusive bodies, mainly salt diapirs, that dissect the area into a series of highs and lows. In the western portion of the Gulf of Mexico, off Mexico, the continental slope consists of some of the largest slump folds in the world. Thousands of square miles of this slope are covered by massive gravity slides. Along certain areas of the U.S. East Coast, such as near Wilmington, Spencer, and Lindenkohl Canyon, large slump areas have been observed on the continental slope.

Offshore activity such as drilling, production platform construction, and the laying of pipelines on or across the continental shelf and eventually out onto and across the continental slope, must take into consideration the various surface and subbottom geological structures, because they will affect the construction and preservation of man-made structures. This chapter presents high resolution seismic profiles of various types of such geological structures.

The word geohazard has become a very popular word within the last decade for those working offshore. A strict definition of the word would be hazards imposed upon us by all the conditions of the world we live in. Obviously the list of geohazards under this definition would be endless. The practical working definition of a geohazard is any geological hazard. This paper will concentrate on several types of geohazards present in the continental margins, they are as follows:

1. Faulting
2. Diapiric structures
3. Channel cuts
4. Gas-charged sediments
5. Blowouts
6. Submarine slumps

The method of presentation will consist of a series of seismic reflection profiles showing the various geological features considered as possible geohazards or engineering constraints in relationship to offshore construction activity.

The seismic reflection profiles used consist of three types. For intermediate resolution and penetration Bolt air guns with 1- and 5-in.^{3} chambers 100 to 200 Hz were used; for high resolution an 800 J mini-sparker 1 to 3 kHz system was used; and for very high resolution seismic reflection profiling a 3.5 kHz system with a resolution of 0.1 to 0.3 m was employed. For lateral surface profiling records a side-scan sonar system of 100 to 200 kHz was used.

SEISMIC INTERPRETATION

The interpretation of a single channel analog seismic record, at best, is an art. A 3.5 kHz seismic record of Eugene Island Block 296 and 305, Figure 1, shows several significant features, most of which without well control could not be interpreted as they are in the profile.

As the search for oil moves offshore, geologists and engineers are concerned with the foundation problems involved in placing expensive platforms in marine environments. The oil industry is most interested in relating shear strength and scour resistance of near-surface sediments to platform design. Foundation borings ranging between 300 to 400 ft are made prior to placing a platform on a lease block. From the use of such borings, zones of very soft clay, firm clay, and stiff clay with sand and silt seams are designated on the section, from data of a platform boring taken at the position indicated. Without this information one could only infer what the sediment conditions would likely be. In Figure 2, a graph showing the water content, plastic limit, and liquid limit of various samples within the section of Figure 1, shows the change from a very soft clay to a firm clay and the appearance of a low water content silt pocket or seam. A similar decrease in water content with a distinct change in the seismic profile record is shown in Figure 3 of a Eugene Island Area record. Note the presence of a pipeline with its characteristic diffraction parabola, and the large gas seep indicated by the large dark area adjacent to the pipeline and extending from the sea floor.

The seismic reflection profile of Eugene Island Area Block 305 in Figure 4 is a typical 40 kJ sparker profile record. The record indicates a regional sand layer throughout this block. A 3.5 kHz seismic record of Ship Shoal Blocks 252 and 267 in Figure 5, shows the top of a regional sand and a section of unconsolidated silty clay overlaying the regional sand layer. None of these profiles indicate the presence of any geohazards or major engineering constraints, but are used to illustrate changing sediment characteristics found on the shelf and the use of borehole data as an aid in interpreting high resolution seismic profiles.

FAULTS

Many fault and fault systems are present in the Gulf Coast geologic province of the U.S. These faults are all normal except for a few uncommon reverse faults that Halbouty and Hardin[2] described as being associated with piercement salt domes. Growth faults and postdepositional faults are found throughout the northern Gulf of Mexico. In addition, grabens and horsts, as well as collapse, radial, and tangential faults are present in the sediments adjacent to piercement salt domes.

The following set of seismic profiles will display the various type of faults found in the continental shelf and slope of the northern Gulf of Mexico. Most of these faults are the shallow expression of very deep-seated faulting within the shelf and slope area.

The air gun profile of Ship Shoal Area Block 311 (Figure 6) shows a large growth fault with a displacement in excess of 100 ft. The emplacement of drilling platforms or other structures on such a fault would be foolhardy.

Such faults act as pipelines or conduits to the surface. Any escaping fluid from a well drilled on such a fault would find its way to the surface. Any movement along

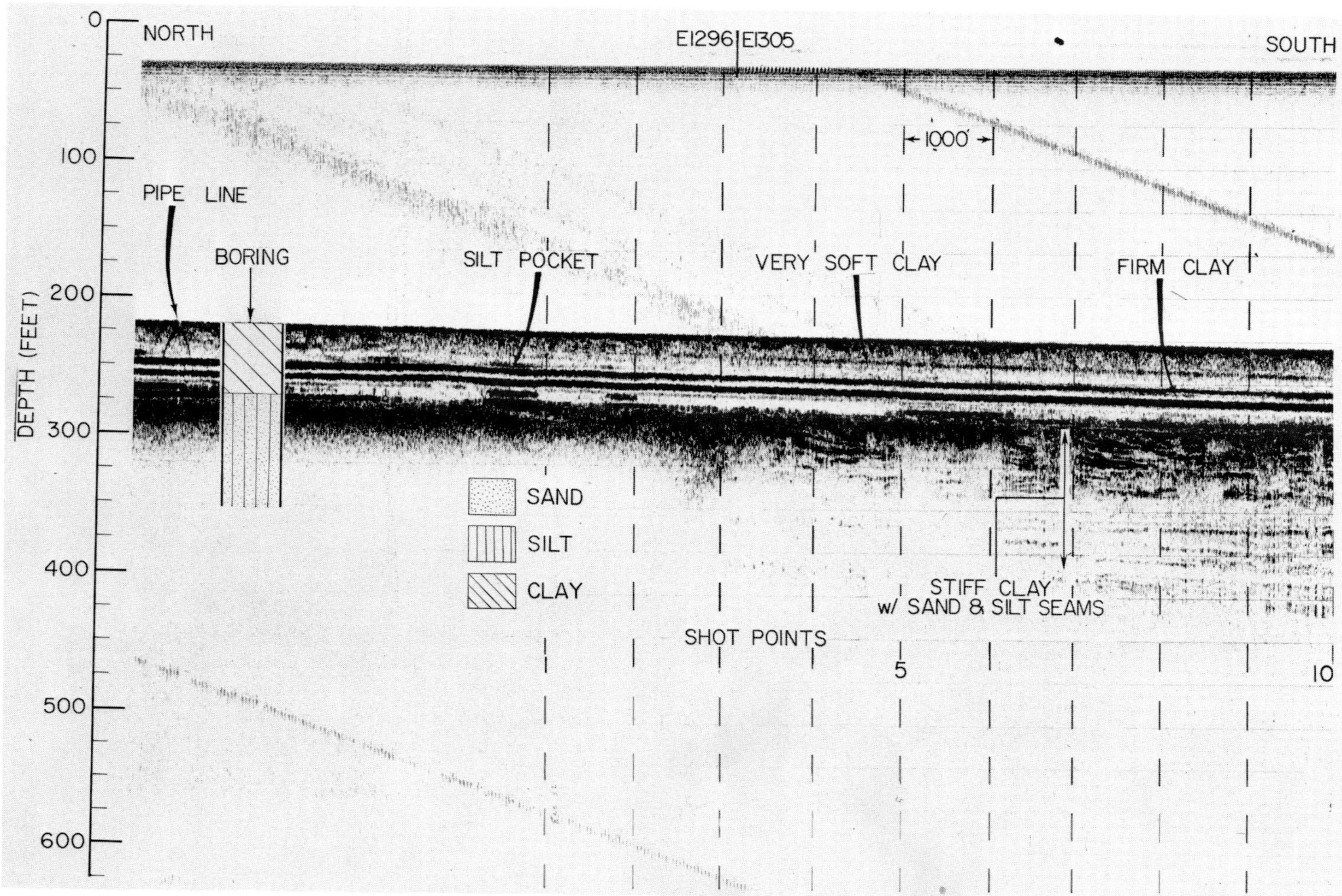

FIGURE 1. 3.5 kHz profile of Eugene Island Block 296-305.

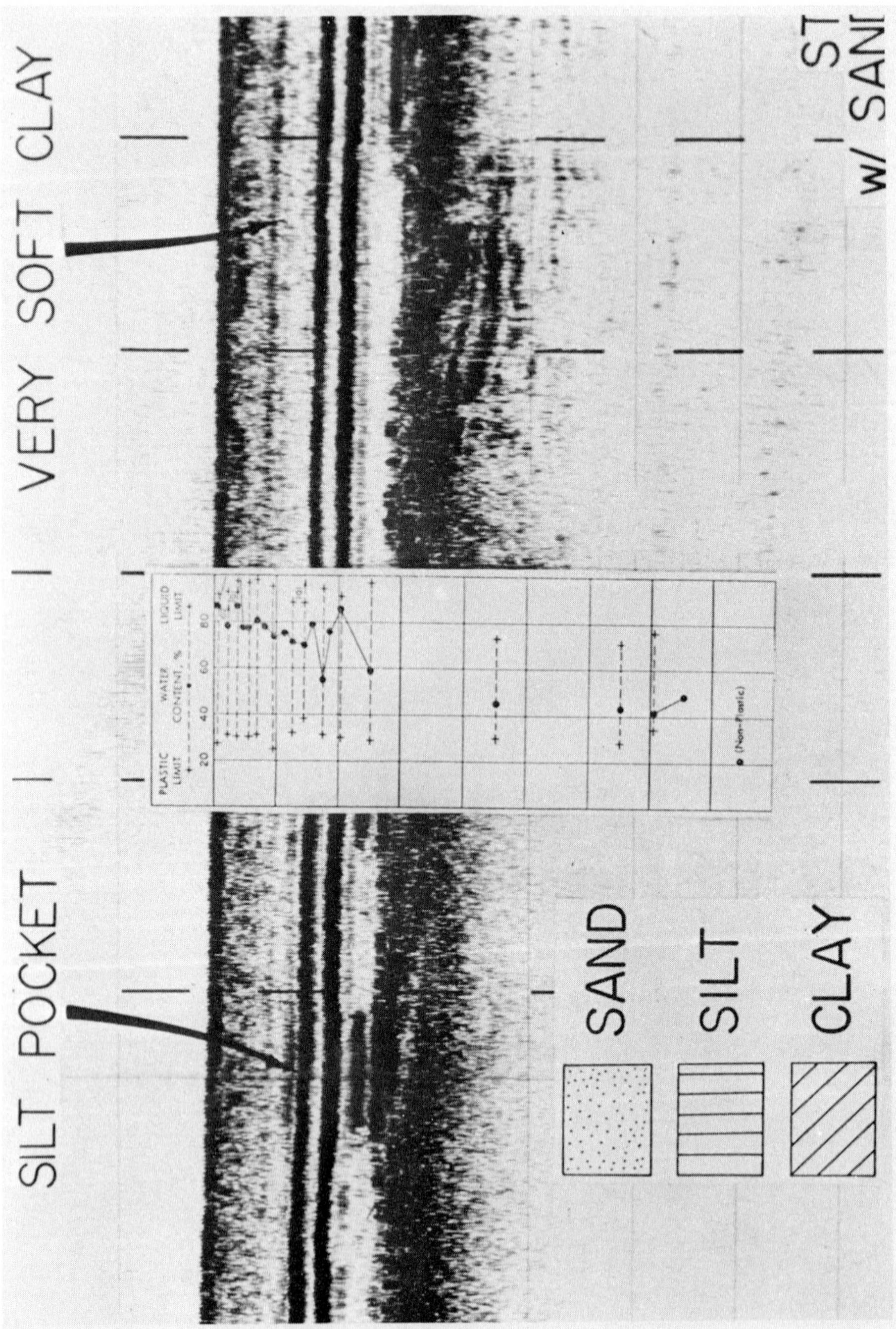

FIGURE 2. 3.5 kHz profile of Eugene Island Block 296-305 relating data from borehole to seismic events.

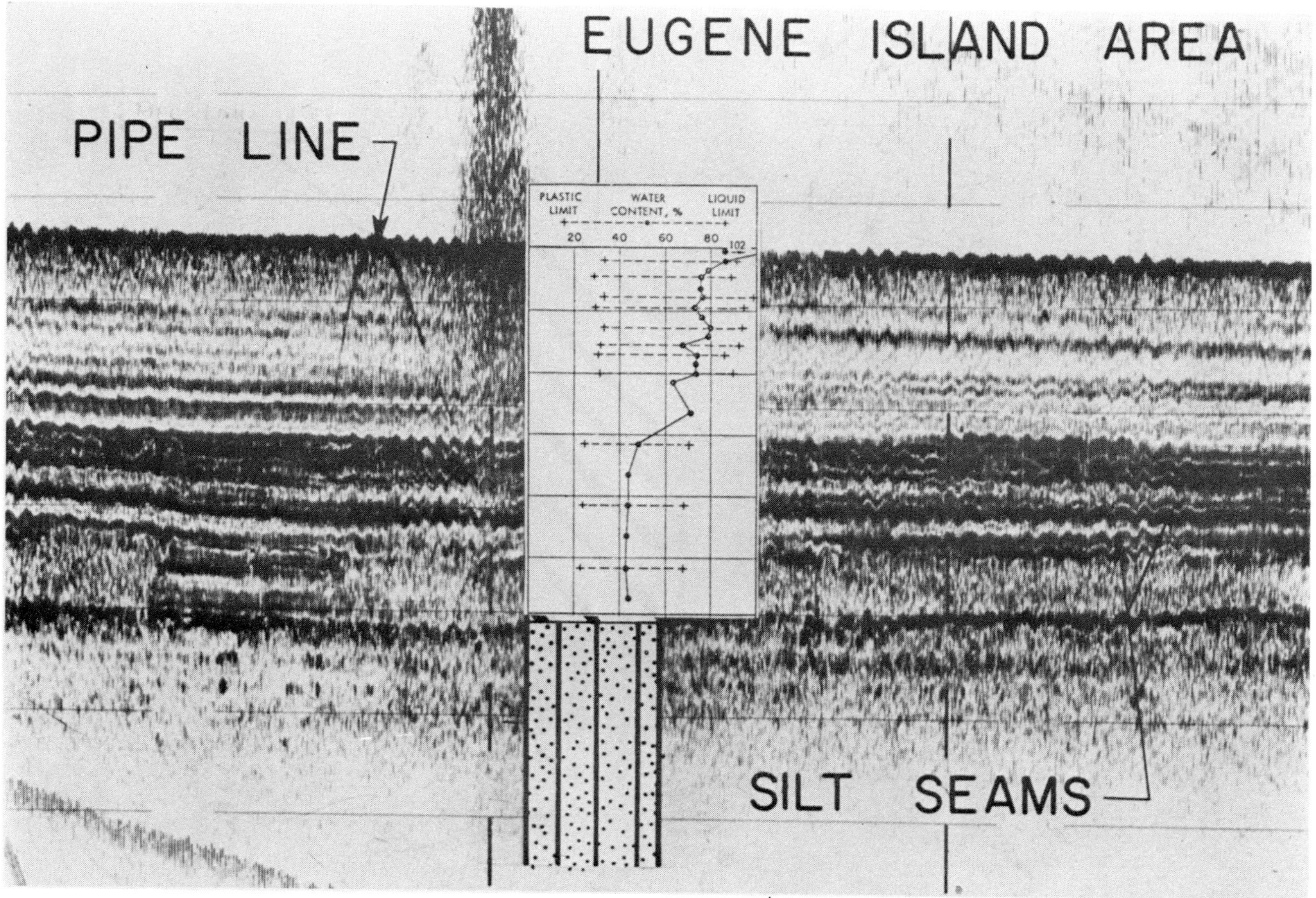

FIGURE 3. 3.5 kHz profile of Eugene Island Area relating data from borehole to various seismic events.

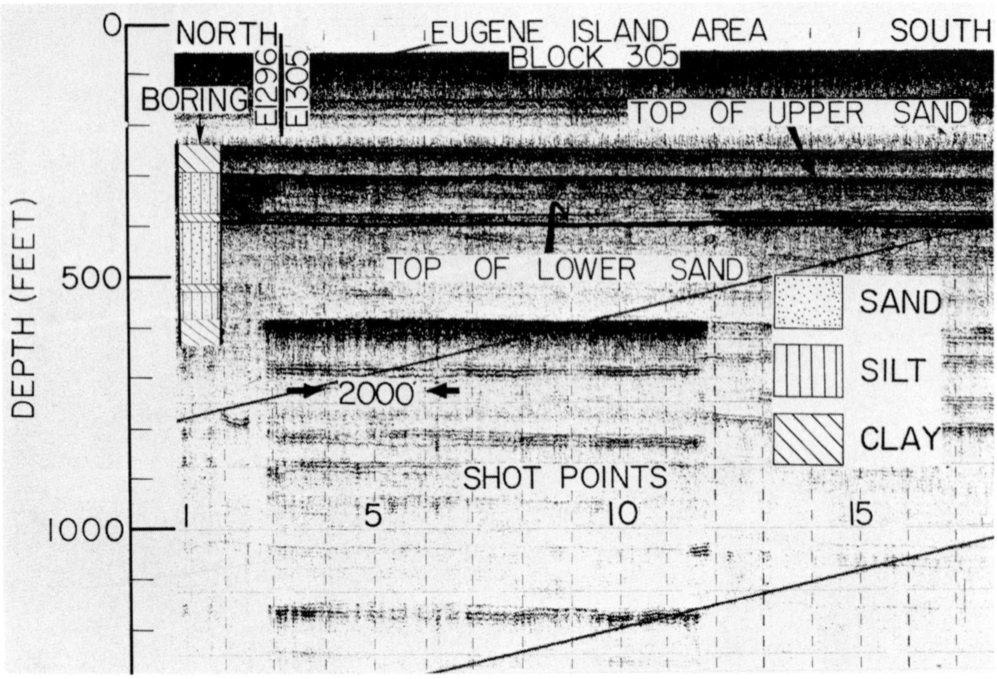

FIGURE 4. Air gun profile in Eugene Island Area Block 305.

the faults could rupture or lead to failure of a bottom supported structure. If the fault extends to the sea floor and is active, a pipeline or other structure placed across such a feature could lose support and fail. A 3.5 kHz seismic record of the same area, Ship Shoal Block 311 (Figure 7) shows that the fault does extend to the sea floor indicating recent displacement. Note the thickening of the sections on the downthrown side of the fault. This is typical of a growth fault. Approximately 80 ft of sand is trapped on the eastern side of the fault, while to the west, only 30 ft is found. This fault pattern is a classic example of the major structure trapping hydrocarbons in the Gulf Coast. Note the rollover on the strata on the downthrown side of the growth fault.

The 3.5 kHz seismic record of South Marsh Island Block (SMI) 127 and 142 (Figure 8) shows a very detailed section of the upper unconsolidated sediment layer of a graben-like structure. This seismic record is similar to a cross-sectional view associated with salt withdrawal basins. The 3.5 kHz records in this area show gas seepage throughout the water column. Literally hundreds of gas seeps are present in the water column. Discontinuities in layering are caused by the vertical migration of gas. The gas appears to be trapped below the regional sand layer, where it accumulates until it is released along fault planes.

The silty clay section of sediments in Figure 8 rests upon a regional sand layer. Immediately above the sand layer, opaque areas of the record indicate the presence of gas-charged sediment pockets. Note also the numerous gas seeps at the top of the record.

The sparker record of West Cameron Area Blocks 629 and 638 (Figure 9) shows a series of faults outlining an uplifted block and small graben feature.

In the Vermilion Area Block 340 (Figure 10) a 3.5 kHz seismic record shows a series of step faults associated with an uplift and an unconformity. In the Eugene Island Area Blocks 308, 315, and 330 (Figure 11) the air gun records shows a small graben

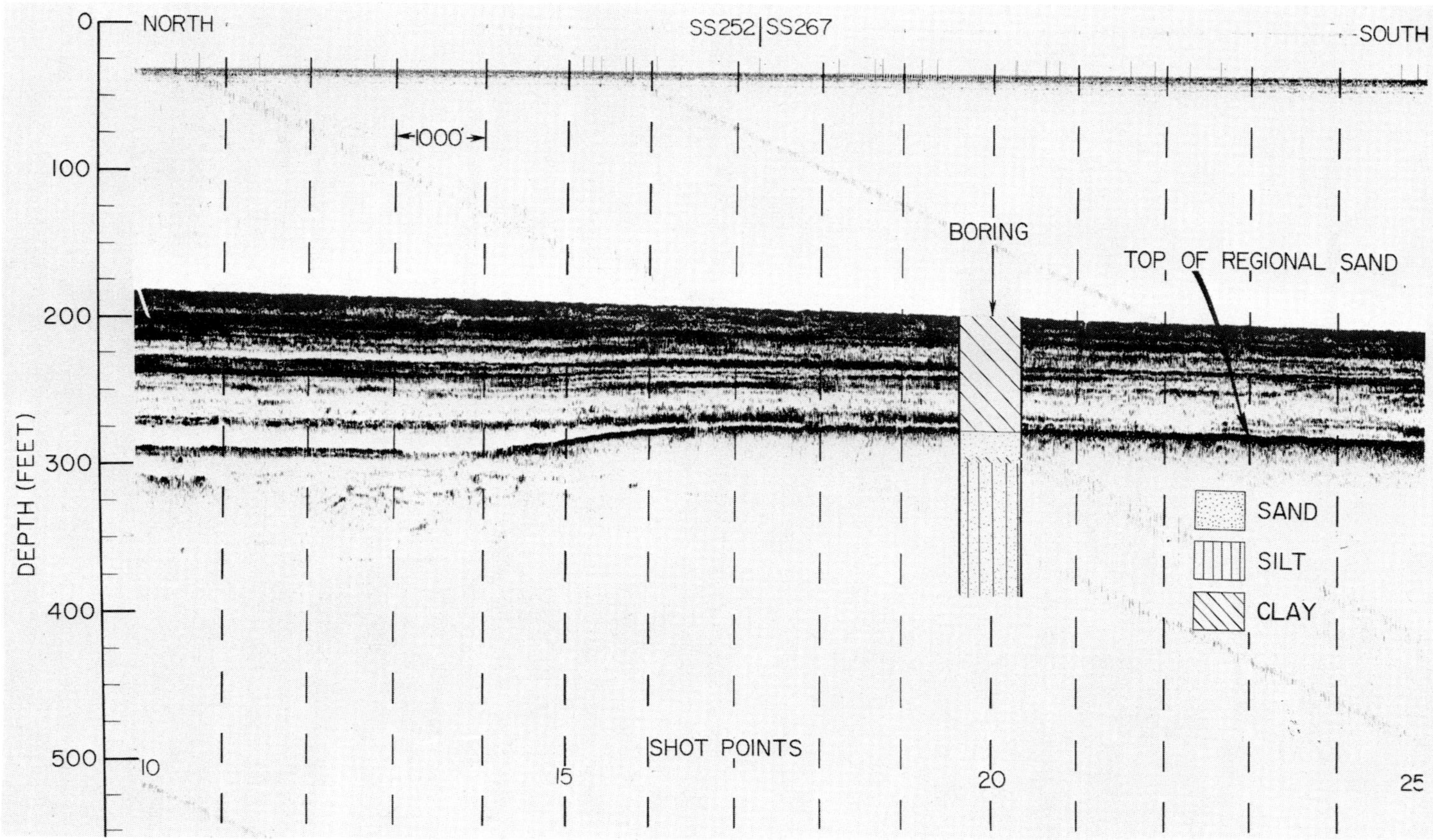

FIGURE 5. 3.5 kHz profile of Ship Shoal Block 252 and 267.

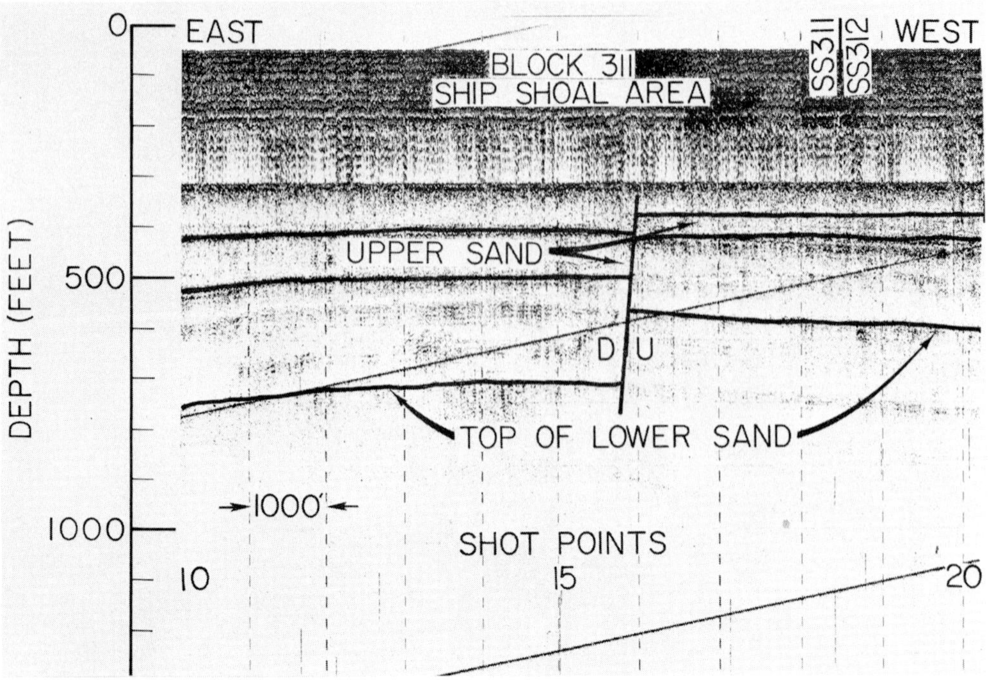

FIGURE 6. Air gun profile Ship Shoal Area Block 311.

formed by two closely spaced faults. The pinch-out of a regional sand layer and the change from very soft clays to firm clays are also designated. Line 46 (Figure 12) displays complex faulting that can occur on the crest of a diapir on the continental slope off Texas. Such faulting may indicate that the salt core of the diapir has been breached by seawater and dissolved salt has caused the collapse of the crest of the structure.

The results of placing a structure or drilling a well on a fault may prove disastrous. A point in case is shown in the sparker record in Figure 13 where a well was drilled (Shot Point 8) on the large fault that extends to the sea floor. After a blowout and near loss of the rig, the well was spotted some distance from any fault within the area.

DIAPIRIC STRUCTURES

A diapir is a dome or anticlinal fold with overlying rocks, which has been ruptured by squeezing out the plastic core material. In sedimentary strata the core material is usually salt or shale.[1] Diapirs are very common subbottom features along the continental shelf of the northern Gulf of Mexico, and the topography of the continental slope is controlled almost entirely by them.

Salt diapirs are the most common piercement structures found on the continental shelf. The air gun record of Ship Shoal Area Blocks 291 and 300 (Figure 14) shows a typical salt diapir or salt dome that has a surface expression on the continental shelf. Associated with such features are a series of faults on the flanks of the dome and an unconformity. These diapiric structures may be very broad as is the case in West Cameron Area Blocks 474-467 (Figure 15), or very narrow, essentially forming a salt spine, such as in the Vermilion Area Block 321 (Figure 16).

An air gun record of Ship Shoal Area Block 303 and 304 (Figure 17) shows a salt

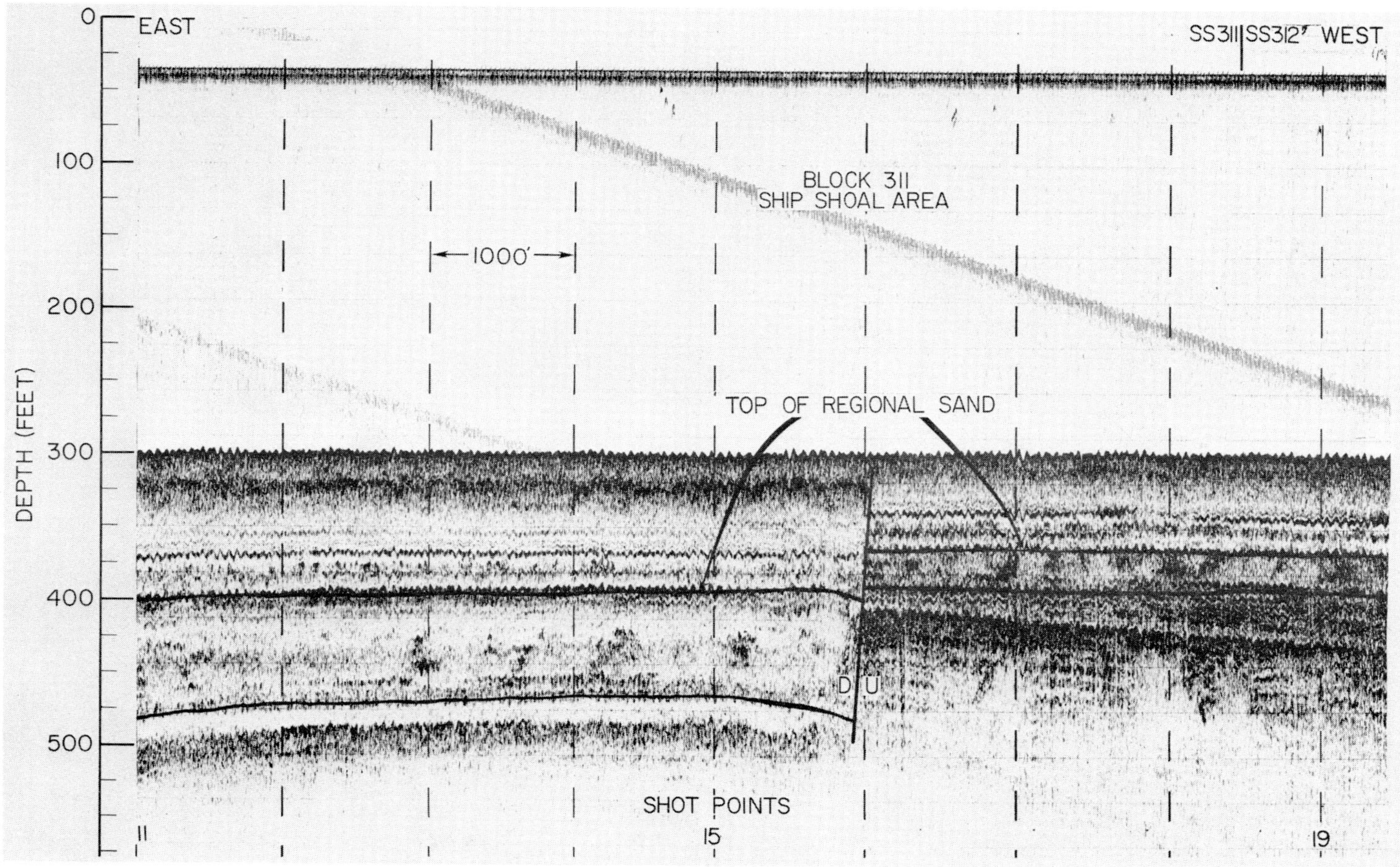

FIGURE 7. 3.5 kHz profile of Ship Shoal Area Block 311.

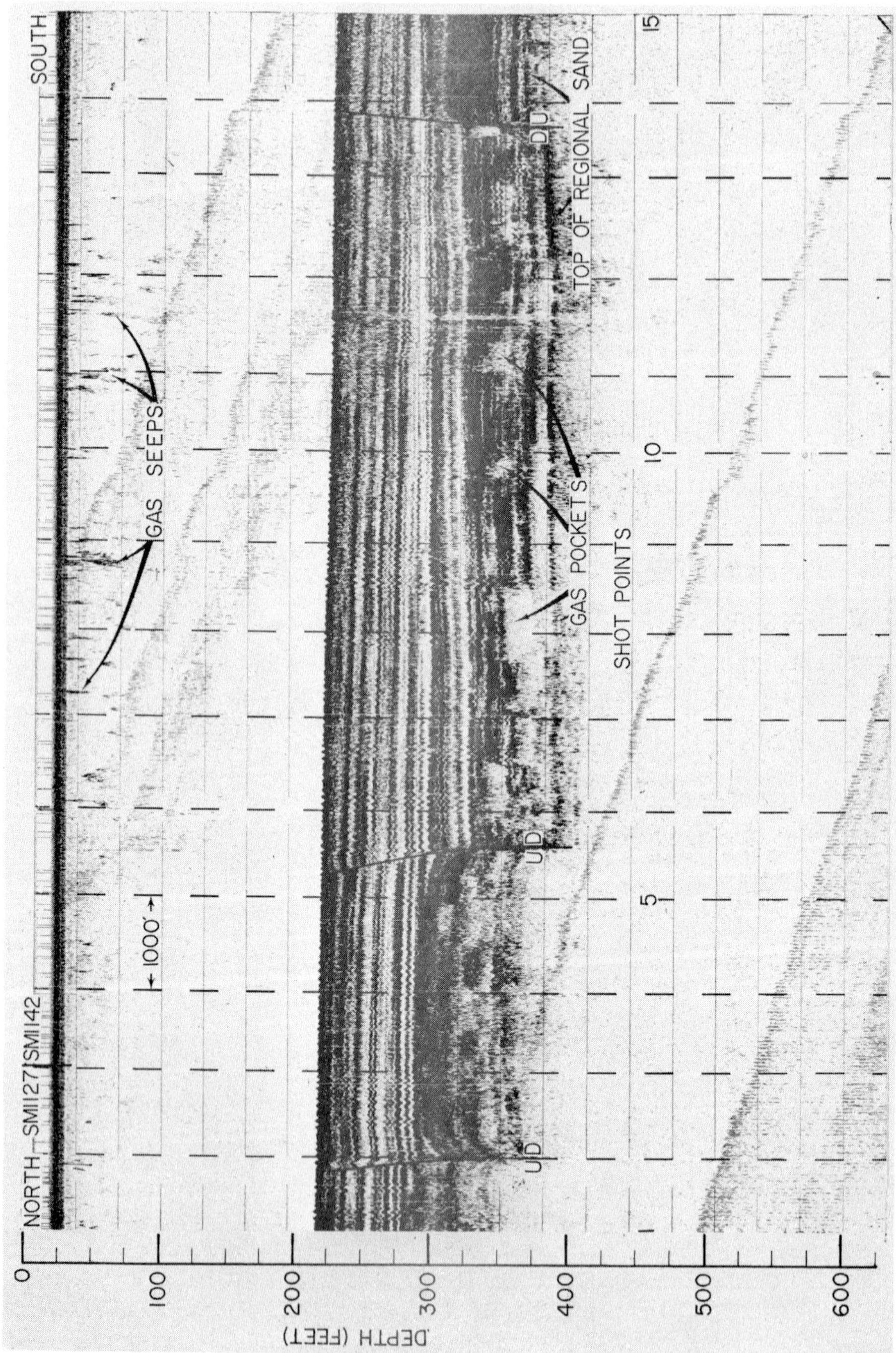

FIGURE 8. 3.5 kHz profile of South Marsh Island Block 127 and 142.

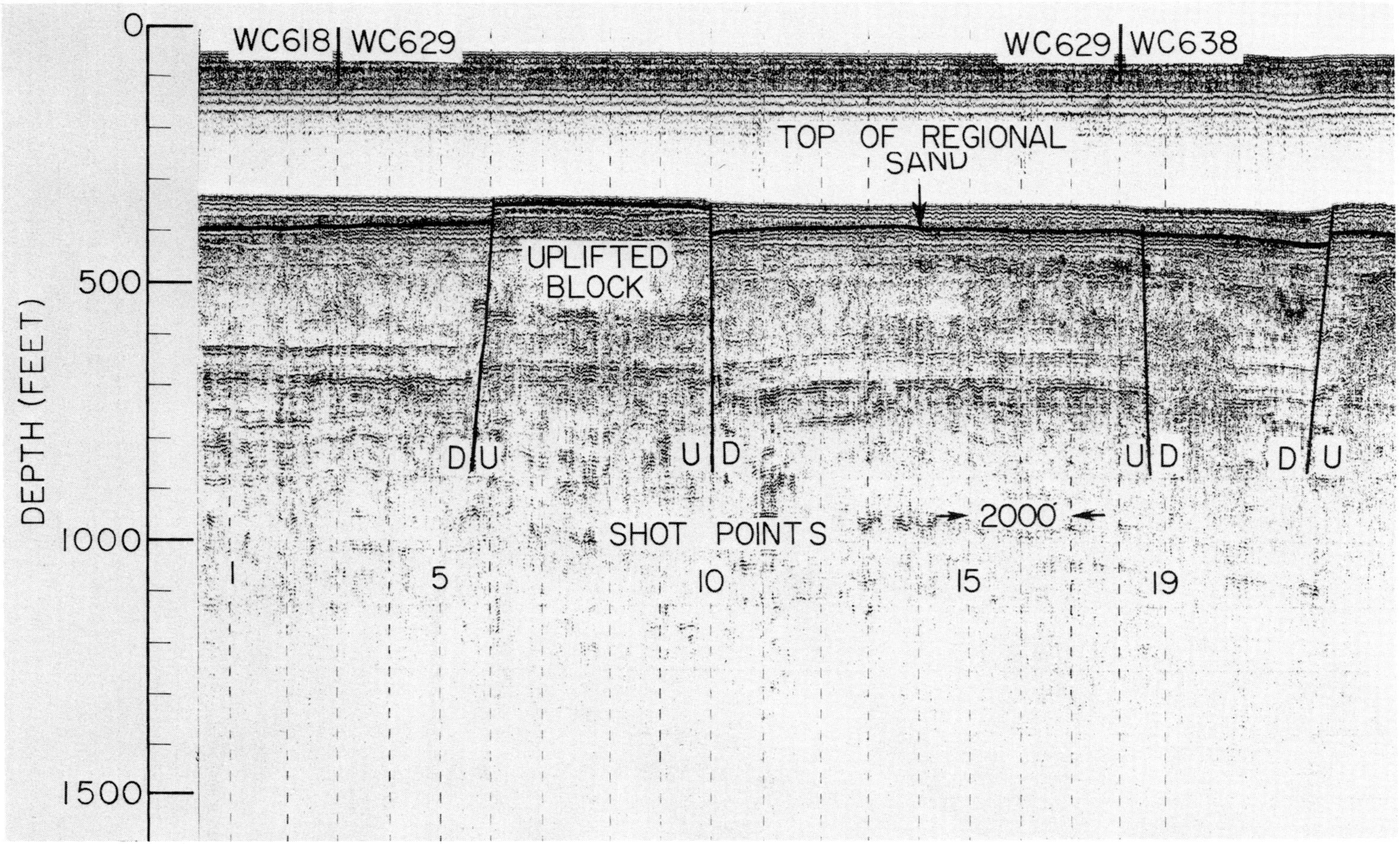

FIGURE 9. Air gun profile of West Cameron Block 629 and 638.

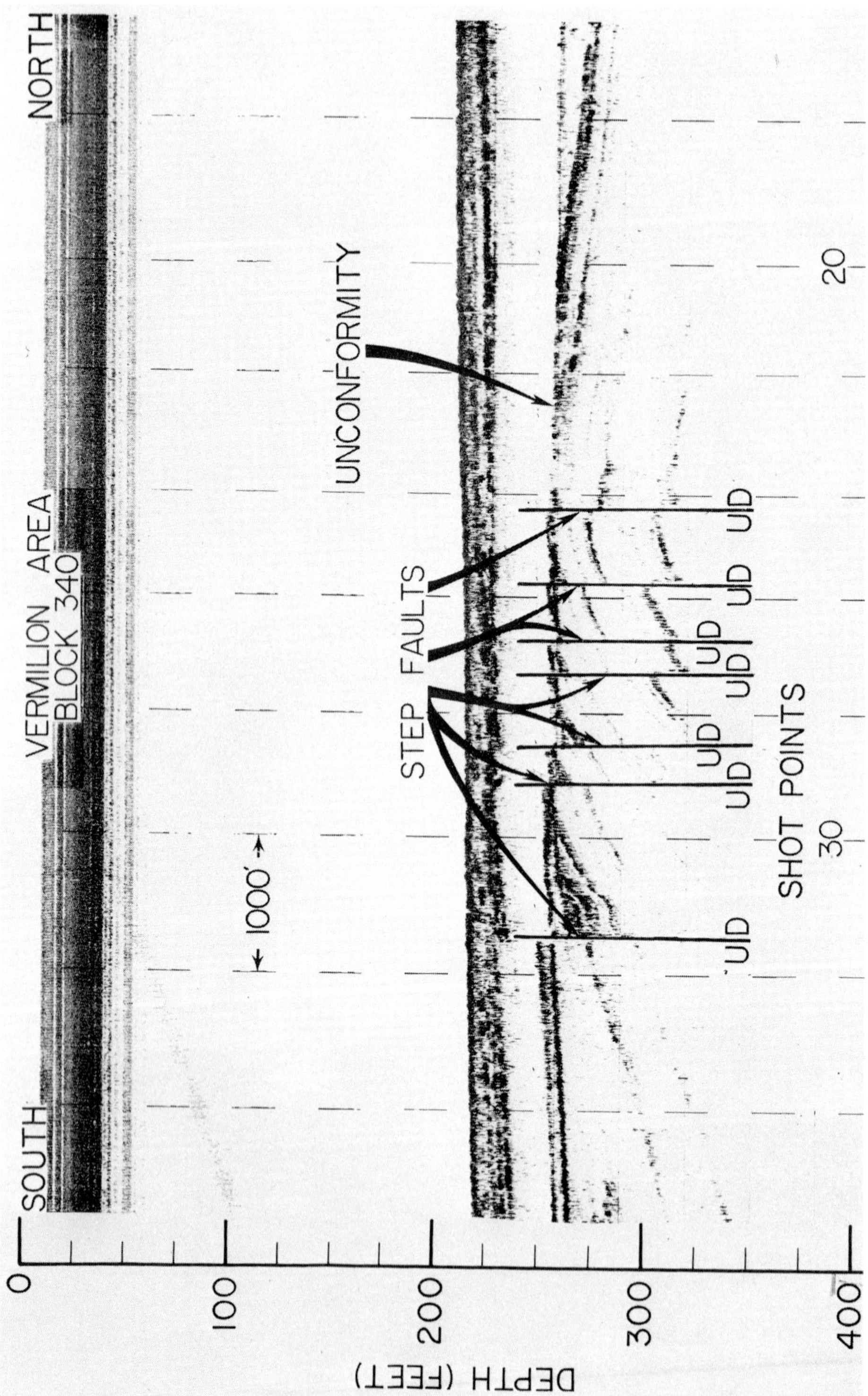

FIGURE 10. 3.5 kHz profile of Vermilion Area Block 340.

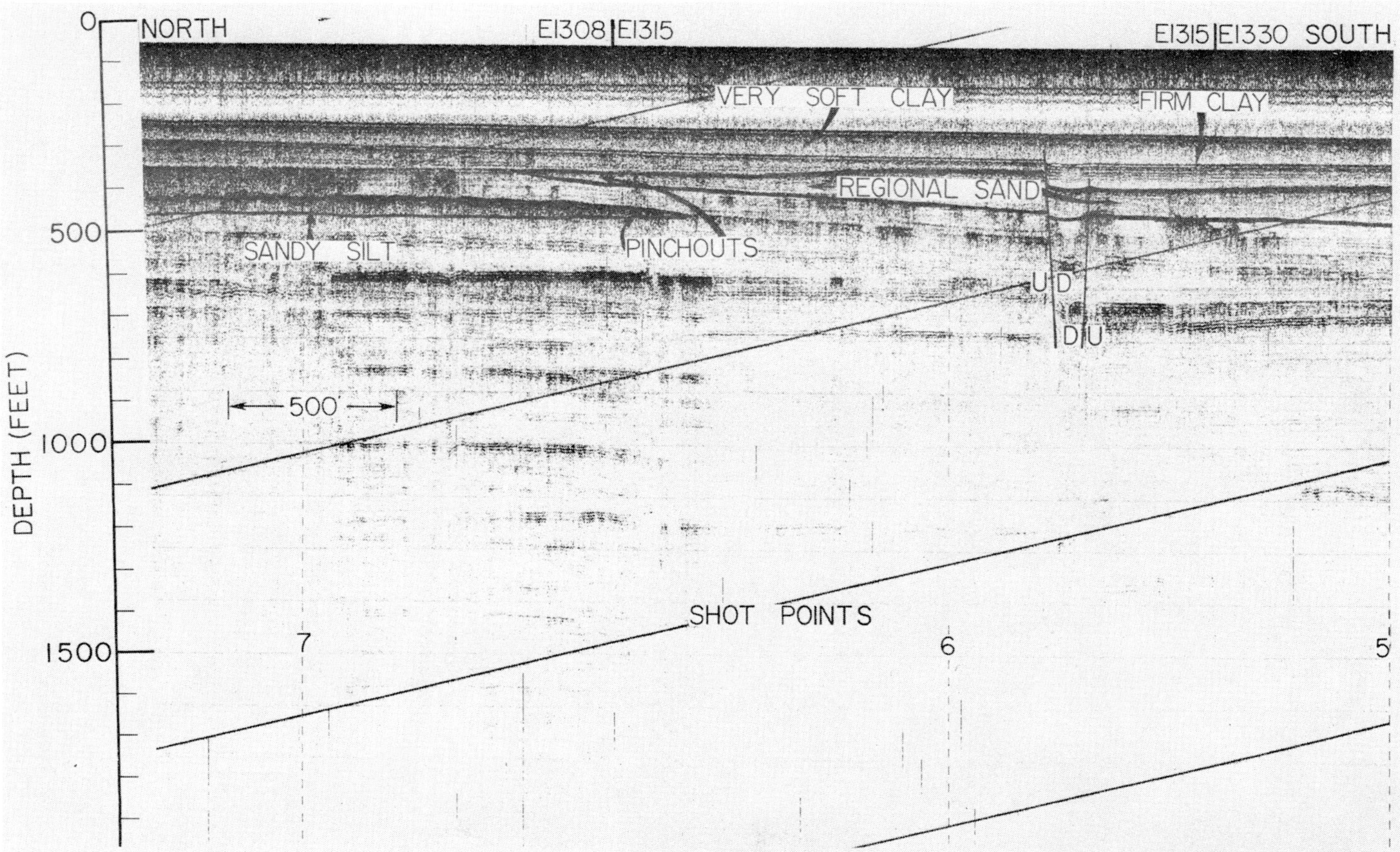

FIGURE 11. Air gun profile of Eugene Island Area Block 308, 315, and 330.

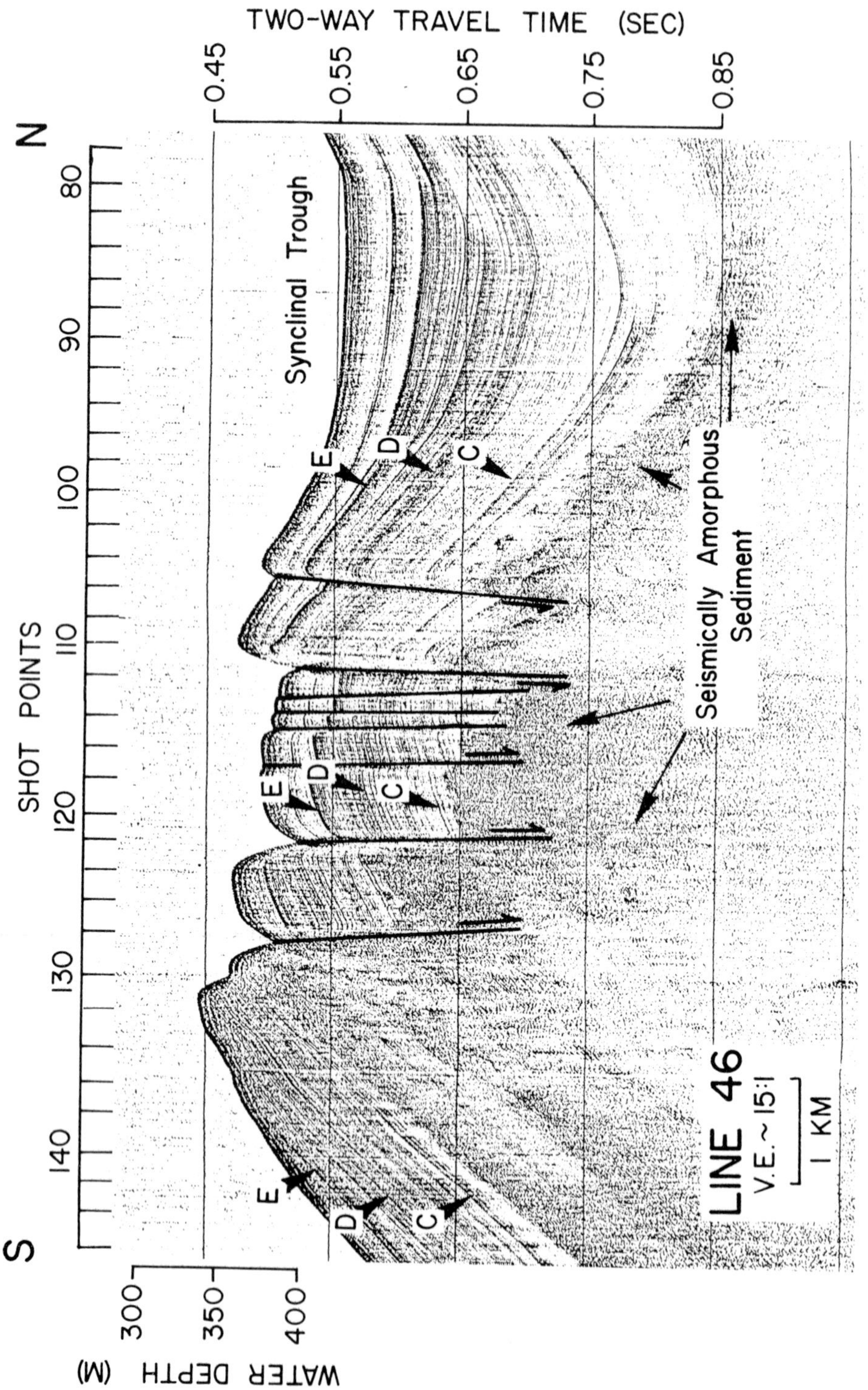

FIGURE 12. Mini-sparker profile of Line 46 on Texas Continental Slope. (From Sidner, B. R., Gartner, S., and Bryant, W., Rep. 77-5-T, Texas A&M University, College Station, 1977. With permission.)

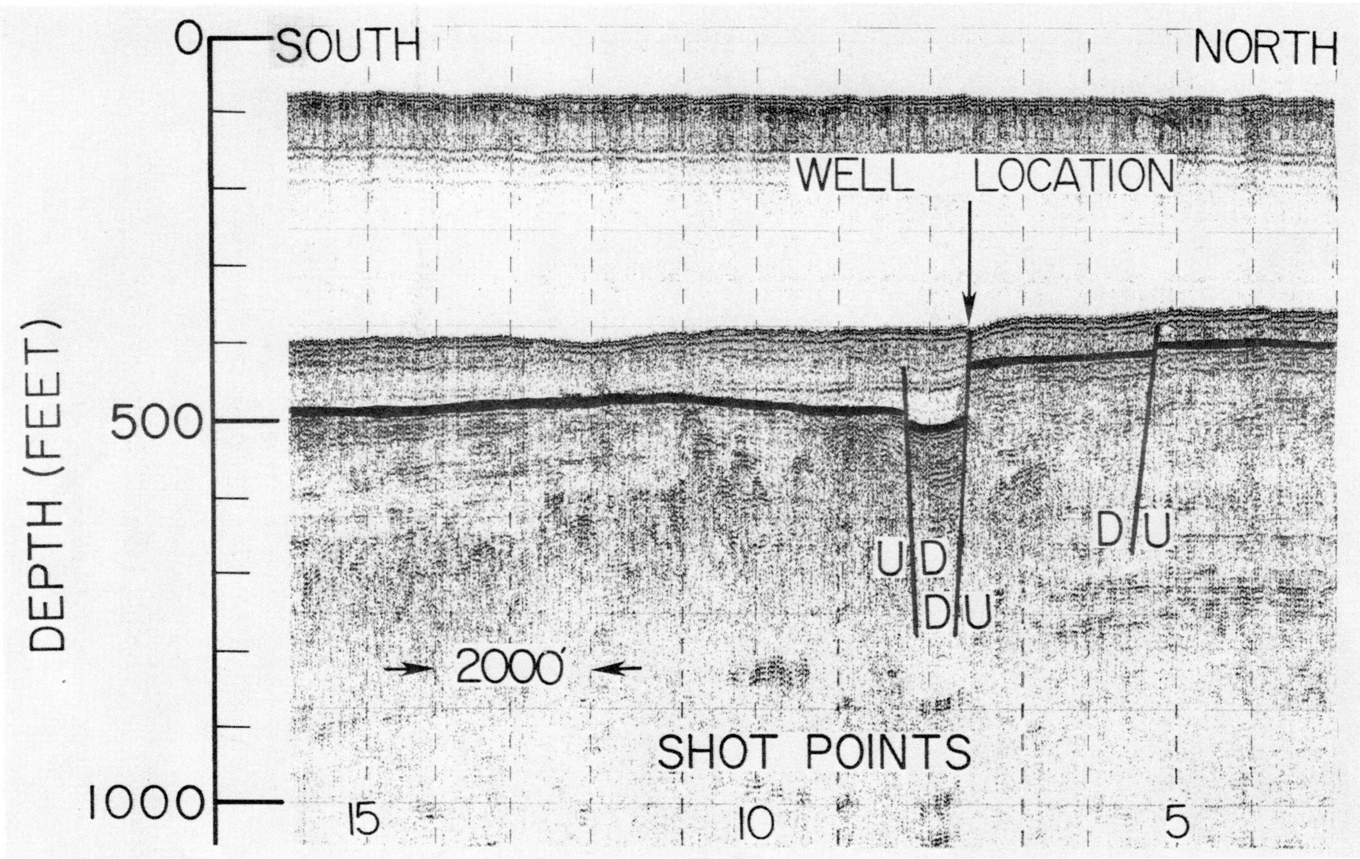

FIGURE 13. Air gun profile of Ship Shoal Area Block 300.

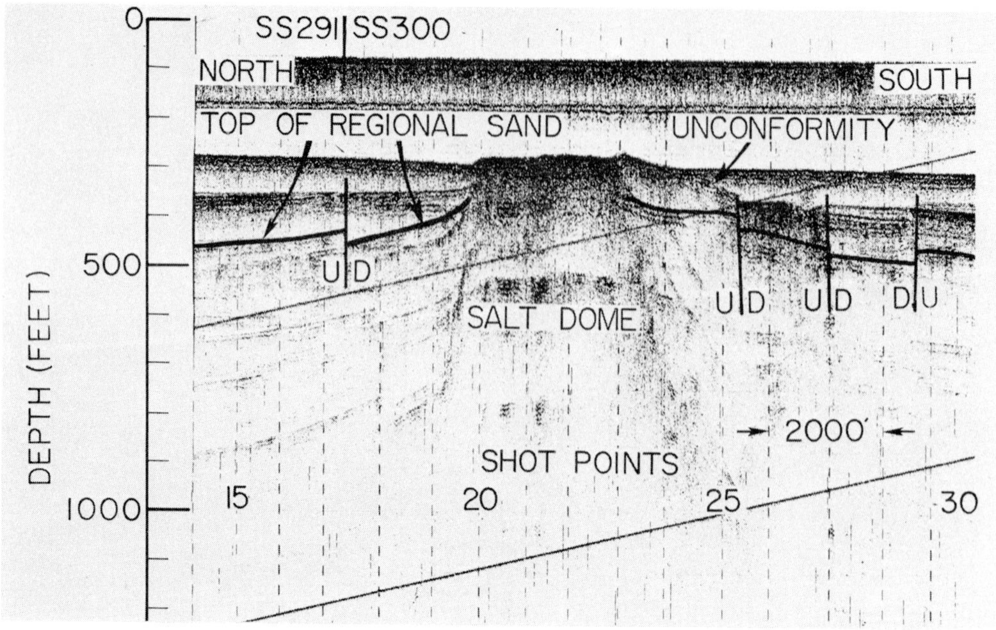

FIGURE 14. Air gun profile of Ship Shoal Area Block 300.

dome where a regional sand layer pinches out on the flanks of the structure, and an angular unconformity caused by the erosion of an uplifted section of sediment. Note the upturned beds at depth on the flank of the diapir. The 3.5 kHz seismic record of the same area (Figure 18) shows a very detailed record of the upper 200 ft of the section. Note the details of the angular unconformity and the regional sand layer pinch-out. Older and very high shear strength sediments are exposed at the sea floor. Numerous gas seeps are noted in the water column over the dome, with the greatest concentration near the crest. The salt dome crops out of the surface with angular unconformities developed on all sides. The 3.5 kHz data illustrate the complexity of the structure in the domal area. Gas pockets, angular unconformities and radial faults are all seen. Small mounds, 12 to 15 ft high, are present over the done where gas is being released into the water column.

The air gun and 3.5 kHz seismic profiles of the East Cameron Area Block 293 (Figures 19 and 20) show a broad eroded diapir. Note the steeply dipping beds forming an angular unconformity on the flanks of the diapir. The upturned beds have exposed older and much stronger sediments than the surrounding surficial sediments. Also note the mud mounds, associated gas seep, and the fairly large depression on the right side of the crest of the diapir.

In contrast to the broad structure of the East Cameron Block 293 a small well-defined salt spine is shown on the air gun records of West Cameron Area Blocks 521 and 544 (Figure 21). The 3.5 kHz records of the structure (Figure 22) shows a series of mud mounds formed by the escaping gas above the salt spine. A diapiric structure in Vermilion Area Block 340 (Figure 23) is present below the sea floor. There is no indications of sea-floor uplifting on the sparker records. Examination of the 3.5 kHz records of this area indicates a slight depression above the sea floor directly above the salt dome. The only sea floor indications of a structure at depth is a mud mound and a large gas seep.

Some of these diapiric structures are large and elevated above the sea floor. The

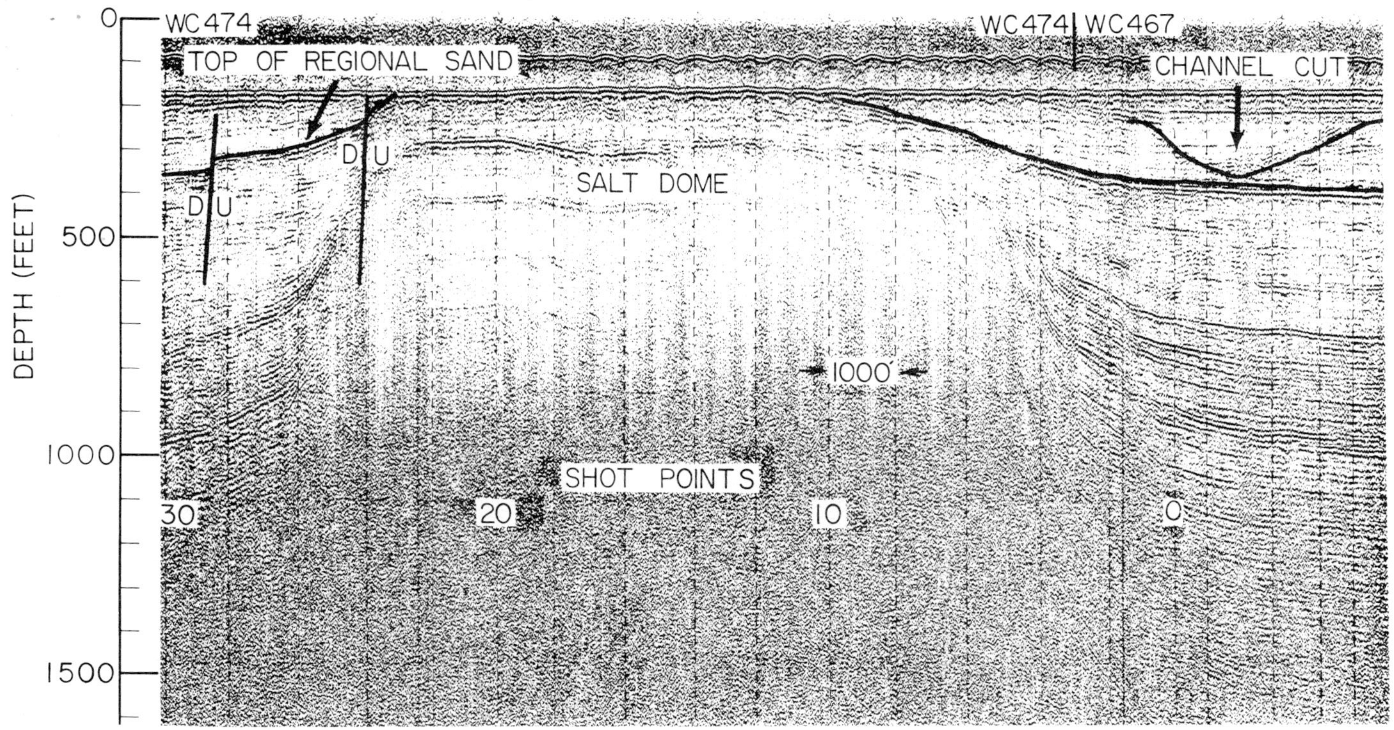

FIGURE 15. Air gun profile of West Cameron Area Block 474-467.

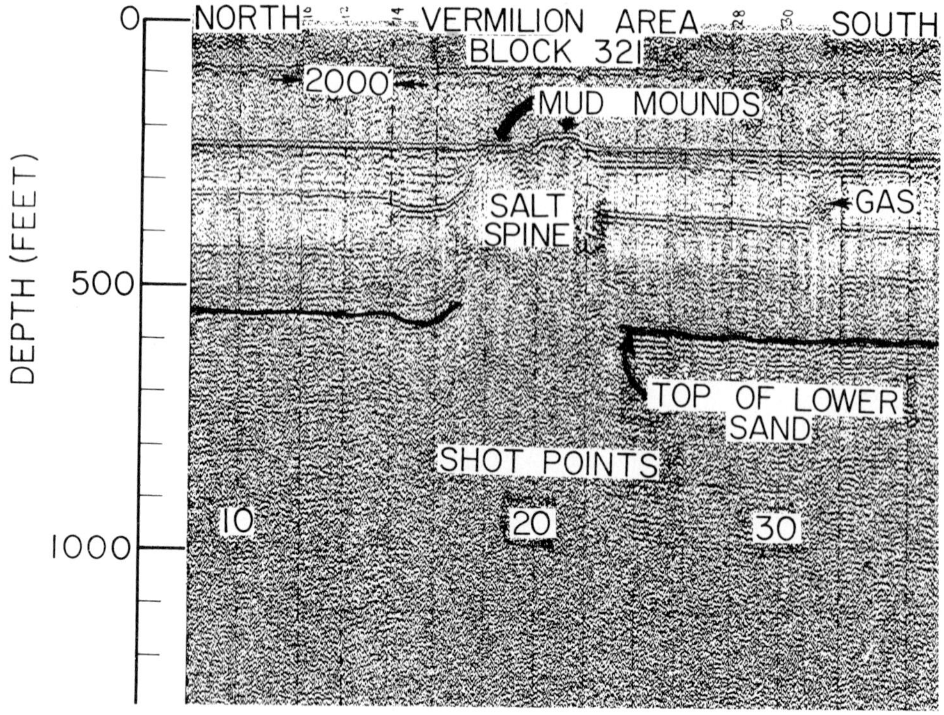

FIGURE 16. Air gun profile of Vermilion Area Block 321.

Eugene Island Area block 356 3.5 kHz record (Figure 24) shows such a feature. In addition to salt diapirs, mud and shale form similar diapiric structures. The West Cameron Area Blocks 648 and 639 3.5 kHz seismic records (Figure 25) show several large submarine mounds. It is suggested that these features are most likely the results of mud diapirism.

An alternate theory on the development of these features is that a series of collapsed faults produced a downfaulted central basin. With a lower sea level, stands would be reworked with a greater amount accumulating on the downthrown side of each fault. These coarse-grained sediments provided a zone through which gas migrated to the surface, creating a mound type feature. These mounds, being subject to subaerial exposure during low stands of sea level, might provide a hard substrate for carbonate growth. As sea levels began to rise, these carbonates would grow upwards, attempting to keep up with the rise. The mounds are found to protrude above the sea floor at various elevations, depending on their growth ability.

The topographic relief of salt diapirs on the continental slope is much larger than those of the shelf. Line 41 mini-sparker profile (Figure 26) shows such a feature located on the upper continental slope off Texas. Note the top of the salt in this record, which is within 150 ft of the sea floor.

The presence of salt diapirs on the continental slope cause extensive oversteepling of the slope on the flanks of the diapiric structure, resulting in slumping of the overlying sediment cover.

The presence of such structures as diapirs in the shelf and slope do not, in themselves, create a geohazard except for the extensive faulting usually associated with the diapirism. The exposure of older and stronger sediments may create trenching problems in pipeline activity. However, the area immediately above one of these diapiric

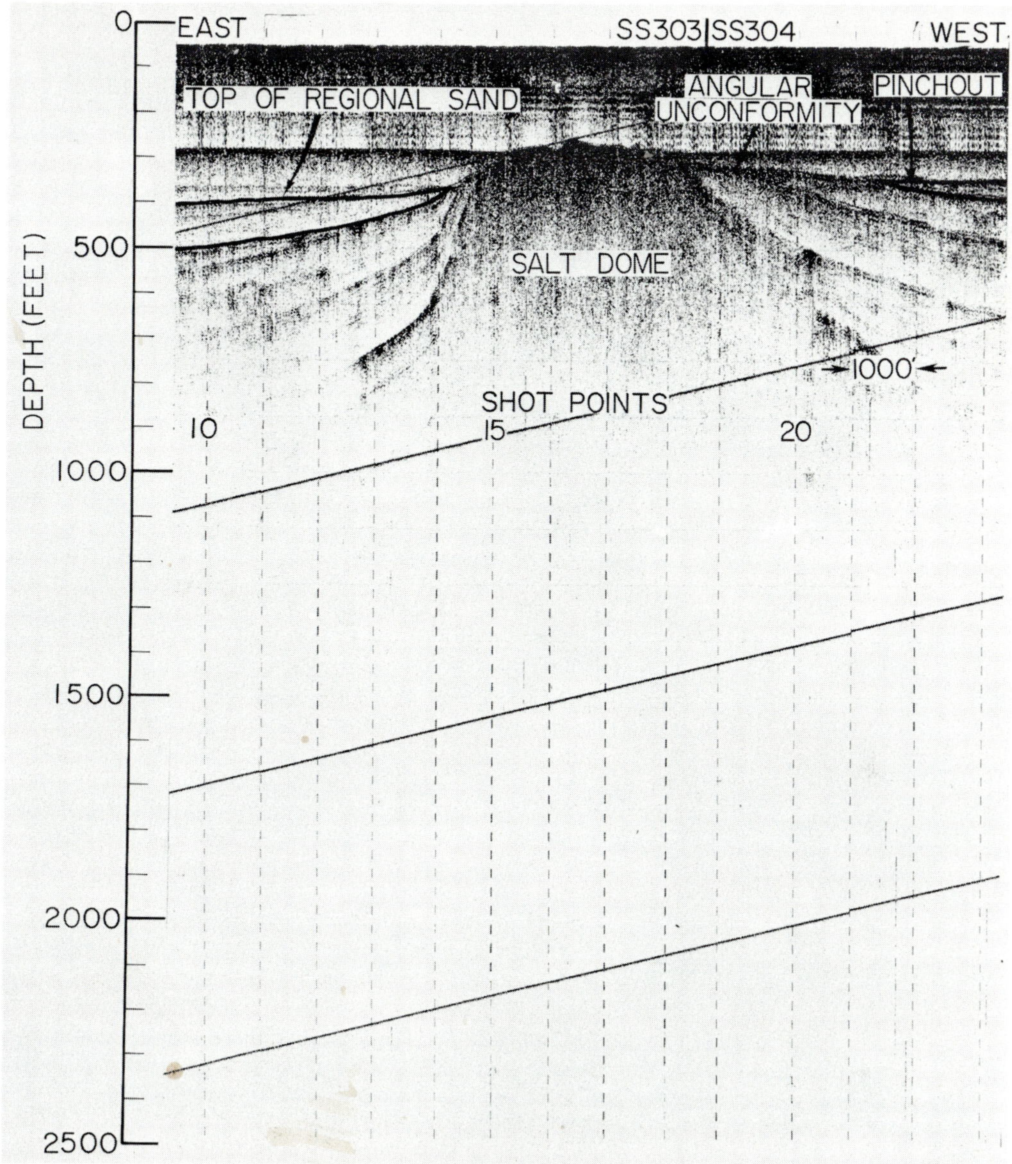

FIGURE 17. Air gun profile of Ship Shoal Area Block 303 and 304.

structures may not be considered stable due to the fact that these features may still be active and growing.

CHANNEL CUTS

Ancient channeling of the sea floor occurs quite often in the continental shelf. During periods of lower sea level most of the shelf area was subaerially exposed to various erosional processes. Periods of rising sea level allowed these channel cuts, formed by various distributaries of rivers and streams, to be filled with sediment that generally was different from the sediment into which the channel was cut. Thus the top section

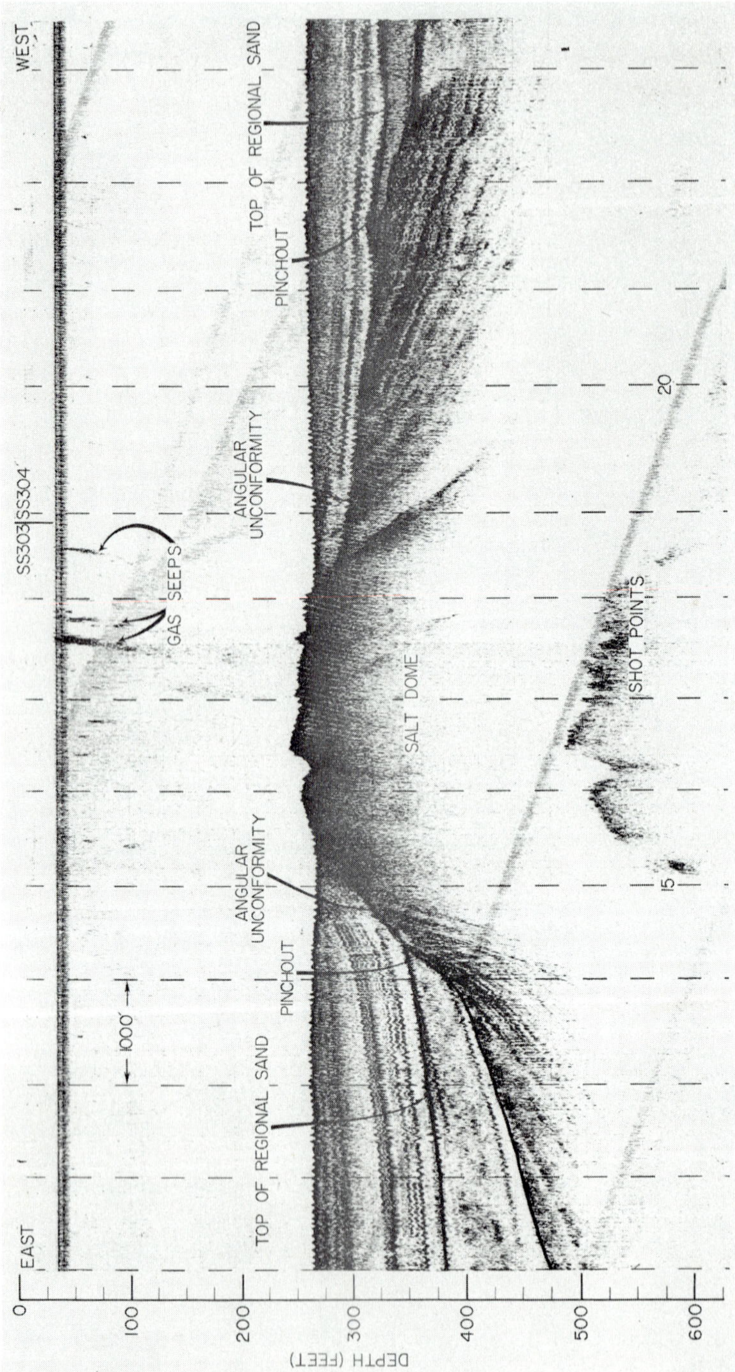

FIGURE 18. 3.5 kHz profile of Ship Shoal Area Block 303 and 304.

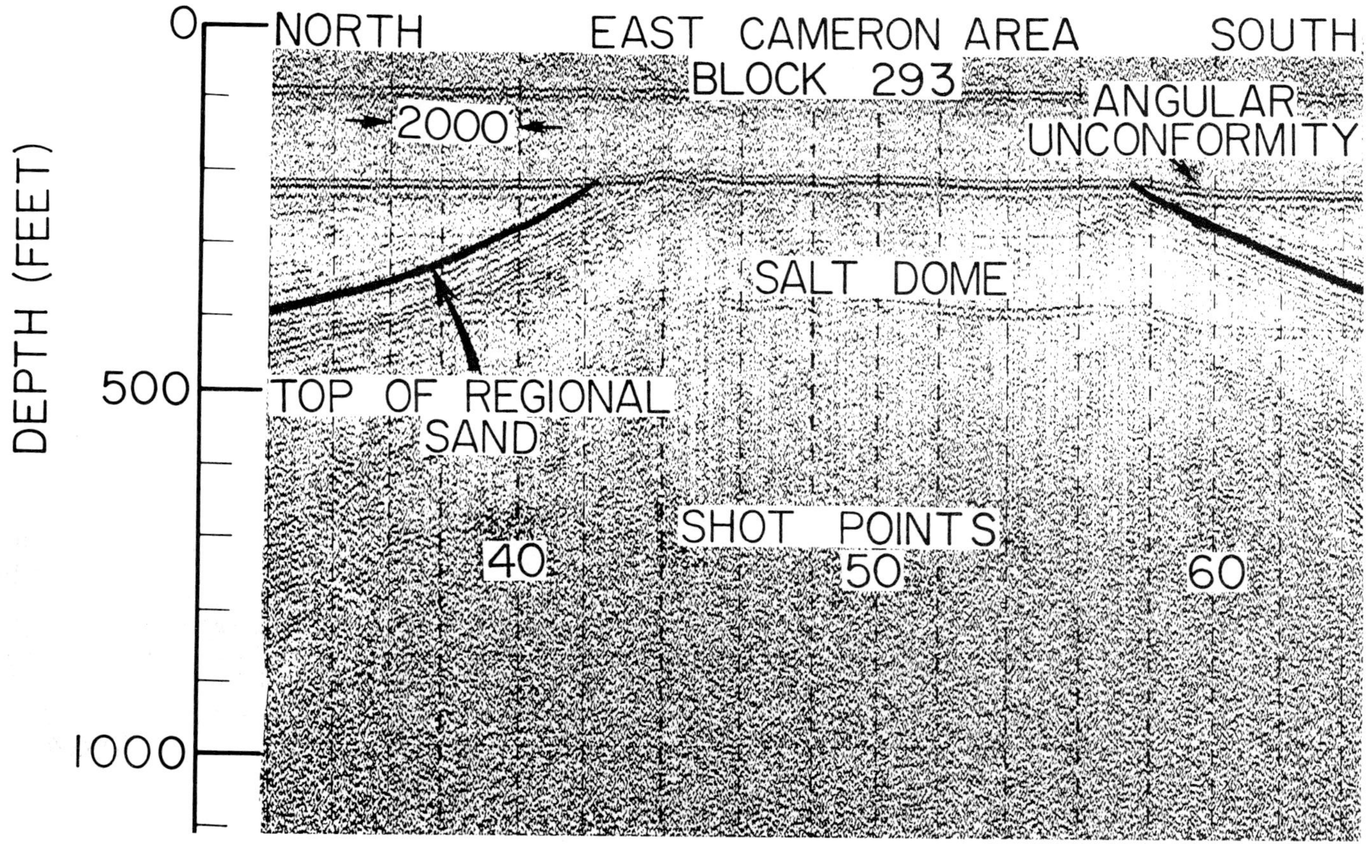

FIGURE 19. Air gun profile of East Cameron Area Block 293.

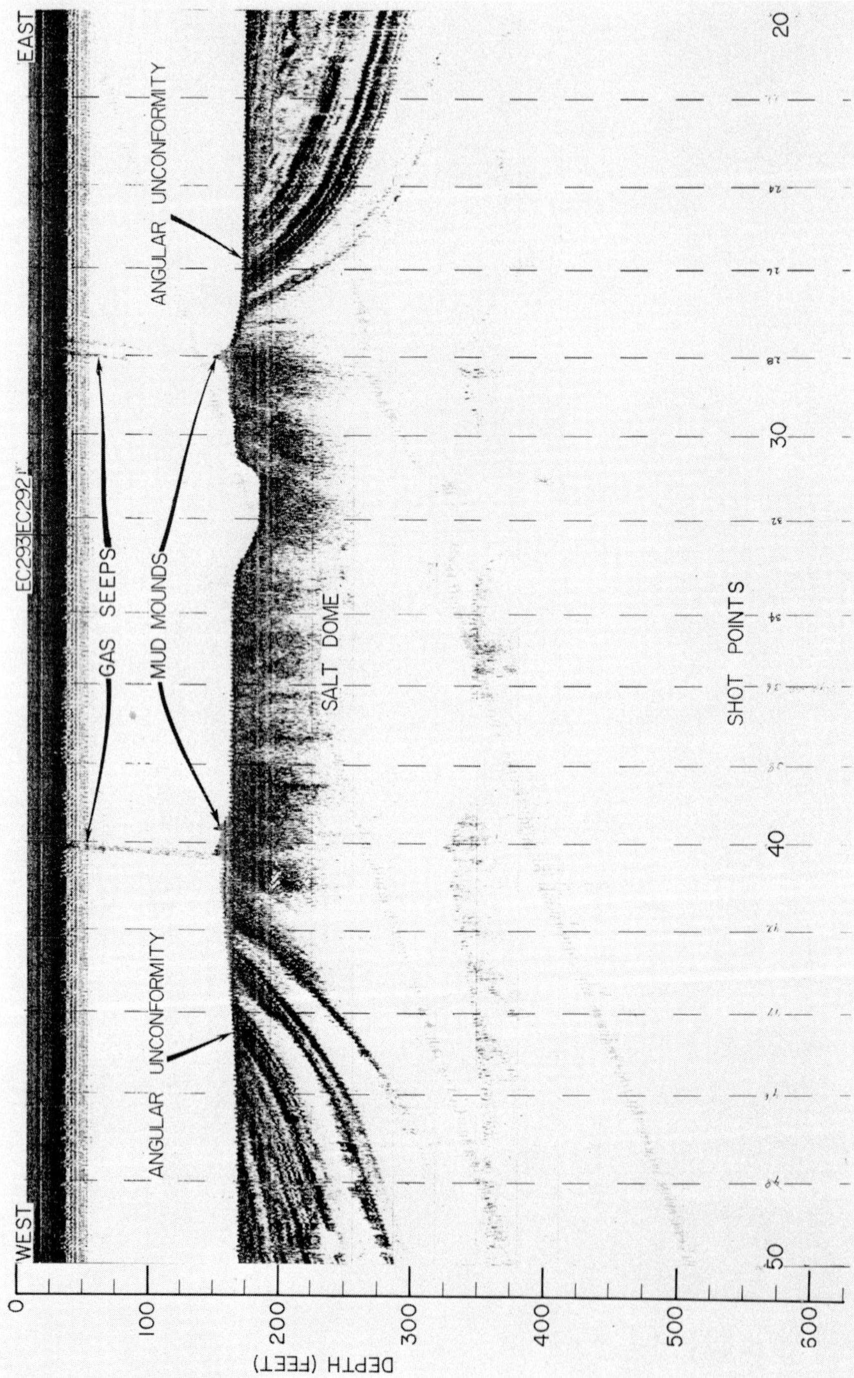

FIGURE 20. 3.5 kHz profile of East Cameron Area Block 293 and 292.

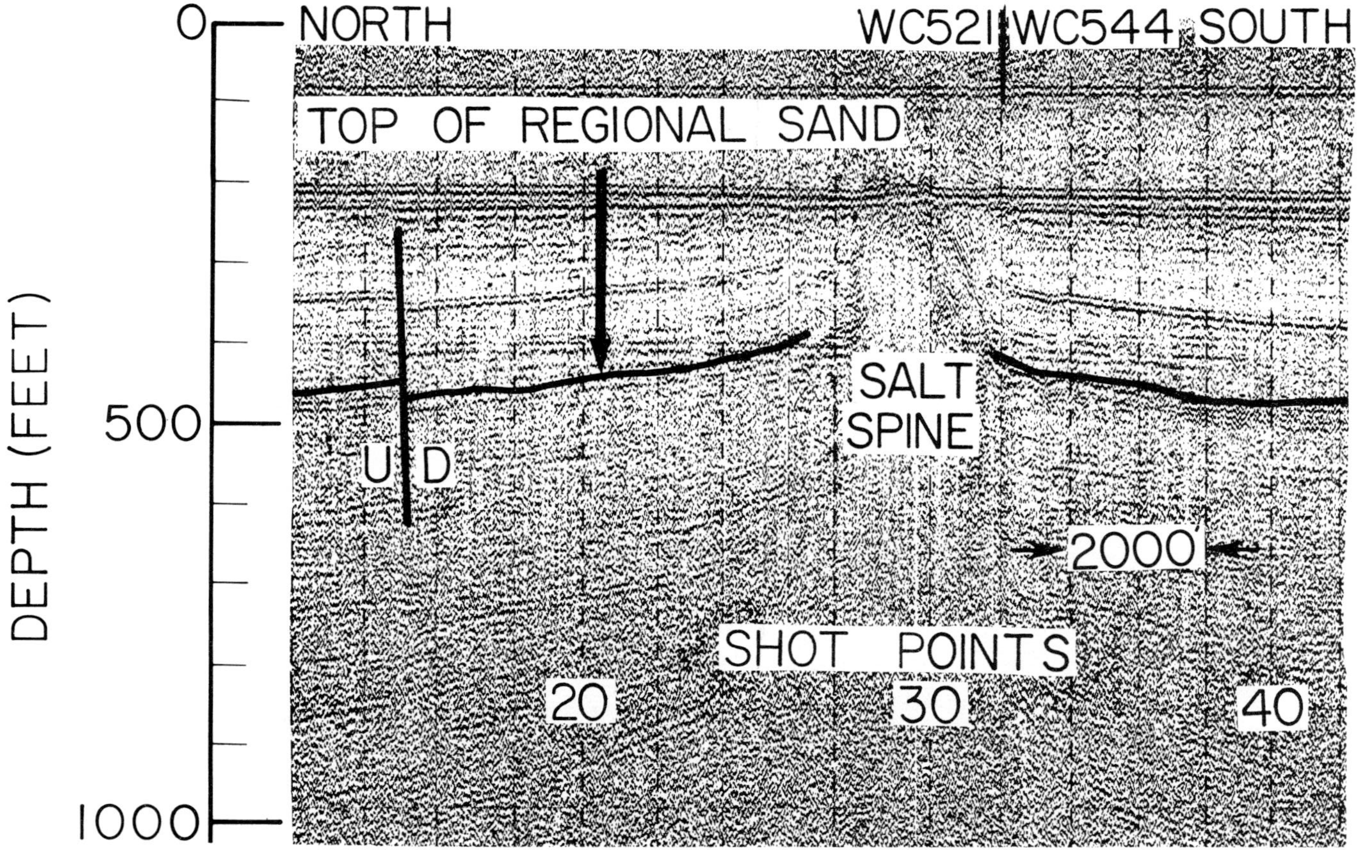

FIGURE 21. Air gun profile of West Cameron Area 521 and 544.

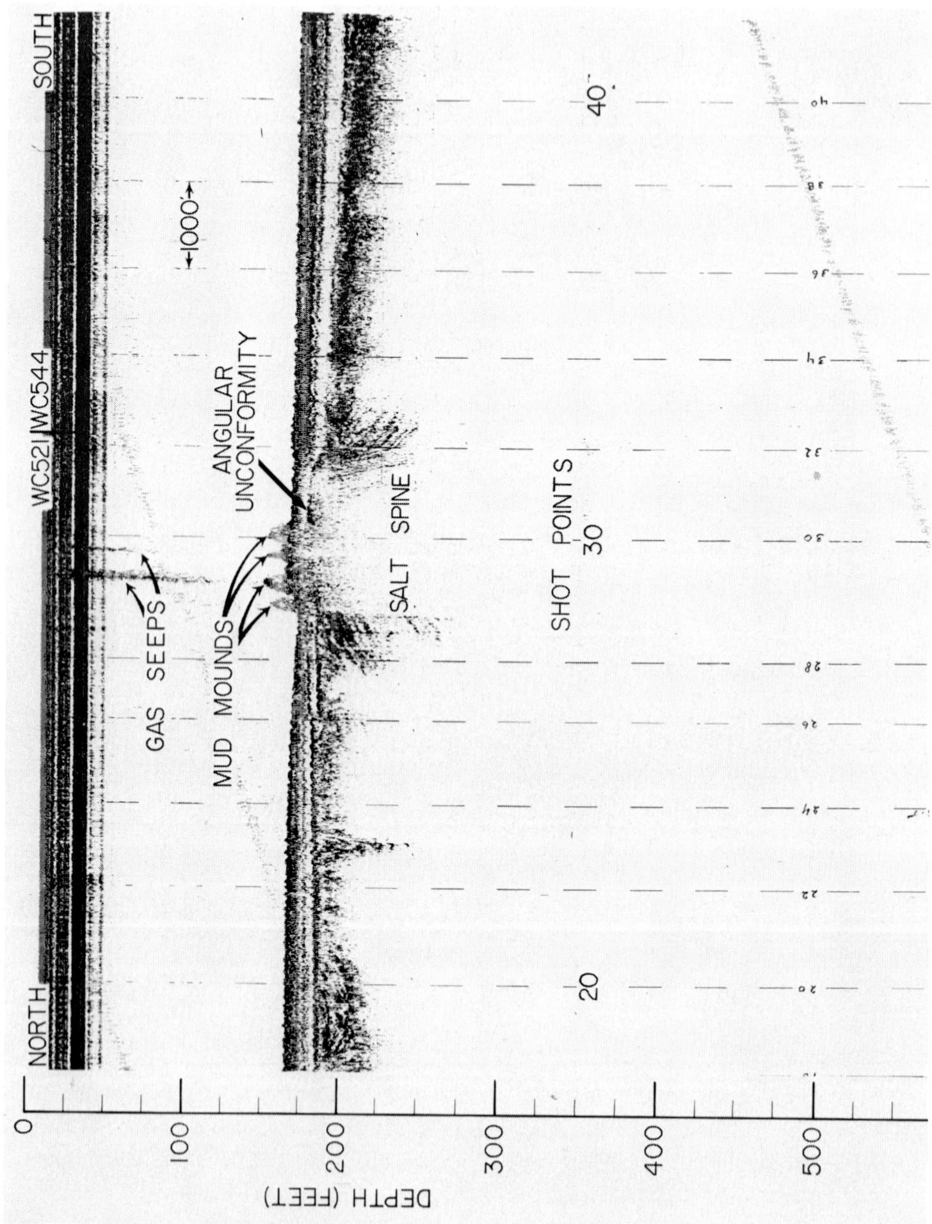

FIGURE 22. 3.5 kHz profile of West Cameron Area Block 521 and 544.

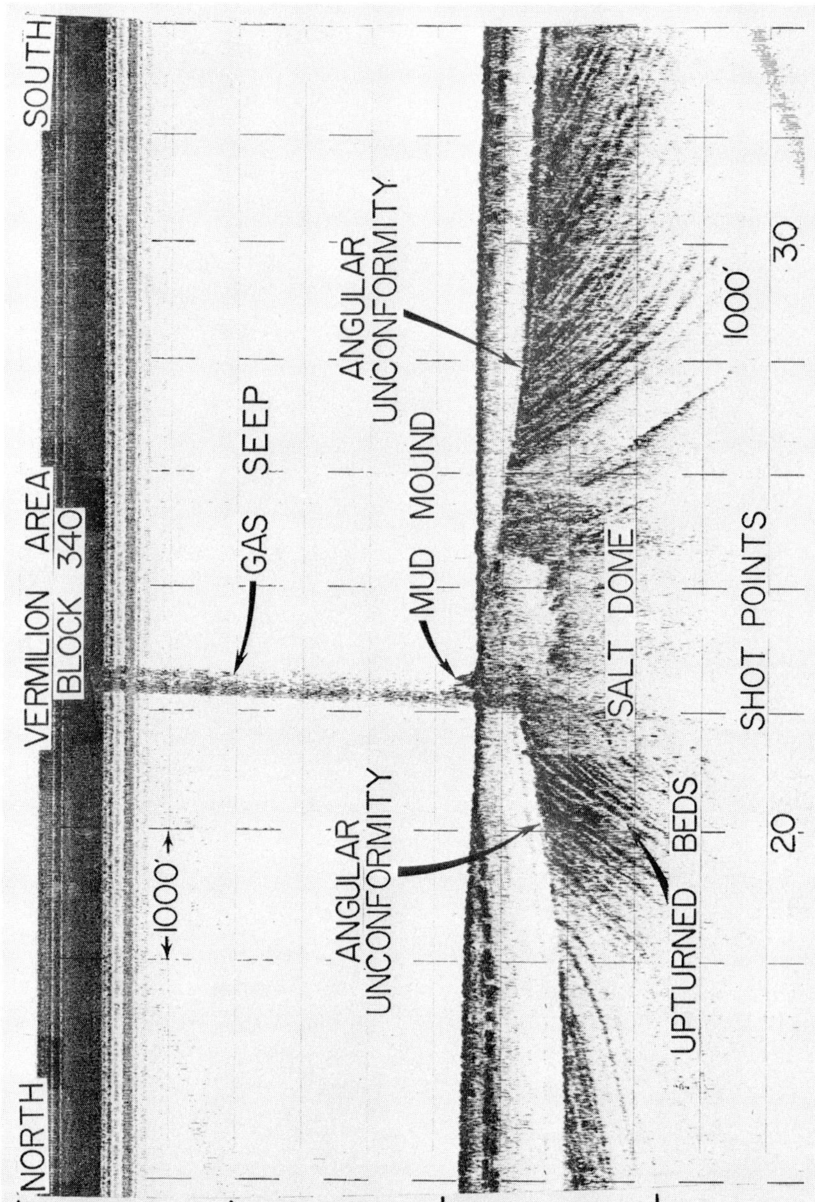

FIGURE 23. Air gun profile of Vermilion Area Block 340.

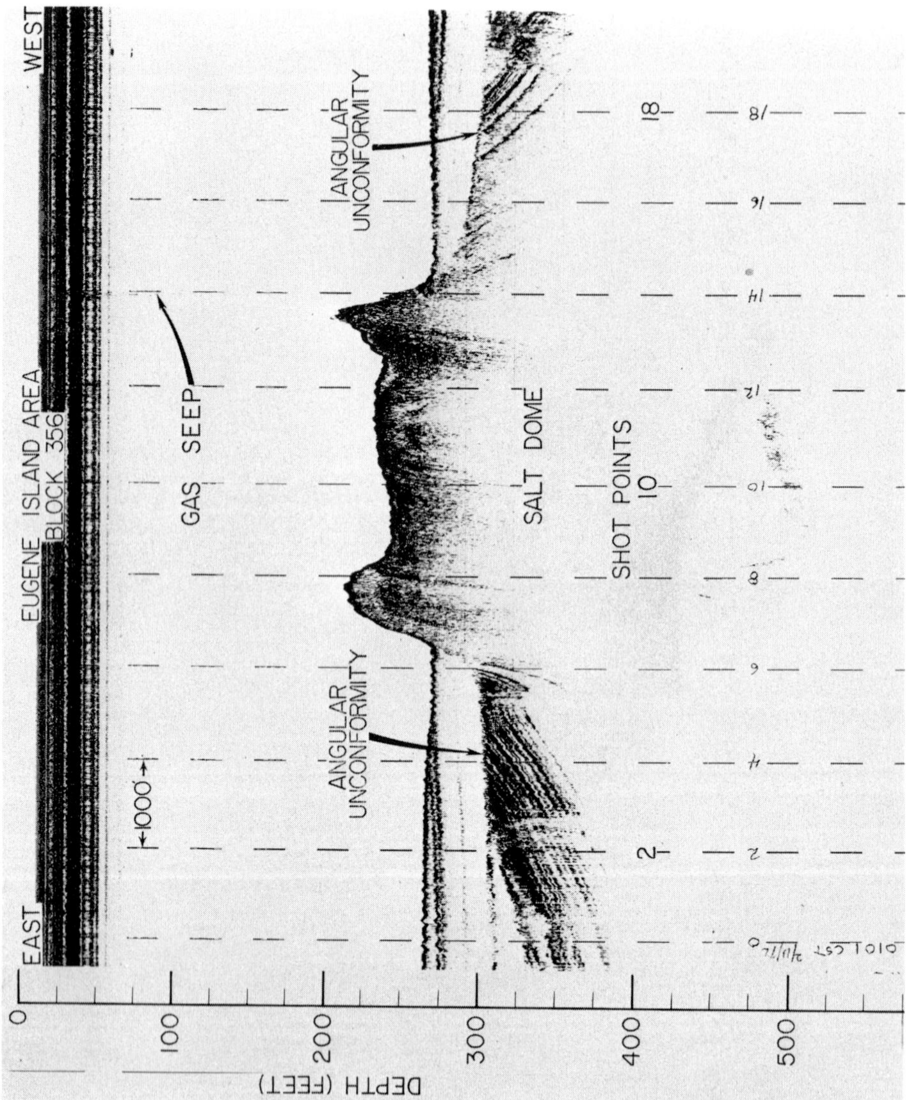

FIGURE 24. 3.5 kHz profile of Eugene Island Area Block 356.

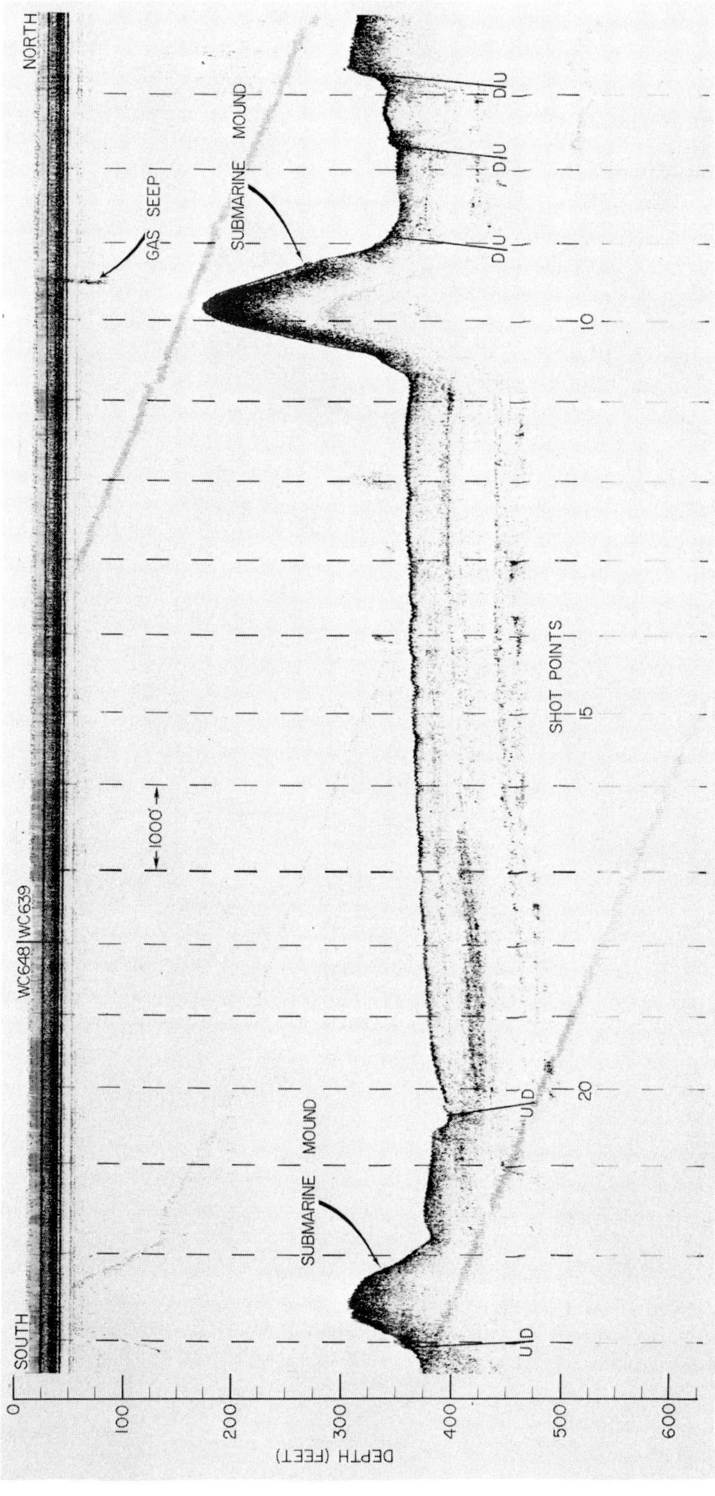

FIGURE 25. 3.5 kHz profile of West Cameron Area Blocks 648 and 639.

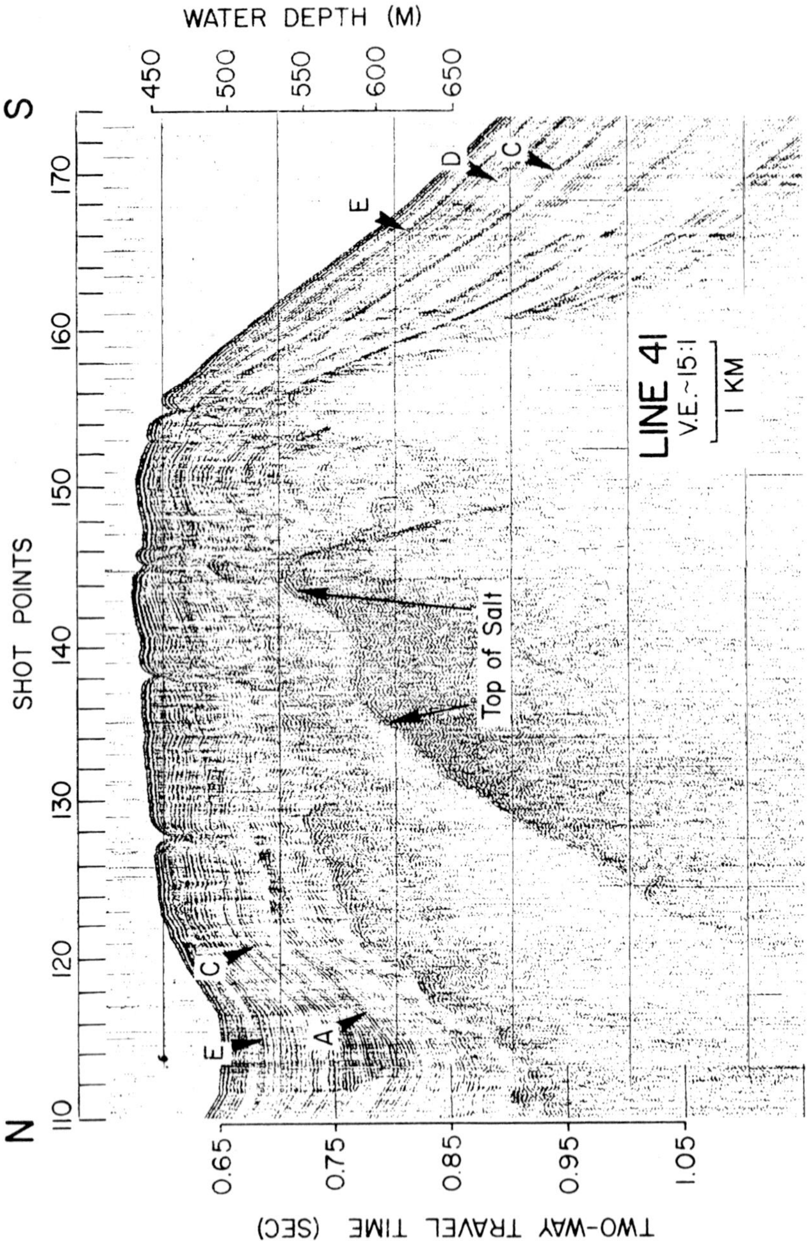

FIGURE 26. Mini-sparker profile of Line 41 off Texas Continental Slope. (From Sidner, B. R., Gartner, S., and Bryant, W., Rep. 77-5-T, Texas A&M University, College Station, 1977. With permission.)

of the sediment cut by a channel is most likely a desiccated sediment horizon, such as the top of the Beaumont clay along the Gulf Coast.

The 3.5 kHz seismic record of West Cameron Area Block 466 (Figure 27) is a typical example of a channel cut-and-fill structure. While channel cuts are not particularly geohazardous, they may indicate desiccated horizons. The very stiff clays associated with such a horizon represent a change in the fill sediment relative to the surrounding host material. Erosional channels filled with porous sand and gravel are further potential hazards to offshore drilling and have cost the oil industry millions of dollars. Circulation is often lost in these zones.

Eugene Island Area Block 298 and 281 3.5 kHz record (Figure 28) shows a large rectangular channel feature. The fill of this structure is most likely sand because no internal structures are present in the fill material. This lack of structure could also be interpreted as indicating gassy sediments, but the extension of a similar feature in other profiles parallel to the one illustrated indicates channeling and not gassy sediments.

A channel cut-and-fill structure that is exposed at the sea floor is shown on the sparker and 3.5 kHz seismic records of Vermilion Area Block 319 (Figure 29). This channel cut is in excess of 7000 ft wide.

It appears that most channel cuts are filled with a sandy sediment. In the Eugene Island Area Block 330 (Figure 30) a channel cut appears to have very straight sides, whereas in reality the channel is curved and most likely filled with sand. Sand has a high reflectivity and sediment strata below are not picked up due to attenuation of the seismic signal.

A very large cut-and-fill structure is shown on the Grand Isle Area Blocks 91 and 92 (Figure 31) air gun record. This cut-and-fill structure is the Old Mississippi Canyon located due southwest of the present-day Mississippi Delta. The erosional processes that created this structure were slumping and water currents.

It is interesting to note that during the 1970s a major archeological constraint was placed on drilling activity in the vicinity of channel cuts because of the possibility of damage to archaeological sites on the banks of an ancient river. During lower sea level the Indians that lived on what is now the present-day continental shelf, favored the banks of old rivers as living places. The major concern of channel cuts in offshore activity is that the infilling sediment most likely represents a different type of sediment than the surrounding material. This is illustrated in Figures 32 where the infilling sediments are sands while the channel was cut into clay sections.

GAS-CHARGED SEDIMENTS

Gas-charged sediments present two problems to offshore activity. The buildup of gas within a sediment increases its excess pore water pressure, which in turn decreases the effective stress of the sediment. A decrease in the effective stress may lead to foundation failure. Gas-charged sediments penetrated by a drill string or foundation pile may cause a blowout, resulting in the loss of a platform or other such structure. The sediments in the continental shelf and slope of the northern Gulf of Mexico are extremely gassy. This is particularly true near the Mississippi Delta area.

Gas-charged sediments may be identified on a seismic record by the appearance of a high amplitude reflecting horizon or an opaque or wipe-out zone. On the air gun record of Eugene Island Area Block 330 (Figure 33) a gassy sediment zone is delimited by a dark seismic horizon between two faults. On the air gun record of Eugene Island Blocks 338 and 330 (Figure 34) gas zones appear as fuzzy nonstructured dark patches on the seismic record. Acoustic voids or wipe-out zones designate the gassy sediments associated with gas vents on the West Cameron Block 629 and 638 (Figure 35) a 3.5

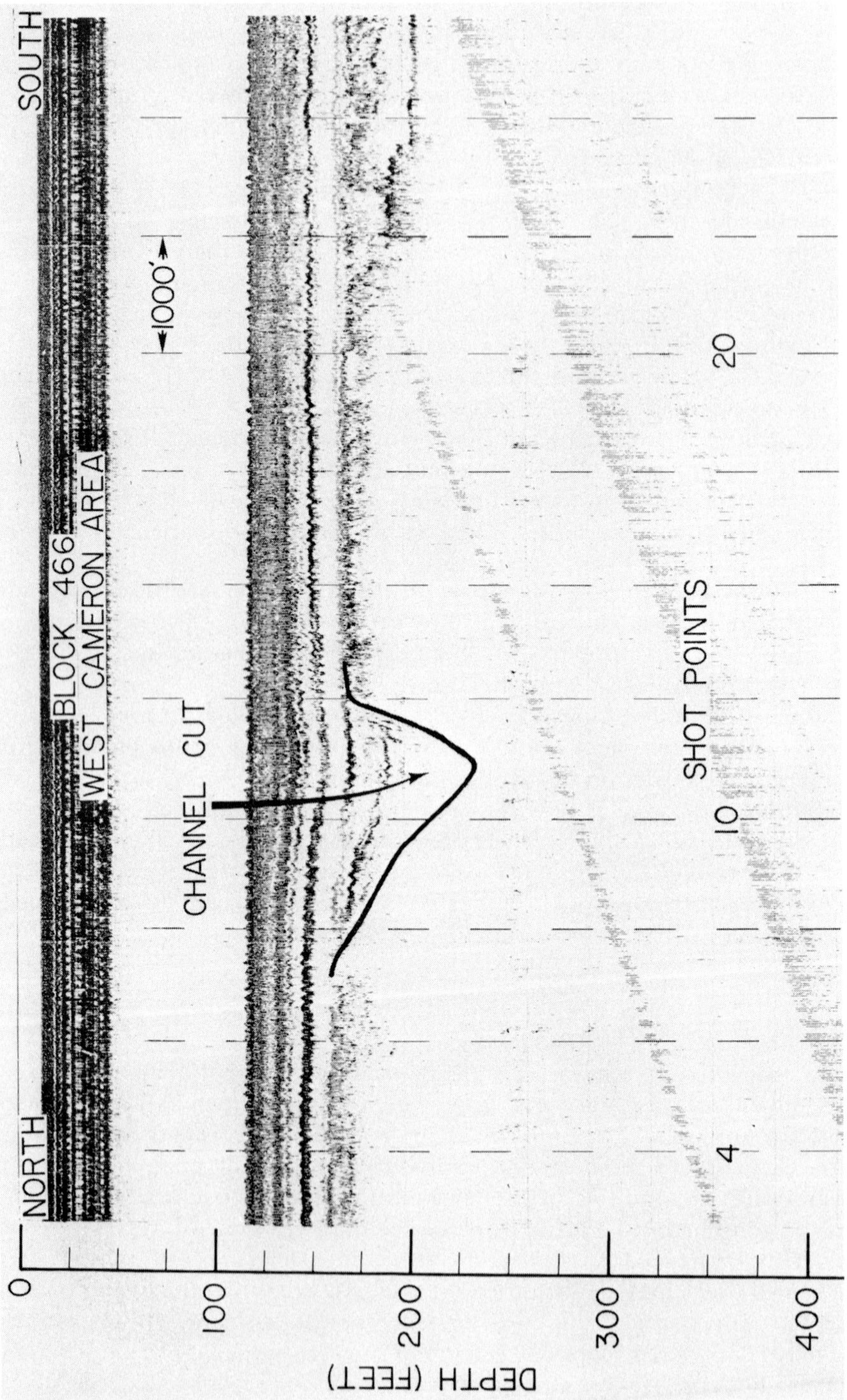

FIGURE 27. 3.5 kHz profile of West Cameron Area Block 466.

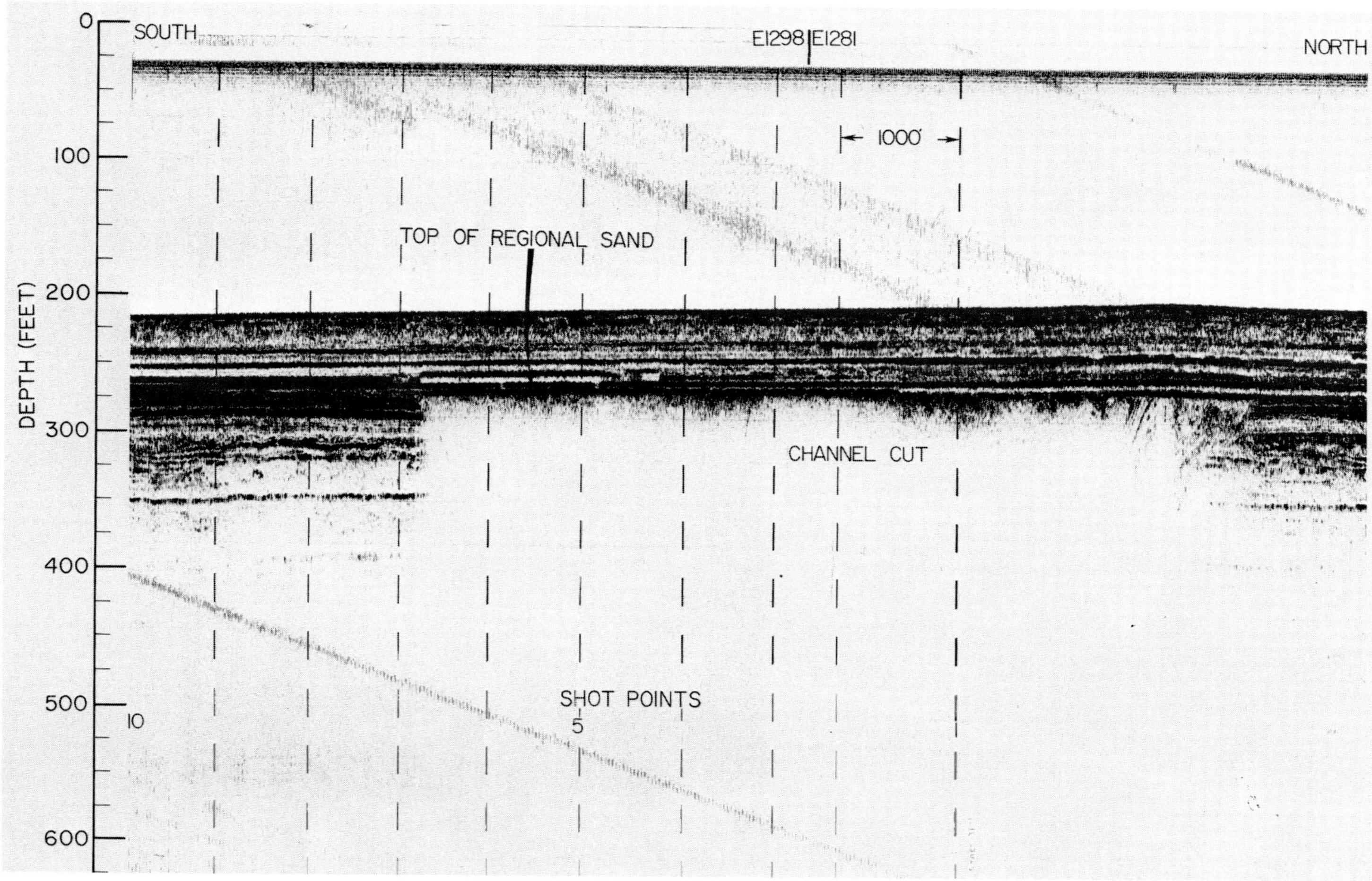

FIGURE 28. 3.5 kHz profile of Eugene Island Area Block 298 and 281.

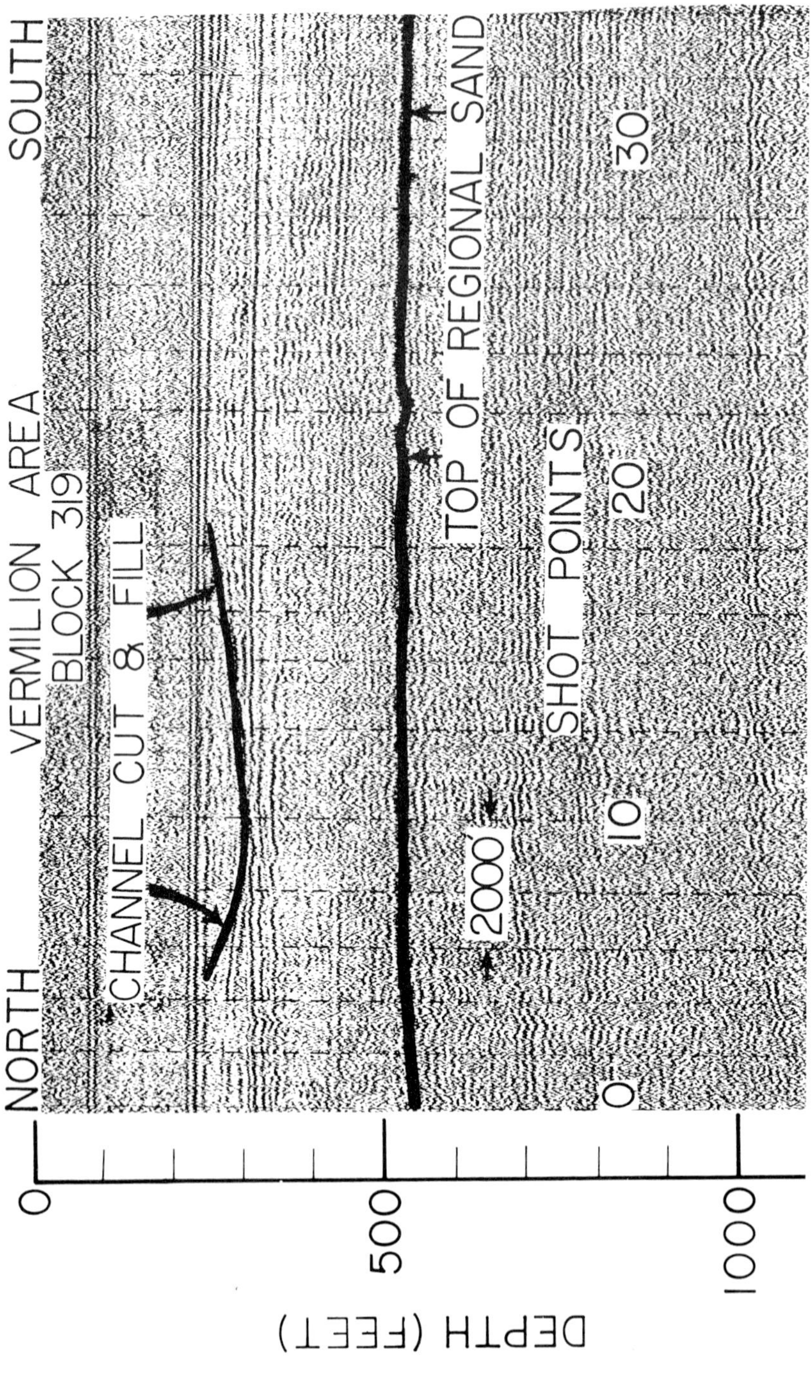

FIGURE 29. Air gun profile of Vermilion Area Block 319.

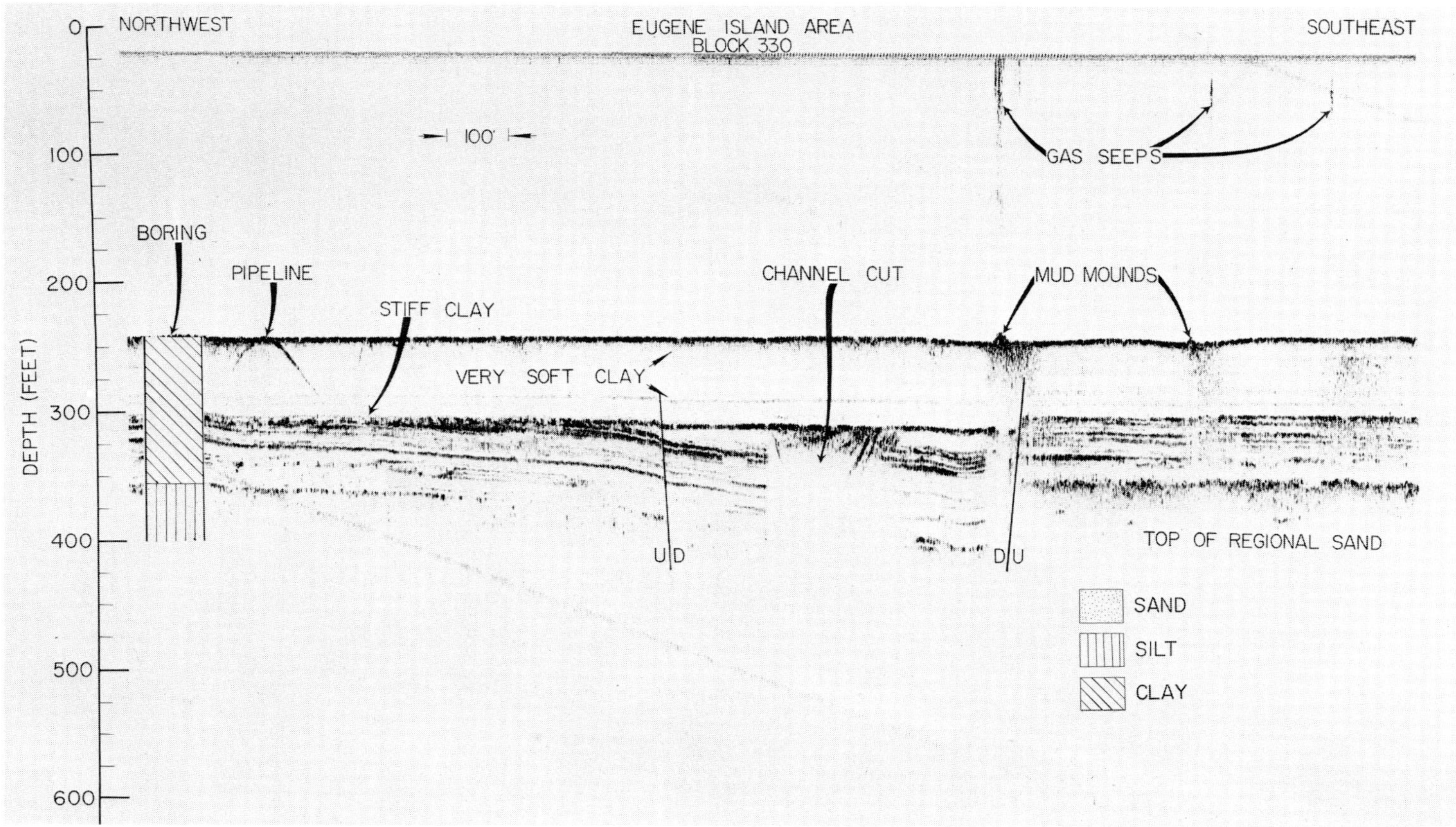

FIGURE 30. 3.5 kHz profile of Eugene Island Area Block 330.

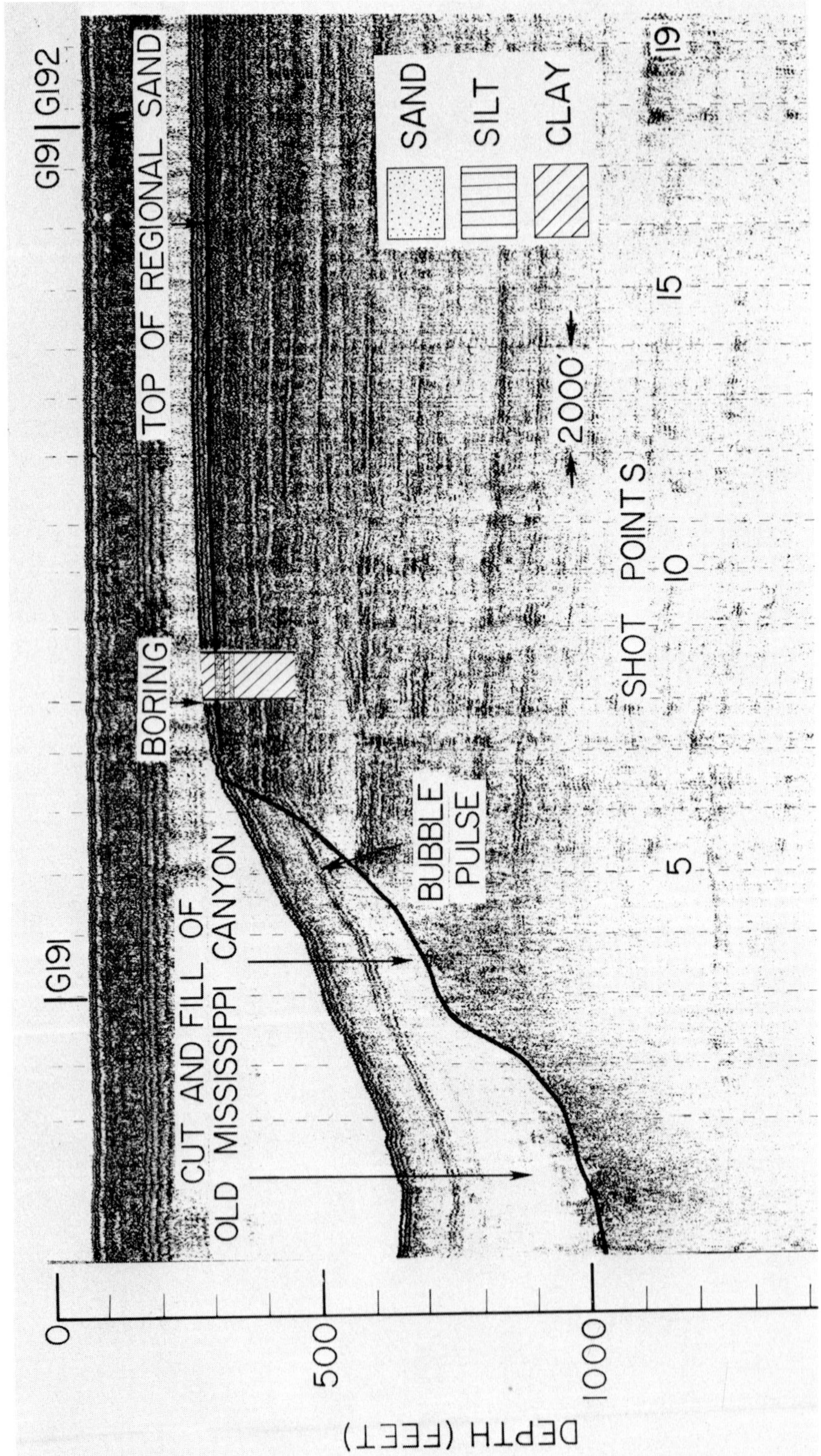

FIGURE 31. Air gun profile of Grand Isle Area Block 91 and 92.

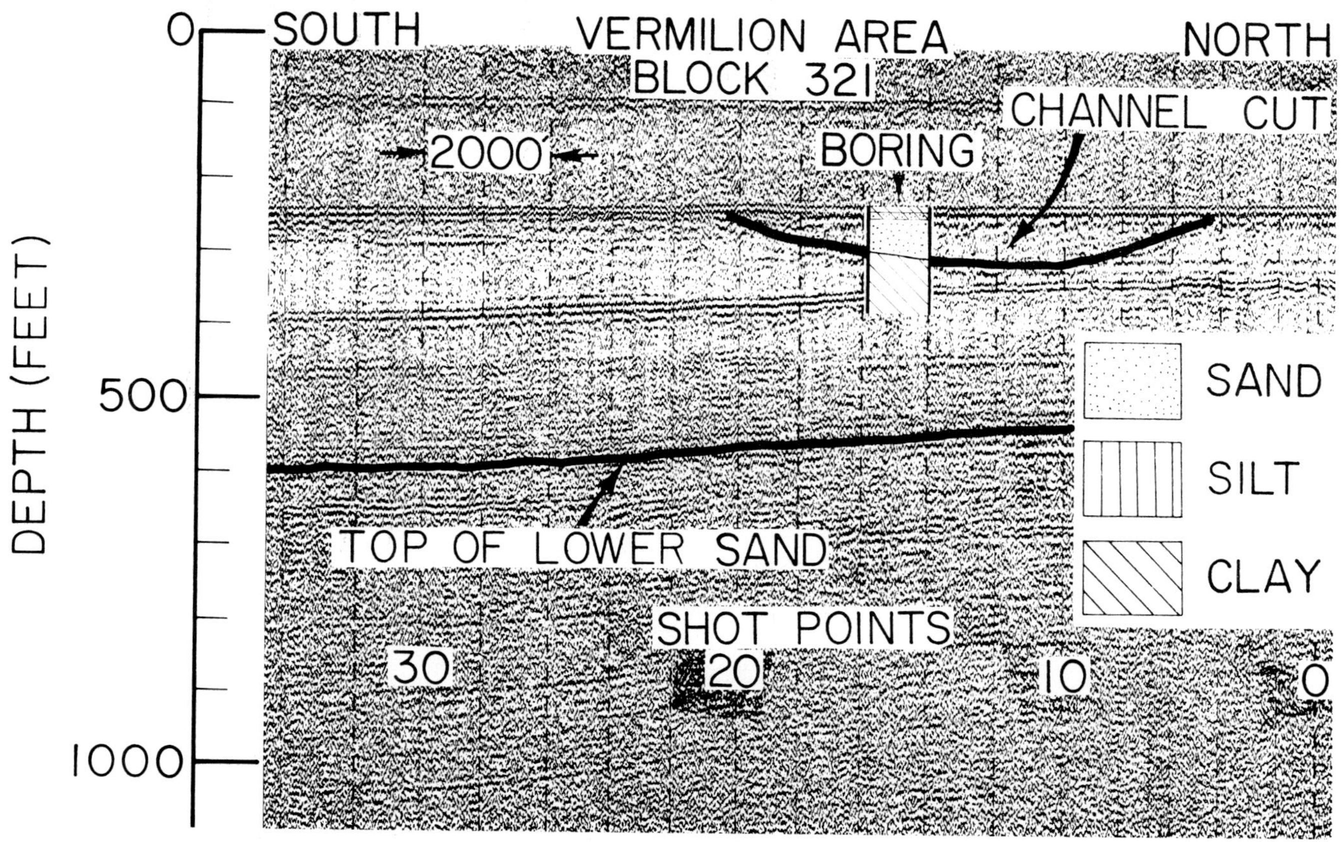

FIGURE 32. Air gun profile of Vermilion Area Block 321.

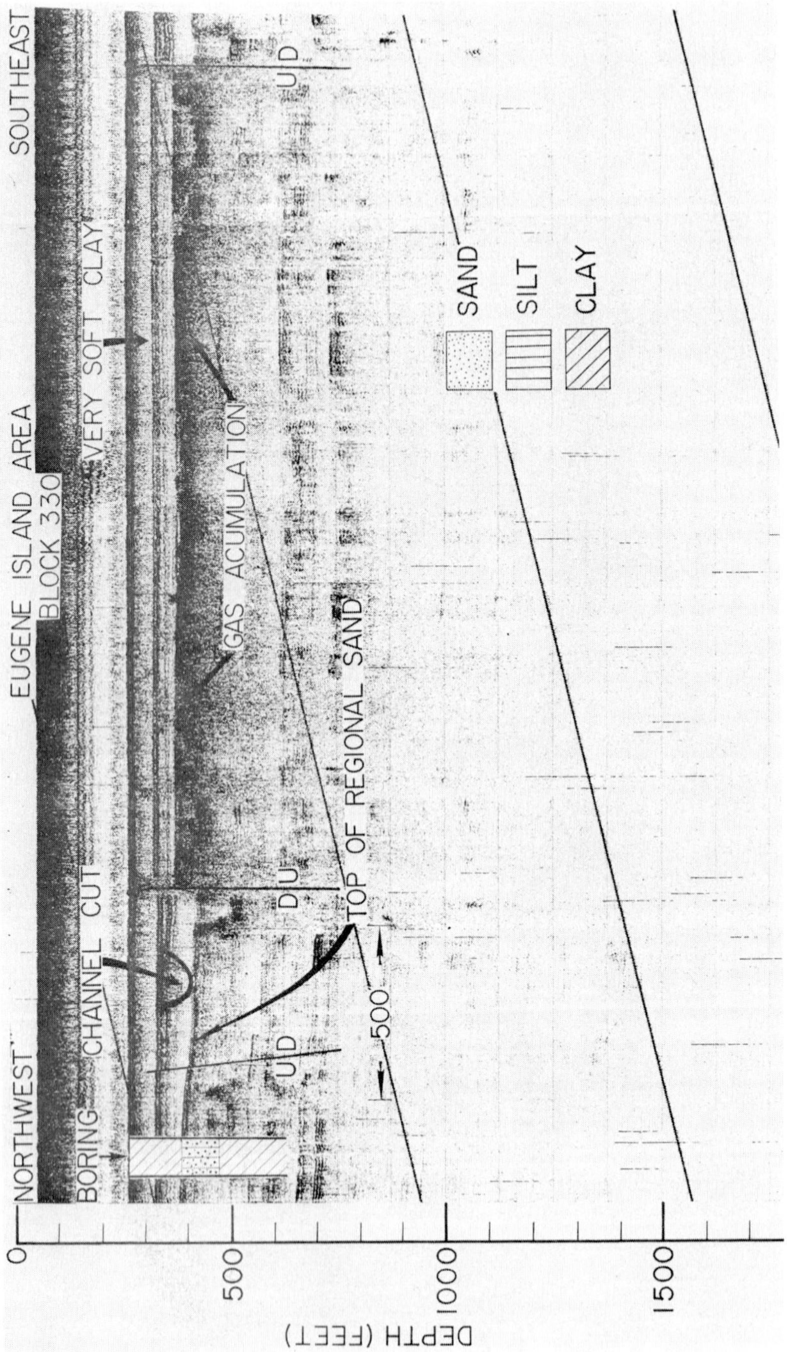

FIGURE 33. Air gun profile of Eugene Island Area Block 330.

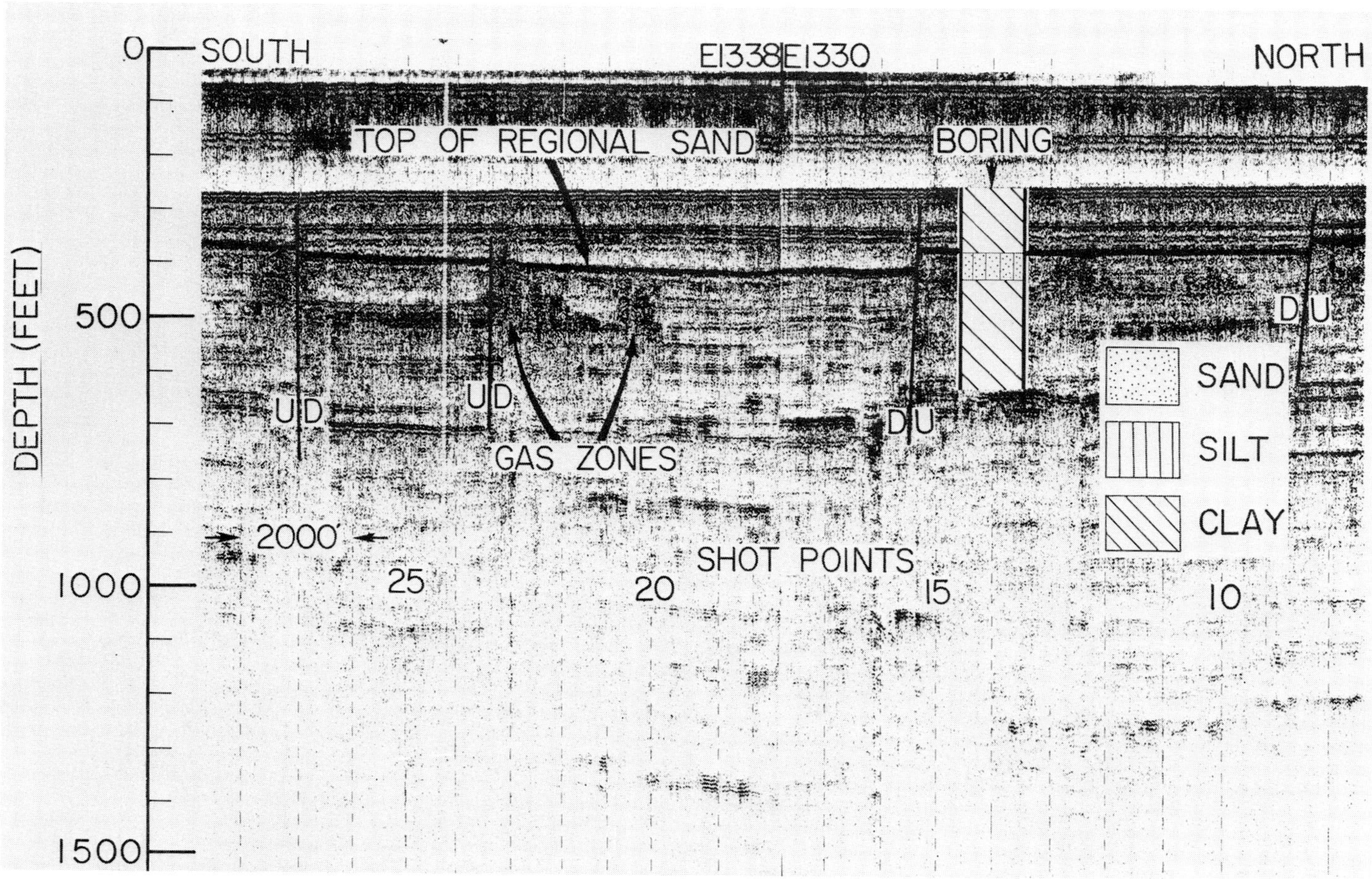

FIGURE 34. Air gun profile of Eugene Island Area Block 338 and 330.

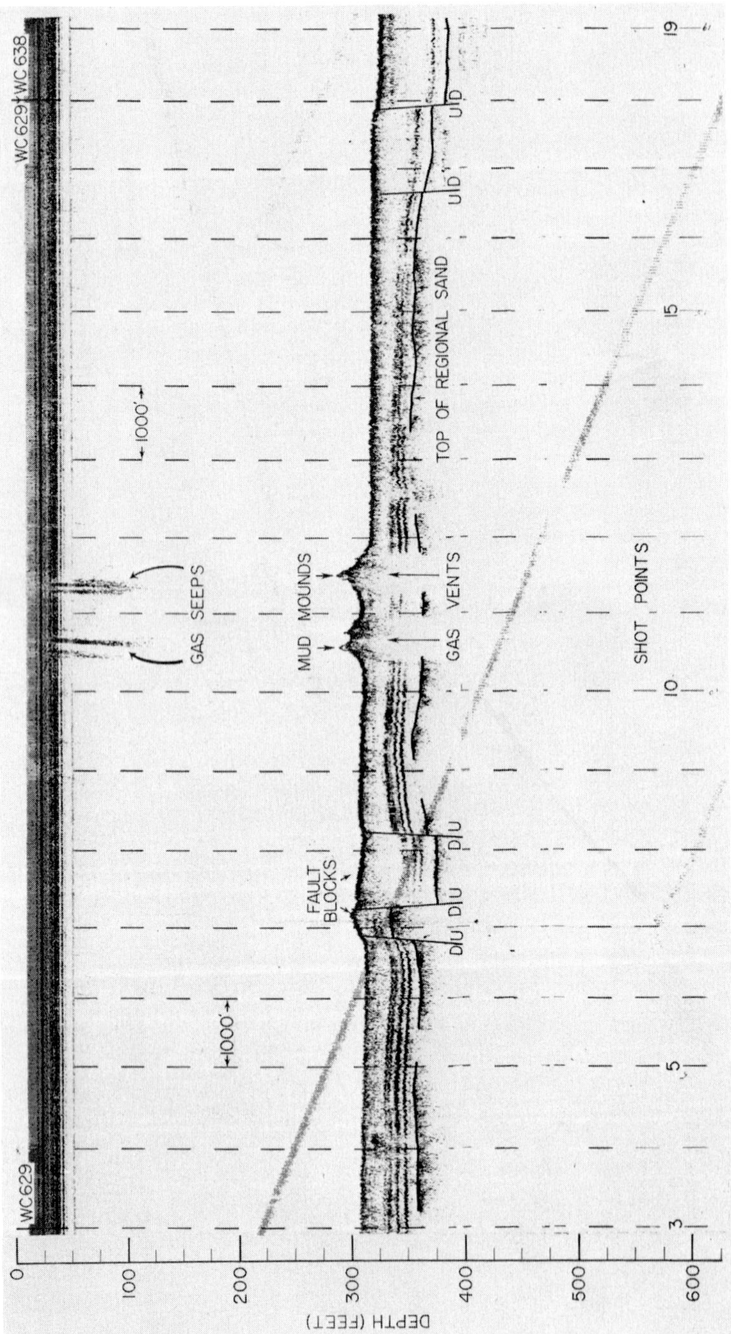

FIGURE 35. 3.5 kHz profile of West Cameron Area Block 629 and 638.

kHz seismic record. Numerous acoustic voids indicate gas zones in the 3.5 kHz seismic record of East Cameron Blocks 263 and 280 (Figure 36).

Mud mounds not associated with a diapiric structure are usually associated with gassy sediment in the underlying sections (west Cameron 629 and 638) (Figure 37). Mud mounds associated with a diapiric structure at depth is shown in Figure 38 of the Eugene Island Area, Block 322 and 323. Faulting, gas venting, and gas seeps in the overlying water column are all associated with petrogenic gases formed at great depths that have escaped via the faults in the area. Figure 39, a profile in Area Block 330 of the Eugene Isle area, shows a 150 ft section of clay on top of a regional sand. Two large gas seeps are in the western portion of the profile. These seeps are not associated with any faults or other structures but appear to emanate from a well-stratified clay section. The seeps are fairly massive and suggest they may be coming from the nearby pipeline. Pockets of gassy sediment are found above the sand layer in a fairly restricted zone within the clay section.

An interesting profile in Block 142 is shown in Figure 40. This seismic record is similar to cross-sectional views associated with salt withdrawal basins. The 3.5 kHz records in this area show hundreds of gas seeps present throughout the water column. Discontinuities in layering are caused by the vertical migration of gas. The gas appears to be trapped above the regional sand layer, where it accumulates until it is released along fault planes.

Figure 41 shows pockets of gas-charged sediments associated wtih the top of a regional sand layer that is offset by a fault that extends to the sea floor. A series of transgressive bars are above the gassy sediment zone. In Figure 42 the top stratum, penetrated by a boring, consists of a 140-ft section of recent marine clay with intermittent silk partings. The upper clay has a water content of about 60% dry weight, in marked contrast to the older, deeper lying, stiff clays where the water content has been reduced to less than 30%. The lower lying clays also have a greater shear strength. A well-defined gassy sediment zone appears at middepth within the profile. These gassy sediments smear the underlying record, which was not indicated as gassy from the borehole information.

A profile taken in the 186 Block of Grand Isle, Figure 43, shows that a velocity difference between gas-charged sediments and the adjoining sediments may create seismic anomalies such as below some gas pockets where the beds are downturned. Although this area is controlled by slumping and the folding of sediments, a question remains as to what effect the velocity contrast between the gas pockets and adjoining sediments has on the structure. Figure 44 shows the escape of gas directly up a fault and into the overlying water column. Shallow gas zones are also prevalent in areas surrounding near-surface piercement salt domes. The abundance of gas in these areas is evidenced by the numerous seeps in the water column. Problems have often been encountered in drilling through angular unconformities on the flanks of the domes. An example is seen in Figure 45 where a blowout occurred (Shot Point 19) in drilling through a shallow angular unconformity after the drilling contractor greatly increased the specific gravity of the drilling mud which broke through to a gas zone. The well blew out for over 6 hr. The danger of gas-charged sediments is not due entirely from the explosive nature of the gas (usually methane) but mostly from the extremely high pressures generated during the gas forming process.

Gas-charged sediments are also found in the continental slope area. Mini-sparker line 44 (Figure 46) shows a section in 350 m of water. Well-stratified sediments are present in the upper section of the record while a chaotic facies is found in the lower one. The dividing line between these two sections is well delineated. Examination of cores from a boring through these two sections showed no change in lithology but did

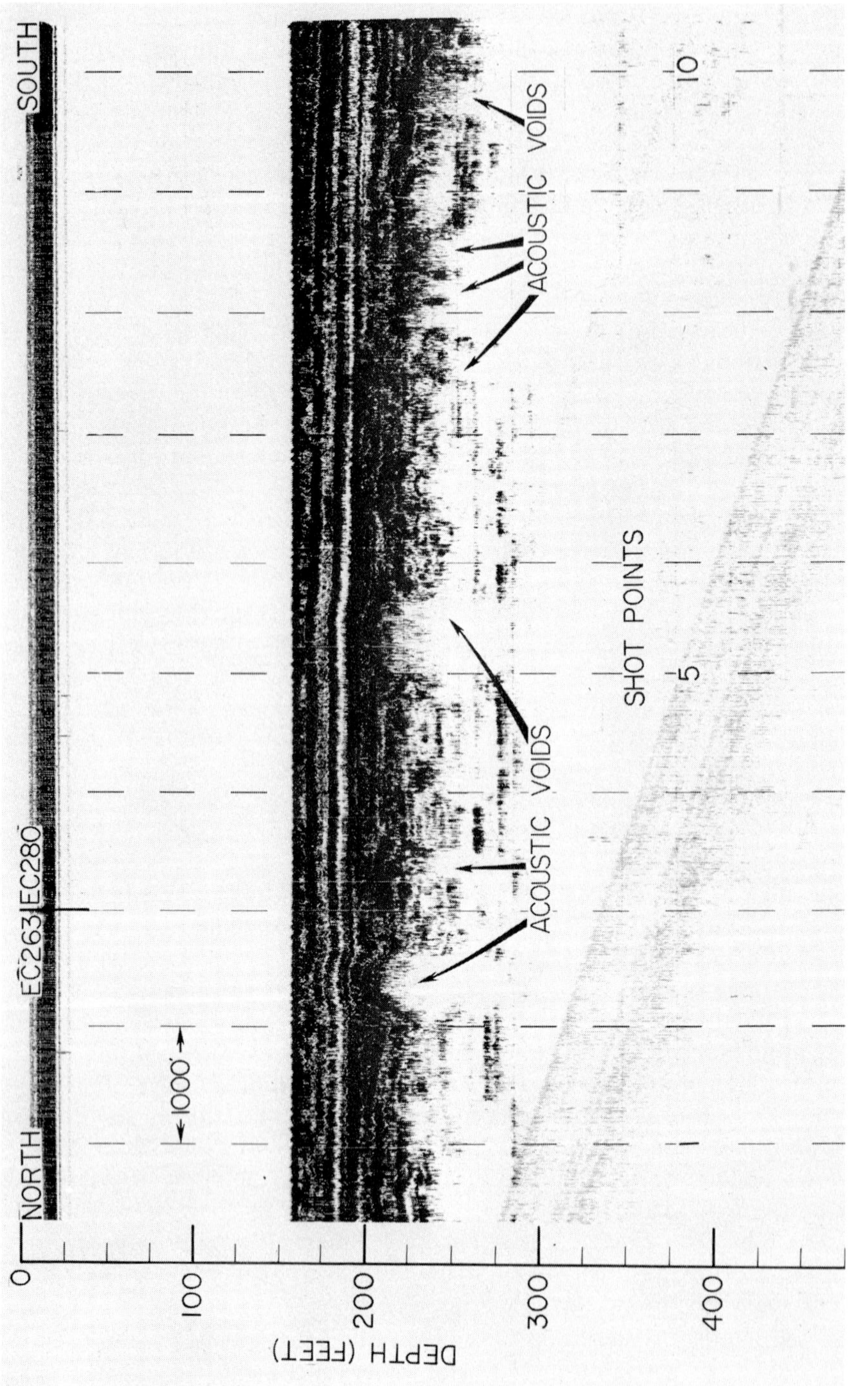

FIGURE 36. 3.5 kHz profile of East Cameron Area Block 263 and 280.

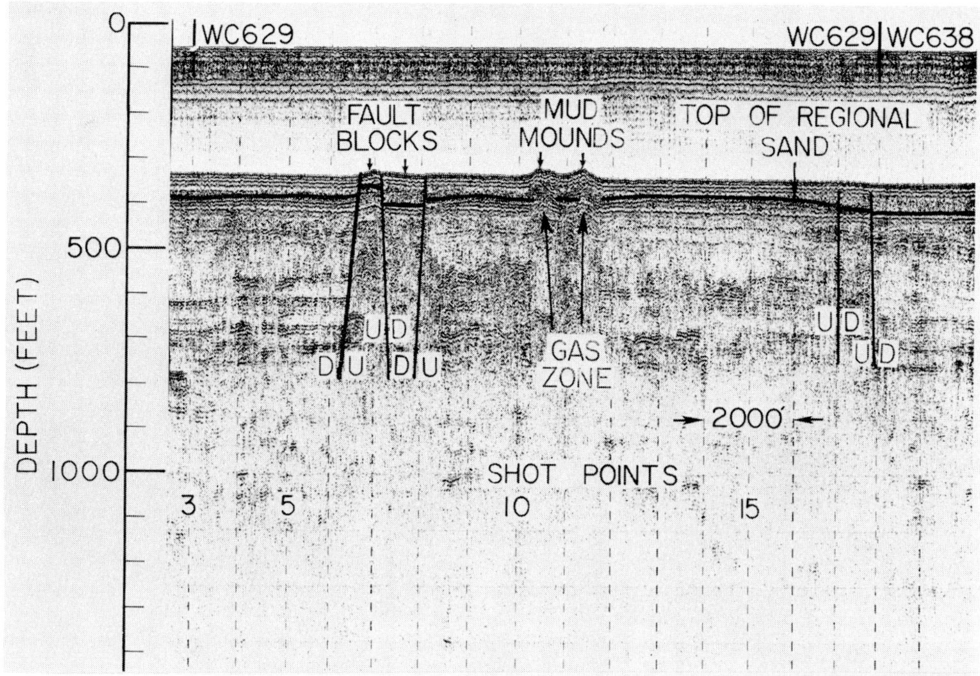

FIGURE 37. Air gun profile of West Cameron Area Block 629 and 638.

show that the sediments in the chaotic facies section had a degassed fabric. A continuation of line 44 downdip (Figure 47) shows that the well-stratified upper section is in the process of becoming gas charged. The difference between the gas-charged sediments of the chaotic facies and the zone designated as gas on the downdip record is the degree of gas present in the sediment. In the wipe-out zone on the record the concentration of gas is higher than in the chaotic facies. Not all acoustic voids or wipe-out zones on a seismic record indicate the presence of gas-charged sediments. Thick sand bodies and oyster reefs will also appear as wipe-out zones.

BLOWOUTS

Gas-charged sediments can and do cause blowouts. A good example of a shallow blowout occurred in the West Cameron Area Block 564 (Figure 48). A foundation boring was being drilled at the position shown on the sparker record (Shot Point 10) of Block 564 at the indicated position. At a depth 350 ft below the sea floor the drillers hit the gas-charged sediment zone indicated on the record. The resulting blowout did not damage the drilling platform, but the tools came out of the hole and it was necessary to delay drilling until the gas had ceased blowing.

The air gun record of South Pass Area Block 78 (Figure 49) shows the results of a large blowout where a platform was destroyed. The crater resulting from the blowout was several thousand feet in diameter and almost 100 ft deep. A large amount of gas was escaping from the crater when the profile was taken. The drilling rig hit a gas pocket at a subbottom depth of less than 700 ft. The gas ignited and the rig collapsed, sinking to the bottom of South Pass Block 78. The 3.5 kHz profile shown in Figure 49 was obtained 5 days after the gas blowout occurred. A crater over 2000 ft wide is seen on this crossing. The water column is literally covered with natural gas seeps over

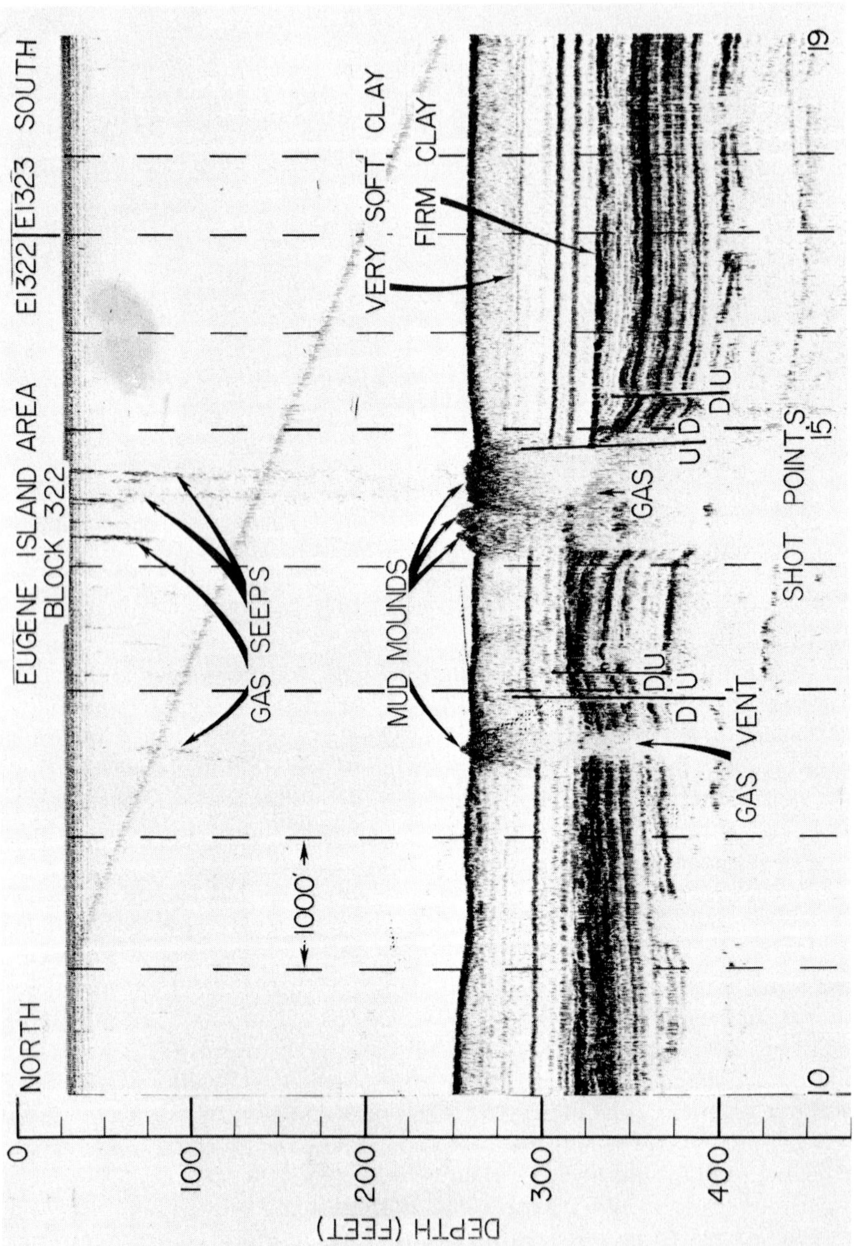

FIGURE 38. 3.5 kHz profile of Eugene Island Area Block 322 and 323.

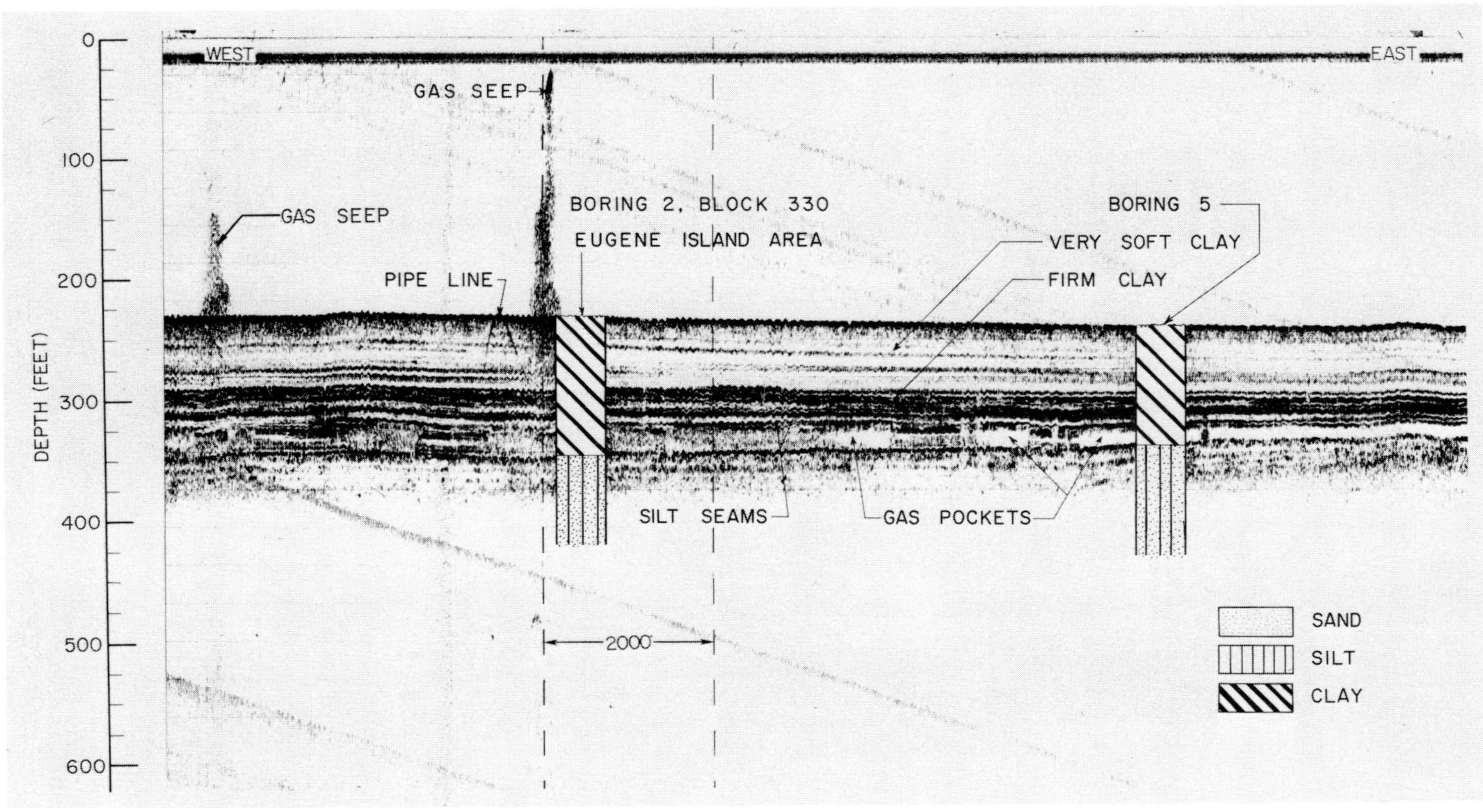

FIGURE 39. 3.5 kHz profile of Eugene Island Area Block 330.

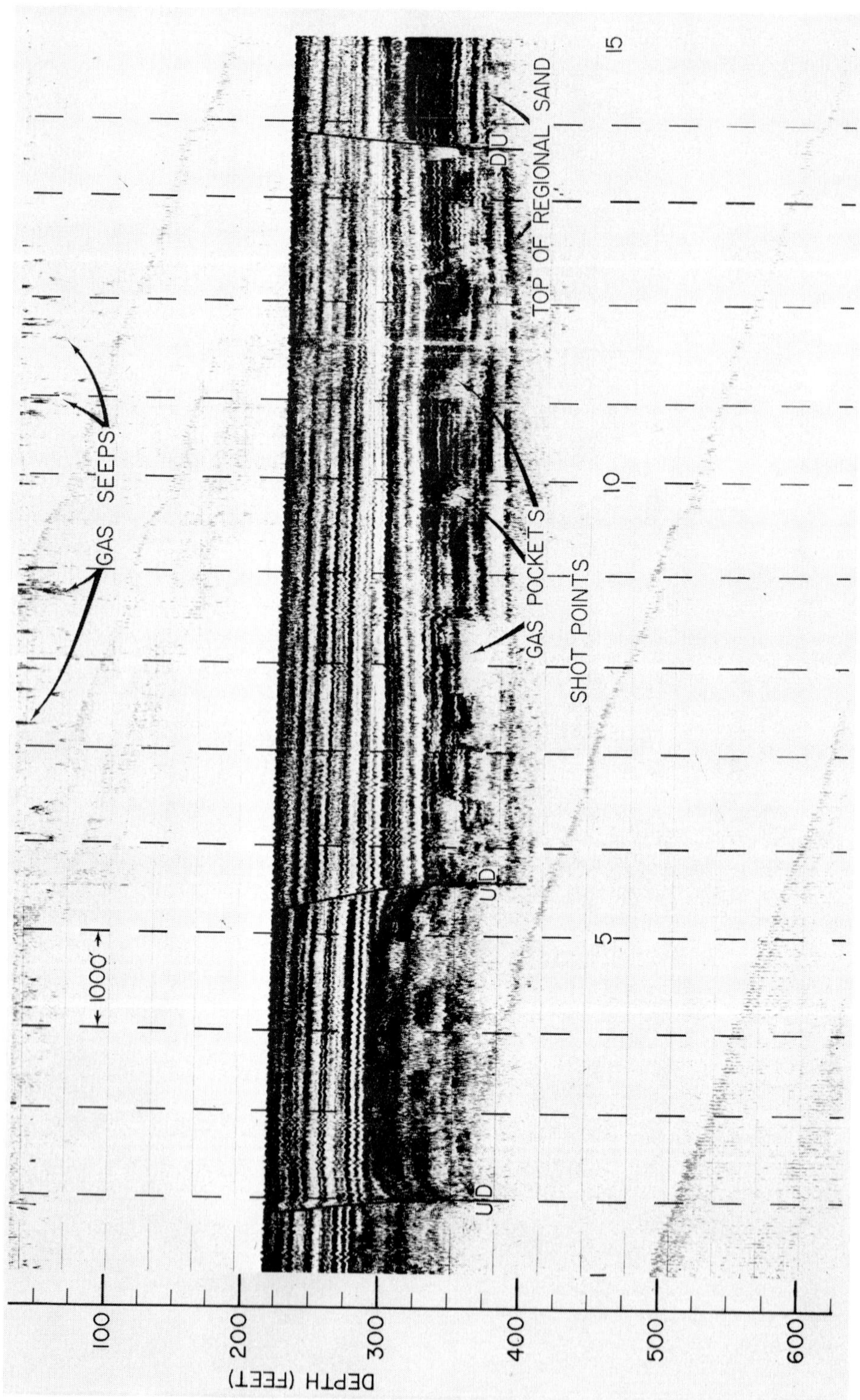

FIGURE 40. 3.5 kHz profile of South Marsh Island 4, Area Block 42.

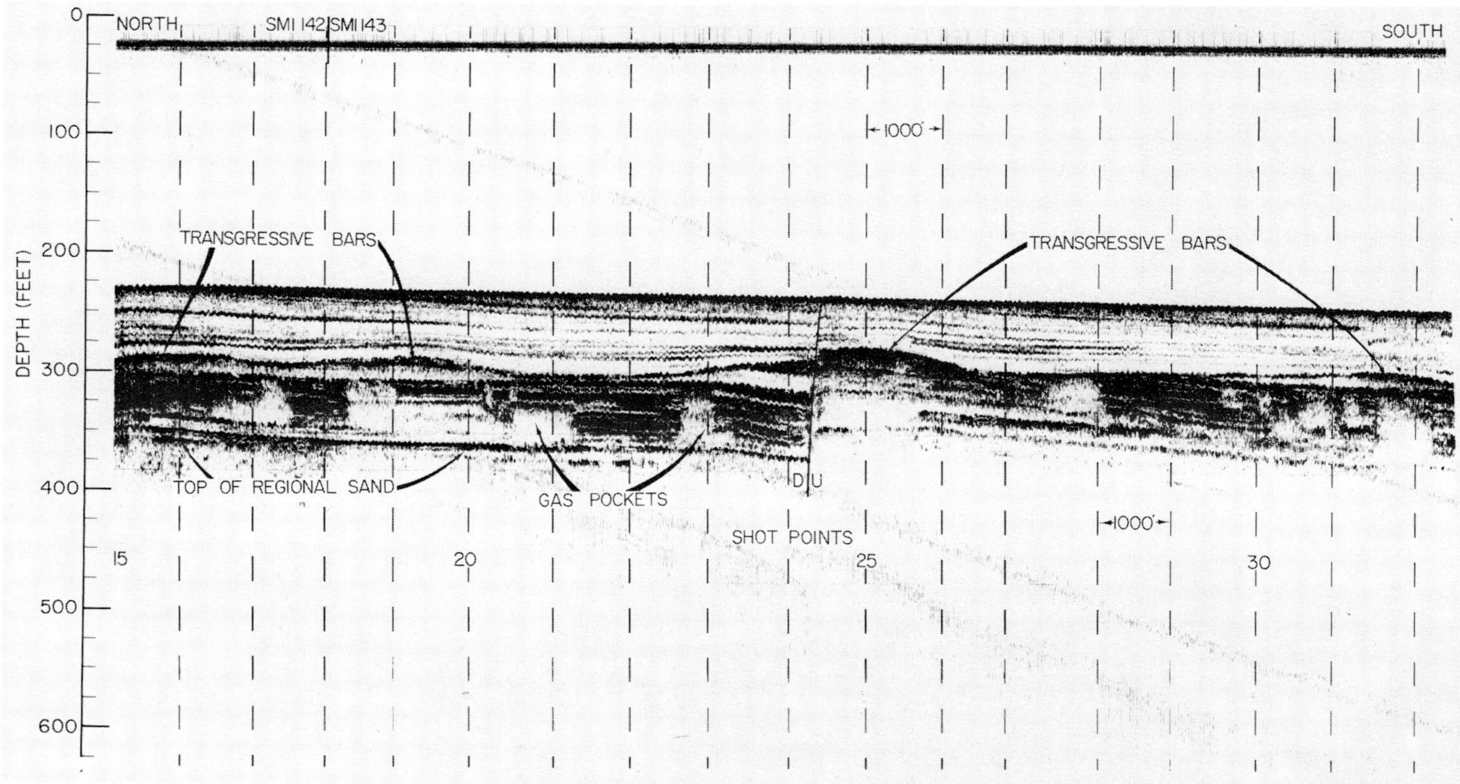

FIGURE 41. 3.5 kHz profile of South Marsh Island Area Block 142 and 143.

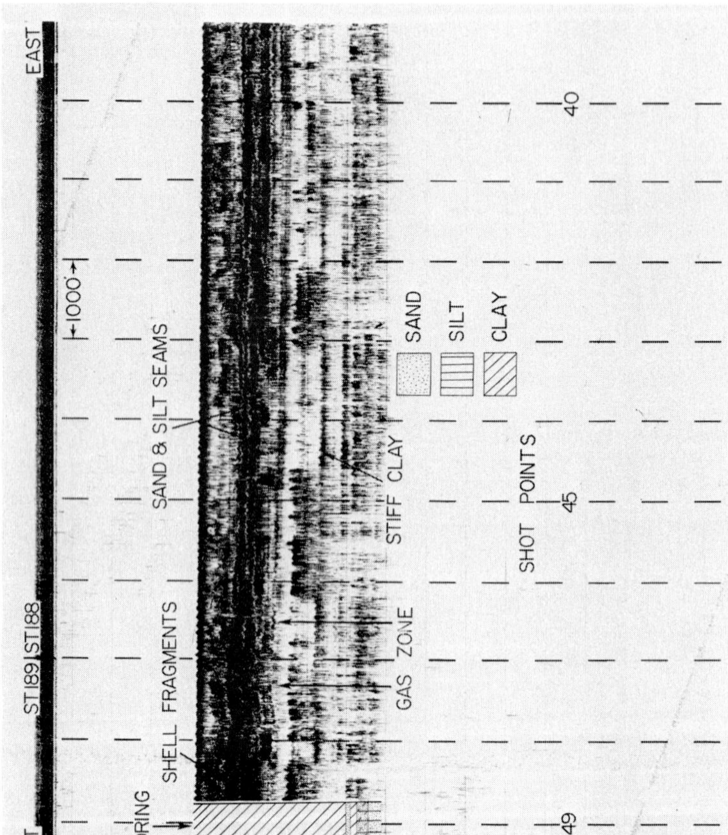

FIGURE 42. 3.5 kHz profile of South Timbalier Area Block 189 and 188.

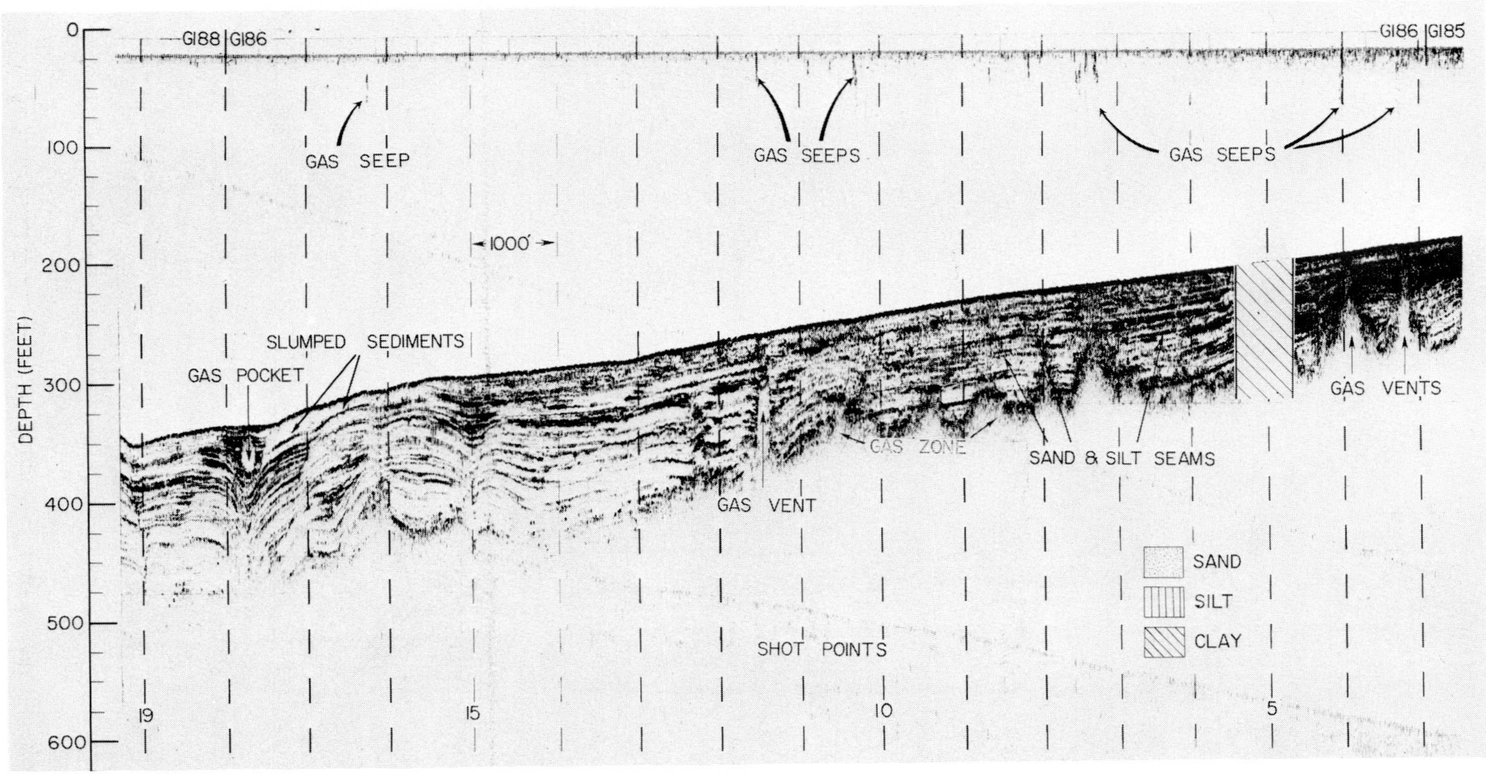

FIGURE 43. 3.5 kHz profile of Grand Isle Area Block 188 and 186.

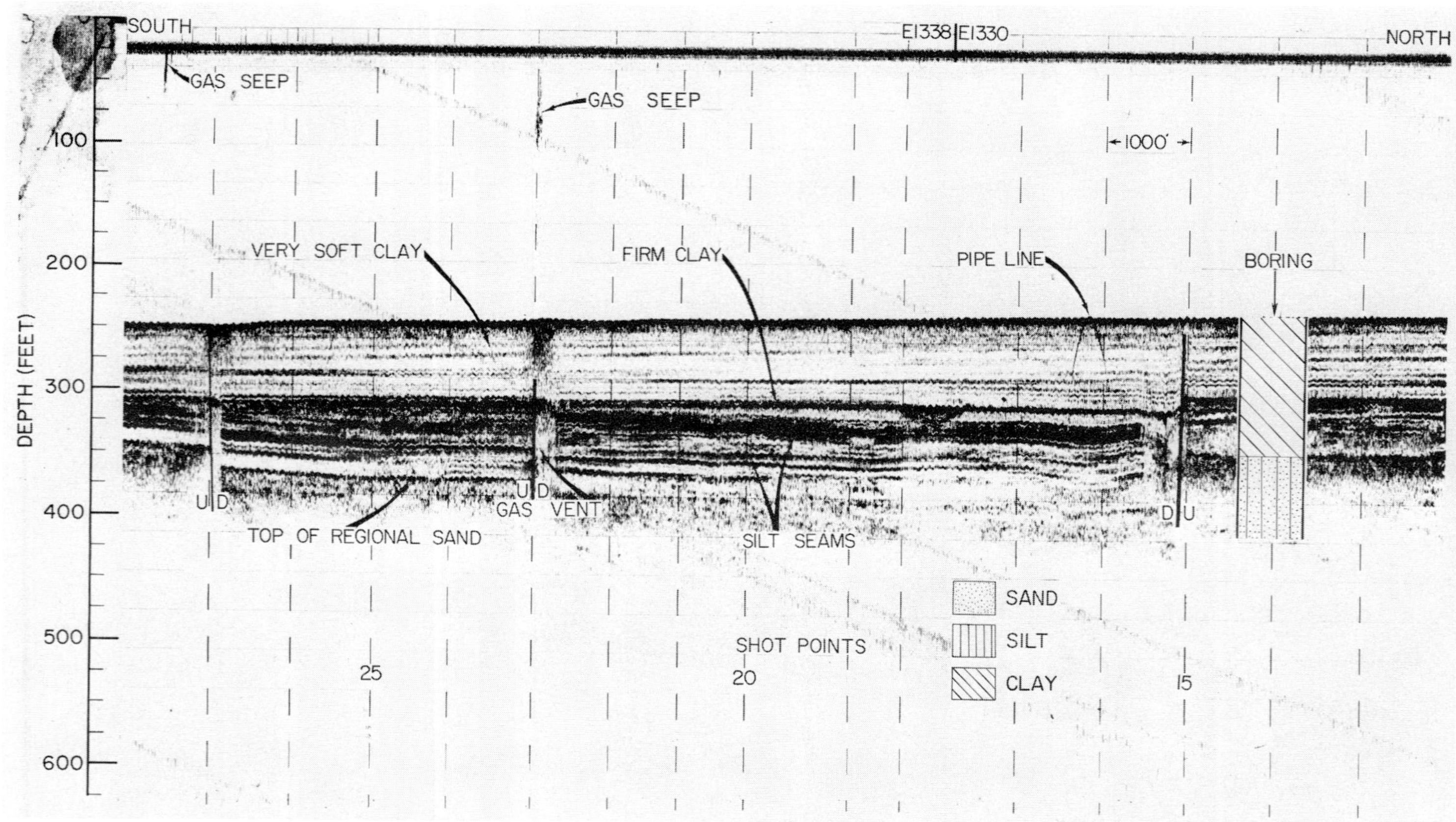

FIGURE 44. 3.5 kHz profile of Eugene Island Area Block 338 and 330.

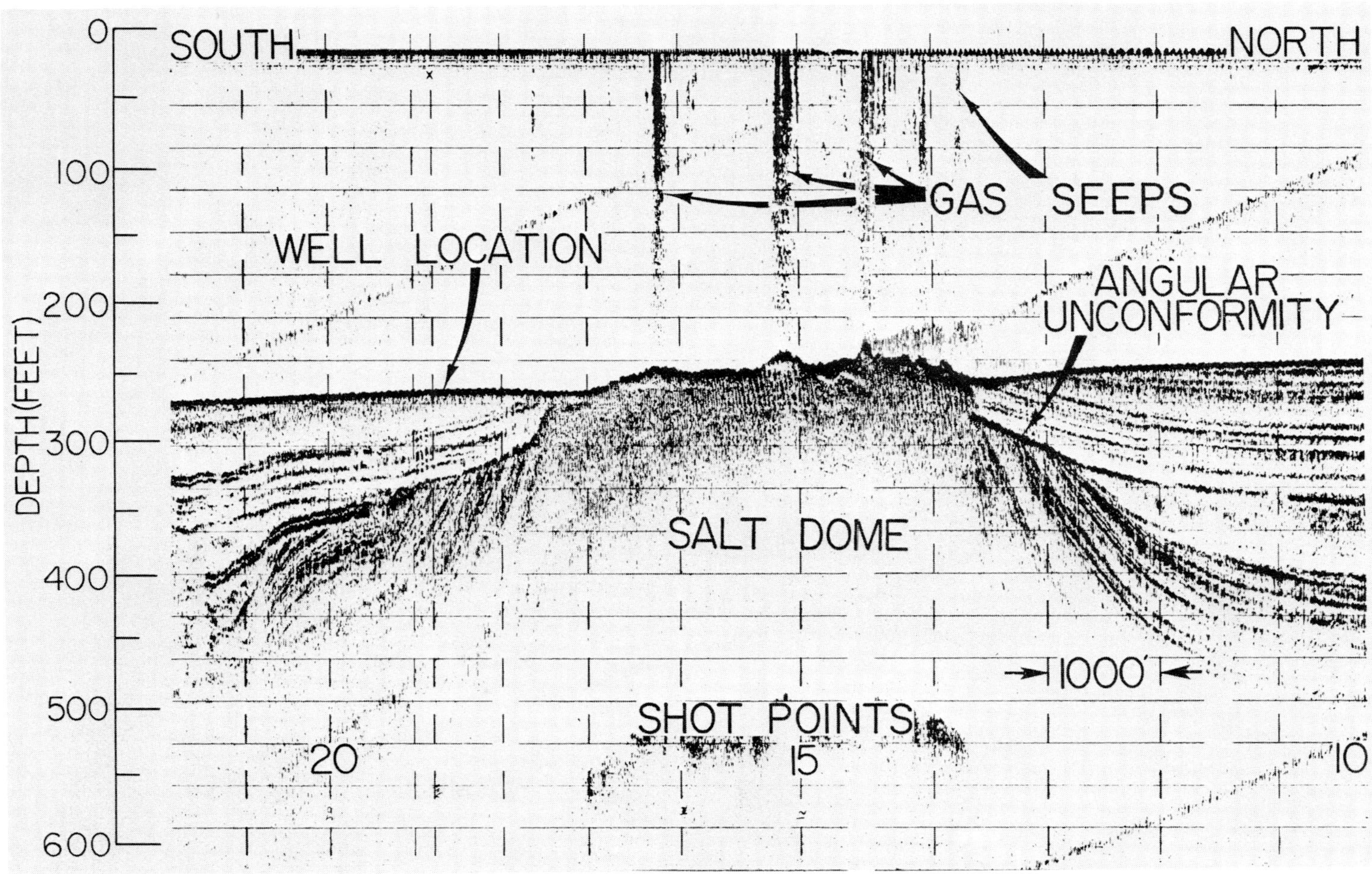

FIGURE 45. 3.5 kHz profile showing well location on flank of a salt dome.

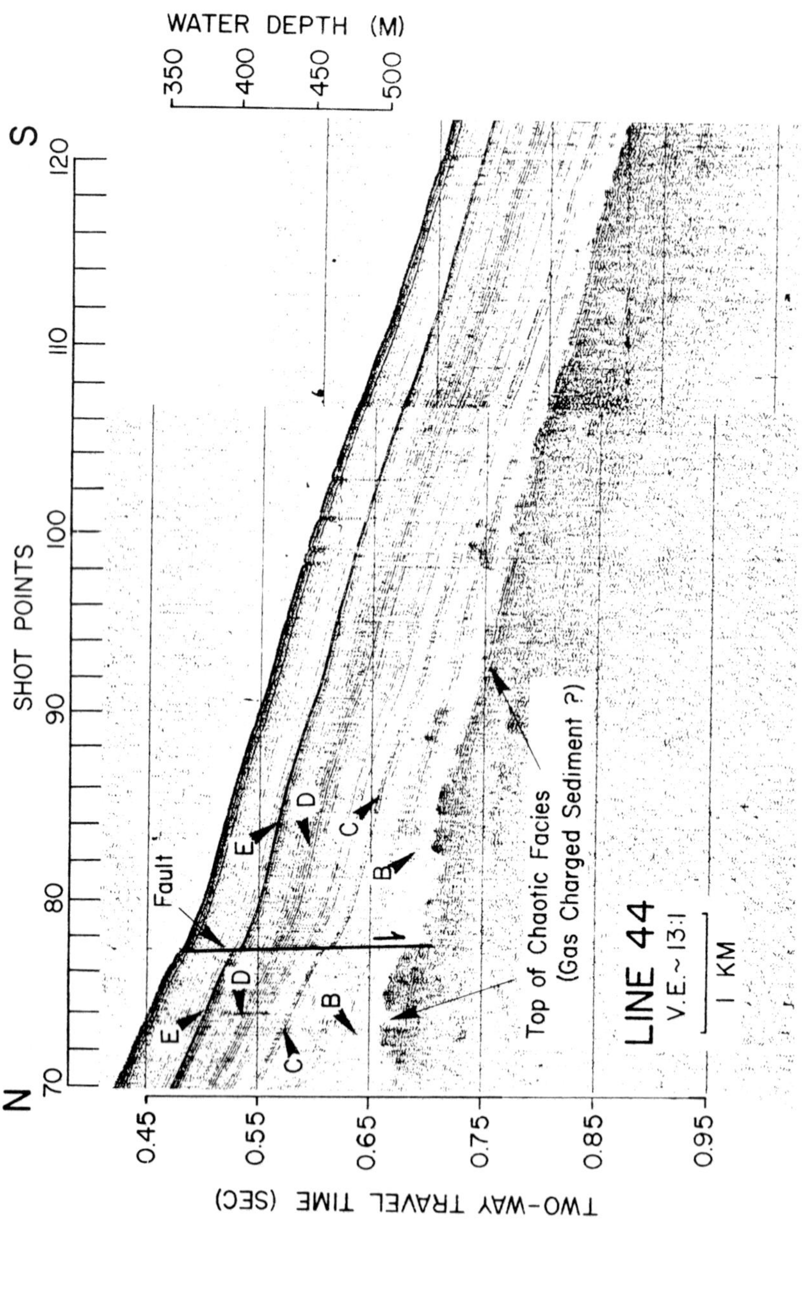

FIGURE 46. Mini-sparker profile of line 44 on Texas Continental Slope (From Sidner, B. R., Gartner, S., and Bryant, W., Rep. 77-5-T, Texas A&M University, College Station, 1977. With permission.)

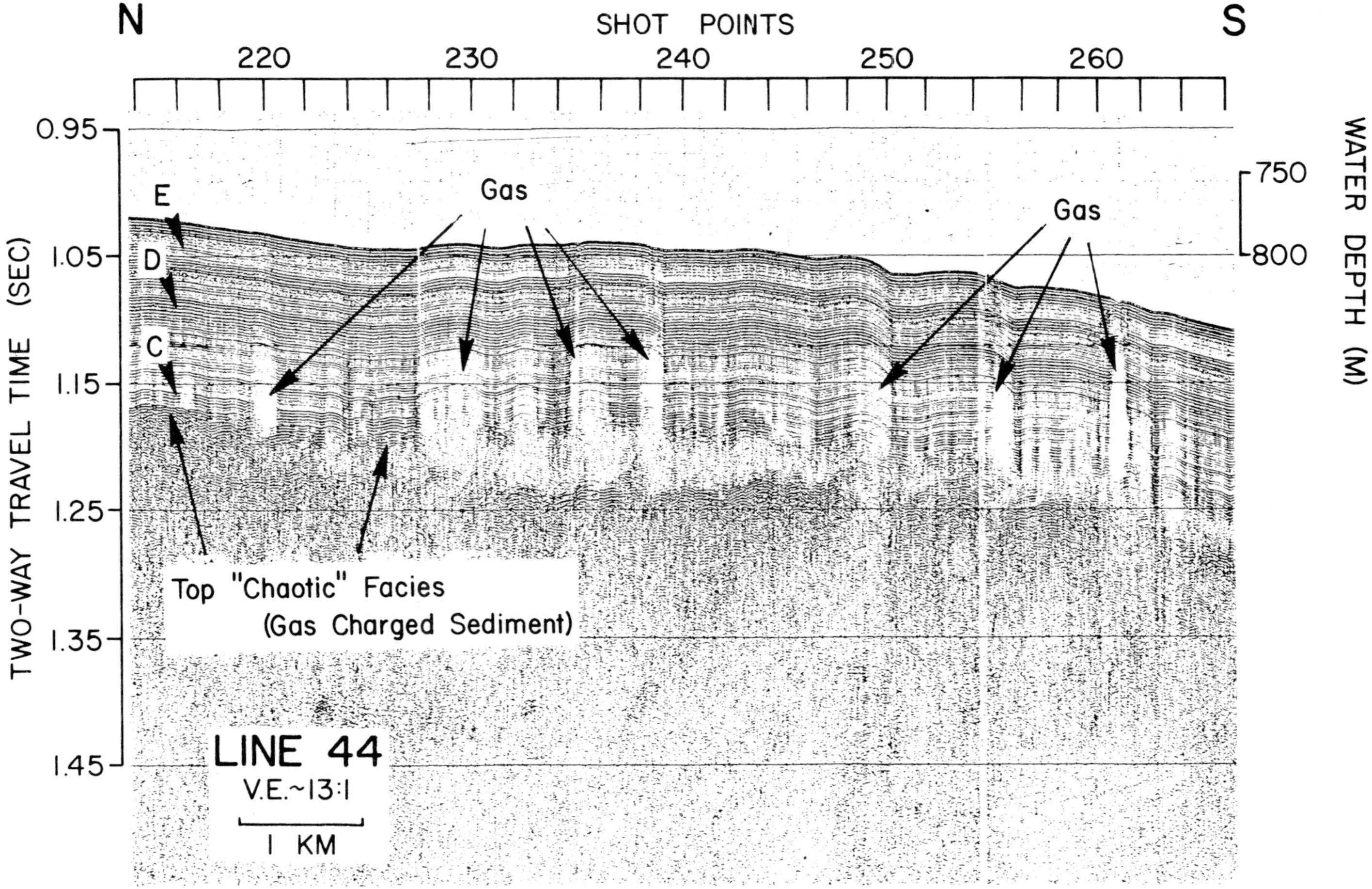

FIGURE 47. Mini-sparker profile of line 44, downdip of profile of Figure 46, on Texas Continental Slope. (From Sidner, B. R., Gartner, S., and Bryant, W., Rep. 77-5-T, Texas A&M University, College Station, 1977. With permission.)

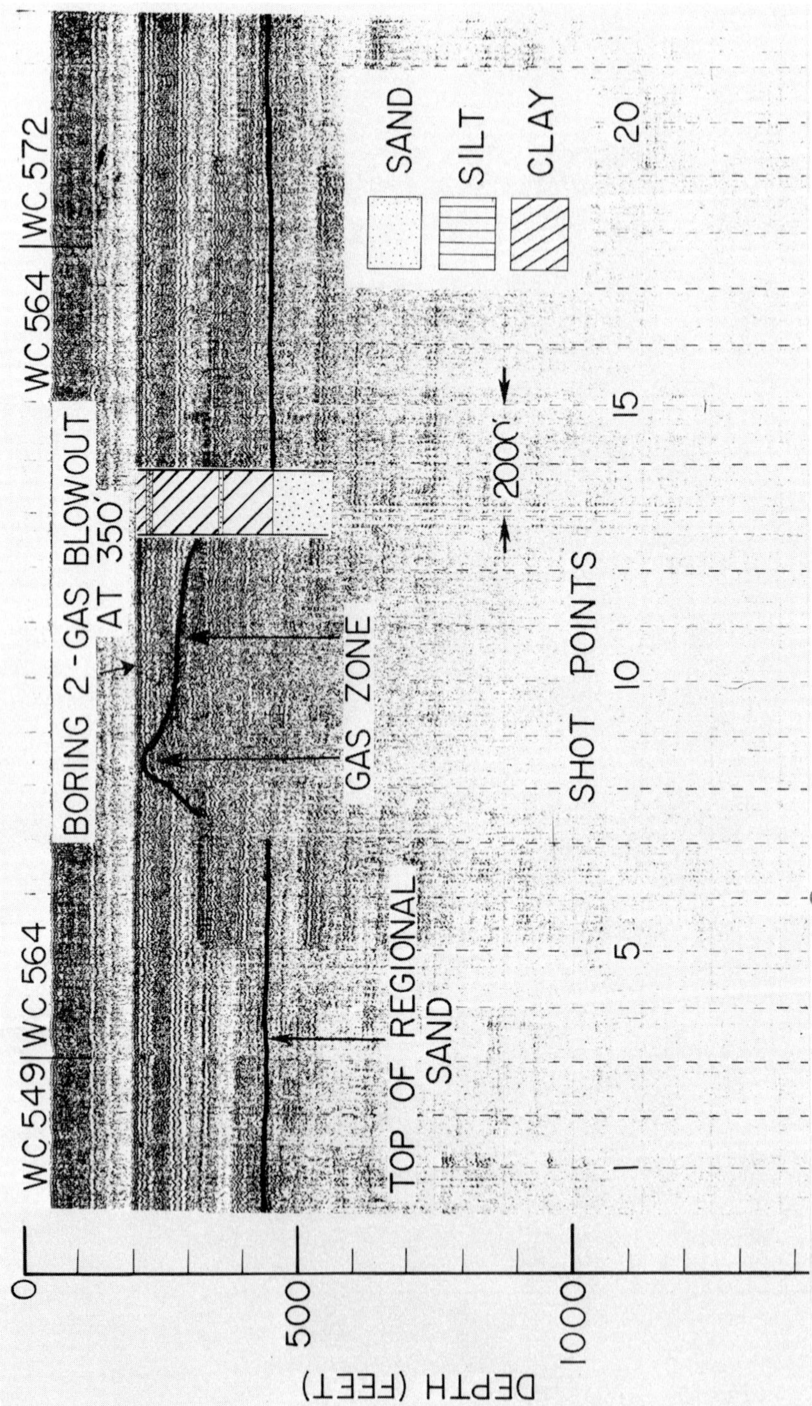

FIGURE 48. Air gun profile of West Cameron Area Block 564.

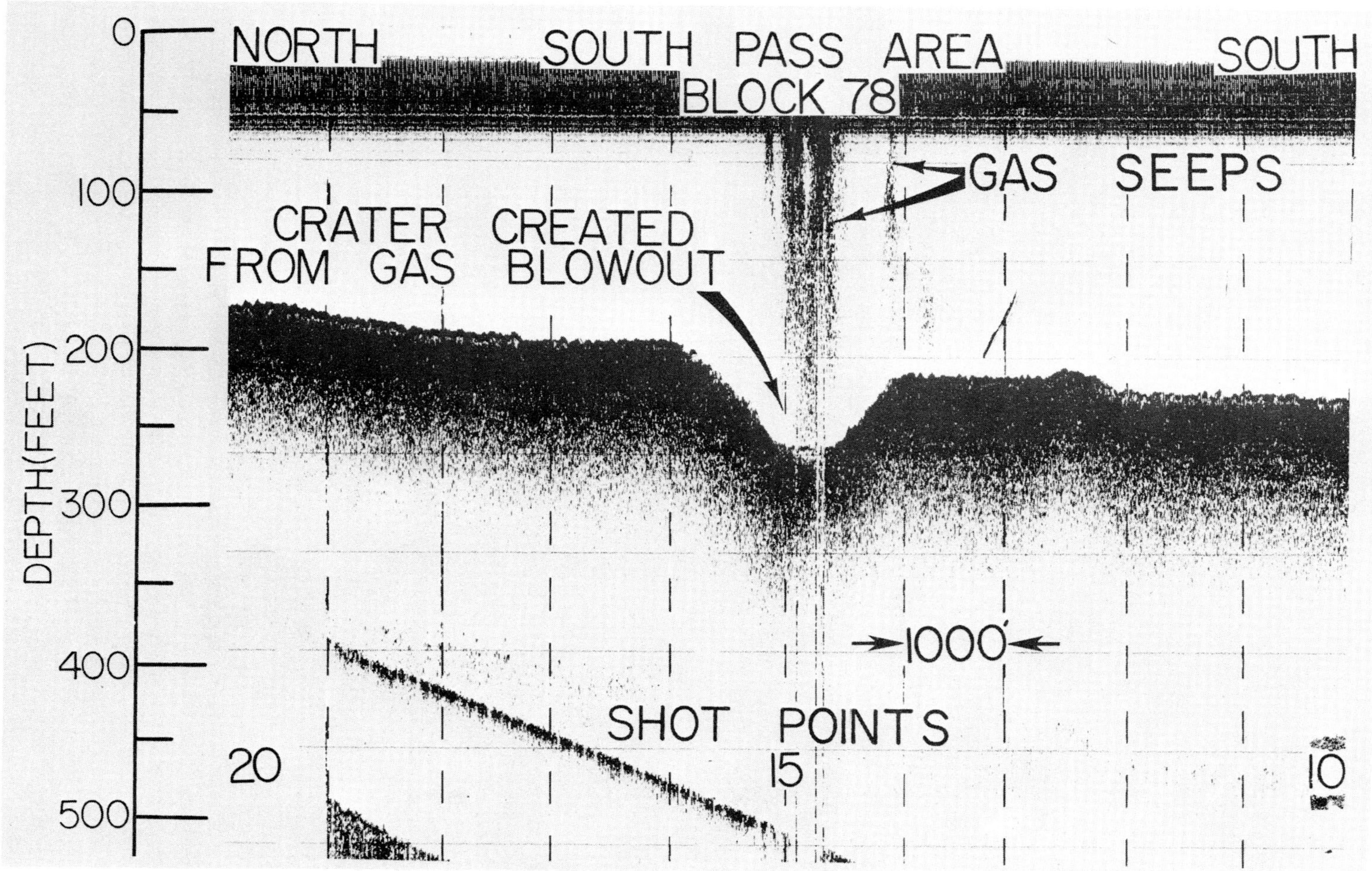

FIGURE 49. 3.5 kHz profile of blowout crater in South Pass Area Block 78.

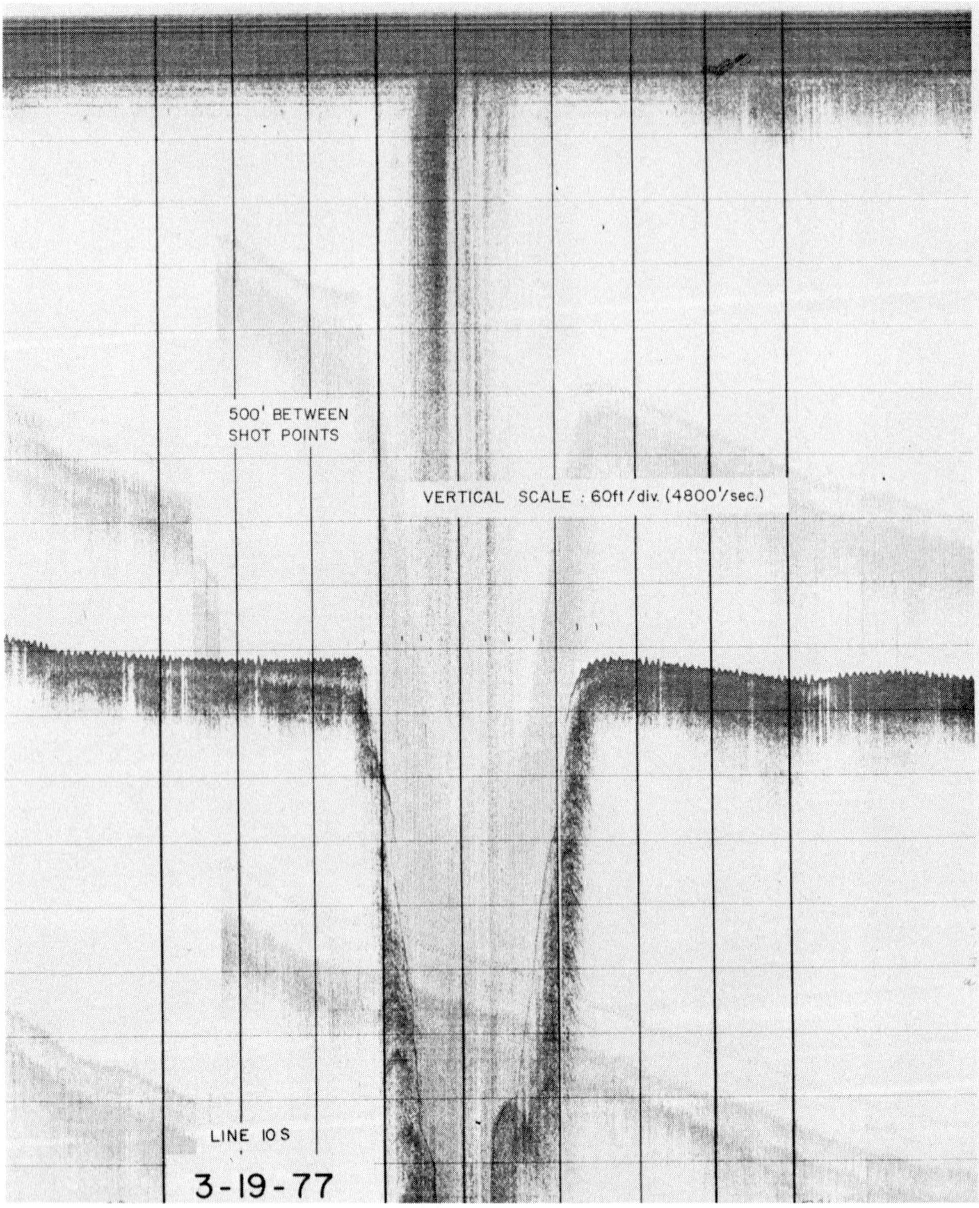

FIGURE 50. 3.5 kHz profile of blowout crater in High Island Area Block A-563.

this crater. The adjoining shallow sediments on the sides of the crater are also saturated with vast amounts of gas at the sea floor, which restricts seismic returns from the underlying layers. It is postulated that the wedge of sediment between the hydrocarbon zone and the rig was indurated by gas, allowing the pressure to increase to a point where the sediment was blown into suspension.

On November 6, 1976, in High Island Block A-563, Pennzoil's A-5 well blew out of control and the entire platform (costing approximately $27 million) sank into the crater formed by the escaping gas and seawater. The gas came from gas-charged sands at a depth of 4200 ft below the sea floor. The 3.5 kHz seismic profile in Figure 50 shows the resulting crater. The crater measured almost 1500 ft across and was 320 ft

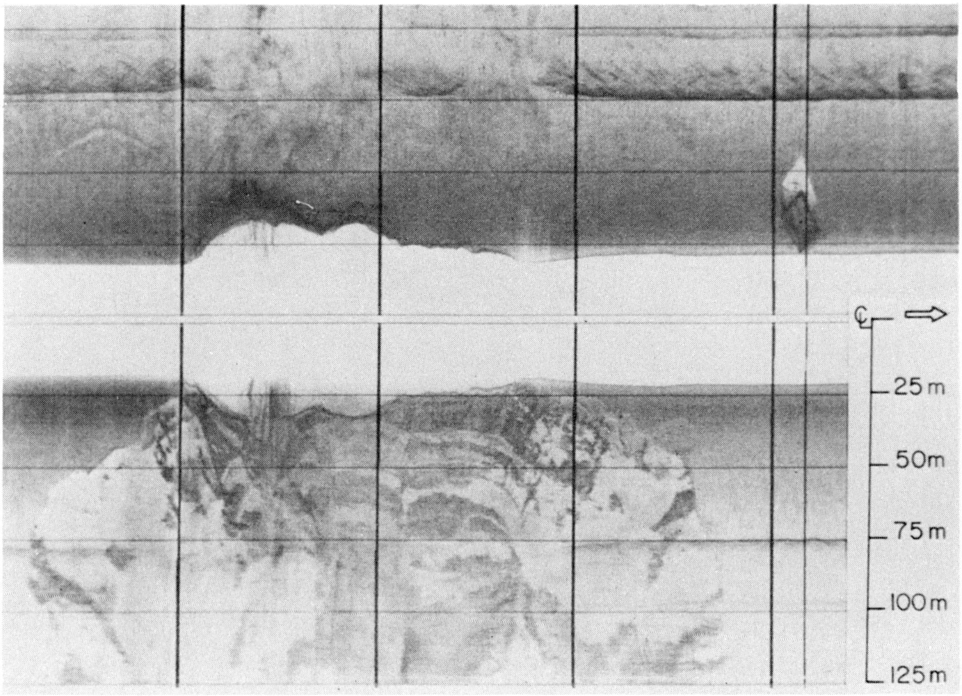

FIGURE 51. Sonograph of blowout crater in High Island Area Block A-563.

deep. An estimated 155 million ft^3 of sediment was ejected from the crater. Figures 51 and 52 show a series of side-scan sonar records of the rim of the crater. The Pennzoil blowout occurred because of a mechanical failure and the results were quite spectacular. Uncontrolled gas-charged sediments are costly, dangerous geohazards.

Gas blowouts can occur naturally at the sea floor creating a crater. A good example of such a crater is shown in the 3.5 kHz seismic record of Ship Shoal Area Block 300 (Figure 53). At Shot Point 27 a small paleo-crater, now covered with sediments, rests on the flanks of a salt dome. The overlying sediments are gassy, indicating that gas is still issuing from the crater.

SUBMARINE SLUMPS

Submarine slumps are very common features whenever the rapid accumulation of sediments occurs on a sloping surface. The deltas of large rivers are such areas and slumps are almost to be expected. Slumping of sediments from a delta area may occur as a gradual process or may take place in a catastrophic manner.

Sea-floor movements caused by slumping or sliding are prevalent in the Gulf area, particularly around the Mississippi Delta. The movements are partially controlled by localized steepening of slopes, storm waves, currents, and creep of steep mud fronts. The recent fill of the Old Mississippi Canyon also produces unstable foundation conditions. The potential hazard of drilling in these unstable sediments, particularly as it relates to the emplacement of platforms, is obvious.

An excellent example of unstable soil conditions is seen in Grand Isle Block 186 (Figure 43). This block is located on the north slope of the east-west trending Old Mississippi Canyon. The dominant structural style of the shallow sediments in this

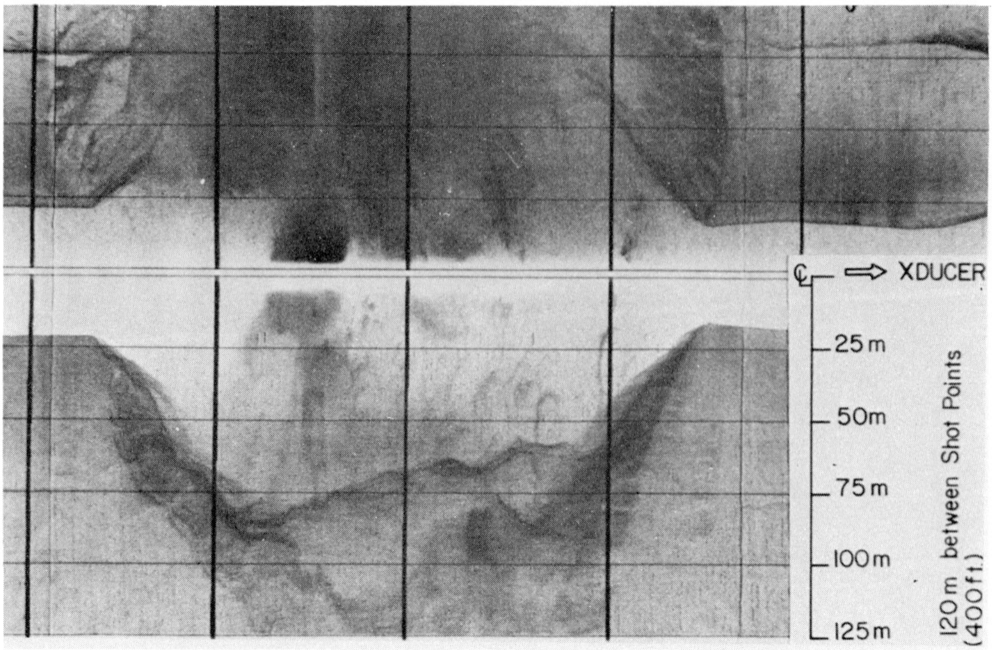

FIGURE 52. Sonography of blowout crater in High Island Area Block A-563.

area includes folded and faulted sediments associated with slumping along the shelf edge. The upper 410 ft of sediment is composed of soft to stiff clay with minor amounts of interbedded silt and sand. The shear strength of these clays is generally equal to, or less than 100 lb/ft^2 at sub-mudline depths down to 70 ft. A prominent salt dome underlying the western portion of this block appears to represent a possible target for drilling. If production were found, then a drilling and production platform that could overcome such instability problems would be vital. However, the oil industry has already lost millions of dollars from platform instability caused by the lateral movements of the seafloor due to slumping.

A typical submarine slump in the process of down-slope migration in the Mississippi Delta area is shown on the first 3.5 kHz seismic record of Figure 54. This slump is approximately 50 to 75 ft thick.

Slumps occurring on steeper slopes than the Mississippi Delta Front are shown in the 3.5 kHz seismic record of Grand Isle Area Block 90 (Figure 55). The slumps are in the section to the north of the contact between the shelf sediments and fill of the Mississippi Canyon.

Slumped material is quite common on the continental slope of the northern Gulf of Mexico. Much of this slumping occurred during the Pleistocene period, when the rivers deposited their loads directly onto the upper continental slope. The diapiric activity in the area of the slope also contributed to slumping due to the oversteepening of the flanks of the diapiric structures, resulting in failure of the sediment cover.

Mini-sparker Line XIII-13 (Figure 56) shows the results of slumping on the upper slope area. The zones designated as chaotic facies, that is contained by well-stratified sediments, are slumped material. Mini-sparker line 49 (Figure 57) shows a syncline, or valley, filled with a chaotic facies which represents slumped material. Some of the largest slumps in the world are found off the east coast of Mexico. An artist's conception of the continental slope off the east coast of Mexico (Figure 58) shows a series of

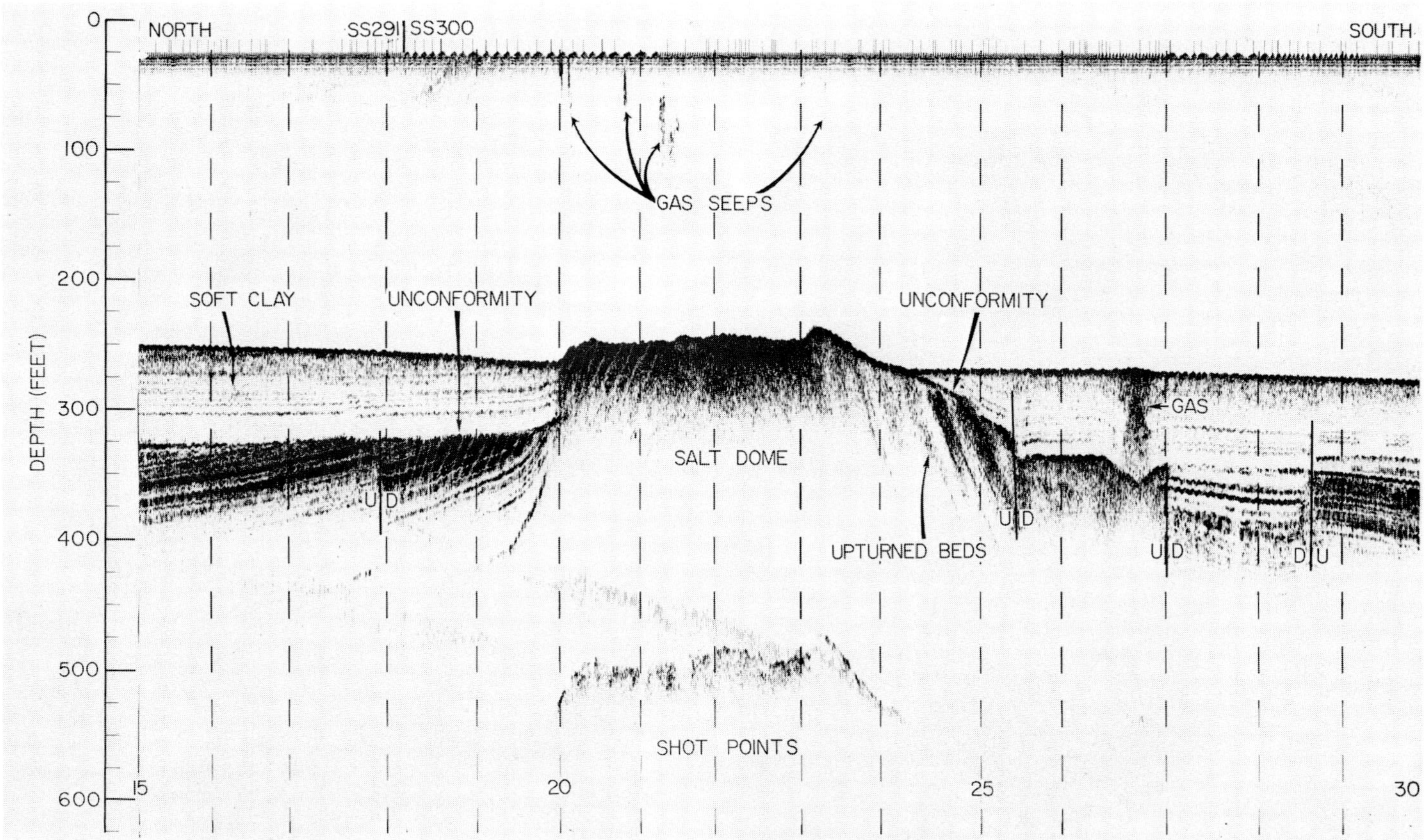

FIGURE 53. 3.5 kHz profile of Ship Shoal Area Block 300 showing Paleo-blowout crater.

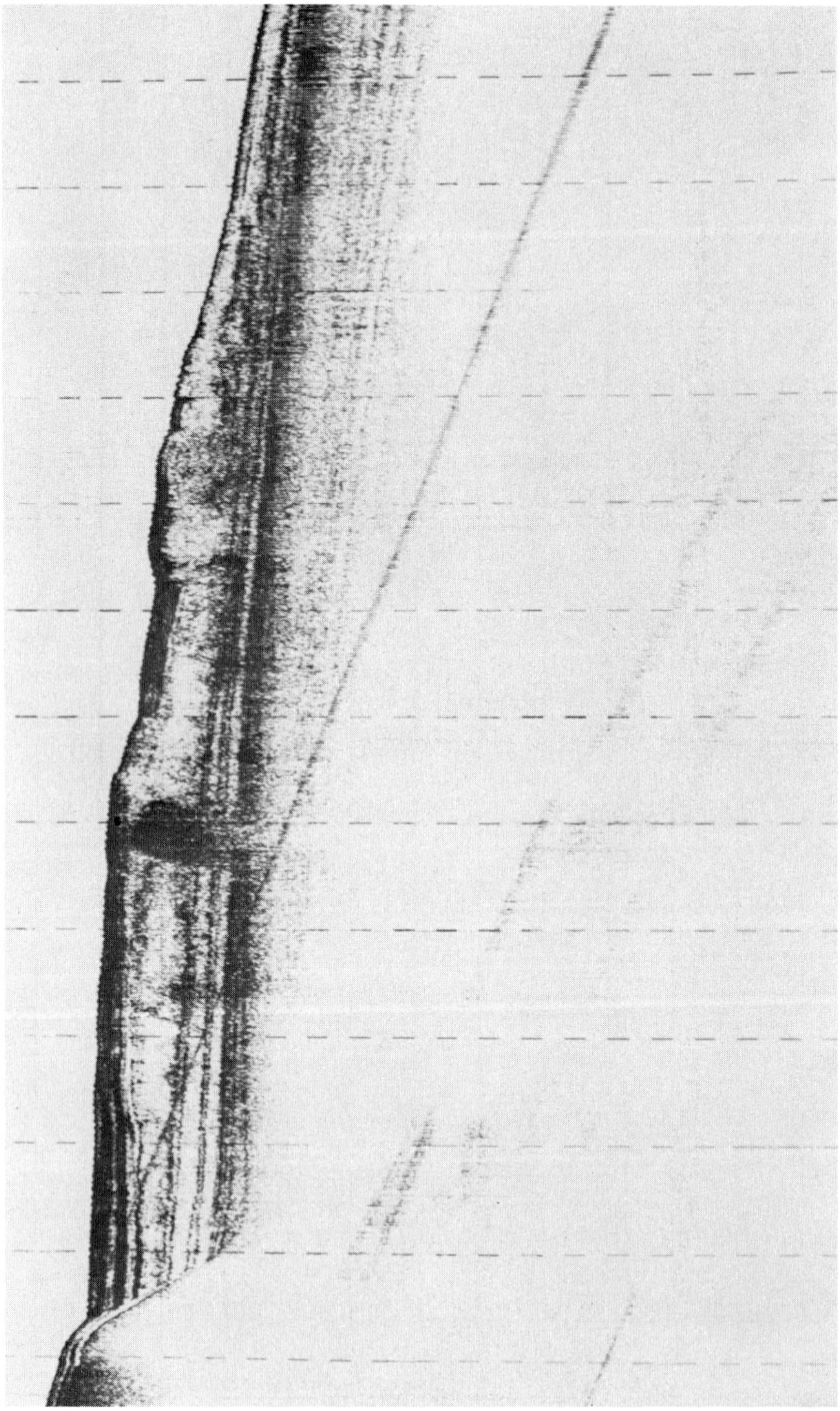

FIGURE 54. 3.5 kHz profile of slump on Mississippi Delta.

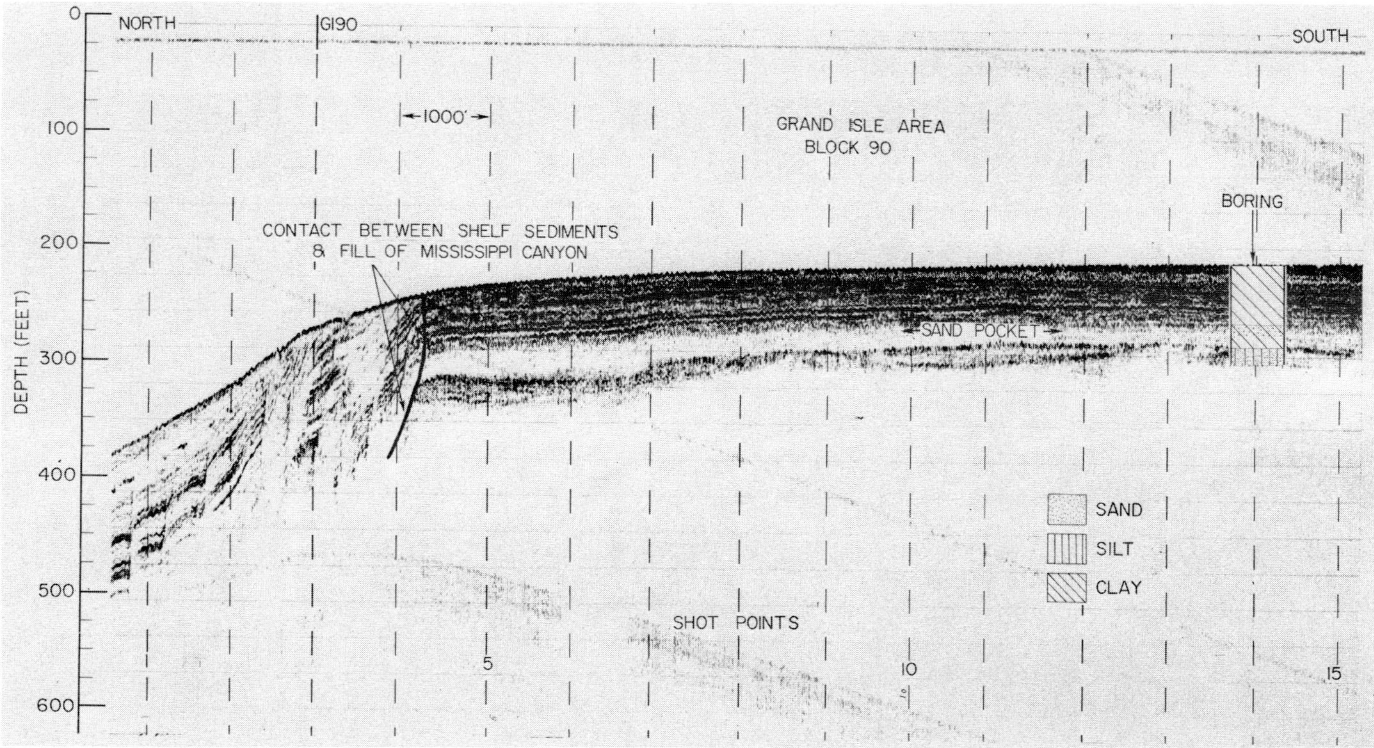

FIGURE 55. 3.5 kHz profile of Grand Isle Area Block 90.

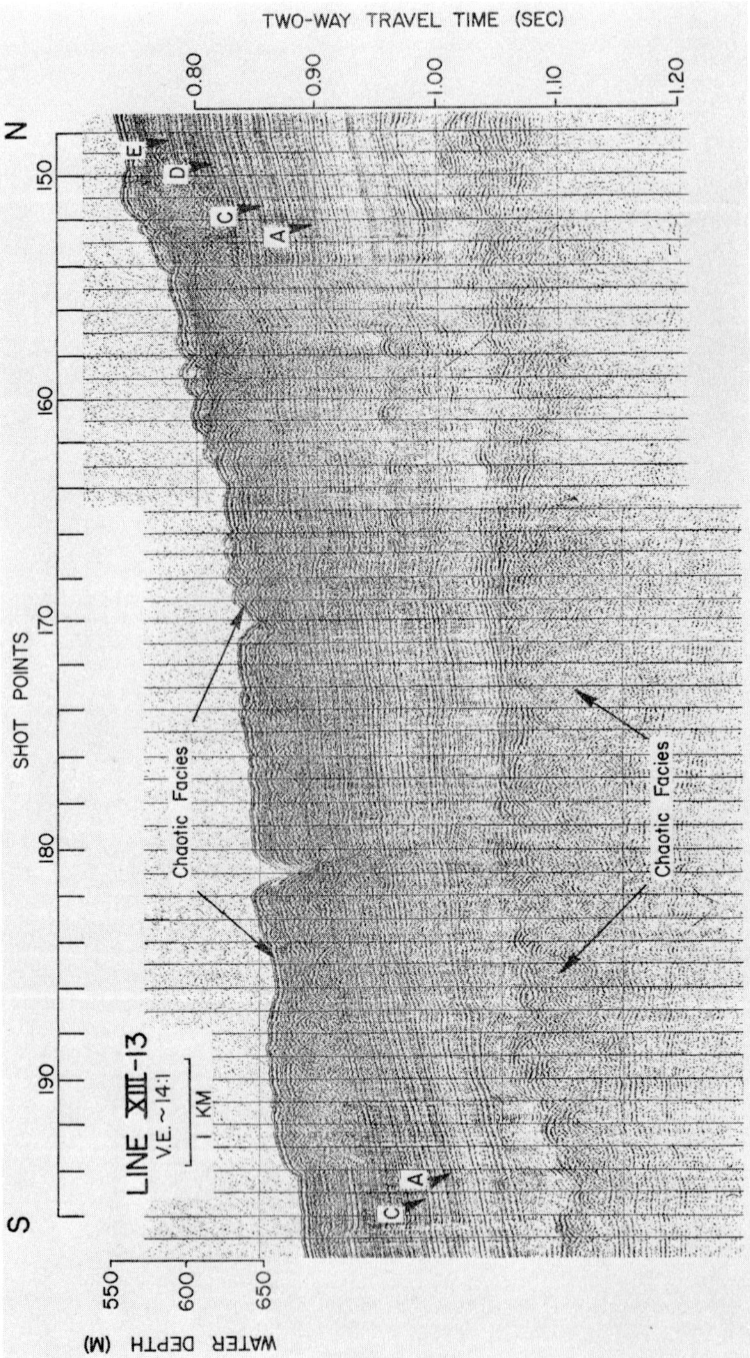

FIGURE 56. Mini-sparker profile of Line XIII-13 on Texas Continental Slope. (From Sidner, B. R., Gartner, S., and Bryant, W., Rep. 77-5-T, Texas A&M University, College Station, 1977. With permission.)

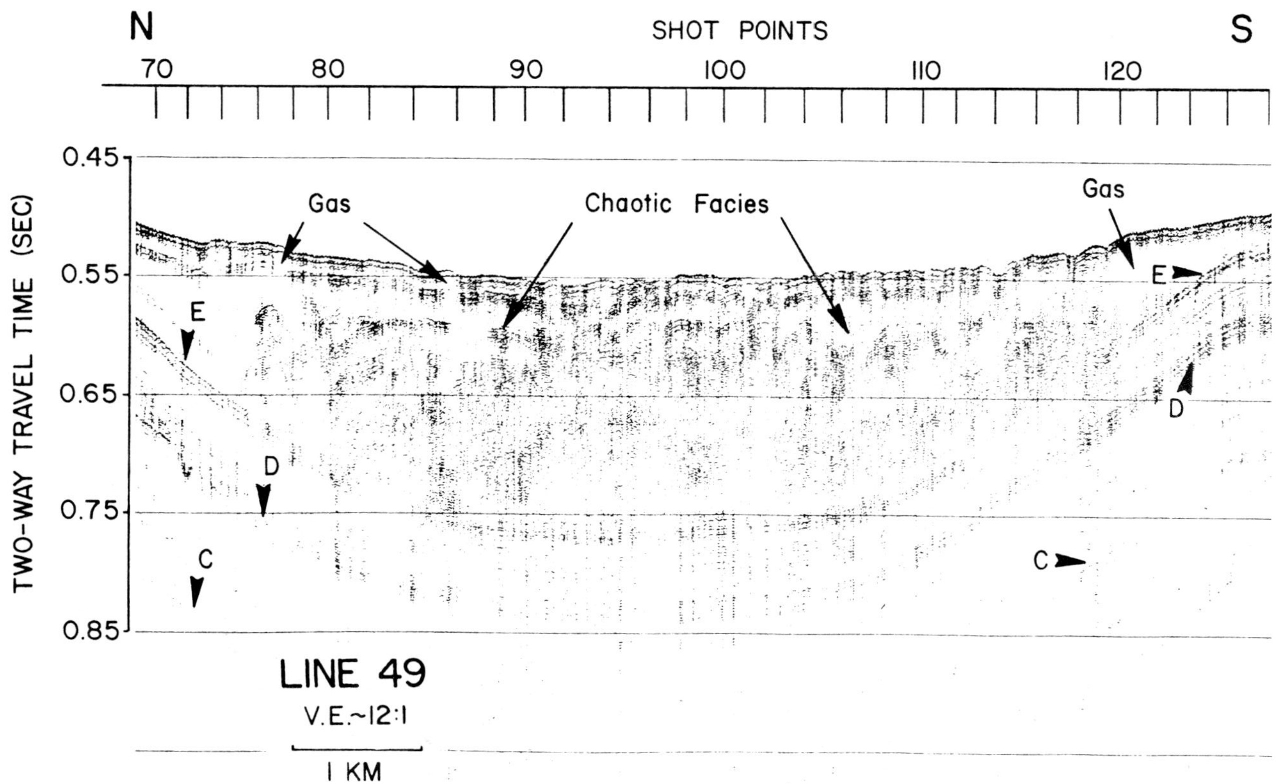

FIGURE 57. Mini-sparker profile of line 49 on Texas Continental Slope. (From Sidner, B. R., Gartner, S., and Bryant, W., Rep. 77-5-T, Texas A&M University, College Station, 1977. With permission.)

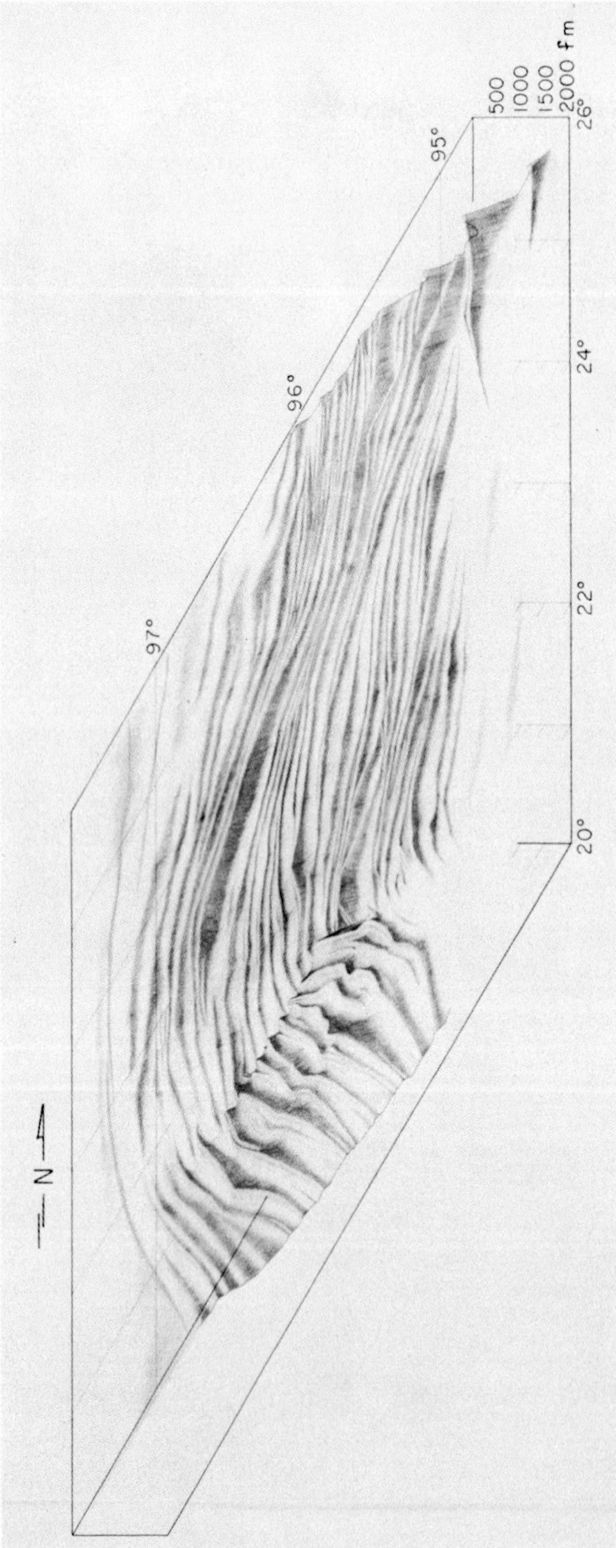

FIGURE 58. Artist conception of gravity slumps on the Continental Slope off Eastern Mexico.

long ridge-like structures extending over 360 mi north and south and 120 mi east and west. A bathymetric profile across these features shows these ridge-like structures to have a relief of almost 2000 ft, and are spaced 5 to 10 mi apart. These features represent the effects of massive gravity sliding and are called the Mexican ridges.

The process of slumping is one of the major geohazards on the continental shelf and slope. In certain cases, such as Chevron's Platform in South Pass Area Block 77, an area of known bottom instabilities, the engineers have constructed a platform to withstand the moveout of a substantial portion of the sea floor. If this is accomplished the geohazard of slumping will turn out to be only an engineering constraint.

REFERENCES

1. **Anon.**, *Glossary of Geology*, American Geological Institute, Washington, D.C., 1972, 805.
2. **Habouty, M. R. and Hardin, G. C., Jr.**, Types of hydrocarbon accumulation of geology of South Liberty Salt Dome, Liberty County, Texas, *AAPG Bull.*, 35, 1939, 1977.
3. **Sidner, B. R., Gartner, S., and Bryant, W.**, Late Pleistocure Geologic History of the Outer Continental Shelf and Upper Continental Slope, Northwest Gulf of Mexico, Oceanography Tech. Rep., 77-5-T, Texas A&M University, College Station, 1977, 131.

GEOPHYSICAL DETECTION OF SUBSEA PERMAFROST

James C. Rogers and John L. Morack

INTRODUCTION

The purpose of this chapter is to introduce the reader to the geophysical methods that have been used to detect subsea permafrost. We also present physical reasons for the existence of subsea permafrost and discuss the implications of its presence. The material is intended to be sufficiently complete to give the reader an understanding of the techniques used by government, industry, and university groups in the delineation of subsea permafrost. Such geophysical investigations have been conducted in the northern hemisphere during the last two decades, and an extensive bibliography exists. We present only some of those materials, principally examples from the literature. However, the interested reader should find sufficient references included here to start a significant investigation of geophysical techniques applied to the detection of subsea permafrost. In many cases the work of others has been included in this chapter, and it is our intention to provide the reader with the opportunity to access these sources.

The accepted scientific definition of permafrost is rock or soil material, with or without included moisture or organic matter, that has remained continuously below 0°C for two or more consecutive years. We are concerned here with the ice-bonded frozen materials, which are a subset of all materials classified as permafrost. In marine environments, material that meets the thermal definition of permafrost is often likely to be free of ice because of infiltration of salt brines. The methods of geophysical detection that will be discussed are relatively insensitive to the temperature of the material, but instead are sensitive to the differing properties between the frozen and non-frozen sediments. (We will use "frozen" and "ice-bonded" interchangeably here.) At the present time, seismic methods have been the most successful in locating subsea permafrost and, therefore, these methods will be emphasized. However, it appears that some electrical and electromagnetic techniques also hold promise of locating subsea permafrost and they will be discussed briefly with the anticipation of improved techniques in the future. At present, no geophysical method has been consistently successful at locating the bottom surface of subsea permafrost. Also, characterization of its physical properties by geophysical methods is poorly developed.

History of Subsea Permafrost Formation

Permafrost has formed beneath approximately 20% of the land mass of the earth over the last several tens of thousands of years. Figure 1, from Ferrians et al.,[1] shows the distribution of land permafrost in the northern hemisphere. The research of Shackleton and Opdyke[2] suggests that the world sea level fell to a minimum level during the last Wisconsin Glaciation about 18,000 years ago. During this period of low sea level, surface temperatures were much like those today and permafrost was formed under much of the present Beaufort Sea continental shelf.

As sea levels rose due to glacial melting, a set of distinctly Arctic processes began to erode the coastline along the Beaufort Sea. Ice-rich Pleistocene sediments subjected to localized thawing developed karst features that are described as thermokarst. Thaw lakes were formed in some of the thermokarst depressions and these lakes were subsequently overrun by the receding coastline, thus forming a highly crenulated shoreline. Along the coast, where the bluffs were composed of ice-rich sediments, a combination of thermal and wave erosion led to a slow disintegration of the coastline. These proc-

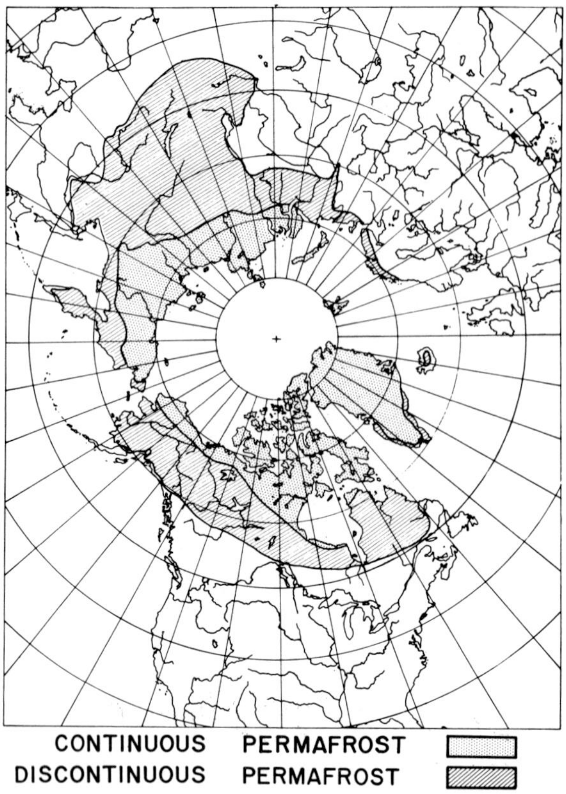

FIGURE 1. The extent of permafrost beneath land in the northern hemisphere as presented by Ferrians, Kachadourink, and Greene.[1] A complete map of permafrost beneath the Arctic Ocean perimeter is not known to exist at this time. (From Ferrians, O. J., Kachadourink, R., and Greene, G. W., *U.S. Geol. Surv. Prof. Pap.*, 678, 1969. With permission.)

esses continue today and are eroding the coastline at an average rate of approximately 1.5 m/year in the Alaskan Beaufort Sea.[3]

Thus, the subsea permafrost that exists today is relict in nature. Models have indicated that thousands of years may be required for its dissipation by thawing after inundation by the relatively warm ocean.[4,5] An idealized vertical section from land to offshore is given in Figure 2. The thickness of permafrost onshore of the Beaufort Sea ranges up to 600 m. Its upper surface is essentially coincident with the land surface, and after inundation, it begins to melt at both the top and bottom. The upward melting of the bottom surface is due to heat flux from the earth's core. The downward melting of the top surface is due to a more complicated process that is presently not entirely understood. It is now clear that the sediment porosity, the interstitial liquid salinity, and the sediment temperature are important in the degradation of relict permafrost. Some of the details of the processes involved are found in the work of Harrison and Osterkamp.[6-8] Elevations in the top permafrost surface are present beneath offshore islands due to the enhanced thermal coupling to the air that the islands provide. While these processes are known to occur, the actual details are not known well enough to predict the depth and thickness of subsea permafrost at a particular site. It is anticipated that specific information on the distribution, depth, and geophysical properties

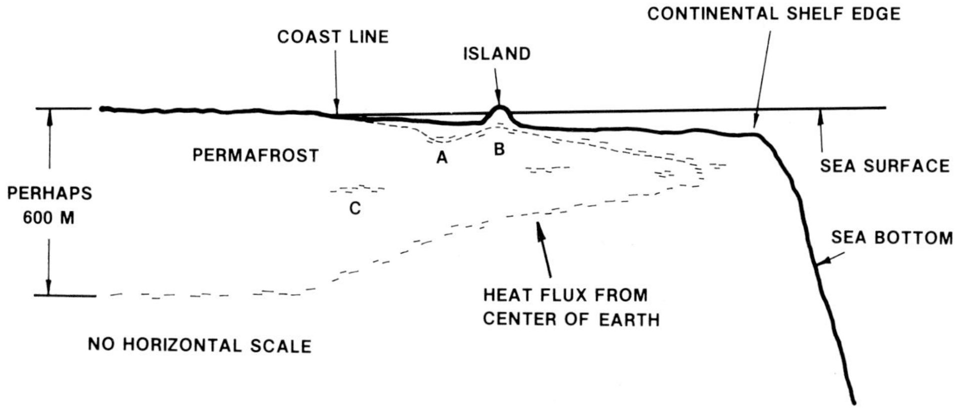

FIGURE 2. An idealized vertical section from land to offshore regions. The depression in the upper permafrost surface (A) is postulated to represent a remnant thaw bulb beneath a river or lake; it was generated before inundation by the sea. Given a mean annual air temperature below freezing, the island (B) acts to extract heat from the sea bottom and preserve relict permafrost. It is expected that this simple model is incomplete in several ways, including possible layering of frozen and nonfrozen materials (C) due to aquifers, freeze-back effects associated with thaw bulbs after inundation, etc. However, it is illustrative of the overall situation along the perimeter of the Arctic Ocean.

of subsea permafrost will help in understanding the complicated physical processes that affect it. A more detailed understanding of these processes, coupled with permafrost distribution information, will also be needed before the complete geological history of the offshore area can be determined.

The discovery of oil and gas beneath the Canadian and U.S. Beaufort Sea shelf, coupled with a need to develop these reserves, has focused a great deal of attention on the distribution and character of subsea permafrost in this area. The development of these reserves will undoubtedly require the construction of bottom-founded structures and buried pipelines, both of which would be adversely affected by any melting and subsequent settling of the permafrost. These considerations, coupled with the high cost of offshore drilling, have given geophysical techniques for locating subsea permafrost a significant impetus during the past few years.

Geophysical Parameters Affected by Ice Bonding

The interpretation of geophysical measurements in the subsea permafrost environment requires a knowledge of the geophysical properties (e.g., seismic velocity, electrical resistivity in the nonbonded and in the ice-bonded subbottom materials).

Changes in the material geophysical properties that can be utilized in this situation are primarily due to the freezing of incorporated water. The degree of change that occurs depends not only on the temperature of the material, but also on moisture content, pore size, pore water chemistry (e.g., salt content) and pressure. For example, in marine sediments the freezing temperature can be lowered several degrees centigrade by the high salt concentrations found in the pore water. In the case of the fine grained sediments such as clay, a considerable fraction of the interstitial water can remain unfrozen below 0°C due to the surface interaction between the material matrix and the water.

Geophysical methods that have been used to investigate subsea permafrost include seismic and acoustic reflection and refraction techniques as well as electromagnetic techniques. All of the seismic and acoustic techniques depend upon the increase in the soil compressional and shear wave velocities that occur as they become ice-bonded. Figure 3 is a plot of compressional and shear wave velocity as a function of tempera-

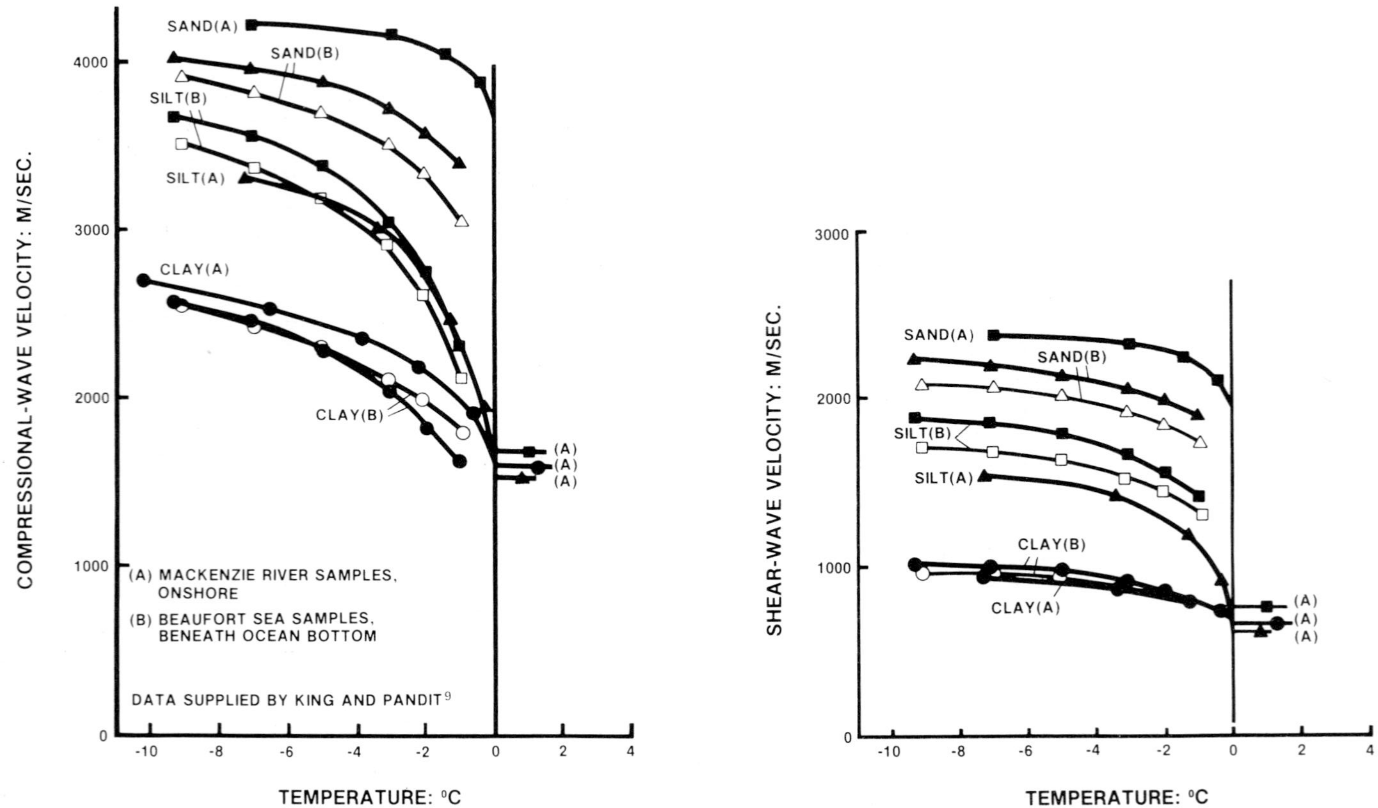

FIGURE 3. Compressional and shear wave velocities as functions of temperature for three general material types as measured by King and Pandit.[9] The Beaufort Sea samples (indicated as (B) in the figure) were to be preserved and their temperatures were always maintained below −1°C. In order to indicate the contrast one might expect between frozen and nonfrozen materials, King and Pandit have shown velocity data obtained for permafrost samples taken from land (indicated as (A) in the figure). These samples were thawed completely to obtain the data above 0°C. A similar behavior is seen for similar material types taken from onshore and from offshore. (From King, M. S. and Pandit, B. I., *Proc. 4th Can. Permafrost Conf.*, National Research Council of Canada, Calgary, Alta., 1981. With permission.)

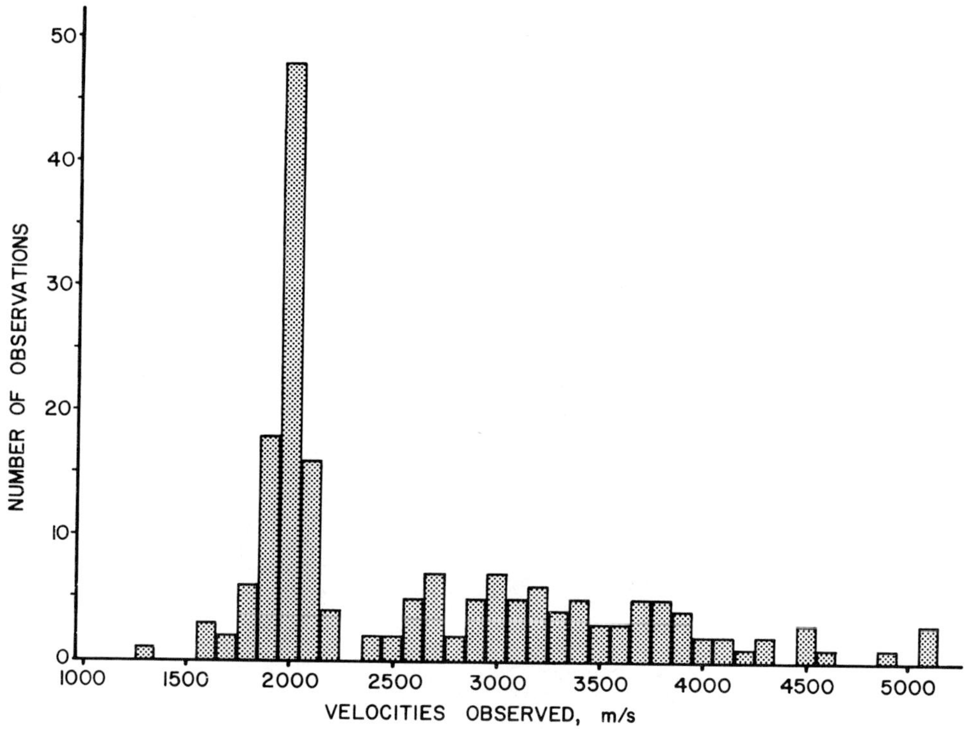

FIGURE 4. Velocities obtained in the Alaskan Beaufort Sea during the course of marine refraction measurements by Morack and Rogers.[13] No effort has been made to eliminate the effects of sloping refractors in the data. The subbottom materials in this area are silty sands with some fine gravel present as well.

ture for several Beaufort Sea subbottom materials as measured in the laboratory by King and Pandit.[9] Appreciable changes in the velocities are observed as the temperature is increased to −1°C, with the effect most noticeable for the silt and clay. Similar freezing effects have been observed on other materials by King.[10] Pandit and King[11,12] have studied the effects of pore fluid salinity on the acoustic and electrical properties of the materials.

Figure 4, a histogram of *in situ* compressional wave velocities in subbottom materials, was obtained by Morack and Rogers[13] using refraction methods with air gun sources near Prudhoe Bay, Alaska. Two distinct velocity groupings are evident from the figure. The large number of velocities in the range from 1550 m/sec to 2250 m/sec are indicative of unbonded bottom sediments. The group of velocities above 2350 m/sec is due to ice-bonded materials. These data were taken within 1 km of shore, and include values that are indicative of onshore permafrost (~ 4000 m/sec) and subsea permafrost (~ 3000 m/sec) in this area. A marked increase in the seismic velocities for ice-bonded materials over that for nonbonded materials can be expected throughout the Beaufort Sea. The data in Figure 4 were obtained over a restricted geographical area and they therefore display a nice separation between the frozen (v > 2300 m/sec) and the nonfrozen materials (v < 2300 m/sec). Actual velocities will vary with thermal conditions and with subbottom material types. However, the contrast required for delineation of frozen and nonfrozen materials is maintained.

Measurements made by electromagnetic techniques at frequencies below 100 kHz are primarily sensitive to changes in electrical resistivity. Figure 5 is a plot of complex resistivity vs. temperatures obtained in the laboratory by King and Pandit.[9] A resistiv-

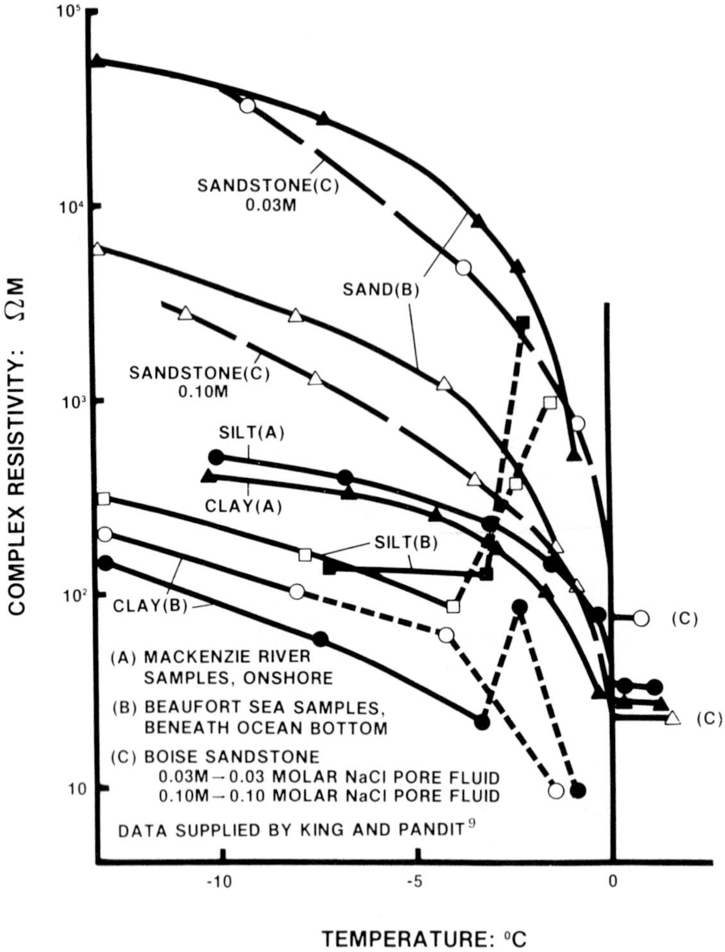

FIGURE 5. Complex resistivity for several materials as a function of temperature as measured by King and Pandit.[9] The Beaufort Sea subbottom samples were maintained at temperatures less than −1°C and other data (sandstone and Mackenzie River delta samples) are shown to indicate the variation in complex resistivity associated with complete thawing. (From King, M. S. and Pandit, B. I., *Proc. 4th Can. Permafrost Conf.*, National Research Council of Canada, Calgary, Alta., 1981. With permission)

ity increase that is associated with increased ice bonding is clearly shown in the figure. The laboratory samples of sand, silt, and clay measured for this figure were the same as those used by King and Pandit[9] to obtain their velocity data shown in Figure 3.

In addition to resistivity effects, electromagnetic measurements made at frequencies above 100 kHz are sensitive to decreases in the permittivity and loss tangent with temperature. These high frequency measurements are, however, limited by their shallow penetration.

The geophysical methods that will be described in the following sections depend upon changes in acoustic velocity and electrical resistivity. Some of the geophysical methods to be discussed have well-developed interpretation procedures (e.g., seismic refraction studies), while others are limited by interpretation methods and geophysical techniques that are just now being developed (e.g., electromagnetic studies). It must be remembered that geophysical data yield only values and distributions of values of the parameters measured. The final step required is correlation of these values with permafrost conditions. Typically the correlations are made with drilling and temperature information. We indicate several of these correlations in the later discussion.

ELECTRICAL TECHNIQUES

Introduction

The use of electrical technique for detection of permafrost beneath land is not uncommon,[14-18] since the contrast in electrical properties between soil in the frozen and nonfrozen state is significant. For example, the resistivity of frozen sand is greater than that of nonfrozen sand by at least an order of magnitude. This contrast has been experimentally demonstrated by a number of measurements,[19-21] some of which included the effects of saline interstitial fluids.[10-12] Recently, samples from beneath the Canadian Beaufort Sea have become available, and good contrast in the electrical properties of these materials has also been shown.[9] Figure 5, from King and Pandit,[9] is illustrative of the results obtained to date. Thus the contrast necessary for successful differentiation of frozen and nonfrozen materials exists, however, the conductivity of seawater complicates the problem significantly beyond that on land. Some efforts have been made to apply electrical methods to the detection of subsea permafrost, and it might be expected that with continued emphasis on offshore resource utilization in the polar region, this tool will be developed further.

Methods

Work has been conducted in the Canadian Arctic on shallow freshwater lakes and in the coastal waters of the Beaufort Sea.[22,23] In polar areas, where the ocean freezes, it is possible to work on the sea ice and use methods generally thought of as applicable only to land surveys. For example, a Schlumberger array has been used to obtain resistivity soundings on lakes and on Kugmallit Bay.

Figure 6, after Scott,[22] shows resistivity sounding data taken approximately 10 km from the eastern side of Kugmallit Bay where the combined thickness of the water column and the sea ice was 4 m. The configuration of the Schlumberger array is shown in the figure. It is seen that the maximum separation of the current electrodes was approximately 2 km and the maximum separation of the receiver electrodes was approximately 30 m. Placement of the electrodes was accomplished by drilling holes in the ice and lowering the electrodes to the ice-water interface.

Interpretation of the sounding data shown was accomplished by fitting a three layer model to the experimental data. The deepest layer in the model was assumed to have infinite resistance and to represent frozen materials. Both the model and the fit of the model calculations to the experimental data are shown. The resistivity data were taken at the location of a drill site and it was possible to compare the permafrost depth determined electrically (46 m) with that obtained by drilling (54 m).

Although the sounding method based upon a Schlumberger array has been demonstrated to be useful, it is somewhat unwieldy and difficult to use in the absence of sea ice. Also, a Wenner array has been towed behind a boat to obtain marine resistivity profiles.[22] Figure 7 depicts the equipment configuration, which consists of nine electrodes towed behind a boat. Two of the electrodes are driven by a 2.5 kVA transmitter which produces a square wave with a 6 sec period. The seven uniformly spaced receiver electrodes provide six channels of data and the overall length of the array is 300 m. The array has been used on a freshwater lake and apparently resistivity contours have been prepared. The receiver electrode pair located farthest from the transmitter elec-

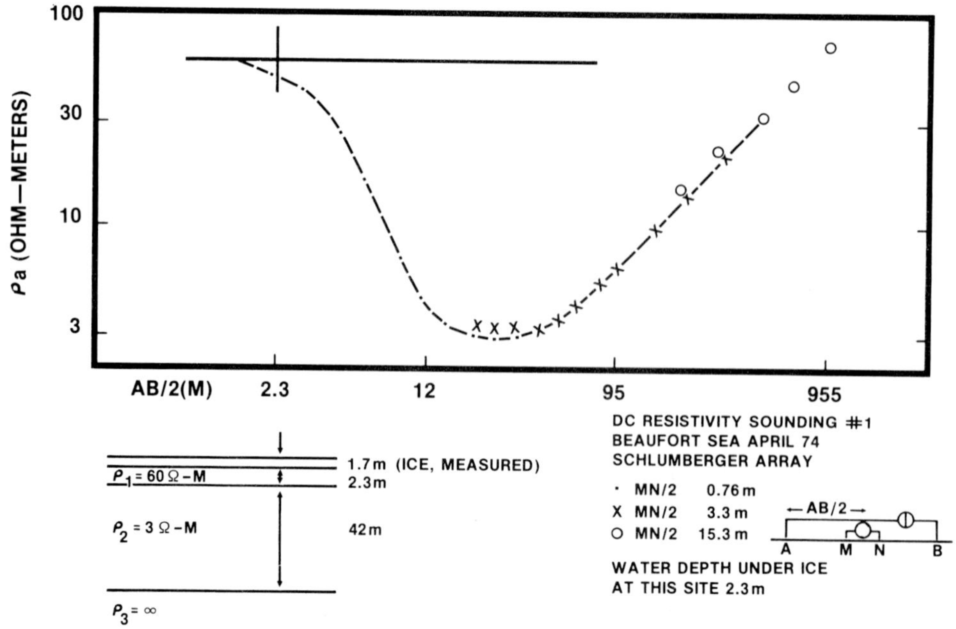

FIGURE 6. DC resistivity sounding data on the Beaufort Sea obtained by Scott.[22] The arrangement of the Schlumberger array is shown along with a layered model. The model parameters were used to calculate a dashed line linking the data points. (From Scott, W. J., Pap. 75-1A Geological Survey of Canada, 1975, 141. With permission.)

trodes provide the deepest resistivity information attainable with the array. While the array has not yet been deployed in seawater, the results obtained on the lake indicate shallow frozen materials extend out from shore beneath the lake.

Other electrical techniques may be useful for studying permafrost beneath the ocean, particularly in coastal areas. Hoekstra and McNeill[23] have discussed time-domain electromagnetic sounding in the detection of permafrost on land and in shallow coastal waters. Also, as will be seen in a later section, it may be that electrical methods can be utilized to estimate the thickness of offshore permafrost.

SEISMIC TECHNIQUES

Shallow Refraction

Seismic refraction profiles require underlying layers that are continuous and have increasing seismic velocity with increasing layer depth. Ice-bonded permafrost overlaid with unbonded materials generally satisfies these conditions. The application of refraction techniques to the detection of subsea permafrost has given the most definitive results in the Beaufort Sea using hydrophone arrays specifically designed for shallow water surveys.[24,26] The distance required between the source and the hydrophones to observe first arrivals from subsea permafrost, and the lower amplitudes of refraction events, requires sensitive hydrophones with good low frequency (< 300 Hz) response. Also, the source must have significant low frequency output to maximize the energy contained in refraction events.

Rogers and Morack[26] have successfully used a system utilizing one to three 40-in.^{3} air guns, operated at approximately 120 bars and suspended 2 m beneath the water surface alongside a 7 m boat. A 260 m- or 480 m-long buoyed hydrophone streamer

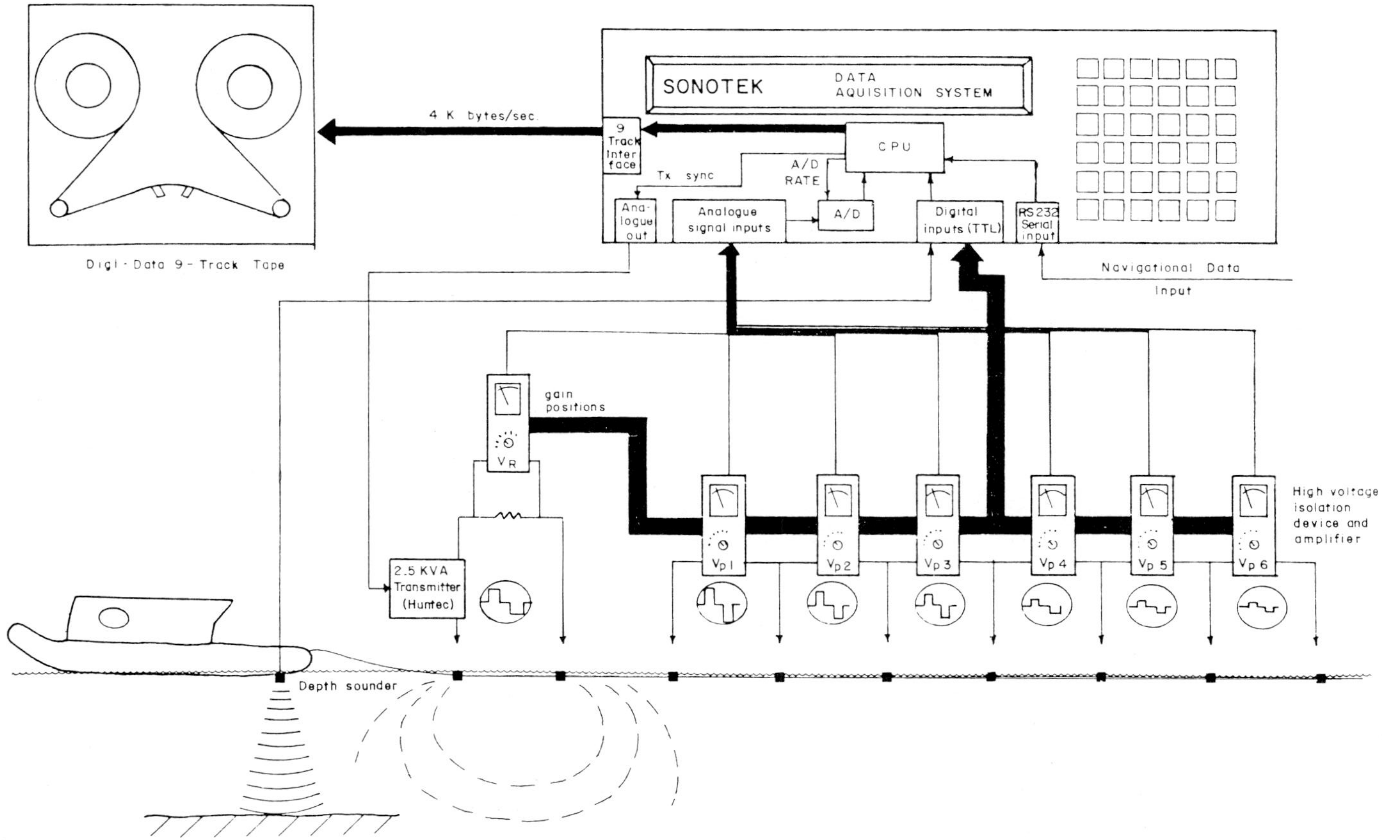

FIGURE 7. Schematic representation of a Wenner array developed by Scott[22] for marine resistivity profiling. One pair of electrodes is driven by the 2.5 kVA transmitter and the seven remaining electrodes are used to obtain six channels of resistivity data. (From Scott, W. J., Pap. 75-1A Geological Survey of Canada, 1975, 141. With permission.)

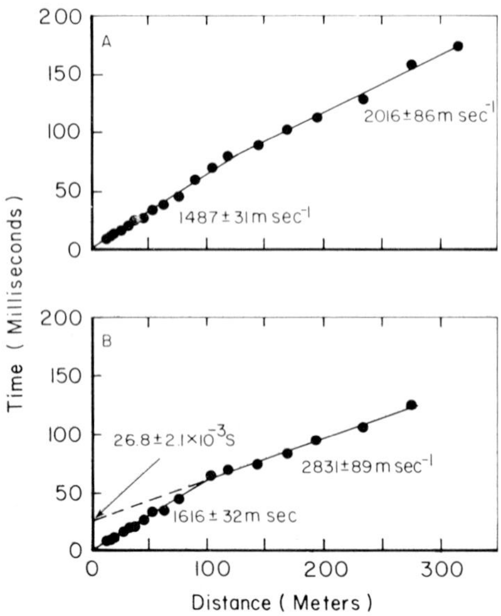

FIGURE 8. Two time-distance plots prepared by Rogers and Morack[26] near Prudhoe Bay, Alaska. The upper plot indicates only nonfrozen materials while the lower plot indicates frozen material approximately 26 m beneath the ocean surface. The error shown is the standard deviation of the straight line fit. (From Rogers, J. C. and Morack, J. L., *J. Geol. Res.*, 85, 4845, 1980. With permission.)

was towed behind the boat with 24 hydrophones suspended approximately 1 m below the surface. A Fourier analysis of the frequencies detected in the refraction signals show that most of the returned energy is centered about 100 Hz. Thus, such sources as explosives, air guns, and gas exploders can be adequate for subsea refraction profiles. Higher frequency sources, sparkers for example, do not have sufficient low frequency output energy for observation of deep refractors.

A typical plot of the arrival time of the first acoustic signal as a function of the horizontal distance between source and receiver is shown in Figure 8. The first part of the curve has an inverse slope of 1616 m/sec, which is the velocity of the compressional wave in the unfrozen bottom sediments. A break in the curve occurs at approximately 100 m from the source and beyond that distance the inverse slope is 2831 m/sec. This is a typical velocity for the refracted wave traveling in ice-bonded permafrost. The depth to the permafrost surface can be calculated as 26 m using the intercepts and slopes of the two straight line segments. Since it is difficult to obtain reversed refraction profiles in the marine environment, most workers interpret refraction data assuming a horizontal refractor. The effects of this assumption are discussed by Rogers and Morack in Reference 26.

The second velocity in the time-distance plot is interpreted to be that of ice-bonded permafrost. This interpretation agrees with data obtained from drill holes in the immediate area and with curves such as Figure 3. Drill hole data in the Prudhoe Bay area has included temperature measurements, penetrometer readings, and some material samples. Reference 27 presents a table comparing permafrost data at drill sites with refraction data.

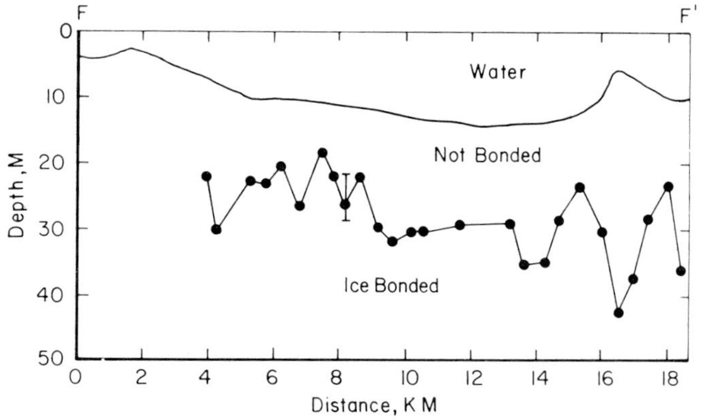

FIGURE 9. Vertical section offshore near Prudhoe Bay, Alaska, prepared from refraction data similar to those shown in Figure 8. The abrupt onset of shallow permafrost is seen approximately 4 km along the line. The error bar shown for the data point at 8.2 km is a combination of systematic and random measurement errors. The maximum penetration depth of the refraction system used was approximately 50 m. It is not known from these measurements whether or not there is marine permafrost beneath the left end of the line or whether it is simply deeper than 50 m. (From Rogers, J. C. and Morack, J. L., *J. Geol. Res.*, 85, 4845, 1980. With permission.)

In general, the correlation between drill hole data and seismic refraction studies has been good. A typical vertical section along a seismic line is shown in Figure 9. This figure shows the interface between the bonded and the unbonded material as determined from refraction studies. A series of several hundred kilometers of seismic lines has allowed determination of the shallow permafrost in the Prudhoe Bay area, as shown in Figure 10.

This kind of study is, at present, capable of forming a detailed distribution map of material velocity and depths to higher velocity refracting layers, which, when correlated with drill hole data, can determine the depth and areal extent of subsea permafrost.

Industry Seismic Reflection Data

The proven offshore oil potential in the Arctic Ocean has stimulated extensive geophysical work in the region. This work has been primarily restricted to reflection techniques with emphasis placed upon reflectors at depths of several seconds. Typically, the seismic sources used might be arrays of air guns with a high energy, low frequency output. Receiver streamer lengths of 2 or 3 km are not uncommon and each channel is the combined output of several hydrophones distributed over an active section several meters long. Separation distances of the active sections might be 100 m or so. The streamers and sources are thus well designed to enhance deep reflected events and first arrival events are usually attenuated. The attractive features of the industry data are the wide geographical coverage and the possibility of greater penetration than is achieved with smaller systems tailored to research or geotechnical tasks.

Hunter et al.,[28,29] with the Geological Survey of Canada, have examined commercial geophysical data gathered for offshore oil exploration. The initial work of 165 km of line tested the feasibility of observing high velocity refractors in the first second of

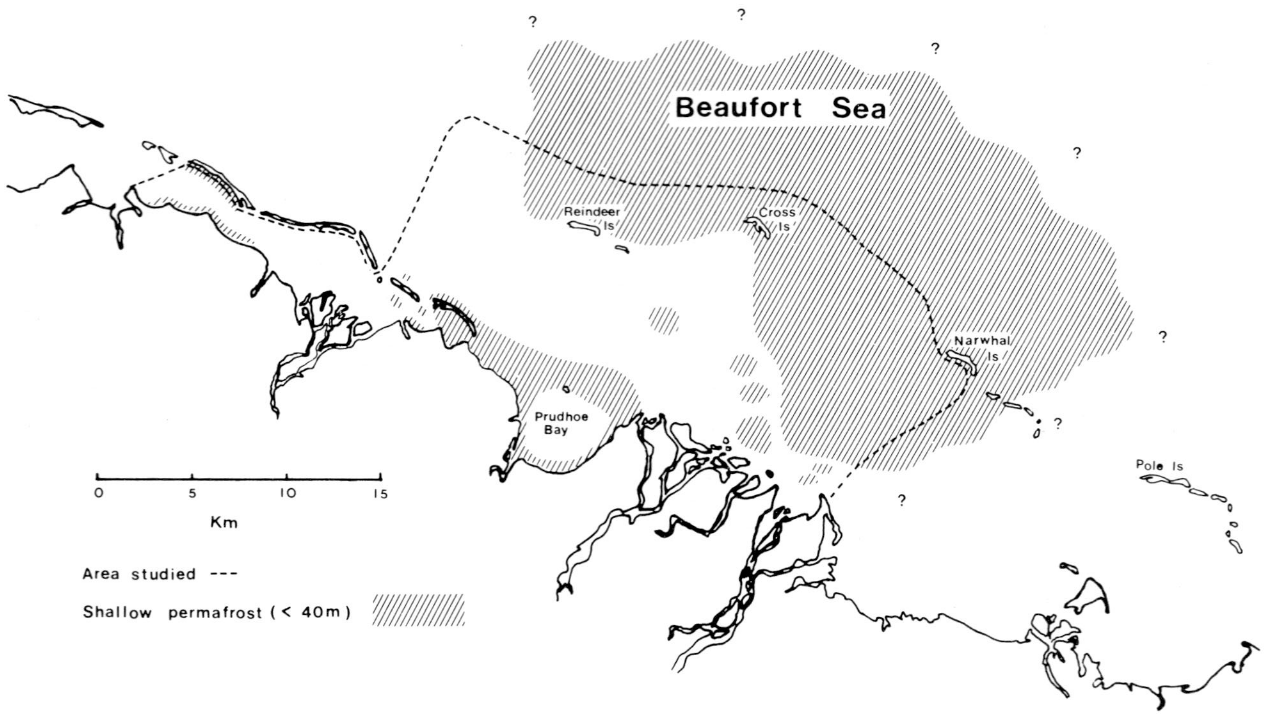

FIGURE 10. Areal distribution of shallow permafrost near Prudhoe Bay, Alaska, prepared by Rogers and Morack.[26] The area south of Reindeer Island is believed to be a location of deep permafrost (greater than 100 m depth to the upper surface). It is speculated that an old river channel may have caused this depression in the permafrost surface. The northern extent of the study area, off Cross Island, was limited by the presence of sea ice which presents a continuing problem to marine activities in the Arctic Ocean. (From Rogers, J. C. and Morack, J. L., *J. Geol. Res.*, 85, 4845, 1980. With permission.)

monitor records. Time-distance plots were prepared for interpretation of all records and ice-bonded permafrost was interpreted to be present for velocities greater than 2500 m/sec. Average depths to the permafrost upper surface were in excess of 100 m beneath the sea surface, in an area where the water depth was approximately 30 m. The maximum depth observed was approximately 250 m and the minimum depths were approximately 80 m.

Subsequent to the initial tests, approximately 40,000 seismic records from the Canadian Beaufort Sea were obtained by the GSC. These consisted of multichannel reflection records recorded on paper without normal moveout corrections, and where feasible, with true amplitude recovery. Thus, refraction events were emphasized at the expense of reflection events. Figure 11 shows a typical data set. Interpretation of the records was done with the use of graphic overlays to obtain refractor velocity estimates and the nonfrozen sediment velocity was obtained by separate high resolution refraction measurements. The nonfrozen sediment velocities used were typically around 1600 m/sec.

Sellmann et al.[30] have also used industry reflection records to interpret frozen materials beneath the Alaskan Beaufort Sea. Their work has been similar in many respects to that of Hunter described above, and they have reported the upper permafrost surface to vary from relatively shallow depths near shore to over 200 m offshore. Velocities over 2200 m/sec have been interpreted to be representative of frozen materials. In addition to classifying materials as frozen or nonfrozen, they have differentiated between high velocity frozen materials and low velocity frozen materials. The former velocities range from 3300 to 4600 m/sec and are believed to represent relatively cold frozen sediments, while the lower velocity group, 2200 to 3200 m/sec, are thought to represent relatively warm frozen sediments. Temperature differences in the permafrost are thought to be indicative of shoreline transgression rates and the subsequent land inundation rates.

Figure 12 indicates the coverage obtained by Hunter et al.[29] in the Canadian Beaufort Sea. The heavy lines, principally northeast of the Mackenzie River delta, indicate the presence of ice-bonded permafrost. In order to increase the resolution over that possible with conventional industry marine reflection systems, higher resolution systems have been constructed. MacAulay and Hunter[31] reported such a system which used a 600 m long streamer with 16 hydrophones per trace. Using data gathered with this system, detailed permafrost profiles have been prepared and Figure 13 is representative of their results. The lines of longitude are indicated on the figure, which gives a good presentation of the variability of the refractor depths and velocities that have been encountered. In this case, the observed velocities were usually significantly greater than the 2500 m/sec ice-bonded interpretation threshold. Shallow high velocity events occurring at intervals along the line are interpreted to be zones of high ice-content material lying at shallow depths below the sea bottom. Such materials can be important considerations in the selection of pipeline routes.

High Resolution Reflection

Shallow reflection work has been performed routinely in the course of marine geological investigations in the Arctic Ocean. In addition to the normal difficulty in interpreting materials responsible for horizons in normal reflection records, there has been difficulty in obtaining good reflections from the upper permafrost surface. One would expect significant energy returned from an acoustic wave normally incident on a nonfrozen/frozen material boundary. The compressional wave velocity contrast is about 1.5 for these materials and the densities of the frozen and nonfrozen materials are approximately equal. (Water expands approximately 10% in volume upon freezing and

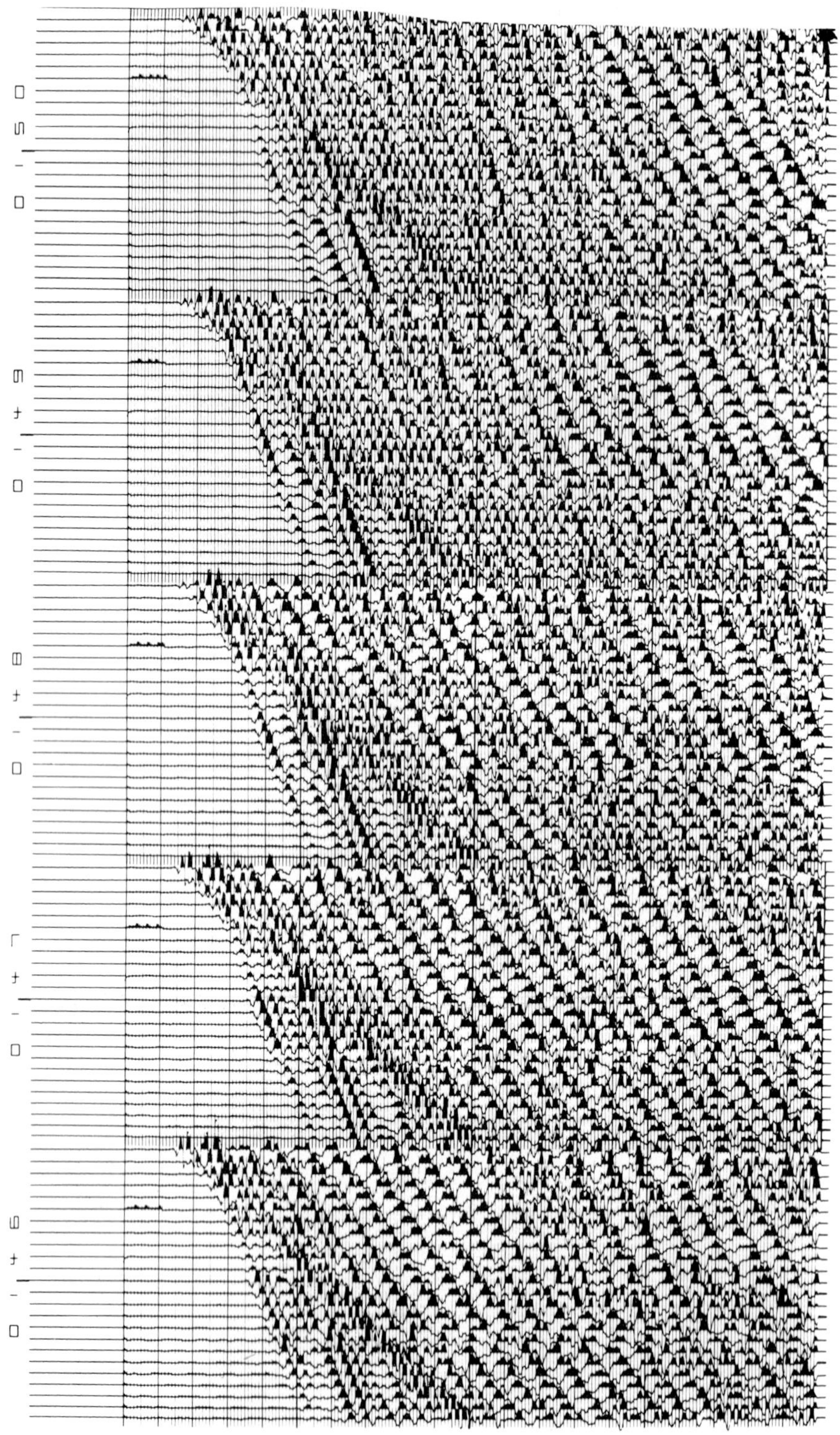

FIGURE 11. Industry seismic records showing refraction arrivals from the upper permafrost surface. Hunter et al.[28] used records of this type to prepare extensive maps of offshore permafrost in the Canadian Beaufort Sea. The vertical scale is time in seconds. (From Hunter, J. A., Judge, A. S., MacAulay, H. A., Gagne, R. M., Burns, R. A., and Good, R. L., Tech. Rep. No. 22, Beaufort Sea Environmental Project, Environment, Canada, 1976. With permission.)

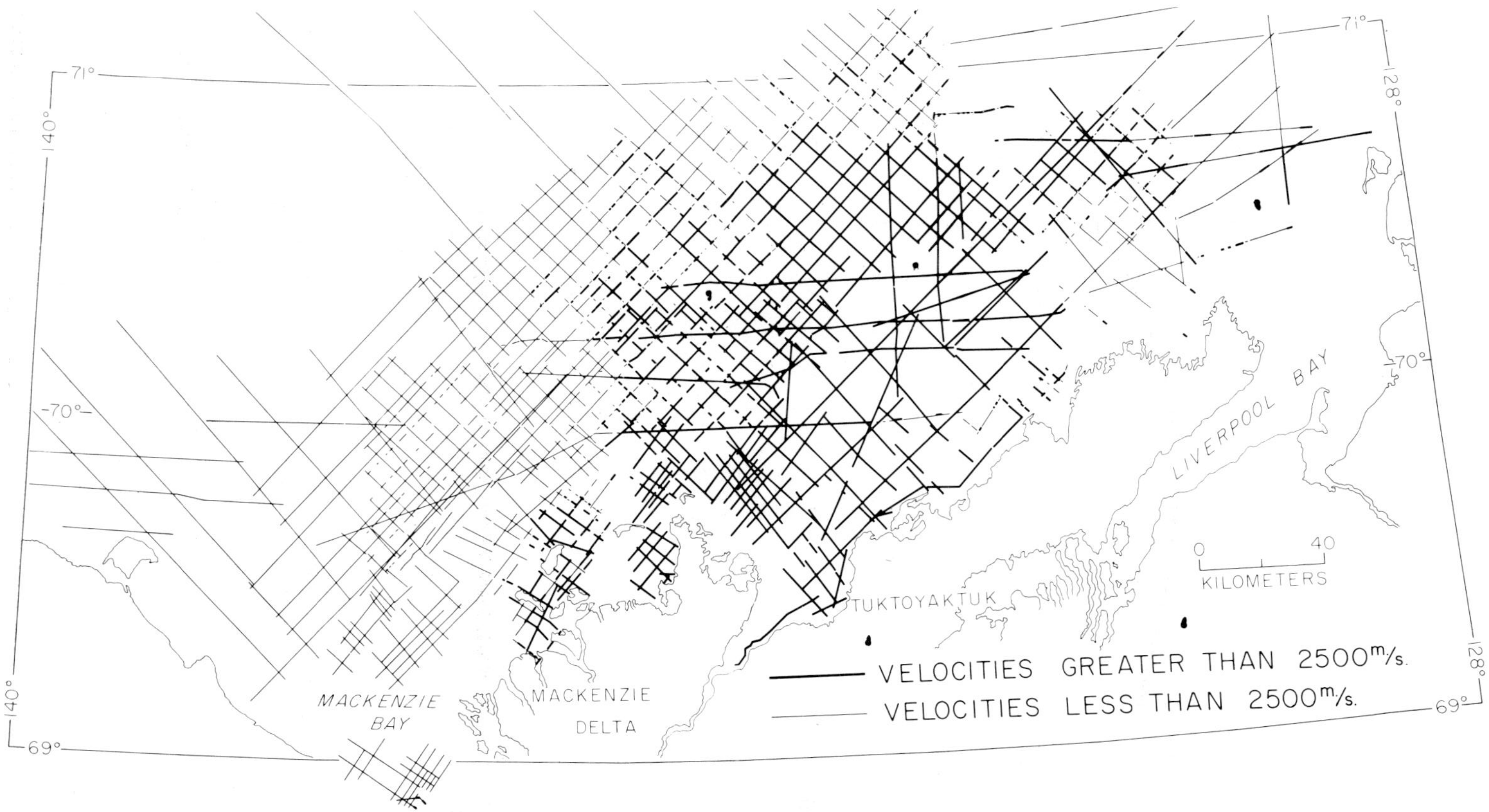

FIGURE 12. The Canadian Beaufort Sea coverage of Hunter et al.[29] obtained using industry marine seismic reflection data. The area to the northeast of the Mackenzie delta is extensively underlain by frozen materials, while to the northwest of the delta these materials are absent. (From Hunter, J. A., Neave, K. G., MacAulay, H. A., and Hobson, G. D., *Proc. 3rd Int. Permafrost Conf.*, National Research Council of Canada, Edmonton, Alta., 1978, 514. With permission.)

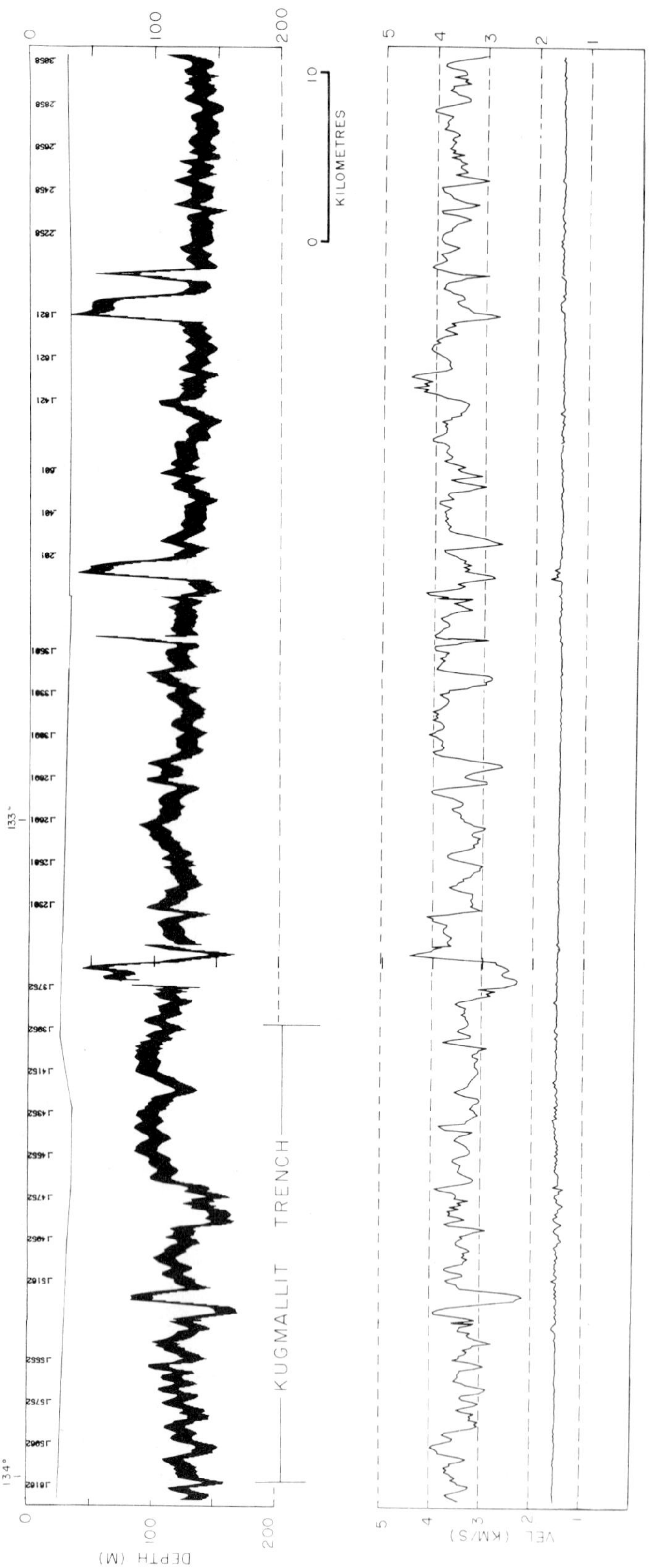

FIGURE 13. A vertical section through the Canadian Beaufort Sea prepared by MacAuley and Hunter.[31] The upper permafrost surface varies between 100 and 150 m beneath the ocean surface. There is considerable variability in the permafrost velocities observed, but they are well above the 2500 m/sec threshold for interpretation of frozen materials. (From MacAulay, H. A. and Hunter, J. A., *Proc. 4th Can. Permafrost Conf.*, National Research Council of Canada, Calgary, Alta., 1981. With permission.)

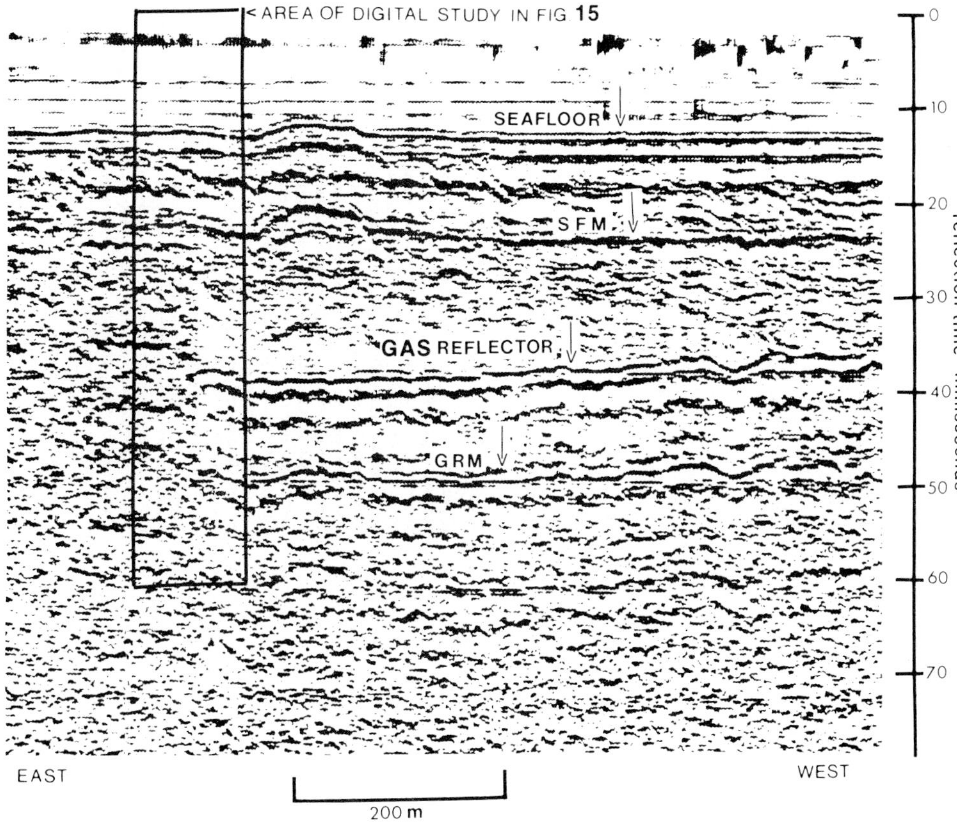

FIGURE 14. Unprocessed high resolution reflection record from the vicinity of Prudhoe Bay, Alaska. Boucher[32,33] correlated the gas reflector with adjacent borehole data. Sea floor multiples (SFM) and gas reflector multiples (GRM) are located about 13 msec below the primary reflectors. (From Boucher, G., *Proc. 50th Annu. Meet.*, Society of Exploration Geophysicist, Houston, Tex., 1980. With permission)

this would represent an upper limit to the volume expansion of the frozen sediments over the nonfrozen sediments.) Thus there is a significant acoustic wave impedance discontinuity expected at the interface.

High resolution reflection work in the Alaskan Beaufort Sea has been reported by Boucher et al.[32,33] One purpose of the work was to apply vertical seismic reflection techniques, with digital signal processing, to the mapping and characterization of ice-bonded material in the offshore environment near Prudhoe Bay, Alaska. The reflection data are recorded on analog magnetic tape and later digitized. A 300 J Uniboom® source at about 0.25 m depth was used with a 20 element, single channel, hydrophone streamer towed at a depth of 1 m. The passband of the system was 150 to 2000 Hz. A typical reflection profile is shown in Figure 14. By correlating the seismic record with a drill log obtained 600 m away[34] (the character of the record shown in Figure 14 was maintained to the location of the drill hole), a shallow gas reflector was identified. This reflector appears as a negative reflector in Figure 15, which is a deconvolved version of the data outlined by the box in the previous figure. Zero-phase wavelet deconvolution was applied to the seismic trace. The recorded wavelet, along with the zero-phase deconvolved wavelet, is shown in Figure 15.

Compressional wave velocities in these sediments were observed by Morack and

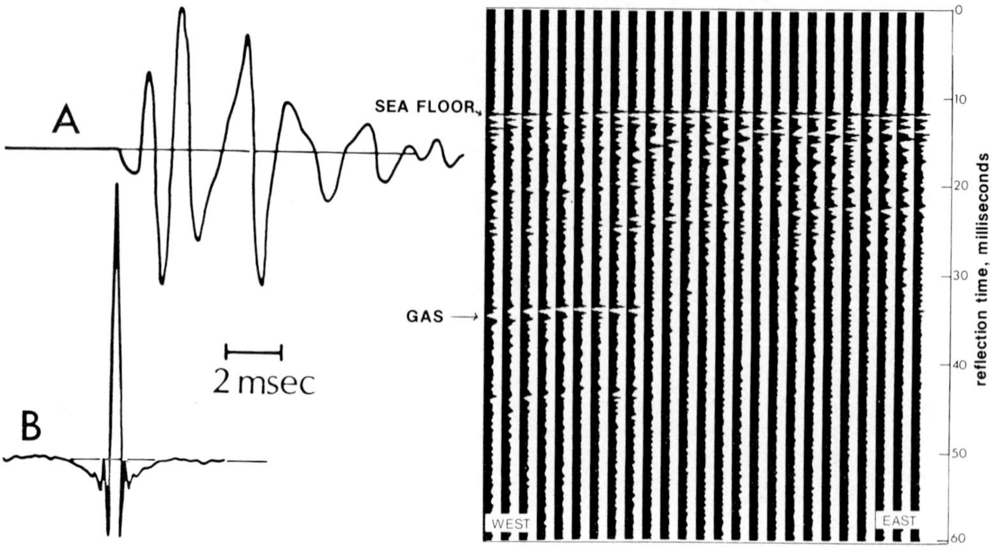

FIGURE 15. Deconvolved data from area within box of Figure 14. The sea floor reflector is seen to be positive while the gas reflector is negative. The received seismic wavelet is shown before (A) and after (B) deconvolution. (From Boucher, G., *Proc. 50th Annu. Meet.*, Society of Exploration Geophysicists, Houston, Tex., 1980. With permission.)

Rogers[13] to be approximately 2000 m/sec. Thus, in the nonfrozen sediments, each 10 msec of reflection time corresponds to a 10 m sediment section. Boucher was able to trace the gas reflector over an area of at least 50 km^2 but nowhere was a reflector that correlated with the depth to the upper ice-bonded permafrost surface identified. Borehole data[34] that were obtained to a depth of 36.5 m indicated possible ice lenses from 25.5 m to 30.5 m with ice-bonded materials starting at 30.5 m. Gas bearing sandy gravels were encountered at 32.5 m and these correlated with the negative reflection coefficient previously mentioned. (All depths are from the ocean surface, the water depth was 7.5 m and tidal effects are negligible.)

Permafrost observation in Prudhoe Bay area boreholes has correlated well with shallow refraction observations of Rogers and Morack.[27] Their refraction data closest to the borehole data used by Boucher is 2.6 km away, and permafrost was observed at a depth of 31.5 m beneath the surface. As the depth of the upper permafrost surface observed by refraction techniques is reasonably well-behaved in the area, and the upper surface gave consistent high velocities, it is believed that there is good corroboration of the drilling data by the refraction data.

It is postulated that the upper permafrost surface is sufficiently rough and perhaps discontinuous, so that little high frequency energy is specularly reflected. The energy received as a refraction event from the permafrost is perhaps an order of magnitude lower in frequency than a typical Uniboom® signal. This difference could account for the different geophysical results.

In addition to the work conducted by Boucher, the U.S. Geological Survey has conducted routine reflection profiles along lines totaling hundreds of kilometers in the Alaskan Beaufort Sea. Sources have included sparkers, boomers, and subbottom profilers, but no positive identification of the upper permafrost surface has been made (Reinmitz et al.[35]). Reasons for the discrepancy between refraction and reflection results in the Alaskan Beaufort Sea remain to be established. However, results in the Canadian Beaufort Sea have been more comparable.

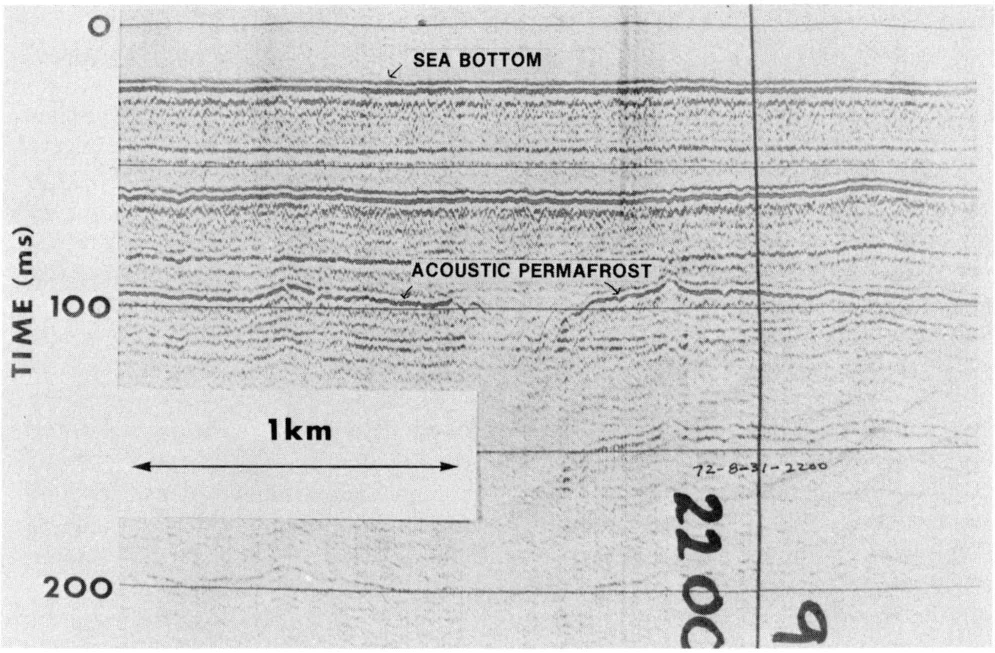

FIGURE 16. A vertical reflection section reported by Blasco and O'Conner[36] in the Canadian Beaufort Sea. The discontinuous reflector shown has been identified as permafrost.(From Blasco, S. M. and O'Conner, M. J., Proc. 4th Can. Permafrost Conf., *National Research Council of Canada,* Calgary, Alta., 1981. With permission.)

Acoustic reflection work conducted in the Canadian Beaufort Sea has been reported by Blasco and O'Conner.[36] Several different reflection systems were examined and varying results were achieved with each. They interpreted and characterized four types of acoustic permafrost: continuous reflectors that conformed to the lithology, continuous reflectors that were not conformable to the lithology, sporadic reflectors associated with a rough upper surface, and isolated strong reflectors interpreted to be permafrost lenses.

Figure 16 is a vertical reflection section discussed by Blasco and O'Conner.[36] They have correlated reflectors identified as permafrost with refraction and drilling data. The data gathering system included two air guns with 5 $in.^3$ and 1 $in.^3$ chambers, a broadband hydrophone receiver, and a graphic recorder. The permafrost that they observed was sometimes coincident with shallow gas, and there is speculation that the gas may be produced by the thermal decomposition of gas hydrates. A later section which discusses gas hydrates shows that seismic velocities in hydrates are probably indistinguishable from those in ice-bonded permafrost. It may be that shallow gas occurring with permafrost can be used as an indicator of the presence of hydrates.

Estimating Permafrost Thickness

Seismic techniques, particularly seismic refraction methods, have been successfully used to map the regional distribution of permafrost and to indicate the depth to its upper surface. Estimating the permafrost thickness is more difficult, and in this case seismic refraction methods cannot be applied due to the velocity inversion that exists at the lower permafrost boundary. It is desirable to estimate offshore permafrost thickness for use in well drilling designs and for static corrections of seismic reflection data.

Given the substantial thickness possible, on the order of 500 m, and the high velocity contrast of frozen and nonfrozen materials, static corrections can be quite important in the interpretation of deep reflection data.

Hunter et al.[37] have discussed three seismic techniques that can potentially be applied to determine permafrost thickness. In permafrost layers that are thin with respect to the signal wavelength, signal attenuation has been used to estimate the thickness of layers less than 10 m. Because of the wavelength limitation of this method and the very long receiver arrays necessary to observe attentuation along permafrost layers, it is not easily applied to thick permafrost. Thicker layers are possibly more amenable to thickness determination using guided wave dispersion theory. In order to use this method it is necessary to clearly identify the guided wave modes of the received signals. The third technique involves observing energy reflected from the lower permafrost boundary. While in theory this method can work, the industry records that were examined by Hunter et al.[37] were not of sufficient quality to identify bottom reflections with certainty.

Estimates of permafrost thickness have been made using industry deep reflection data.[38,39] The technique of MacCallum and Card,[39] which is applicable to estimating lateral variations in permafrost thickness, depends upon identifying laterally continuous seismic reflections from well below the suspected permafrost zone. The interpretation is made using the stacking velocity for the underlying layer. Anomalies in the stacking velocity are fitted in an interactive manner by adjusting the thickness of the permafrost layer. Figure 17, from MacCallum and Card,[39] illustrates results obtained with this method.

Another method, described by Baird,[40] involves lowering seismic detectors into wells drilled near the ocean edge. A series of seismic data sets were then gathered as the source was moved sequentially farther from the location of the well. After analysis of seismic travel time, a model was prepared indicating the variations in permafrost thickness with increasing distance from shore.

Card[41] has used an array of shots on a grid over 2 km × 3 km in size. The grid, located a few kilometers from shore in the Beaufort Sea, was roughly centered on a well which provided downhole records. These were used to obtain the depth to the bottom of the permafrost. A line of surface geophones was used to obtain the depth to the upper permafrost surface in the center of the grid.

Although we have restricted our previous discussion to estimating the thickness of a permafrost model composed of only one layer, this is only the simplest possible model and not necessarily the best representation of offshore permafrost.

Neave et al.[42] observed the tops of two permafrost layers in parts of the Canadian Beaufort Sea. Their velocity data were divisible into an upper high-velocity group, with a top 60 to 100 m below sea level, for which the seismic records showed an attenuated, intermittent signal; and a lower group (130 to 200 m below sea level) for which their records showed a strong continuous signal. Using the attenuation of the upper layer signals they interpreted the layer to have a thickness of 10 to 30 m. It is presumed that the thickness of the lower layer is limited by the maximum depth expected for permafrost in the region, perhaps 600 to 700 m. Thus, the thickness of the lower permafrost layer may be estimated.

Electrical methods have been used to estimate permafrost thickness on land.[17,43,44] These measurements have been made on permafrost which is of the order of 100 m thick, but little work has been reported in the offshore environment. Hoekstra and McNeill[23] have suggested that, in shallow water locations, a time domain electromagnetic sounding technique could be used to determine the thickness of deep permafrost.

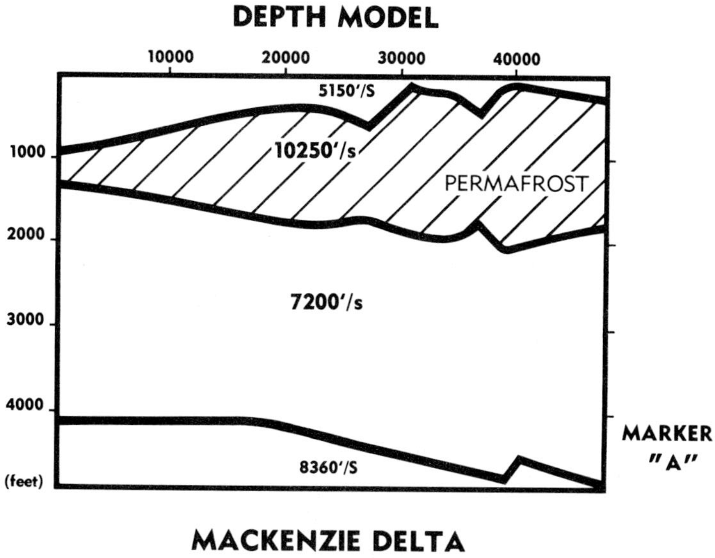

FIGURE 17. Permafrost thickness variation as determined by MacCallum and Card[39] using industry seismic reflection data. Their model thinned simultaneously from top and bottom, a fact that accounts for the symmetry of the permafrost cross section. They observed that this model fit the data rather well. (From MacCallum, R. and Card, J., *J. Can. Soc. Exploration Geophys.*, 1978, 55. With permission.)

Detection of Hydrates

The subsea environment, that has been shown to host frozen materials or permafrost, according to the definition used here, is also suitable for the maintenance of stable clathrate hydrates. These consist of a frozen water and gas mixture existing at temperatures below 0°C and at pressures greater than atmospheric pressure. The gas might be methane, propane, or some other naturally occurring gas. Several papers have been presented on this subject, as the presence of hydrates can affect drilling procedures in the exploration for offshore petroleum resources.[45-48] Potential problems encountered during drilling through hydrates are associated with freeing the gas present by an influx of heat during drilling.[45,48] Utilization of this feature for the production of gas occurring in the hydrate form has been discussed, since it is speculated that these reserves contain significant quantities of gas.[47]

Geophysical detection of subsea hydrates depends upon the contrast of their seismic or electrical properties with that of their host materials. Pandit and King[49] have performed laboratory tests of propane gas hydrates. The compressional wave velocity for a typical hydrate specimen changed from approximately 3200 m/sec to approximately 2100 m/sec as the temperature was raised from −2°C to +1 °C. The decrease in velocity might be due to the melting of excess ice (10 to 20%) in the hydrate specimen. Thus, good seismic contrast exists. However, the range of compressional wave velocities observed for hydrates is quite like that for sand or silt (see Figure 3A). It appears, therefore, that compressional wave velocities cannot be used to distinguish gas hydrates from other ice-bonded permafrost. Preliminary studies suggest that the elastic attenuation in hydrates is greater than in ice at comparable temperatures and pressures and, therefore, it is possible that attenuation might be used to infer the presence of hydrates.[49] Hunter et al.[29] speculated on the possibility that gas hydrates might be responsible for the anomalous velocities that they observed in industry seismic data, but no confirmation has yet been made.

CONCLUSIONS

The study of offshore permafrost is a relatively new endeavor and the application of geophysical techniques to the problem has been common only since the mid-1970s.[25,50,51] Presently, seismic techniques have been the most successful; the regional distribution and depth to the upper surface of marine permafrost have been successfully determined in many areas.[26,31] Additional work is needed in characterizing the permafrost thickness and to determine the permafrost variability.

Offshore resource exploration has been, and will probably continue to be, a prime motivation for studies of this interesting aspect of polar regions. It is expected that many of the geophysical techniques applied to permafrost on land[52] will be modified for marine investigations. Electrical methods, which have recently been more common on land,[53-56] may find application in marine work. Whatever the technique used, geophysical data will continue to aid in the study of processes along the Arctic coast. These processes, including the effects of barrier islands, are closely related to the offshore permafrost in existence today.[57-63] We note that there are a great deal of additional data available that have not been discussed here. While Vigdorchik[64] has compiled an extensive bibliography of the Soviet literature on permafrost, our attention has been restricted largely to the Canadian and American literature.

REFERENCES

1. **Ferrians, O. J., Kachadourink, R., and Greene, G. W.,** Permafrost and Related Engineering Problems in Alaska, *U.S. Geol. Sur. Prof. Pap.* 678, 1969.
2. **Shackleton, N. J. and Opdyke, N. D.,** Oxygen isotope and paleomagnetic stratigraphy of equatorial Pacific core V28-238: oxygen isotope temperatures and ice volumes on a 10^5 year and 10^6 year scale, *Quaternary Res. (N.Y.),* 3, 39, 1973.
3. **Hopkins, D. M. and Hartz, R. W.,** Coastal Morphology, Coastal Erosion, and Barrier Islands of the Beaufort Sea, Alaska, open File Rep. 78-1063, U.S. Geological Survey, Washington, D.C., 1978.
4. **Lachenbruch, A. H.,** Thermal effects of the ocean on permafrost, *Geol. Soc. Am. Bull.,* 68, 1515, 1957.
5. **Mackay, J. R.,** Offshore permafrost and ground ice, southern Beaufort Sea, *Can. J. Earth Sci.,* 9, 1550, 1972.
6. **Harrison, W. D. and Osterkamp, T. E.,** Theoretical models for subsea permafrost, in *Proc. 3rd Int. Conf. Port and Ocean Eng. Under Arctic Conditions,* Burrell, D. C. and Hood, K. W., Eds., University of Alaska Institute of Marine Sciences, Fairbanks, Alaska, 1976, 1011.
7. **Harrison, W. D. and Osterkamp, T. E.,** Heat and mass transport processes in subsea permafrost, I: an analysis of molecular diffusion and its consequences, *J. Geol. Res.,* 1978.
8. **Osterkamp, T. E. and Harrison, W. D.,** Subsea permafrost regime at Prudhoe Bay, Alaska, *J. Glaciology,* 19, 91, 1977.
9. **King, M. S. and Pandit, B. I.,** Some seismic, electrical, and thermal properties of subseabottom permafrost samples from the Beaufort Sea, in *Proc. 4th Can. Permafrost Conf.,* National Research Council of Canada, Calgary, Alta., 1981.
10. **King, M. S.,** Acoustic velocities and electrical properties of frozen sandstones and shales, *Can. J. Earth Sci.,* 14, 1004, 1977.
11. **Pandit, B. I. and King, M. S.,** Influence of pore fluid salinity on seismic and electrical properties of rocks at permafrost temperatures, in *Proc. 3rd Int. Conf. Permafrost,* National Research Council of Canada, Edmonton, Alta., 1978, 553.
12. **Pandit, B. I. and King, M. S.,** A study of the effects of pore-water salinity on some physical properties of sedimentary rocks at permafrost temperatures, *Can. J. Earth Sci.,* 16, 1566, 1979.

13. **Morack, J. L. and Rogers, J. C.**, Marine refraction measurements of permafrost characteristics, in *Proc. 4th Can. Permafrost Conf.*, National Research Council of Canada, Calgary, Alta., 1981.
14. **Hoekstra, P.**, Electromagnetic methods for mapping shallow permafrost, *Geophysics*, 43, 782, 1978.
15. **Arcone, S. A., Sellmann, P. V., and Delaney, A. J.**, Shallow electromagnetic geophysical investigations of permafrost, in *Proc. 3rd Int. Conf. Permafrost*, National Research Council of Canada, Edmonton, Alta., 1978, 501.
16. **Davis, J. L., Scott, W. J., Morey, R. M., and Annan, A. P.**, Impulse radar measurements on permafrost near Tuktoyaktuk, N. W. T., *Can. J. Earth Sci.*, 13, 1584, 1976.
17. **Daniels, J. J., Keller, G. V., and Jacobson, J. J.**, Computer-assisted interpretation of electromagnetic soundings over a permafrost section, *Geophysics*, 41, 752, 1976.
18. **Koziar, A. and Stangway, D. W.**, Permafrost mapping by audio frequency magnetotellurics, *Can. J. Earth Sci.*, 10, 1539, 1978.
19. **Olhoeft, G. R.**, Electrical properties of permafrost, in *Proc. 3rd Int. Conf. Permafrost*, National Research Council of Canada, Edmonton, Alta., 1978, 127.
20. **Olhoeft, G. R.**, The Electrical Properties of Permafrost, Ph.D. thesis, University of Toronto, 1975.
21. **Katsube, T. V., Wadleigh, M., and Erickson, R.**, Electrical Properties of Permafrost Samples, Pap. 76-1C, Geological Survey of Canada, Ottawa, 1976, 83.
22. **Scott, W. J.**, Preliminary Experiments in Marine Resistivity Near Tuktoyaktuk, District of Mackenzie, Pap. 75-1A, Geological Survey of Canada, 1975, 141.
23. **Hoekstra, P. and McNeill, D.**, Time domain electromagnetic sounding for deep permafrost, in *Proc. 4th Can. Permafrost Conf.*, National Research Council of Canada, Calgary, Alta., 1981.
24. **Hunter, J. A. M.**, The application of shallow seismic methods to mapping of frozen surficial materials, in *Proc. 2nd Int. Conf. Permafrost*, National Academy of Sciences, Washington, D.C., 1974.
25. **Hunter, J. A. M. and Hobson, G. D.**, A seismic refraction method to detect subbottom permafrost, in *Beaufort Sea Symposium Proceedings*, Reed, J. C. and Sater, J. E., Eds., Arctic Institute of North America, San Francisco, 1974, 401.
26. **Rogers, J. C. and Morack, J. L.**, Geophysical evidence of shallow nearshore permafrost, Prudhoe Bay, Alaska, *J. Geol. Res.*, 85, 4845, 1980.
27. **Rogers, J. C. and Morack, J. L.**, Beaufort seacoast permafrost studies, in Environmental Assessment of the Alaskan Continental Shelf, U.S. National Oceanographic and Atmospheric Administration, Rockville, Md., 1980.
28. **Hunter, J. A., Judge, A. S., MacAulay, H. A., Gagne, R. M., Burns, R. A. and Good, R. L.**, The Study of Frozen Seabed Materials in the Southern Beaufort Sea, Tech. Rep. No. 22, Beaufort Sea Environmental Project, Environment Canada, Victoria, B.C., 1976.
29. **Hunter, J. A., Neave, K. G., MacAulay, H. A., and Hobson, G. D.**, Interpretation of subseabottom permafrost in the Beaufort Sea by seismic methods, Part I: Seismic refraction methods, in *Proc. 3rd Int. Permafrost Conf.*, Vol. 1, National Research Council of Canada, Edmonton, Alta., 1978, 514.
30. **Sellmann, P. V., Chamberlain, E., Delaney, A., and Neave, K. G.**, Delineation and Engineering Characteristics of Permafrost Beneath the Beaufort Sea, Outer Continental Shelf Environ. Assessment Program, National Oceanic and Atmospheric Administration, Washington, D.C., 1980.
31. **MacAulay, H. A. and Hunter, J. A.**, Detailed seismic refraction analysis of ice-bonded permafrost layering in the Canadian Beaufort Sea, in *Proc. 4th Can. Permafrost Conf.*, National Research Council of Canada, Calgary, Alta., 1981.
32. **Boucher, G., Reinmitz, E., and Kempema, E.**, Seismic evidence for an extensive gas-bearing layer at shallow depth, offshore from Prudhoe Bay, Alaska, *Cold Reg. Sci. Technol.*, 4, 63, 1981.
33. **Boucher, G.**, Digital high-resolution study of shallow frozen and gas-bearing layers, offshore Prudhoe Bay, Alaska, in *Proc. 50th Annu. Meet., Society of Exploration Geophysicists*, Houston, Tex., 1980.
34. U.S.G.S., Geotechnical investigation Beaufort Sea 1979, Ref. #AK17718, Natl. Geophys. Solar-Terrestrial Data Cent. U.S. Geological Survey, Boulder, Colo., 1979.
35. **Reinmitz, E., Wolf, S. C., and Rodeick, C. A.**, Preliminary Interpretation of Seismic Profiles in the Prudhoe Bay Area, Beaufort Sea, Alaska, Open File Rep. No. 548, U.S. Department of the Interior, Washington, D.C., 1972.
36. **Blasco, S. M. and O'Conner, M. J.**, Acoustic characteristics of subsea permafrost in the southern Beaufort Sea, in *Proc. 4th Can. Permafrost Conf.*, National Research Council of Canada, Calgary, Alta., 1981.
37. **Hunter, J. A., Neave, K. G., MacAulay, H. A., and Hobson, G. D.**, Interpretation of subseabottom permafrost in the Beaufort Sea by seismic methods, Part II, estimating the thickness of the high velocity layer, in *Proc. 3rd Int. Conf. Permafrost*, National Research Council of Canada, Edmonton, Alta., 1978, 521.
38. **Acheson, C. H.**, Estimating permafrost thickness using permafrost velocities, in *Proc. Symp. Permafrost Geophys., Tech. M-124*, Scott, W. J. and Brown, R. J. E., Eds., National Research Council of Canada, Ottawa, 1979, 101.

39. **MacCallum, R. and Card, J.,** Velocity modeling: the interpreters' magnifying glass, *J. Can. Soc. Exploration Geophys.,* 1978, 55.
40. **Baird, R. J.,** A cross sectional view for permafrost feathering out under sea ice, in *Proc. Symp. Permafrost Geophys., Tech. M-119,* Scott, W. J. and Brown, R. J. E., Eds., National Research Council of Canada, Ottawa, 1977, 114.
41. **Card, J. D.,** Permafrost in the Mackenzie Delta: a detailed study around the SUN BVX et al. UNARK L-24A well site, in *Proc. Symp. Permafrost Geophys., Tech. M-124,* Scott, W. J. and Brown, R. J. E., Eds., National Research Council of Canada, Ottawa, 1979, 125.
42. **Neave, K. G., Judge, A. S., Hunter, J. A., and MacAuley, H. A.,** Offshore Permafrost Distribution in the Beaufort Sea as Determined from Temperature and Seismic Observations, Pap. 78-1C, Geological Survey of Canada, Ottawa, 1978, 13.
43. **Arcone, S. A., Sellmann, P. V., and Delaney, A. V.,** Effects of Seasonal Changes and Ground Ice on Electromagnetic Surveys of Permafrost, CRREL Res. Rep., U.S. Army, Washington, D.C., 1979.
44. **Scott, W. J. and Mackay, J. R.,** Reliability of permafrost thickness determination by DC resistivity soundings, Tech. M-119, National Research Council of Canada, Ottawa, 1977, 25.
45. **Judge, A.,** Natural gas hydrates in northern Canada, in *Proc. 4th Can. Permafrost Conf.,* National Research Council of Canada, Calgary, Alta., 1981.
46. **Goodman, M. A. and Franklin, L. J.,** Thermal model of new concept of hydrate control during drilling, in *Proc. 4th Can. Permafrost Conf.,* National Research Council of Canada, Calgary, Alta., 1981.
47. **McGuire, P. L.,** Methane hydrate gas production by thermal stimulation, in *Proc. 4th Can. Permafrost Conf.,* National Research Council of Canada, Calgary, Alta., 1981.
48. **Weaver, J. S. and Purcell, D.,** *In situ* hydrates under the Beaufort Sea shelf, in *Proc. 4th Can. Permafrost Conf.,* National Research Council of Canada, Calgary, Alta., 1981.
49. **Pandit, B. I. and King, M. S.,** Compressional and shear wave velocities of propane gas hydrates, in *Proc. 4th Can. Permafrost Conf.,* National Research Council of Canada, Calgary, Alta., 1981.
50. **Rogers, J. C., Shapiro, L. H., Gedney, L. D., and VanWormer, D.,** Nearshore permafrost in the vicinity of Pt. Barrow, Alaska, in *Proc. 3rd Int. Conf. Port and Ocean Eng. Under Arctic Conditions,* Fairbanks, Alaska, 1975, 1071.
51. **Rogers, J. C. and Morack, J. L.,** Geophysical investigation of offshore permafrost, Prudhoe Bay, Alaska, in *Proc. 3rd Int. Conf. Permafrost,* Vol. I, Edmonton, Alta., 1978, 560.
52. **Scott, W. J., Sellman, P. V., and Hunter, J. A.,** Geophysics in the study of permafrost, in *Proc. 3rd Int. Conf. Permafrost,* Vol. II, National Research Council, Edmonton, Alta., 1978, 93.
53. **Scott, W. J., Campbell, K. J., and Orange, A. S.,** EM pulse survey method in permafrost, Tech M-113, National Research Council of Canada, Ottawa, 1974, 92.
54. **Scott, W. J.,** VLF Resistivity (Radiohm Survey), Agricola Lake Area, District of Mackenzie, Pap. 75-1A, Geological Survey of Canada, 1975, 223.
55. **Hoekstra, P., Sellmann, P. V., and Delaney, A. J.,** Ground and airborne resistivity surveys of permafrost near Fairbanks, Alaska, *Geophysics,* 14, 641, 1975.
56. **Sellmann, P. V., Delaney, A. J., and Arcone, S. A.,** An electromagnetic geophysical survey at an interior Alaska permafrost exposure, CRREL Spec. Rep. SR-79-14, U.S. Army, Washington, D.C., 1979.
57. **Nummendal, D., Owens, E. H., and Harper, J. R.,** Sediment transport processes and coastal variability on the Alaskan North Slope, in *Proc. 17th Int. Coastal Eng. Conf.,* Sydney, Australia, 1980.
58. **Morack, J. L. and Rogers, J. C.,** Seismic evidence of shallow permafrost beneath islands in the Beaufort Sea, Alaska, *Arctic* (in press).
59. **Sellmann, P. V.,** Regional distribution and characteristics of bottom sediments in Arctic coastal waters of Alaska, review of current literature, CRREL Spec. Rep. SR-80-15, U.S. Army, Hanover, N.H., 1980.
60. **Harper, J. R., Owens, E. H., and Wiseman, W. J., Jr.,** Arctic beach processes and the thaw of ice-bonded sediments in the littoral zone, in *Proc. 3rd Int. Conf. Permafrost,* Vol. I, National Research Council, Edmonton, Alta., 1978, 195.
61. **Hopkins, D. M.,** Coastal processes and coastal erosional hazards to the Cape Krusenstern archaeological site, Open File Rep. 77-32, U.S. Geological Survey, Menlo Park, Calif., 1977.
62. **Smith, P. A. and Hopkins, D. M.,** Offshore permafrost studies and shoreline history of Chuckchi and Beaufort Seas as an aid to predicting offshore permafrost conditions, Outer Continental Shelf Environ. Assessment Program, National Oceanographic and Atmospheric Administration, Washington, D.C., 1979.
63. **Vigdorchik, M. E.,** *Arctic Pleistocene History and the Development of Submarine Permafrost,* Westview Press, Boulder, Colo., 1980.
64. **Vigdorchik, M. E.,** *Submarine Permafrost on the Alaskan Continental Shelf,* Westview Press, Boulder, Colo., 1980.

ACOUSTIC CLASSIFICATION OF THE SEABED

N. G. Pace

INTRODUCTION

Acoustic techniques provide a rapid, remote method for the examination of the sea floor and the subbottom. Echo sounders have long been used to provide bathymetric information along the line of the ship track. Subbottom profilers, which are essentially similar to echo sounders but are usually operated at lower frequencies, can provide data from which the subbottom interfaces between materials of different acoustic impedance can be inferred. Some acoustic sources are designed to provide acoustic returns from a line of the seabed normal to the ship track. These so-called side-scan sonars thus provide information on a wide swath of the sea bottom as the ship moves.

This chapter is concerned with the information pertaining to the surficial sediments of the sea floor that can be obtained acoustically. Common acoustic sources produce a short burst of narrow-band acoustic energy which is directed into a limited solid angle or beam towards the seabed. The sudden change in the acoustic properties at the water/sediment interface causes a proportion of the incident energy to be reflected. In addition, any roughness of the water/sediment interface introduces scattering of the incident energy. Usually, but not always, the receiver is the same transducer as the source, and in such cases parameters which are important in determining the received signal level are the normal incidence reflection coefficient and the backscattering strength[1] of the interface. The latter is a measure of the incident energy scattered into the direction of incidence. Thus, the acoustic energy returned to the receiver from the sea bottom depends on the acoustic impedance of the sediment, the interface roughness, and the geometry of the situation. In addition to data on acoustic impedance and interface roughness, by using some specific techniques it is possible to obtain remotely data on the velocity[2] of longitudinal sound waves and on the attenuation[3] of sound in the sediment, the latter as a function of both frequency and penetration depth.

In practice, the acoustic data collected even from closely spaced positions above the same sea bottom exhibit a statistical spread. In addition to mean values of specific data there are many instances, to be reviewed below, in which measures of their statistical nature are also treated as data.

The activity of sediment classification can be viewed in two ways. In one, it is a case of examining the extent to which the acoustic data can provide features that are of value in discriminating one area of sea bottom from another and can be approached via clustering algorithms. In the other, the process is really one of sediment identification and requires extensive bottom truth data[4] so that correlations may be established between the acoustic data and the physical parameters of the sediments.

REFLECTION COEFFICIENT AND SEDIMENT PROPERTIES

Where the sea floor appears smooth in terms of the acoustic wavelengths, and when the top sediment layer is sufficiently thick that interference between reflections from the water/sediment interface and subbottom interfaces does not occur, simple models[5] of the reflection process are possible. At normal incidence, the ratio of reflected to incident acoustic intensity is

$$R = (\rho_1 c_1 - \rho_2 c_2)^2/(\rho_1 c_1 + \rho_2 c_2)^2$$

where ϱ and c are, respectively, the density and velocity of sound; and the subscripts 1 and 2 refer, respectively, to sediment and water. A number of studies of the relationships between physical properties of sediments such as porosity and grain size and the acoustic reflectivity are available.[6-10] In general, the reflectivity bears an inverse relationship to porosity. High porosity is associated with fine grained sediments and low porosity with coarse grained ones; so that clays tend to be poor reflectors while sand tends to be a good reflector.

Measurement of acoustic reflection coefficients of the water/sediment interface at angles of incidence other than normal require the acoustic source and receiver to be separated and the practical arrangements become more complicated and expensive.[2] However, at off-normal angles of incidence the reflection coefficient is more sensitive to changes in sediment parameters than it is at normal incidence.[11]

SURFACE ROUGHNESS EFFECTS AT NORMAL INCIDENCE

In the context of sediment classification studies of the sea bottom, interface roughness concerns the small-scale departures from a plane interface rather than the large-scale variations, which are sometimes called the geography or topography of the sea bottom. The latter can be mapped using time of arrival data obtained with echo sounders, whereas acoustically characterizing the small-scale roughness or texture of the sea bottom which can bear a close relationship with the sediment descriptors such as particle size,[12-14] is a much more demanding task as will be described below. If a large number of the scatterers associated with the roughness are within the area of the bottom which contributes to the return signal at any one time, then that roughness may be described as texture rather than topography.

A rough interface can be described by the rms height relative to a level such that the mean height is zero together with the autocorrelation function of the interface heights.[15-17]

There is an enormous volume of literature on the subject of scattering of waves from rough surfaces.[18-20] It is not the purpose of this article to review such work, but to outline ways in which the acknowledgment of interface roughness affects the sediment classification exercise. The discussion will be initially restricted to the case of normal incidence and that of a coincident source and receiver.

If $P_i(t)$ represents the time domain pressure incident on the interface, and $P_r(t)$ that returned to the receiver, then various quantities may be introduced.

1. The coherent return is the ensemble average $\langle P_r(t) \rangle$, obtained by averaging the returns as the source/receiver combination moves horizontally at a fixed distance from the mean plane of the water/sediment interface. If the interface is smooth, then the coherent return has the same time waveform as $P_i(t)$ and an amplitude which varies as the inverse square of the range to the interface. As the interface becomes rough, the energy in the coherent return reduces and its time domain form becomes, to a good approximation, the convolution of the probability distribution of the interface heights, considered in terms of the time-delays they introduce, with the incident waveform.[19-21] Thus, in principle the time stretching of the coherent return provides information on the interface height statistics.
2. The coherent intensity is

$$I_c(t) = \langle P_r(t) \rangle^2$$

3. The total returned intensity I(t) is the sum of the coherent and incoherent intensities

$$I(t) = I_c(t) + I_{inc}(t)$$
$$= \langle P_r^2(t) \rangle$$

It is useful to note that in most practical situations the energy ($\propto_o I(t) \cdot dt$) received back from a totally reflecting rough interface is, on average, the same as would be returned from a plane interface.[22]

4. The incoherent intensity

$$I_{inc}(t) = \langle (P_r(t) - \langle P_r(t) \rangle)^2 \rangle$$

represents the received intensity due to the scattering of the incident energy into the direction of incidence. The time domain form of the incoherent intensity is dependent upon the duration of the incident pressure waveform $p_i(t)$, the beam pattern of the acoustic transducer, the height statistics of the sea bottom, and the angular dependence of the sea bottom backscattering strength.[23-24] For a conventional transducer, the dependence of the incoherent intensity on the range r of the source/receiver combination from the sea bottom varies between r^{-2} and r^{-3}, depending on whether the insonnified patch on the sea bottom is limited in area by the beam pattern or the transmitted pulse length, respectively.[25]

If the received intensity I(t) has a "tail" and is thereby of significantly longer duration than the transmitted waveform, such a tail may be attributed to the incoherent component since elongation of the signal beyond that due to the roughness amplitude must be due to signals arriving from off-normal directions.[26] However, confusion can arise if there is shallow layering in the sea bottom. An elaborate treatment[26] of the incoherent tail interprets its form in terms of the information it contains concerning the interface characteristics.

A number of authors[14-20] have discussed the scattering of acoustic waves from a rough interface in terms of the mean scattered intensity as a function of σ/λ, the ratio of the rms interface heights to acoustic wavelength. For $\sigma/\lambda \ll 1$, the normal incidence backscattered return is coherent and determined by the reflection coefficient. The mean backscattered intensity is again independent of frequencies for $\sigma/\lambda \gg 1$, although its magnitude is usually much reduced. The intermediate range, which is centered on

$$\sigma/\lambda \simeq \frac{1}{2\pi}$$ [13]

is characterized by a complicated dependence[15] of the mean scattered intensity on frequency, this being the region bridging the coherent and incoherent regimes. Various authors[13,27-29] have exploited the usefulness of the frequency dependence of the mean scattered intensity in this region for remotely determining interface roughness. In addition to mean values, parameters of the rough interface influence the statistics of the backscattered signal. At normal incidence the distribution varies from Gaussian for the coherent ($\sigma/\lambda \ll 1$) returns to Rayleigh for the incoherent ($\sigma/\lambda \gg 1$) returns.[18] A succinct way of expressing this change in the form of the statistics is found in the coefficient of variation v, which is the ratio of the standard deviation to the mean value of the signal. Experimental studies[12,13] have shown that a useful correlation exists between the value of v and the interface roughness expressed as the ratio σ/λ. In the case of wideband width systems, the fluctuations between successive return signals may be described in terms of their cross correlation or coherence functions.[30]

NORMAL INCIDENCE CLASSIFICATION STUDIES

A number of laboratory studies[13,27,31] have examined the information content of signals returned from artificial and real sediments. Such studies have concentrated on sensing the small-scale roughness or texture of the water/sediment interface and its relation with the sediment mean particle size using wide bandwidth acoustic pulses. The coefficient of variation of the return signal amplitude in a narrow frequency band has been shown to be well correlated with the ratio of sediment mean particle radius to the acoustic wavelength.[12,13] If wideband width pulses are used at sea to provide data directly on sediment particle sizes, then narrow beam widths are called for in order to avoid topographic effects. To this end, the transient parametric array[32] has recently been employed[12,33] both in the laboratory and in a small-scale sea experiment to provide an acoustic pulse with a bandwidth of 80 kHz centered on 80 kHz and a 3 dB energy beam width of ±2°. Coherent processing of the returns of such pulses has provided the mean particle size in the very coarse sand-pebble range of sediments. A small-scale experiment[33] with the same acoustic source at sea has separated sea bottoms of mud, medium sand, granules, and large pebbles one from another on the basis of measures of the time-spreading of the returned intensity waveform.

Although laboratory experiments can be useful in identifying fruitful approaches to remote sediment classification, the practicalities of a method are only fully appreciated following full-scale sea trials. Accounts of such sea experiments in the literature can, on the whole, be divided into two classes: those which employed gated CW pulses and those which employed large bandwidth pulses. In the former category, the signals returned from the seabed can be analyzed for their mean amplitude and its statistical spread, and for the form of the signal tail, while many more possibilities are presented when wideband width pulses are employed.

GATED CW PULSES

A review of progress in acoustic classification of sedimentary sea bottoms up to 1966[34] concluded that the dominant sediment property which determines the acoustic return amplitude is the porosity, a conclusion fully supported by more recent work.[6-10] A milestone in sediment classification studies from the point of view of scale, instrumentation, and success is provided by the work of Breslau.[35] A simple and useful theory is used to predict relationships between the reflection coefficient and sediment porosity together with the relations between sediment velocity, density, and porosity. The measurements of the bottom loss (= 10 log R) showed a good correlation with both sediment porosity and the percentage of silt and clay in the sediment. There now exists in the literature[6-10] a large amount of data concerning the interrelation between physical and acoustic properties of sediments.

An early study[36] that acknowledged to good effect the variability of acoustic returns from the sea bottom, employed a 14 kHz echo sounder, and reported that sand, silt, clay, and glacial till could be separately identified. The classification was based on the shape of the sea bottom interface as well as the relative degree of compaction of the bottom as interpreted from the depth to which sound penetrated the bottom.

In an account[37] of the measurement of the seabed reflectivity at the high frequency of 75 kHz, the importance of the interface roughness is discussed together with the dependence of the returned signal level on range to the sea bottom. The importance of the experimental geometry in determining the partition of energy between the coherent and incoherent components of the return signal has been pointed out[38] and the present understanding of the range dependence of signals returned from rough interfaces has recently been discussed.[39]

Principal component analysis has been used[40] to analyze the variability in shape and amplitude of the returns of a 7.5 kHz signal incident normally on the seabed. Using cluster analysis,[41] sediment distribution maps were produced which bore very good comparison with other observations of the same sediment regimes.

WIDEBAND WIDTH PULSES

Apart from the small-scale experimental studies[12,13,27,33] of the use of wide bandwidth pulses for sediment classification, the majority of work reported in the literature concerns the application of the Huntec® Deep Towed Seismic System (DTS).[42] This system consists of an impulsively driven plane piston transducer, which produces a high intensity broadband acoustic pulse with a usable bandwidth extending from about 1 kHz to beyond 10 kHz. Several avenues of analysis have been followed in evaluating the returns due to the incidence of such a pulse (0.3 msec) on the seabed. One approach[43] calculates the magnitude and statistical variation of the energy contained in two different time windows about the first return from the seabed. The windows are chosen so that most of the coherent energy is contained in the first window while energy in the second window provides a measure of the energy scattered at the water/sediment interface together with that returned both coherently and incoherently from shallow depths within the sediment. Changes in r_1 and r_2, the acoustic reflectivities obtained as the square root of the ratio of reflected to incident energy for the two time windows, can define a change in the material of the sea floor. Further, it was reported that a degree of correlation exists between the r_1 and the median grain size of the sediments.

Another approach[44,45] evaluates, using two methods, the amount of coherent signal present in the seabed returns. This, being related to interface roughness, is of value as roughness may in turn be associated with particular sediment types.[14] The first method, based on the stability of the coherent returns with small spatial translations of the source/receiver combination, calculates the maximum value of the normalized cross correlation function between successive returns. High values of this parameter are expected for both smooth bottoms and smooth, finely layered bottoms, and low values are expected for rough bottoms. The second method estimates the coherent energy within a single return by normalizing the return energy contained in a time window equal to the transmitted pulse length, with the energy contained in a time window of somewhat longer duration. This measure of coherence has a high value for smooth bottoms and a low value for both smooth, finely layered bottoms and rough bottoms. Taken together, the two parameters provide a powerful tool for sediment classification.[45]

This time domain approach to estimating measures of return signal coherence has been extended by the same authors to the frequency domain.[46,47] Three frequency domain coherence functions are employed, namely, the coherence function, the reflecting coherence function, and the scattering coherence function. Only two of these are independent. The coherence function is used to estimate the overall coherent energy between successive returns and is related to the Fourier transform of

$$\langle P_r^i(t) P_r^{i+1}(t) \rangle$$

where the superscripts i and i + 1 indicate successive returns. The reflecting coherence function is used to estimate the reflected component of the returned energy and is related to the Fourier transform of $\langle P_r(t) \rangle$. The scattering coherence function is used to estimate the amount of coherent scattered energy between successive returns and is related to the Fourier transform of

$$<(P_r^i(t)P_r^{i+1}(t) - < P_r(t)>)>$$

The estimation of the coherence functions is complicated by the fact that allowance for topography of the sea bottom has to be made. The effect on the data of the bottom topography is manifest as simple shifts in the time of the first return and also in a complicated manner that is dependent on the area of the sea bottom insonnified. The coherence functions are interpreted in terms of scattering theory,[48,49] and the estimates of bottom roughness thus obtained compare with estimates obtained by other means.

The data collected by the Huntec DTS[42] have also been presented in the form of "sonograms"[50] that depict the returned signal intensity as a function of both time and frequency, in an easily assimilated form. The attenuation and scattering effects of the seabed, and the directional response of the source, are frequency dependent and sonograms can be interpreted in terms of these effects. The attenuation of sound in the surficial sediments is an important parameter which should prove to be of value in remote sediment classifications.

SIDE-SCAN SONAR

The area coverage of the sea bottom is restricted when returns from normally incident signals are used. With an eye on the rapid area coverage possible with side-scan sonar, the remainder of this article examines the usefulness of such a technique for sediment classification studies. In the side-scan method, use is made of the backscattering, at low grazing angles of incidence to the seabed, of relatively high-frequency acoustic waves radiated in a fan-shaped beam from a transducer contained in a towed body. The fan beam is oriented so that the wider angle is in the vertical plane and the narrower angle in the horizontal plane with the axis of the beam normal to the ship track. The display of the returns is usually realized as a series of closely spaced intensity-modulated lines on a paper recorder. In this way a two-dimensional picture of the seabed is built up. The interpretation of side-scan sonar records by experienced observers can supply information on the nature of the sea bottom in terms of geological features, bottom topography, and sediment types.[51-56] However, it is in the latter category that the most subjective opinions are offered. The acoustic return from a grossly flat sedimentary sea bottom due to an acoustic pulse incident at a low grazing angle depends on the bottom roughness, and thus, in part, on sediment particle size.[52] Hence the distribution of signal levels over an area of side-scan record will contain information on the statistics of the bottom roughness which, in turn, gives to the usual side-scan display the appearance of texture.

A study[57] that draws on methods of texture quantification[58] has explored the potential of using a number of textural features to characterize small areas of side-scan record. These can produce, via a pattern recognition procedure, classification decisions in the context of sedimentary sea bottom identification.

A fresh departure in the study of the off-normal incidence backscatter of sound from rough boundaries for the purpose of sediment classification is found in a laboratory based study[59] of the power spectra of individual return signals. The frequency separation of "resonance peaks" found in the power spectra of backscattered signals is being studied and assessed for its value as a characerizing property for rough surfaces, and thus for sediment classification.

Related to sediment classification is the determination of bottom topography: ongoing programs[60,61] have, as their goal, the rapid quantitative measurement of sea bottom topography that special arrangements of side-scan sonars can, in principle, provide.

REFERENCES

1. **Urick, R. J.,** *Principles of Underwater Sound,* 2nd ed., McGraw-Hill, New York, 1975.
2. **Bell, D. L., Porter, W. J., and Westneat, A. S.,** Progress in use of acoustics to classify marine sediments, in *Proc. IEEE Int. Conf. Eng. Ocean Environ.,* Seattle, September 25 to 28, 1973, 354.
3. **Dodds, D. J.,** Attenuation estimates from high resolution subbottom profiler echoes, in *Bottom Interacting Ocean Acoustics,* NATO Conf. Ser. IV, Vol. 5, Kuperman, W. A. and Jensen, F. B., Eds., Plenum Press, New York, 1980, 173.
4. **Chari, T. R.,** Instrumentation problems related to determining geotechnical properties of ocean sediments, *IEEE J. Oceanic Eng.,* OE-3, 120, 1978.
5. **Mackenzie, K. V.,** Reflection of sound from coastal bottoms, *J. Acoust. Soc. Am.,* 32, 221, 1960.
6. **Smith, D. T. and Li, W. N.,** Echo sounding and sea floor sediments, *Mar. Geol.,* 4, 353, 1966.
7. **Faas, R. W.,** Analysis of the relationship between acoustic reflectivity and sediment porosity, *Geophysics,* 34, 546, 1969.
8. **Hamilton, E. L.,** Reflection coefficients and bottom losses at normal incidence computed from Pacific sediment properties, *Geophysics,* 35, 995, 1970.
9. **Hamilton, E. L.,** Prediction of deep sea sediment properties: state of the art, in *Deep Sea Sediments, Physical and Mechanical Properties,* Inderbitzen, A. L., Ed., Plenum Press, New York, 1974, 1.
10. **Smith, D. T.,** Acoustic and mechanical loading of marine sediments, in *Physics of Sound in Marine Sediments,* Hampton, L., Ed., Plenum Press, New York, 1974, 41.
11. **Newman, A. K.,** Computer Analysis of Oblique Acoustic Reflection for Ocean Sediment Identification, Pap. No. 1791, Offshore Tech. Conf., 1973.
12. **Pace, N. G. and Ceen, R. V.,** Unpublished data, 1981.
13. **Williams, J. P.,** Quantitative Sediment Identification Using Remote Acoustic Techniques, Ph.D. thesis, University of Wales, Cardiff, 1978.
14. **Drapeau, G. and King, L. H.,** Surficial geology of the Yarmouth-Browns Bank map area, in *Marine Science,* Pap. 2, Information Canada, Queens Printer, Ottawa, 1972.
15. **Horton, C. W., Hempkins, W. B., and Hoffman, A. A. J.,** A statistical analysis of some aeromagnetic maps from the north western Canadian Shield, *Geophysics,* 29, 582, 1964.
16. **Beckmann, P. and Spizzichino, A.,** *The Scattering of Electromagnetic Waves From Rough Surfaces,* Pergamon Press, Elmsford, N.Y., 1963, chap. 4-6.
17. **Bass, F. G. and Fuks, I. M.,** *Wave Scattering from Statistically Rough Surfaces,* Pergamon Press, Elmsford, N.Y., 1979, 32.
18. **Bass, F. G. and Fuks, I. M.,** *Wave Scattering from Statistically Rough Surfaces,* Pergamon Press, Elmsford, N.Y., 1979.
19. **Berry, M. V.,** The statistical properties of echoes diffracted from rough surfaces, *Phil. Trans. R. Soc., Ser. A,* 273, 611, 1973.
20. **Ishimaru, A.,** *Wave Propagation and Scattering in Random Media,* Vol. 2, Academic Press, New York, 1978, chap. 21.
21. **Roebuck, I.,** Scattering of short acoustic pulses from a rough surface, *J. Sound Vib.,* 12, 383, 1970.
22. **Berry, M. V.,** The statistical properties of echoes diffracted from rough surface, *Phil. Trans. R. Soc., Ser. A,* 273, 635, 1973.
23. **Brown, G. S.,** The average impulse response of a rough surface and its applications, *IEEE J. Oceanic Eng.,* OE-2, 67, 1977.
24. **Hammond, D. L., Mennella, R. A., and Walsh, E. J.,** Short pulse radar used to measure sea surface wind speed and S.W.H., *IEEE Trans. Antennas Propag.,* AP-25, 61, 1977.
25. **Moore, R. K. and Williams, C. S.,** Radar terrain return at near-vertical incidence, *Proc. IRE,* 45, 228, 1957.
26. **Berry, M. V.,** The statistical properties of echoes diffracted from rough surfaces, *Phil. Trans. R. Soc., Ser. A,* 273, 611, 1973.
27. **Pace, N. G.,** Sediment identification using acoustic techniques, in *Proc. Ultrasonics Int.,* London, March 24 to 26, 1975, 245.
28. **Clay, C. S. and Leong, W. K.,** Acoustic estimates of the topography and roughness spectrum of the sea floor southwest of the Iberian peninsula, in *Physics of Sound in Marine Sediments,* Hampton, L., Ed., Plenum Press, New York, 1974, 373.
29. **Clay, C. S., Medwin, H., and Wright, W. M.,** Specularly scattered sound and the probability density function of a rough surface, *J. Acoust. Soc. Am.,* 53, 1677, 1973.
30. **Cochrane, N. A. and Dunsiger, A. D.,** Sediment roughness characteristics measured by broadband spectral analysis of acoustic echoes, in Proc. 1st Can. Conf. on Mar. Geotechn. Eng., Calgary, 1979, 140.
31. **Gurcan, M. K.,** Target Identification Using a Wideband Sonar Transducer, M.Sc. (Qual.) thesis, University of Birmingham, 1981.

32. **Berktay, H. O.**, Possible exploitation of nonlinear acoustics in underwater transmitting applications, *J. Sound Vibr.*, 2, 435, 1965.
33. **Pace, N. G. and Ceen, R. V.**, *In situ* Sediment Classification Using an Acoustic Technique, Rep., School of Phys., Bath Univ., England, 1980.
34. **Smith, D. T. and Li, N. N.**, Echo sounding and sea floor sediments, *Mar. Geol.*, 4, 353, 1966.
35. **Breslau, L.**, Classification of sea floor sediments with a shipborne acoustical system, in Symp. "Le Petrole et la Mer", Sec. 1, No. 132, Monaco,
36. **King, L. H.**, Use of conventional echo sounder and textural analyses in delineating sedimentary facies, *Can. J. Earth Sci.*, 4, 691, 1967.
37. **Bezdek, H. F.**, Reflection of high frequency sound at normal incidence from the ocean bottom, *J. Geophys. Res.*, 78, 3390, 1973.
38. **Clay, C. S.**, Coherent reflection of sound from the ocean bottom, *J. Geophy. Res.*, 71, 2037, 1966.
39. **Mikeska, E. E. and McKinney, C. M.**, Range dependence of underwater echoes from randomly rough surfaces, *J. Acoust. Soc. Am.*, 63, 1375, 1978.
40. **Milligan, S. D., LeBlanc, L. R., and Middleton, F. H.**, Statistical grouping of acoustic reflection profiles, *J. Acoust. Soc. Am.*, 64, 795, 1978.
41. **Anderberg, M. R.**, *Cluster Analysis for Applications*, Academic Press, New York, 1967.
42. **Hutchins, R. W., McKeown, D. L., and King, L. H.**, A deep-tow high resolution seismic system for continental shelf mapping, *Geosci. Can.*, 3, 95, 1976.
43. **Parrott, D. R., Dodds, D. J., King, L. H., and Simpkin, P. G.**, Measuremnt and evaluation of the acoustic reflectivity of the sea floor, *Can. J. Earth Sci.*, 17, 722, 1980.
44. **MacIsaac, R. R. and Dunsiger, A. D.**, Ocean sediment properties using acoustic sensing, in *Proc. 4th Int. Conf. Port and Ocean Eng. Under Arctic Conditions*, 1978, 1074.
45. **Dunsiger, A. D. and MacIsaac, R. R.**, Broadband seismic data used for sea floor sediment classification, *Proc. IEEE Oceans*, 521, 1978.
46. **Cochrane, N. A. and Dunsiger, A. D.**, Sediment roughness characteristics measured by broadband spectral analysis of acoustic echoes, in *Proc. 1st Can. Conf. Mar. Geotech. Eng.*, Calgary, 1979, 140.
47. **Cochrane, N. A. and Dunsiger, A. D.**, Seabed roughness characterization by broadband acoustic echo sounding, in *Proc. 5th Int. Conf. Port and Ocean Eng. Under Arctic Conditions*, Trondheim, 1979, 877.
48. **Clay, C. S. and Leong, W. K.**, Acoustic estimates of the topography and roughness spectrum of the seafloor southwest of Iberian peninsula, in *Physics of Sound in Marine Sediments*, Hampton, L., Ed., Plenum Press, New York, 1974, 373.
49. **Clay, C. S. and Medwin, H.**, *Acoustical Oceanography*, John Wiley & Sons, New York, 1977.
50. **Dodds, D. J.**, Attenuation estimates from high resolution subbottom profiler echoes, in *Bottom Interacting Ocean Acoustics*, Plenum Press, New York 1981, 173.
51. **Clay, C. S., Ess, J., and Weisman, I.**, Lateral echo sounding of the ocean bottom on the continental rise, *J. Geophys. Res.*, 69, 3823, 1964.
52. **Wong, H. K. and Chesterman, W. D.**, Bottom backscattering near grazing incidence in shallow water, *J. Acoust. Soc. Am.*, 44, 1713, 1968.
53. **Chesterman, W. D.**, The ocean floor, *Contemp. Phys.*, 15, 501, 1974.
54. **Chesterman, W. D. and Heaton, M.**, Sonar studies of the continental shelf, Proc. Ultrasonics Int. Conf., Brighton, 1977, 1.
55. **Sanders, J. E., Emery, K. O., and Uchupi, E.**, Microtopography of five small areas of the continental shelf by side-scanning sonar, *Geol. Soc. Am. Bull.*, 80, 561, 1969.
56. **Belderson, R. H., Kenyon, N. H., Stride, A. H., and Stubbs, A. R.**, *Sonographs of the Seafloor*, Elsevier, Amsterdam, 1972.
57. **Pace, N. G. and Dyer, C. M.**, Machine classification of sedimentary sea bottoms, *IEEE Trans. Geosc. Electron.*, GE-17, 52, 1979.
58. **Haralick, R. M., Shanmugam, K., and Dinstein, I.**, Textural features for image classification, *IEEE Trans. Syst. Man Cyber.*, SMC-3, 610, 1973.
59. **Gurcan, M. K.**, Target Identification Using Wideband Sonar Transducers, M.Sc. (Qual.) thesis, University of Birmingham, 1981.
60. **Denbigh, P. N.**, A bathymetric side-scan sonar, in Proc. Ultrasonics Int., Graz, 1979, 321.
61. **Cloet, R. L. and Heaton, M. J. P.**, Recent advances in seabed surveying in Bath University, Proc. Soc. Underwater Technol./Hydrograph. Soc., London, January, 1980.

OCEAN BOTTOM SEISMOMETRY

B. D. Loncarevic

INTRODUCTION

Ocean bottom seismometry is an experimental geophysical technique that employs seismic sensors implanted in, resting on, or located near the ocean bottom. When Ewing et al.[1-3] and Bullard and Gaskell[4] initiated this technique in the mid-1930s the ocean-covered areas of the Earth's surface were geological "terra incognita." The state of knowledge had not advanced much since the CHALLENGER Expedition 60 years earlier and the 1913 statement of de Launay[5] was still valid:

" . . . The Geology of the bottom of the Seas is still very poorly known to us. . . . It is the almost unexplored field where one may hope some day to find the key to problems that terrestrial geology elsewhere proposes to us. . . . It may be hoped that some day, instead of limiting the work to the epidermis, we shall penetrate them . . . deper and still deeper . . . ".

The progress of marine geophysics in the 50 years since Ewing's pioneering experiments has been spectacular. Marine seismology has played a key role, leading our advance in knowledge.

Development of the ocean bottom seismometry has gone through three phases.

Phase I — The early work was interrupted by World War II. After the war, other geophysical techniques were developed and research focused on deep ocean basins where existing OBS techniques could not yet be used.

Phase II — The second phase was stimulated by progress in the negotiations on the Nuclear Test Ban Treaty, which created a need for the detection of clandestine underground nuclear explosions. An expectation of a lower level of background seismic noise on the ocean bottom and more uniform geographical distribution of monitoring observatories led to development (mostly commercial) of several test instruments in the early 1960s.[6-16] The expectations of this program were not realized: the noise at ocean sites was not substantially lower than at the best continental sites,[17-26] and the instruments were expensive, complex, and unreliable.

Phase III — The third phase started in the late 1960s and continues to this day. The great success of the current OBS research is due to a successful exploitation of the latest advances in microelectronics, which make it possible to construct small, reliable, self-contained, and relatively cheap instruments. This application of latest technology has enabled at least 20 research groups to become involved in OBS design and use: in Canada (2), France (2), Germany (1), Japan (2), and U.K. (3), U.S.A. (9) and U.S.S.R. (1). A listing of these groups, together with their addresses in early 1981, is given in Appendix 1. Distribution of the experiments carried out by these groups is shown in Figure 1.

The bibliography accompanying this review is reasonably complete up to mid-1981 with the exception of cruise reports and certain technical reports issued by individual institutions. The results of the OBS investigations are reported through primary research journals. The five most frequently cited journals have published over 55% of the contributions listed in the bibliography: *Journal of Geophysical Research* (and other publications of the American Geophysical Union) (29%), *Geophysical Journal of the Royal Astronomical Society* (21%), *Seismological Society of America Bulletin* (13%), *Marine Geophysical Researches* (28%), and *Bulletin of the Earthquake Research Institute, Tokyo* (10%). There are no books or monographs dealing with ocean

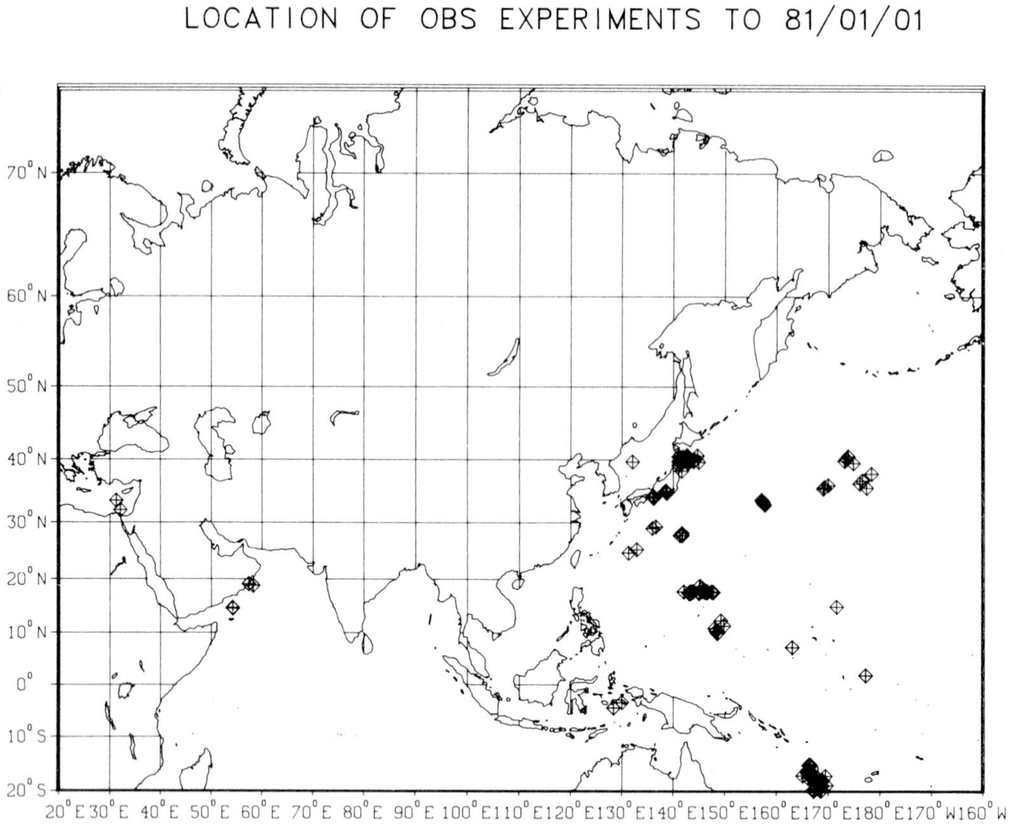

A

FIGURE 1. Location of launch sites of 735 successful OBS experiments as reported by Bedford Institute of Oceanography (BIO-44 stations), Hawaii Institute of Geophysics (HIG-183), Massachusetts Institute of Technology (MIT-12), Institute of Oceanographic Sciences (Wormley, IOS1-110, Blacknest-46), Lamont-Doherty Geological Observatory (LDGO-29), Woods Hole Oceanographic Institution (WHOI-64), Earthquake Research Institute, Tokyo (TOK-55), University of Washington (UOW-28), Institute for Geophysics, Hamburg (IFGH-12), Centre Océanologique de Bretagne (COB-43), Pacific Geoscience Centre (PGC-10), and Marine Science Institute, Texas (TEX-99). Another 18 BIO stations north of 80°N and 1 MIT station south of 25°S could not be shown on the map. (A) Stations in the eastern hemisphere. (B) Stations in the western hemisphere.

bottom seismology. A comprehensive review has been published recently by Whitmarsh and Lilwall.[27] Other reviews have been published by Bolt,[28] and others.[29-31]

Ocean bottom seismometers (OBSs) are not the only instruments for seismic research at sea. Two-ship seismic work was the standard technique up to the mid-1960s. Free-floating or moored sonobuoys (either self-recording or transmitting by radio to the ship) are still occasionally used. The main advantages of OBSs (in order of importance) are

1. Fixed location of the receiver on the ocean bottom
2. Simultaneous recording of several components, so that shear waves and particle motion can be identified
3. Ease of deployment of arrays, which maintain a fixed geometry during an experiment

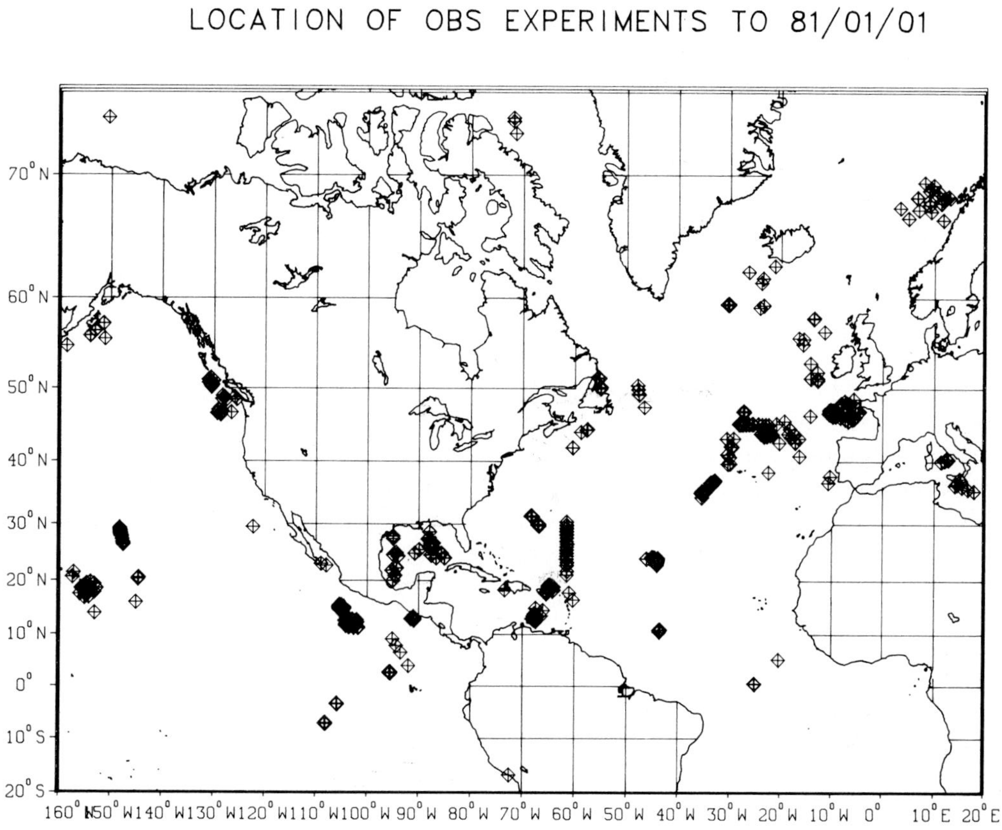

FIGURE 1B.

4. Extension of the length of seismic refraction lines well beyond the usual radio range of transmitting sonobuoys
5. Attenuation of the sea-surface wave-induced noise, which increases the signal-to-noise resolution of seismic sensors
6. Geometry of travel paths that is advantageous for the study of the fine structure of the upper crust
7. The capability to study the elastic properties of near surface sediments by sound sources located on the bottom
8. Cost-effectiveness in terms of ship time and the amount of explosives required for an experiment

Some of the disadvantages of the OBS could, and probably will, be overcome through better design in the future. Listed in their order of importance, these disadvantages are

1. A finite chance (estimated at 10 to 20%) that an instrument may be lost or that an experiment may be ruined through a malfunction that cannot be detected until after the completion of the experiment
2. Uncertainty of the exact bottom conditions and coupling at any OBS site
3. A tradeoff between the width of the pass-band and the recording time
4. Limited dynamic range (except in digital instruments)

5. Lack of information to control an experiment (data can be examined only after the instrument is recovered)

The number of successful launchings in the last decade is estimated to be around 1000. The cost per launching can be estimated using recent participation of CSS Dawson in Project LADLE.[32] The ship was away from home port for 36 days and the seismic work represented about 80% of her program. Ship operating costs are $7000/day; thus the total ship-time cost was 36 × 0.80 × 7000 = $203,000. A total of 30 tons of explosives were used at a cost of $30,000. The cost of each launch is small, not more than $100. Because of the finite chance of losing an instrument, a capital depreciation of 10% is assigned to each launch. At an estimated total cost for an OBS of $10,000, this represents $1000/launch. There were 20 launches, so the total operating costs were $22,000, and the total project cost was $255,000. (This does not include the salaries of the scientific staff, the overhead of institute administration or the cost of data analysis and publication.) The cost/launch is then calculated at $12,500. Assuming that this type of analysis is applicable to other expeditions and other scientific institutions, the worldwide investment in ocean bottom seismic work is of the order of $1.3 million/year.

INSTRUMENTATION

The variety of possible instrumental configurations available for ocean bottom seismometry is illustrated in Figure 2. In all these configurations, a sensor (or a sensor package) must be on or below the ocean bottom. These configurations fall into two classes.

Instruments with sensors connected by wire to the sea surface and then directly to the ship, to the ship via radio transmitting buoy, or to a recording buoy, represent **Class I**. These instruments belong to an older technology and are being phased out. Their advantages are that the results of a seismic experiment can be made available while the experiment is in progress and they have a wide frequency band and a greater dynamic range. Their disadvantages outweigh those advantages: substantially greater cost of equipment, complex and time-consuming launch and recovery procedures (weather dependent and depth limited), and the fact that extending the length of seismic lines with them is very difficult. These instruments were important in the development of OBS techniques, but as they are unlikely to be of any further importance, they will not be discussed in this review.

Class II are the self-contained instruments capable of autonomous operation over several hundreds of hours, independent of any surface support. These instruments can be further subclassified according to the number and type of sensors, to the recording technique used, or to the type of release mechanism employed.

Ocean bottom seismometers have been adequately described in the literature by their designers.[33-65,175] In reviewing this literature, the most striking impression is of the variety of configurations employed to achieve presumably similar scientific results. In the absence of a general consensus regarding the optimum characteristics for an OBS, each group proceeded to design its own instrument. The final appearance of these instruments therefore reflects the differences in emphasis of the type of experiment for which the instrument was designed, and the financial resources and technical expertise available to each group.

According to the nature of experiment, the instruments can be divided into two types: instruments for monitoring natural phenomena (earthquakes, microseisms, sea-bottom ambient noise) and instruments for controlled seismic source experiments.

FIGURE 2. Different OBS configurations as implemented by: (1) Bedford Institute of Oceanography; (2) U.S. Geological Survey; (3) Marine Science Institute, University of Texas; (4) University of Washington; (5) Massachusetts Institute of Technology; (6) University of California at Santa Barbara; and (7) Lamont-Doherty Geological Observatory. (Photograph taken less # in June 1978 during Lopez Island experiment[168] by D.E. Heffler.)

The first type should have a capacity for prolonged operation (several months) on the sea bottom. Because of the inherent limitation of the magnetic tape recording medium, this means that the recording is usually intermittent, and triggered by some device that detects signal arrivals.[57,73,88] The sensor package includes one vertical and two horizontal geophones. A broad frequency response, preferably below 2 Hz, is required. Absolute calibration in terms of ground motion is desirable if earthquake magnitudes are to be calculated.

The second type, intended primarily for controlled source seismology, need not have a capacity for prolonged operation as a seismic refraction profile can be easily completed in 2 to 7 days. Multiple geophones are an advantage but are not essential, and several instruments use a single hydrophone. Response below 4 Hz is less important than an ability to detect high frequency (300 to 2000 Hz) water waves, which are often used to determine the direct distance between the source and the receiver. Finally, these instruments need to be relatively cheap and simple to use, as multiple launchings and recoveries during any expedition are likely.

An ocean bottom seismometer consists of eight principal components (Figure 3). A typical instrument at the time of launching is shown in Figure 4.

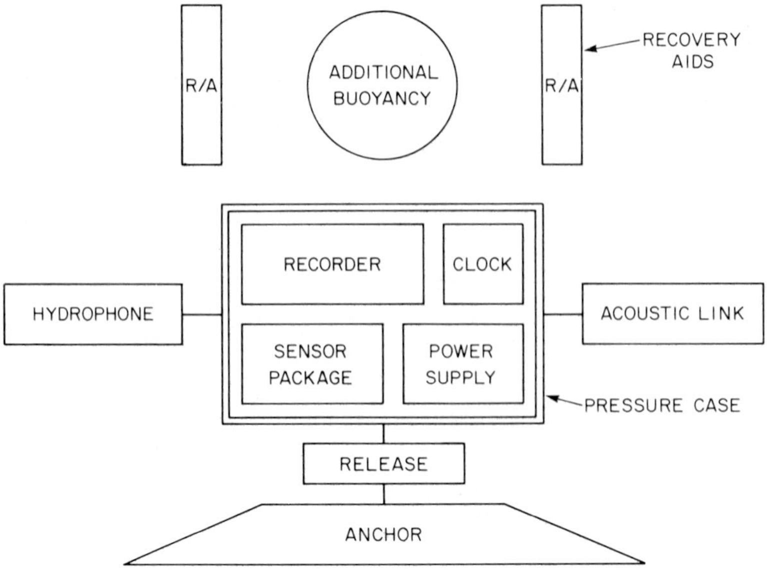

FIGURE 3. Principle components of an OBS instrument assemblage.

FIGURE 4. An OBS being launched over the side in Baffin Bay. This instrument was designed and built at the Atlantic Geoscience Centre, Bedford Institute of Oceanography. Some of the principal components shown in Figure 3 can be easily identified.

Sensor Package

Hydrophones

Hydrophones are pressure sensitive transducers that convert the ambient pressure fluctuations to electrical voltages utilizing the piezoelectric effect. They are not sensitive to acceleration and therefore were used for all seismic work from ships and sonobuoys prior to the advent of the OBS. They still represent the most common seismic sensor for work at sea.

The heart of the transducer is a disc or small cylinder of special ceramic material. There may be an oil-filled cavity in the cylinder to increase its sensitivity. The frequency response is usually flat from DC to 5000 Hz. Superior high frequency response is important to detect waves traveling directly from the sound source to the receiver through the water. The sensitivity range of hydrophones in common use today ranges from 50 V/Pa (PZT-4 ceramic discs, 5 × 5 cm diameter — used by IOS) to 450 V/Pa (OAS Inc. Model E-2SD used by HIG). The higher sensitivity ceramics have lower capacitance and thus require charge-coupled amplifiers, and a compromise between the gain of the first stage of amplification and its lower corner frequency.

The hydrophone is susceptible to hydrodynamic flow noise[176] and has to be mounted inside a perforated protective housing.[163] The electrical connection to the OBS electronics is through a high pressure bulkhead connector, which must be selected and handled with care to avoid intermittent failures.

Geophones

Geophones are velocity-sensitive transducers, usually of a moving coil variety with an external damping resistor. The most commonly used geophones are manufactured by Geospace Corp. (U.S.A.), Model HS-1, with a lower corner frequency of 4.5 Hz. Sensitivity depends on coil resistance, and ranges from 16 V/msec for coil resistance of 215 ohm to 75 V/msec for coil resistance of 960 ohm. Other types of geophones used include Sprengthener Model S-6000 (2 Hz, R_{coil} = 10,000 ohm, 300 V/msec, Mark Products Model L1-G (2 Hz), Mark Products Model L-4-3D (1 Hz geophone with a separate calibration coil), and a special seismometer developed for the Ranger Lunar Program (1.8 Hz).

Horizontal and vertical geophones are usually identical electrically but the vertical geophone has a spring to cancel out the acceleration due to gravity. The leveling of horizontal geophones is quite important. An error of 1° produces an acceleration bias of 0.0175 g, which overloads a geophone with a wide dynamic range. To achieve self-leveling within a 1° angle, the geophones are mounted inside a two-axis gimbaled frame, or are hanging like a pendulum from a two-axis gimbal support, or, in the case of the University of Washington instrument, are mounted on a small boat floating on a thin film of silicon oil inside a hemispherical pressure case. In all cases, two problems exist: (1) electrical contact through the gimbals, and (2) the degree of sticktion of the gimbal bearings. The most reliable way to solve the first problem is to use a fine wire conductor, but this limits the degree of rotation of the housing relative to the gimbals. Other solutions have been to use gold-plated slip rings or mercury-wetted point contacts. The second problem is due to the conflicting requirements that bearings be sufficiently free to allow accurate settling of geophones to true vertical, yet must be tight enough to allow enough friction for the gimbals and the instrument housing to act as a rigid body at seismic frequencies. The solution has been to slightly misalign supports in order to increase friction, to clamp the gimbal mounting once the leveling has been completed, and to depend on slip ring friction or to use heavy silicon oil for damping. A simple calculation by Lister[55] shows that for the University of Washington instrument, the viscosity of silicon oil should be greater than 360 cP.

The calibration of the geophone sensitivity as a function of frequency is important to evaluate the absolute magnitude of ground motion, for comparative measurements at different sites, or when comparing measurements made with different instruments. The sensor alone can be calibrated by moving it through a known displacement, for example on a shaking table, or by applying an electrical transient to the coil and measuring back emf. The latter can be done effectively through a separate calibration coil. A sophisticated scheme for broadband calibration using random binary sequence of pulses was proposed recently[66] and will be implemented in the latest instrument being developed at Scripps Institution of Oceanography (SIO).[57]

The calibration of the geophone does not provide the full information about the response of the instrument. That response is governed by the characteristics of the amplifiers, filter, and recording system used. Even more importantly, when the instrument is sitting on the nonrigid ocean floor, the mechanical coupling with the sea floor becomes a part of the transfer function. This problem has been the subject of numerous studies and is discussed further in this chapter in the section entitled ''Operating Environment''.

Internal Time Reference

The basis of seismological research is the measurement of velocity of propagation of elastic waves through different geological formations. The seismic velocities of interest are in the range of 2 to 8 km/sec. It is desirable to determine these velocities with an accuracy of 1:1000, or a timing accuracy of 0.01 sec. To obtain this accuracy over a period of 10 days (or 864,000 sec) requires an absolute clock stability of $1{:}10^7$. The short-term stability of this order is possible, but this stability cannot be achieved over a long-term in portable instruments with a reasonably low power consumption. The typical drift rate of OBS clocks is 2 to 5 msec/hr. After 250 hr of operation this becomes 0.5 to 1.25 sec, and is thus significant. It is important to rate the clock before launching and after recovery against a shipboard standard such as a cesium clock (itself rated against satellite or radio time signals). The assumption is that the drift between the two calibrations is constant and thus a simple linear interpolation is used for correction. This assumption is probably not valid. Most of the thermal and mechanical shock happens during brief handling of the instrument assembly on deck and over the side of the ship; while on the sea floor the instrument is in a constant-temperature environment. It is thus desirable to rate the clock while the instrument is on the bottom. This will be possible with those instruments incorporating an acoustic command/response link to the survey ship.

Oscillators and the associated electronics are usually powered by a separate supply in order to increase the reliability of the time reference. The power consumption is 60 mW to 750 mW. This power is usually interrupted during the pre- and post-launch checkup; this is undesirable as the transients disrupt the crystal stability. A better method would be to switch to a shipboard standby power supply without even a short-term interruption of power supply to the crystal oscillator.

The time reference is a frequency from a stable oscillator which must be converted to a coded signal in order to read real time. The simplest way is to use frequency dividers to obtain regular pulses at 1 sec intervals. Decoding then consists of counting the pulses since the reference time at the beginning of the experiment. This is an unsatisfactory method, since an error of one count anywhere on the record causes an error in interpretation of 1 sec and that is 200 times greater than the expected accuracy. A modern approach is to use low power digital electronics (in some cases incorporating microprocessors) which count the oscillations and produce timing pulses. The pulse train is encoded (in minutes, hours, and days) for recording with the seismic data.

Data Recording

Despite the various merits of different methods of data recording (and a great future promise of mass storage devices such as bubble memories) the magnetic tape has become a universally accepted medium for data storage in an OBS. The choice of the medium and method of recording should be based on eight factors: frequency response, dynamic range, length of recording time, number of parameters to be recorded, efficiency (physical volume of recording medium per parameter), reliability (e.g., availability of off-the-shelf commercial units), physical size and weight, and power consumption.

Gilbert[67] discussed a number of considerations for a good design of oceanographic equipment. Regarding the recording method, he gives a useful table showing that the efficiency (in terms of cm^3 volume per parameter; smaller the number, more efficient the medium) of FM 0.635 cm (¼ in.) width magnetic tape is 0.1, of digital tape of the same width is 1.1, and of punched tape is 13 (roughly a ratio of 1:10:100).

The magnetic tape recorders are either reel-to-reel (mostly using 0.635 cm tape) or cassettes. In some cases, a commercial product can be used with minimum modification (e.g., UHER R-L or UHER 4000 IC recorder). In other cases, a commercial tape deck is extensively modified, or a completely new design is undertaken.[68-71] A very sophisticated special purpose recorder was recently designed and built by Davis and Ambuter[72] at USGS, Woods Hole, Mass.

Data can be recorded on magnetic tape in digital form, as modulated frequency (FM), or directly in the form of varying magnetization (DR). The quality of reproduction and the amount of tape used decreases in the order given. Lister and Lewis[55] thoroughly discuss the merits of different methods. The basic tradeoff is between the quality of reproduction and the economy of tape consumption. Their conclusion is that for an equivalent signal quality, the amount of tape required for 100 hr of recording with digital, FM, and DR recording is 2970 m, 1125 m, and 360 m, respectively (a ratio of 8.25:3.13:1).

The main disadvantage of DR is a restricted dynamic range (35 to 40 dB) of recording. This can be overcome in two ways: by recording at two separate levels, thus doubling the number of channels required, or by employing some method of signal compression on recording and expansion on playback.

Three methods of signal modification are available: automatic gain control (AGC), logarithmic encoding, and bipolar square rooting. None is fully satisfactory and some investigators interested primarily in the onset time of seismic signals (first arrivals) ignore the problem. Others employ a full digital method of recording; often an automatic or programmed gain ranging scheme is used to extend the dynamic range to 100 dB. With digital recording, the continuous recording time with a reasonable length of magnetic tape is about two orders of magnitude less than the typical length of a launching. Therefore, some scheme of intermittent recording must be employed.

Intermittent recording can be achieved in two ways. A predetermined table of recording intervals can be stored in the OBS to correspond with a proposed shooting schedule (this severely limits the flexibility of the ship's program). The second approach is to implement an event trigger to sense the arrival of a seismic signal and switch on the tape recorder. Since event triggers respond to the large signal, which arrives up to 15 sec after the small but important first arrivals, a continuous buffer storage is required.

Such a trigger usually depends on the comparison between average noise over a short-term (1 sec) and a long-term (up to 1024 sec = 17 min). Algorithms for implementing such a trigger are discussed by Prothero[63,88] and by Moore et al.[57] The major problem with an event detector is that the setting of the trigger level must be such that no significant events are missed, yet there should be no false alarms due to fish bumps, whale whistles, passing ships, etc.

When using magnetic tape at sea, great care should be exercised to ensure that it is properly handled. At no time should the tape be exposed to excessive heat or humidity. The tape should be mounted with care so that reels are properly secured and all the slack to take-up spool eliminated. Cassettes should be properly centered and tape should be run back and forth to the end of the spool to eliminate any sticking between layers. Finally, the recording heads should be frequently cleaned and demagnetized to eliminate the buildup of permanent magnetization which degrades signal quality and amplitude.

Power Supplies

As a self-contained instrument, an OBS must include all the power supplies necessary for its operation. The overall objective of extending the recording time on the bottom can be achieved in several ways: by decreasing the power consumption through a careful selection of components and design, by increasing the volume (and therefore the weight) of the batteries carried by the package, by increasing the efficiency (i.e., electrical capacity to volume ratio) of batteries, etc. Remarkably low power consumption can be achieved by taking advantage of the advances in silicon chip technology. The reported total power consumptions (in recording mode) of several OBSs designed by different groups are between 0.11 W (University of Oregon)[48] and 3.75 W (Scripps Institution of Oceanography).[63]

The types of batteries that can be used in an OBS are shown in Table 1.

Although the most "efficient" batteries are lithium cells, other batteries must also be considered. Alkaline cells are the cheapest, and most readily available, but have a limited shelf life and reduced capacity at low temperatures (by about 25% between room temperature and the ocean floor temperature). Lithium cells are efficient but expensive, and there have been cases of violent gas discharges. Nickel-cadmium batteries are rechargeable and therefore reusable, but unless a careful log of use of each cell is kept, their reliability after several discharge cycles becomes questionable; furthermore, there is a potential problem with corrosion due to leakage and degassing. On balance, most designers have opted for alkaline cells that are discarded after each experiment.

Almost all current OBS designs employ multiple power supplies; different functions are powered independently and the risk of a single component failure disabling the whole instrument minimized. This is particularly important for release circuitry, which ought to be the most reliable part of an instrument. To increase reliability, a low voltage alarm can be incorporated; it allows for a premature, fail-safe release in case of battery failure.

Release

In order to sink, an OBS must have negative buoyancy at the time of launching. In order to float back to the surface on completion of the experiment, it must release a ballast (excess) weight to acquire positive buoyancy. The release mechanism consists of an electrical command subassembly and a mechanical device for separating the capsule from its anchor weight.

The easiest way of generating the electrical release command is to have an internal timer counting the elapsed time up to a preset value. This would correspond to the end of the experiment when the recovery ship is above the launch position and ready to retrieve the instrument. The elapsed timer may be set to count a certain number of time units (seconds, minutes, or hours) or, in a more sophisticated instrument, it may compare the clock time with a preset future time value, and generate the release command when the two times coincide.

Table 1

Battery type	Capacity	
	Wh/kg	Wh/cm^3
Lithium	386	197
Mercury or silver oxide	110	137
Alkaline	66—99	33—49
Carbon-zinc	44	33
Nickel-cadmium	26—35	33—49
Sealed lead-acid	30	13

An external acoustic command sent by a surface ship can also generate the electrical release signal. The acoustic command may be a special time sequence of shots or a transmission of a coded acoustic signal. In either case, the acoustic receiver in the OBS must have a discriminating decoder to prevent accidental release due to spurious noise in the ocean.

A third method, which may be used as a backup to either of the first two methods, is to release when a certain special condition is encountered: low power supply voltage, intrusion of salt water, high angle of repose on the ocean bottom, end of recording tape, etc. The main advantage of this type of backup release command is to prevent the loss of the instrument or of experimental data.

The oldest and simplest release mechanism is a soluble link between the ballast weight and the instrument capsule. Soluble links have been made from a variety of materials, ranging from sugar cubes to magnesium rods. In principle, these are "elapsed timers" with a low accuracy because the solution rates are dependent on pressure, temperature, seawater flow, etc. For longer periods (say above 12 hr) the uncertainty of release time may be as great as 50% of the expected value. Under those conditions, a soluble link may be useful only as a last resort backup device.

The most widely used release mechanism in the oceanographic community is an electrically fired explosive charge which may cut a wire, drive out a pin, release a latch, etc. This device is reliable but requires a substantial power source and special precautions when transporting and handling explosive charges on board ship. It is a brute force method suitable for releasing substantial assemblies, such as a string of moored current meters. The main disadvantages of electric bolts are that they require watertight electrical connections that are difficult to test before a launch.

A mechanical lever driven by an electrical motor is used by at least three groups.[33,48,56] There are usually two arms held down by a latch that is operated through a gearbox; unlatching of either arm will result in release. While these devices are imaginative in design, they are difficult to construct (because of the high-pressure waterproof mechanical linkages required), expensive, and add weight to the capsule payload that must then be carried back to the surface.

University of Washington OBSs[55] use an electromagnet in their hold-down assemblies. Using 20 mA at 12 V and a simple 10:1 mechanical advantage, the electromagnet provides a hold-down force of about 70 kg. The advantage of the system is that it is fail-safe: if the battery voltage fails, the electromagnet releases. Since there is a constant power drain, if the clock fails the instrument will eventually release when the hold-down battery is exhausted.

Institute of Oceanographic Sciences (IOS)[53] OBSs use a pyro-release, which consists of a small diameter tube of solid propylene capped at its open end by a disc sealed by an O-ring. The two parts are held together by an alloy rod around which is packed a special powder. When electrically ignited, the powder burns with intense heat causing the rod to melt. The combustion gases produced equalize the pressure in the device with the external hydrostatic pressure and allow the two parts to separate and thereby operate the ballast release linkage.

An ingenious device that uses a vacuum chamber as a hold-down has been described by Heffler and Locke.[74] It has several advantages: it uses little power, all its critical components must withstand only a dry partial vacuum rather than a high pressure marine environment, and if its chamber leaks, the instrument will release prematurely. The disadvantages are the need to test at high pressure, the high cost of machining the vacuum chamber, its weight on retrieval, and the skill required to arm the release.

The most promising mechanisms are based on electrolysis. In these, a simple mechanical arm, with a mechanical advantage of about 20:1, holds the instrument package down to the anchor plate. The other end of the arm is held fast by a loop of thin (0.5 mm) stainless steel wire. On command, a current of 100 mA flows through the wire and decomposes it in about 6 sec thereby releasing the instrument. The cost and weight of this release mechanism are insignificant; it is simple and easy to assemble and because of that, ought to be highly reliable.

Anchor Weight

The anchor weight must be sufficient to sink the instrument but not so great as to strain the assembly or create handling difficulties on a moving ship. It should have a simple shape and be inexpensive to manufacture; it should establish a firm and competent contact with the ocean floor; and its shape should contribute to a smooth hydrodynamic flow during descent.

Concrete anchors have been used in oceanographic work because of their low cost and the ease with which complicated concrete shapes may be formed. Because of the low specific density of concrete, these anchors must be heavy in air, which is a distinct disadvantage on board ship. The long-term behavior of concrete slabs under pressure is not known. Once cast, concrete anchors cannot be easily modified on board ship to accommodate additional devices, etc. University of Washington OBSs use a ''flowerpot'' design for an anchor, the best example of taking the full advantage of the construction medium.[55] The anchor has the shape of a truncated cone of moderate side angle (12 cm thick, 300 kg weight). This gives the unit fast and stable descent characteristics, so that they land not more than a few tens of meters from the launch position in 3 km ocean depths.

Steel is the other material most often used for ballast weight. Part or all of the frame supporting the capsule is constructed so that it is detachable and remains on the bottom after release. For kinematic stability, the frame is supported on three legs, usually fitted with foot pads that prevent it from sinking too deeply into soft sediments. As an alternative to three point support, some instruments use a flat steel plate that is sometimes fitted with pins or prongs to grab the bottom and thus prevent sliding.[79] A steel plate on a simple steel frame anchor that is not too complicated to manufacture has a definite advantage over concrete anchors unless a very fast descent is required. Typical descent speeds for instruments fitted with steel plates or steel frames are 0.5 to 1.2 m/sec. In contrast the ''flowerpot'' concrete anchor descends at 3 m/sec.

Buoyancy

Once released, the capsule must have sufficient positive buoyancy to ascend to the surface. Instruments in spherical capsules usually have enough reserve buoyancy; those in cylindrical pressure cases require additional floats.[78] This additional buoyancy can be in the form of one or more glass spheres or specially shaped syntactic foam. The latter is a matrix of glass spherules imbedded in epoxy, a convenient material to handle onboard ship and one that can be shaped to fit the instrument capsule. There are indications that this material may crack at depths of 2000 m and over, and is to be avoided for deep ocean launchings.

Recovery Aids

Once the instrument capsule returns to the surface it must be located by the recovery ship. If there are navigational uncertainties at the time of launch or recovery, or strong subsurface currents, the capsule may be some kilometers away from the ship. Because OBSs have a low profile they are be very difficult to spot from the bridge. Therefore recovery aids (RA) are essential if high rate of instrument loss is to be avoided.

Acoustic Links

An acoustic link between the ship and an OBS on the ocean floor greatly facilitates experimental work. The link can be passive (one-way only) or active (two-way).[75] In the passive link, a self-contained acoustic source transmits regular pulses or "pings" in the 10 to 12 kHz frequency band. The pinger can be programmed to be switched off after landing (to prevent overload of the hydrophone by pings) and switched on again before release (to allow recovery ship to "home" on the OBS position);[42,65] also the pinging rate can be changed to indicate the release command.[45] A two-way acoustic link permits for sending commands from the ship to the OBS.[33] The most important of these commands is the release command, that places the recovery ship in full control, and thus adds a great operational flexibility. Other possible commands are to change a gain setting or to switch the tape recorder on or off. The OBS pinger is inactive until interrogated by the ship. Depending on the sophistication of the design, the pinger may respond with a variety of status information: gain settings, clock offset, battery levels, number of events recorded, length of tape remaining, etc.

Other Recovery Aids

The most common RA is a flashing light. The device in common use today (Model SF-500, manufactured by Ocean Applied Research Co., U.S.A.) has a pressure switch to turn the light off at a depth greater than 10 m; the batteries are conserved until the unit resurfaces again after it is released. Its typical lifetime is 100 hr of flashing. The light can be easily seen on a clear night from a distance of 5 to 10 km unless the sea is very choppy. The flashing light is less useful during the daytime and the Arctic summer.

The second most commonly used RA is a radio beacon (Model ST-200, Ocean Applied Research Co., U.S.A.). This device operates in the citizens band (27 MHz) and has a pressure-activated switch also. A hand-held receiver is fitted with a loop antenna for direction finding. This transmitter has a range of 10 to 20 km and is equally useful during day or night recovery.

If the OBS assembly is fitted out with any kind of a mast, a flag and a radar reflector may be attached. These may be a mixed blessing. While it may assist the search under certain conditions, radar reflectors or flags will increase the drag so that strong winds could drive the OBS beyond the search area. Finally, it has been proposed that once on the surface, the instrument could be made to release fluorescent dye. In practice, this is not very useful for searches by ship but may be worthwhile if an aerial search is conducted.

Another simple and effective RA is a retrieve line. This is a length of buoyant rope, uncoiled as the instrument lifts off from the bottom.[45] This line floats on the surface and can be easily picked up by a grapnel thrown from a ship.

EXPERIMENTAL PROCEDURES

Preplanning

Before an expedition sails, the general "time-budget" must be drawn to insure that all the available ship time and other resources are used to the greatest benefit of the experiment. The time-budget will take into account: steaming time to the site of the experiment, local survey around a launch site, instrument assembly time, launch and descent time, steaming time between sites, time for preparation of charges, shooting schedule, additional data collection (gravity, magnetics, seismic reflection profiling) along the refraction line, recovery time, and steaming time back to port. Additional constraints on the time-budget are that certain operations should be confined to certain periods of the day (e.g., shooting should be during the daylight hours, recovery may be preferred at night when it is easier to see a flashing light, etc.), the crew should not be required to work overtime, and contingency time be provided for bad weather, emergencies, or equipment failures. If the cruise is a long one, then a port call should be scheduled at 3- to 4-week intervals for crew R & R.

The effect of weather on operations is very difficult to plan and control. One way of coping is to plan on a basis of reduced ship speed, say 18 to 20 km/hr, rather than on ship's normal cruising speed, which may be 22 to 24 km/hr. Another method is to make an allowance in the cruise plan of 5 to 10% for bad weather. Similarly, in planning each step, a conservative estimate should be made based on the estimate of the longest required time for each operation. Of course if one is lucky and there is no time lost due to weather, etc., one may end up with a surplus of time. In that case, the cruise may return a day or two early, or a lower priority experiment may be undertaken.

In laying out a launch and recovery schedule, a plan must be fairly detailed. An example of a scheduled relaunching of five OBS along a 240 km refraction line is shown in Figure 5. The OBS used in this example had a preset time release, so that the ship had to be on the recovery site at a specified time, and the search time could not be extended beyond the allocated time window or she would be late for the recovery of the next instrument. In the experiment, there were 20 launchings and 19 recoveries; the cause of the one loss, as is the case with all losses when there is no evidence, is unknown.

Instrument Checkout

Because of variations in construction and assembly details, procedures for final checkout are different for each instrument. Among the checks that need to be made are the following: battery voltages and connectors, recorder transport and take-up spool tension, clock rate calibration, secondary or backup clock calibration, programmer or microprocessor check, acoustic command or pinger operation, amplifier gain settings and sensor response. It is desirable to carry out these tests as close to the launch time as possible. On the other hand, the pressure case containing electronics usually ought to be closed in the laboratory some hours before the launch (to prevent excess moisture filling the inside air space). The trend now is to conduct the final checkout through a bulkhead connector by the direct interrogation of the on-board computer.[57]

Final Assembly

The final assembly takes place immediately prior to launch, usually on deck and near the crane or other lifting device that will put the instrument over the side. (One varient is Prothero's instrument, which is assembled in a portable deck laboratory and

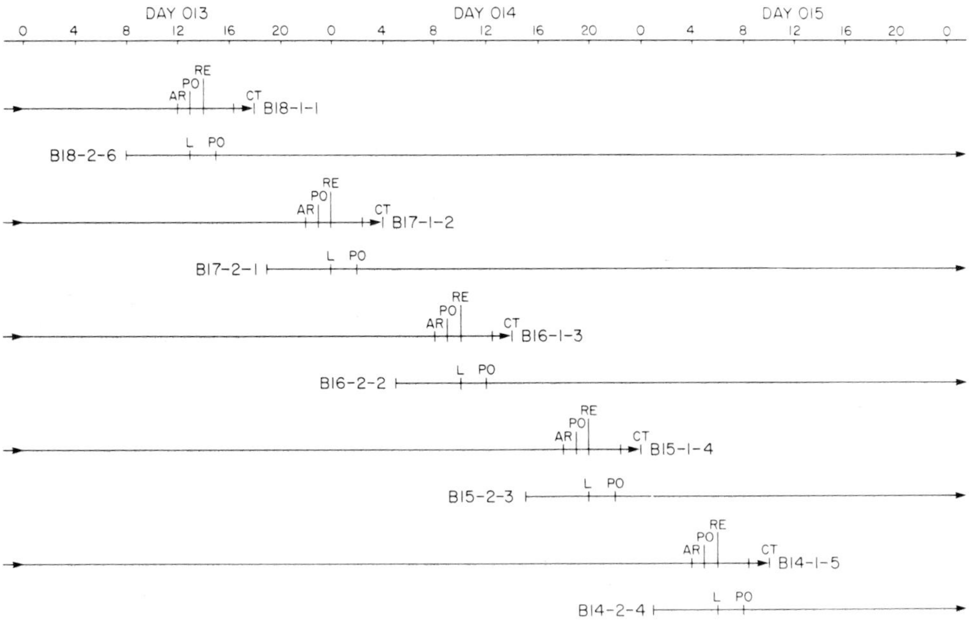

FIGURE 5. A ship schedule for occupying and launching five OBSs at 60-km spacing between stations in 6000 m of water depth. The release times were preset, allowing for no departures from the schedule. Stations were designated B14 to B18. AR: time of arrival at a station; PO: pinger on; RE: release from the bottom; CT: contingency time for surface search; L: launching of replacement OBS; PO: pinger off.

travels in and out of the laboratory on rails.) Every step in the final assembly is essential since any one error could lead to a nonrecovery. The following need to be checked: integrity of the buoyancy containers; mechanical reliability of all clips, fasteners, clamps, etc.; operation of recovery aids (radio beacon, flashing light, flag, radar reflector, etc.); anchor weight release, etc. During the final assembly on deck of instruments incorporating microprocessors it may be desirable to switch the ship's radar to standby to prevent an accidental erasure of a memory location.

Over the Side

The exact procedure for launching an instrument over the side depends on the type, size, and weight of the final assembly, the lifting arrangements on the deck of the ship, the experience of the bosun and his crew, and on weather conditions. The procedure should be carried out smoothly and quickly, otherwise there is a danger that the instrument may hit the side of the ship due to pendular swinging if the end of the crane boom is high above the water. If the final assembly consists of several components connected by flexible linkages[33] even greater coordination is required from the deck crew so that all the components go over the side in the proper order without any snags or delays.

The moment the instrument commences a free descent from surface to the bottom, the station data must be recorded: time, position of the ship, and depth of water, instrument number and characteristics, recovery aids used, weather and wave conditions, etc. If the instrument has a pinger, the descent can be observed on the echo sounder record and the exact time of bottom contact observed. As mentioned earlier, the descent rate for most instruments is 0.5 to 1.2 m/sec. Once the instrument settles on the bottom, the ship can proceed to the next station. If the OBS is without a pinger,

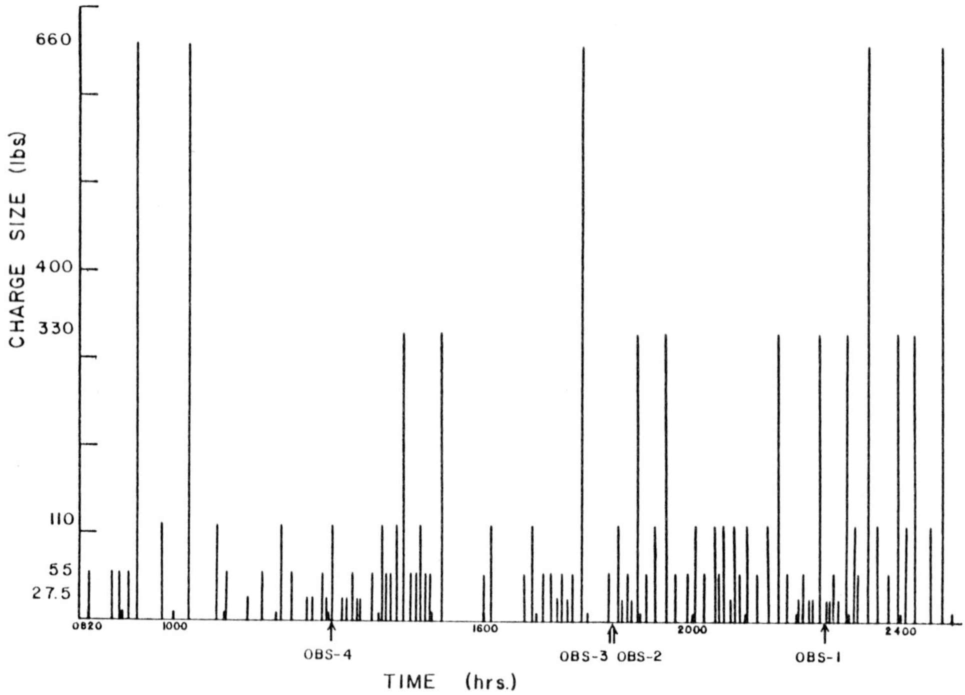

FIGURE 6. An example of spacing and charge size distribution along a 165-km long refraction line.

then it may be necessary to wait an equivalent descent-plus-ascent time period to make sure that the anchor weight did not prematurely release on hitting the bottom.

Acoustic Energy Sources

Once the instrument array is deployed, the ship may continue with other research programs if the instruments are to be left in place to monitor natural phenomena. If it is a controlled source experiment, the ship can commence to shoot. When discussing the acoustic energy sources, several topics must be covered: shooting pattern, use of repetitive sources or explosives (including special problems with very large charges), firing of charges, shot-time recording, and safety precautions.

Shooting Pattern

The shooting pattern describes the position and size of charges relative to the elements of an OBS array. The simplest pattern is a straight line along the length of the array, with equally spaced, equal charges. If the OBSs have horizontal geophones of unknown azimuth orientation, it may be necessary to shoot a cross line across each instrument to calculate this azimuth. For a study of a layer anistropy in a horizontal direction, a circular pattern of shots is required. The diameter of the circle will depend on the depth of the layer studied.

For a reversed refraction profile with most of the instruments concentrated near the two end points, an optimum shooting pattern should have variable-sized charges. When the shooting ship is near one end of the line, frequent small charges are needed to define upper crustal layer arrivals at the nearby instruments, while less frequent but much larger charges are required to define deeper layer arrivals at the far-end instruments. As the ship approaches the midpoint of the line, intermediate size charges are used, to be received at both ends. A typical shooting pattern for an OBS line in 4000 m of water is shown in Figure 6.

There are a number of tradeoffs that have to be considered when designing a shooting pattern, assuming that the quantity of available explosives is limited and that the minimum ship speed is set by the safety distance between the charge and the ship.

Increasing the number of shot points will give more points in the time-distance graph, but if this means dividing the total quantity of explosives and decreasing the size of individual charges the result will be indistinct and it will be difficult to read first arrivals. When a charge explodes, reverberations will be set up, depending on the charge size, bottom conditions, depth and shape of the ocean basin, etc. The next shot must wait about 4 to 10 min for this "noise" to be attenuated. The shooting ship usually travels at 14 km/hr (8 kn) so that the minimum spacing becomes 1 km. Seismic work at sea is an environmental issue and may lead to confrontation with fishing interests, native peoples, and others. Particular concern is expressed when the proposed work is near fishing banks or in areas visited by marine mammals. The evidence that seismic shooting may lead to large scale fish-kill is inconclusive and concerns appear to be overstated.

Repetitive Sound Sources

Nondestructive repetitive sound sources are often used for the definition of the sedimentary and upper crustal layers. There are several types of these sources: air guns, boomers, flexichock, gas exploders, etc. Air guns, with capacities of either 16.4ℓ (1000 in.^{3}) or 33ℓ (2000 in.^{3}) are the most frequently used devices. They are usually fired at 1-min intervals, though this rate may depend on the compressor capacity. The pressure of compressed air is 12 to 20 MPa. The maximum range with air guns is about 25 km, depending on the absorption in the bottom sediments. Although the total energy of an air gun "bang" is equivalent to less than 1 kg of explosive, the energy is concentrated in the lower end of the spectrum so that excellent arrivals are recorded at close ranges as can be seen from an example in Figure 7.

Explosive Charges

Explosives must be used whenever longer ranges and deeper penetration (to and below Mohorovic discontinuity) are required.[162,164] Some experimenters have had access to surplus military explosives (such as depth charges) though most use commercial explosives (various types of ammonium nitrate gels). Three trade names frequently mentioned in the literature are Hydromex® T3 (CIL), Tovex® (Dupont), and Geophex® (ICI). These explosives are quite safe as they require a special detonation booster to ignite. Some aluminum oxide should be added during manufacture to allow for detonation under hydrostatic pressure. They can be packaged in a variety of shapes and containers. A common form is as 12.5-kg plastic "sausages". Charges up to 500 kg can be easily made up by piling together a number of sausages on a pallet. One or more boosters are imbedded in the center of the package together with a firing cap.

Very Large Charges

One of the advantages of OBS seismology is that very long refraction lines (longer than several hundred kilometers) are possible. To penetrate to these ranges requires charges of several thousand kilograms and this has necessitated developing procedures for firing very large charges at sea.

Smaller charges can be easily tipped over the side while the ship steams away. This procedure is not practical with large charges. A large charge may break up on hitting the water or it may sink too deep while the ship is steaming away to a safe distance. Also, it may be desirable to fire simultaneously an array of very large charges.[76] The practice is to make up large charges on board, the size depending on the lifting capacity of the crane. Before the charge is lowered over the side, a long length of primacord (a

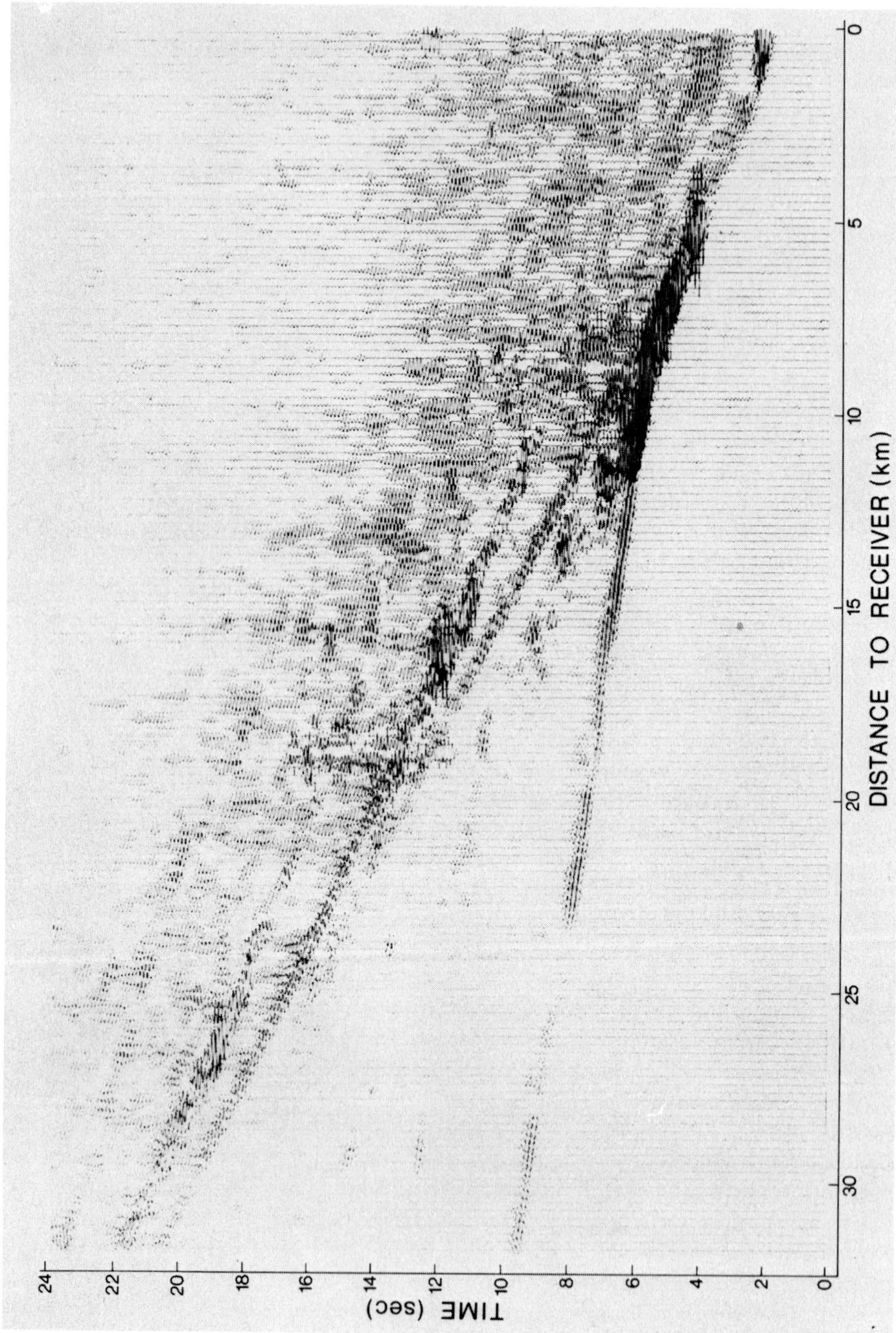

FIGURE 7. Time-distance plot of seismic arrivals from a 33-ℓ air gun in 6000 m of water. This plot was enhanced by polarization filtering of combined arrivals on vertical and horizontal geophones.

fast-burning fuse) is attached to the detonators, interspersed throughout the volume of the charge. The charge, weighing perhaps 1000 kg, is lowered over the side attached to a sufficient number of floats so that it can be maintained at the desired depth. The primacord is also attached to the floats, and the whole string is allowed to drift away from the ship, or is pulled away by a motorboat (for example a Zodiac rubber dinghy). Several such strings can be made up, and when all are in the water, the ship sails away to a safe distance. The boat then attaches an electric detonator cap to a free end of primacord on each floating charge and connects to them one end of the electric firing line. The other end is attached to a blasting machine in the motorboat. The blasting machine is triggered by a radio signal from the ship. This procedure is safe and has been used to fire explosive charges of up to 5000 kg. The intense shock wave, which could cause damage to a rigid ship, does not harm a rubber boat. The weather conditions must be good (winds less than 3 m/sec) to allow a small motorboat to work among the floating charges. In cold seas this work is uncomfortable and may be difficult. A diagram of an assembled floating charge is shown in Figure 8.

Firing of Charges

The firing of a charge can be either by a slow burning fuse (approximately 2 cm/sec) or by an electrical detonator. The length of the fuse is determined by the depth at which the charge should explode to give a maximum positive interference from the collapse of the gas bubble and the reflection from the sea surface. The depth varies with the charge size and is in the range of 60 to 120 m. The sinking rate depends on the density and geometry of the charge; it is of the order of 1 m/sec.[77] The sinking rate can be easily determined by attaching a simple float (for example a plastic garbage bag) to the charge with a known length of rope (say 50 m) and noting the interval between the time when the charge is thrown over the side and time when the float is pulled under by the sinking charge. The fuse burns faster under hydrostatic pressure.

The fuse length giving a sufficient burning time at the surface must be increased by about 30 to 50% to give an equivalent burning time while the charge is sinking to its desired depth of detonation. Once the fuse is lit, the charge must be quickly dropped over the side as there is no safe way of stopping the fuse from burning. With experience, a routine can be worked out so that the expected explosion or "shot-time" can be predicted with an accuracy of ±5 sec.

An alternative to fuse firing is to use electric detonators and an electric firing line. As the ship steams away from a dropped charge, the firing line is paid out. When the ship is at a safe distance and the charge is at the desired depth, a blasting machine sends a high voltage pulse (discharge from a bank of condensers) to fire the detonators. This moment can be recorded with a shot-time accuracy of a few milliseconds. The paying out of the shot line must be smooth, as any sudden jerks could cause a break in the continuity of the electrical circuit. The line must have a good quality insulation to prevent leakage and short circuit through seawater. Precautions must be taken against accidental firing by stray electrical pulses picked up by the firing line acting as a receiving antenna. All radio transmitters on the ship must be off-the-air during the shooting program. The line must have sufficient mechanical strength to withstand the strain of retrieval by a high-speed winch in preparation for the next shot.

Shot-Time Recording

In order to determine accurately the length of time for the travel of acoustic energy from source to the OBS, the origin, or shot time must be determined with an accuracy better than a hundredth of a second. This is easy with electrical firing. The burning time of fuses is not reliable and the shot time must be calculated from the recorded

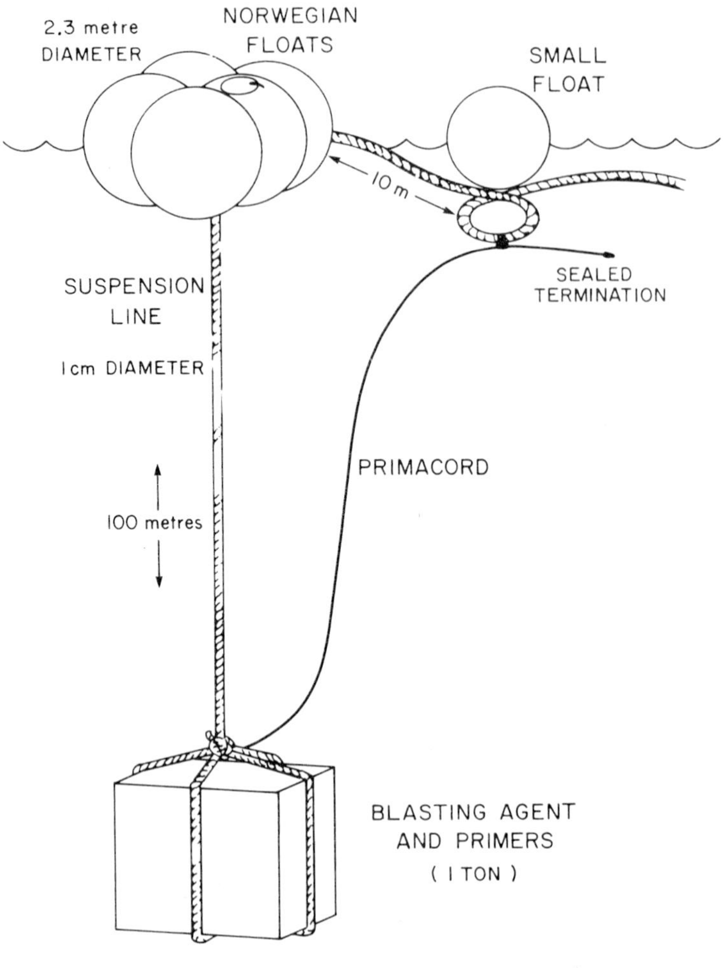

A

FIGURE 8. Method of rigging very large charges as practiced by Bedford Institute of Oceanography. (A) Suspension of individual charges. (B) An array of five 1-ton charges.

time of the arrival at the ship of the direct wave from the detonation. This arrival time is detected by a low sensitivity geophone firmly attached to the metal hull of the ship or by a low sensitivity hydrophone towed just below surface on the side of the ship. A shooting log must be kept of the following information at each detonation: shot point number, position of the ship and depth of water, speed and course of the ship, length of fuse, fuse burning time, charge flight time (i.e., time between dropping of the charge over the side and the detonation), estimated depth of charge (from an assumed sinking rate), shot time (to nearest second), magnetic tape counter, and the total amount of explosives used.

The arrival time and the absolute time reference are recorded on a two-channel graphic recorder and a backup magnetic tape. The chart speed of the graphic recorder must be 0.25 m/sec or greater to make possible scaling of the record with the required precision. The shot-time as recorded must be corrected for the distance of the charge from the shooting ship based on information in the shooting log. The horizontal dis-

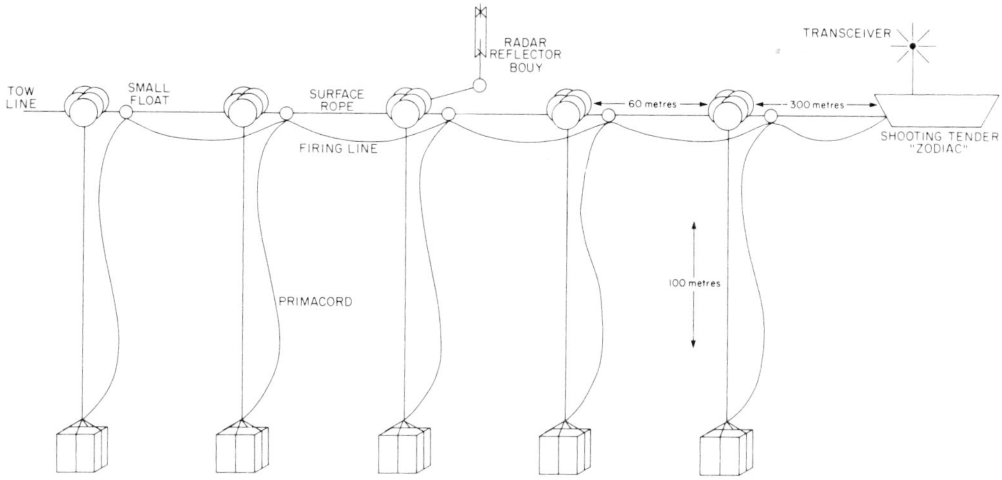

FIGURE 8B.

tance is calculated by multiplying the speed of the ship (V) by the flight time (FT). The vertical distance is the depth of the charge based on the estimated sinking rate (SR). The correction, which must be subtracted from the recorded shot-time, is

$$\text{Corr} = ((V*FT)^2 + (SR*FT)^2)^{1/2}*WV^{-1} = (FT/WV) * (V^2 + SR^2)^{1/2}$$

where WV is the velocity of sound in seawater. This correction is usually less than 0.4 sec.

Safety Precautions

Before a ship sails to carry out a seismic program, a number of permissions must be obtained from fisheries and environmental protection agencies, defense authorities, and external affairs (if the proposed program is to be in waters beyond the national jurisdiction). It is advisable to find out what other seismic exploration programs may be underway in the area in order to avoid interference. Proper Notices to Mariners should be issued at least 24 hr prior to the commencement of shooting.

Although Hydromex® and Tovex® are quite safe to handle, the preparation of charges and firing must be done by properly qualified and certified shooters. There should be at least two members on a firing team. Unauthorized personnel should not be allowed near the explosives and smoking should not be allowed. Fire hoses should be made ready to prevent any accidental spread of fire to the magazine. The explosives magazine must be constructed to the standards of ship insurers. Boosters, detonators, and caps must be stored in a separate magazine. A safe distance must be maintained at all times between the ship and the detonations. This distance depends on the charge size, the ship size, the depth of detonation, and the depth of water. An empirical formula for the minimum safe distance in deep water is

$$D = SF * W^{1/2} * (1 + \sin \alpha)$$

where D = distance to the ship in kilometers, W = weight of charge in kilograms, d = depth of detonation in meters, and α = angle subtended by the charge, i.e., sin α

$= d/(D^2 + d^2)^{1/2}$ and SF is a factor dependent on a ship. For a typical research ship of 80 to 100 m in length (about 2500 t) the factor is SF = 30.

Additional precautions may be necessary in shallow water, where the major shock may be due to the reflections from the sea floor and the simple distance formula is not valid. The author participated in 1968 in a seismic experiment when serious damage was done to the ship due to constructive interference between the 4th and 5th bounces of reflected shock waves in a water depth of less than 100 m, although the ship was about twice the "safe" distance away according to the simple formula.

Recovery Procedures

If the OBS has a pinger, the ship should be in the recovery area 1 to 2 hr before the release. Once the recovery position is reached, the ship must wait till the capsule reaches the surface, observing the ascent as a pinger trace, if one is available. If the recovery aids are operating and the weather conditions are favorable, the search is easy and the instrument can be recovered and returned on board within 10 to 20 min of reaching the surface.

If there is a two-way acoustic link, the ship reaches OBS position whenever it is desired to recover the instrument and issues the necessary commands. A timed backup release is usually employed as well. It is difficult to decide on an appropriate time allowance that will not prematurely release an OBS into bad weather or ice conditions, and still not cause an excessive wait if the acoustic system fails.

The problems arise when there is no pinger to indicate the release and ascent, and no recovery aid signals at the expected time of surfacing. Whether, where, and how to search must be quickly decided by taking into account the cost of the instrument and the value of the data recorded as compared to the cost of ship-time that will be used on search. If additional information about sea currents, possible uncertainties in navigation, etc., are available, they may indicate the direction in which to search. If the search is strictly visual, it is important to have the largest possible number of observers at the flying bridge, since the ability to spot a small object on the surface of the sea varies greatly from person to person.

If the ship is called away on a mercy mission, suffers a mechanical breakdown, or encounters ice or severe storms, the instrument may be lost unless the ocean currents carry it to a friendly shore. (There have been several instances of return of instruments lost at sea, indicating the importance of clearly labeling the owner's name and address on the capsule and inside the pressure case.)

When the instrument is first approached by the ship, the log book should record the time, position and depth, states of recovery aids and pinger, wind and sea state, and other conditions that may be helpful in future planning.

Post-Recovery

When the instrument capsule is returned on board, it should be first examined for physical integrity, damage from handling, landing on the bottom, fish bites, etc. Next, it should be cleaned and washed with fresh water to minimize seawater corrosion. The pressure case should be brought into the laboratory and allowed to warm up to room temperature to prevent condensation on the electronic components and the magnetic tape.

On opening the pressure case, the first check is for the presence of salt water indicating possible leaking end caps, bulkhead connectors, etc. A quick visual inspection of tape recorder should show if the tape transport operated properly.

The internal clock must be calibrated next against a shipboard reference. The offset between these two times, compared to the offset determined prior to launching, determines the clock drift, which becomes one of the corrections to be applied during the data reduction. Next, the battery levels are measured and the power drain recorded. Finally, a quick, qualitative examination of the magnetic tape is carried out to confirm the recording of seismic signals at expected arrival times.

If all the checks are satisfactory, the instrument can be refurbished and prepared for next launching. The total time for refurbishing is between 2 and 6 hr.

OPERATING ENVIRONMENT

The environment in which an OBS must operate is particularly harsh and requires a rugged design. On deck, the instrument is exposed to a corrosive atmosphere that may be particularly active in a humid, tropical climate. While handled on board and over the side, the instrument assembly is often subjected to mechanical shock and vibration.

The terminal velocity during the free fall descent depends on the hydrodynamic shape and the drag forces. A considerable stress could be exerted on the frame and various components by the drag forces. The differential pressure between the top and the bottom of the capsule may be substantial and may overstress the release system. Violent gyration may be experienced during the descent, shaking the internal components which may cause problems with the geophone leveling mechanism, magnetic tape transport, batteries, etc. The landing of a free-falling assembly causes impact forces, which may be large if the bottom is a rock outcrop.

The pressure at operating depths (6000 m) is 69 MPa (10,000 psi). Any pressure case must be designed to withstand this pressure or it will fail.[78] Bulkhead connectors that provide for electrical feed through the pressure case present a special technological problem not yet solved to everyone's satisfaction. The greatest danger to an instrument from leaks occurs when it is first submerged and before the hydrostatic pressure seals all caps. Such partial flooding may result in reduced buoyancy and may cause the loss of an instrument.

In the laboratory the instrument is at room temperature, while on the sea floor the temperature is near 0°C. All the circuitry must be designed to operate efficiently at this temperature, which fortunately is very stable. The low temperature reduces the rate of corrosion but the high pressure increases cathodic corrosion between dissimilar metals.

Noise

The movement of bottom water can cause a noise to be recorded by an OBS through two mechanisms: Karman vortex shedding,[163] and current pressure directly against an instrument resting on soft sediment.[165,166] Not much is known about the bottom currents but frequent scour marks around pebbles on photographs of the sea floor indicate that bottom currents with speeds of 10 cm/sec or greater can be present almost anywhere in the ocean.

In Karman shedding, vortices of water spin away downstream from obstructions and exert a force on the object. The vortices spin off at well-defined intervals, resulting in a harmonic force on the body. The frequency of the force is proportional to the current speed and inversely proportional to the diameter of the body. The shedding is nonlinear in that the vortices do not begin to spin off until a critical speed is reached (about 10 cm/sec, dependent on instrument). The motion resulting from these forces can be amplified if the shedding frequency is near a natural resonant frequency of the structure.

In soft sediments, an instrument with a circular cross section about 1 m in diameter, centered about 0.5 m above the sea floor, may tip about 5×10^{-6} m (10^{-5} rad) in response to a 10 cm/sec current. A small fluctuation in current speed, with additional perturbations due to vortex shedding, may produce objectionable noise on horizontal geophones. The magnitude of currents probably changes over a period of a few days depending on "ocean weather" (analogous to weather on dry land) with alternate periods of calm and considerably variable currents.

Many OBS recordings show evidence of short, impulsive signals of 0.5 to 3.0 sec duration that occur at irregular intervals. Evidence has been presented[167] that these signals are of biological origin based on the following observations: (1) the distribution of these signals on instruments launched in depths shallower than 1000 m shows a 24 hr periodicity, while there is no periodic pattern on instruments at sites deeper than 1000 m (where there is no visible light), (2) the frequency of occurrence of signals is similar to the vertical distribution of biomass in the ocean, i.e., they appear most frequently on OBS instruments launched in shallow depths, (3) biological material ("fish eggs") has been found attached to several OBS units upon recovery. How the creatures interact with the OBS is not clear. Perhaps fish swimming around an OBS bump into it or bite it, or crabs and worms burrow under it to cause irregular settling and tilting.

Bottom Coupling

Soft sediments can act as dissipative mechanical springs.[168] Coupling through soft sediments can modify the response to ground motion as does a low pass filter.[172,173] High buoyancy tends to counteract this effect. Depending on the configuration of the instrument, the effect of bottom coupling can seriously distort a recorded seismic signal thus reducing the value of an OBS experiment.[79] First order theory,[169,170] assuming a damped harmonic oscillator, suggests that increasing the surface area in contact with sediment increases the coupling frequency and the damping. This effect forces the OBS motion to be closer to that of the sediment. In addition, more sediment and water are likely to be entrained with the OBS, thus causing a drop on the coupling coefficient C, and improved response to frequencies above the coupling frequency.

Adjusting the density of the package to be slightly greater than that of seawater at depth also decreases the value of C, decreasing the effect of coupling. This makes the instrument more sensitive to high frequencies and reduces resonant amplification. Special care must be taken, in this case, to keep a low profile and good contact with the bottom to reduce cross-coupling and current problems.

The resonant frequency and coupling coefficient depend on the stiffness of the sediments on which the OBS is resting. The response of an OBS then will be site specific and some method of *in situ* calibration may be highly desirable. In order to achieve a desirable low ratio of package density to bearing area, several recent OBS designs incorporate an external geophone package that can be emplaced away from the main instrument capsule.[171] Experimental evidence shows that this arrangement will return superior records; there are however serious design problems in ensuring that recoveries will be reliable.

DATA REDUCTION

On recovery of an OBS, the raw data must be processed through a number of steps before meaningful geophysical interpretations can begin. A schematic data flow is shown in Figure 9. Stream "A" (on the left, Steps 3 through 7) represents the preliminary examination of data, usually carried out on board ship. Stream "B" (Steps 9 through 13) represents a more thorough and painstaking process that may take several months or years for completion.

The playback of raw data (Step 2) is carried out with a speed-up factor of between 10 and 20 times. This saves time during the data reduction, but it also shifts the fre-

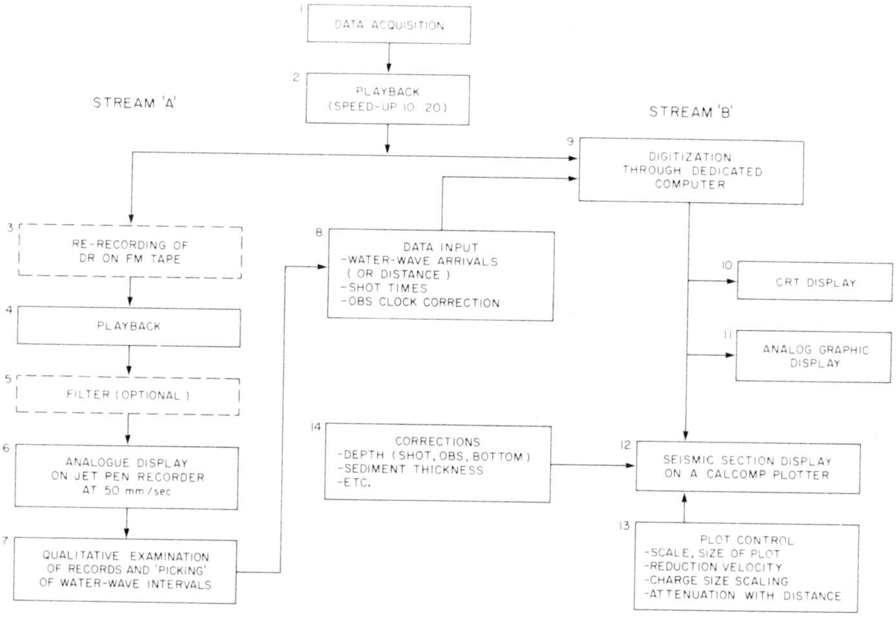

FIGURE 9. A flow chart of typical steps encountered in data reduction of OBS observations.

quency band towards higher frequencies. If the experiment was designed to study seismicity, the whole raw data tape needs to be examined. In a controlled-source experiment only the events corresponding to the arrivals from shots need be rerecorded (Step 3) for further examination. This rerecording allows for a slowdown on further playback (Step 4) so that the frequency band is shifted down, close to the original, and is within the response band of the analog recorder (Step 6). An optional filter (Step 5) may be included in the data stream if it is necessary to improve signal-to-noise ratio.

The display of data in analog form as a wiggly trace (Step 6) is important for recognition of arrivals of various phases. The times of direct water wave arrival are picked off at this time (Step 7) if they are to be used for calculating the distance between shots and receivers. The first arrivals also may be picked at this time in order to construct preliminary time-distance diagrams. This allows for construction of approximate velocity-depth structures that can guide the size and spacing of shots for further seismic experiments in the area.

During Step 8, log books are checked to make sure that all the parameters and corrections are tabulated and available for further detailed data reduction. The most important of these data are listed in Figure 9: distance between each shot and OBS, table of corrected shot times, and corrections to the OBS clock. A systematic error in any of these three parameters will lead to a distortion of the structural model that will be hidden, since the data set will appear self-consistent.

Data reduction through Stream "B" is usually carried out at the home laboratory and may require dozens of iterations before an acceptable seismic cross section is constructed. In modern practice, Steps 9 and 12 are automated. The degree of success is attested by high quality of results reported in recent literature.

The reason that Steps 9 through 12 are tedious and time consuming is that a number of corrections must be applied at this stage. The effect of these corrections is to shift seismic profiles either along distance (x) or time (y) axis, or both.

The depth of water and the sediment thickness usually vary along any seismic profile shot in an ocean basin. It is desirable to eliminate these variations by selecting a datum level and then applying corrections for each shot-receiver path. The datum level is conveniently taken as the average depth of water in the area studied. These corrections give us the travel times that would have been measured if: (1) the sea bottom under shot was at the datum level; (2) the OBS was located at the datum level; (3) the shot was fired at the datum level; and (4) the layer with sediment velocity was replaced by a layer with basement velocity. Because the ray path of acoustic energy enters the sea floor at an angle away from the vertical, the correction (1) has a horizontal component which decreases the shot-receiver distance.

The use of these corrections can be illustrated using typical values for P waves in the North Atlantic:[177]

Average water velocity	$= V_W = 1.54$ km/sec
Bottom water velocity	$= W = 1.55$ km/sec
Sea floor datum	$= t_D = 3.629$ sec
Sediment velocity	$= V_S = 1.8$ km/sec
Basement velocity	$= V_B = 4.0$ km/sec
Refractor velocity	$= V_R = 8.0$ km/sec
Water depth under shot	$= t_W$ (one-way sec)
Water depth under OBS	$= t_{OBS}$ (one-way sec)
Depth of shot beneath sea surface	$= h_{shot}$ (km)
Sediment thickness	$=$ TS (two-way sec)

1. Correction for sea floor topography under a shot is given by:

$$ST = \frac{-\left(t_D - \frac{h_{shot}}{W}\right)}{\sqrt{1 - \left(\frac{W}{V_R}\right)^2}} = -3.6963 + 0.6701\, h_{shot}$$

and the horizontal correction by:

$$\Delta X = \frac{-\left(t_D - \frac{h_{shot}}{W}\right) \cdot \frac{W^2}{V_R}}{\sqrt{1 - \left(\frac{W}{V_R}\right)^2}} = -1.0675 + 0.1935\, h_{shot}$$

2. Correction to move the receiver to the datum level:

$$TC_1 = \left(t_{OBS} - t_D\right)\left\{\left(\frac{V_W}{V_B}\right)\sqrt{1 - \left(\frac{V_B}{V_R}\right)^2} - \sqrt{1 - \left(\frac{W}{V_R}\right)^2}\right\}$$

These corrections can be positive or negative, depending on whether the OBS depth is above or below the datum level.

3. Correction to move the shot to the datum level:

$$TC_2 = \left(t_W - t_D\right)\left\{\left(\frac{V_W}{V_B}\right)\sqrt{1-\left(\frac{V_B}{V_R}\right)^2} - \sqrt{1-\left(\frac{W}{V_R}\right)^2}\right\}$$

the above two corrections, (2 and 3), can be conveniently combined to give the total datum correction (for our example) of:

$$TC = 0.6455\,(2t_D - t_W - t_{OBS})$$

4. Correction for sediment thickness under a shot (or under a receiver):

$$\Delta t_S = \frac{-TS}{2}\left\{\left(\frac{V_S}{V_B}\right)\sqrt{1-\left(\frac{V_B}{V_R}\right)^2} - \sqrt{1-\left(\frac{V_S}{V_R}\right)^2}\right\} = -0.2923\ TS$$

Once a self-consistent, corrected, and relatively error-free data set is available, data interpretation will be undertaken — often requiring further reprocessing through Steps 9 to 12. The aim of this final stage of data processing is to construct a most plausible structural model to which the data can be fitted. A variety of interpretational techniques may be used. Of particular value is ray tracing, tau-p method, and construction of synthetic seismograms.[81] Kennett[80] has given a useful introduction to these techniques, together with an extensive bibliography.

RESULTS OF OBS RESEARCH

A wide range of geophysical problems has been studied with the OBSs during the past dozen years or so.[146] To review in detail the results of this research would require a review of our present knowledge of the seismic structure of the ocean basins — a task beyond the scope of the present chapter. Only a brief sketch of the main areas of research can be given here. For more information, the interested reader must consult the original literature. Whitmarsh and Lilwall[27] give a useful overview. Another recent summary is given by Spudich and Orcutt.[199,200]

Studies of Natural Seismicity

Studies of Earthquakes

A worldwide, uniform distribution of observing stations is important for the detection, location, and identification of earthquakes. Since 71% of the Earth's surface is covered by water, it is clear that OBS have an important role to play. Deployment of OBS is particularly important in seismically active areas which are offshore, as for example off Vancouver Island and off Japanese islands. With only a shore-based network, errors in hypocentral location of 20 to 50 km were quite common.[178,179]

To observe a teleseismic event on an OBS, the instrument must be in place and recording at the time of the earthquake, and the local sea-bottom noise must be below the signal level. Except for a deployment of 5 OBS for 8 months near Kodiak Island by the Marine Science Institute of the University of Texas,[180,181] most launchings to date have lasted a month or less. It is thus not surprising that the number of detections reported in the literature is rather small.

Among the events discussed in the literature are those detected by OBS near Bermuda,[31] in the Middle America Trench,[137] off Japan[85-87] and in the Pacific near Midway, Wake, and Eniwetok Islands.[161] Over 80 events recorded on Pacific islands were detected by offshore hydrophones (prior to 1971) and not by OBS.[89] Nevertheless, the number of events recorded is an indication of the detectability possible with OBS were they to be deployed over a long enough period. There has been more success with the detection of earthquakes off Japan, off the west coast of Canada, and off Alaska. The work off Kodiak Island is of particular interest as it involves strong motion instruments,[54] and is designed to study the long-term seismicity.[111,180,181] This experiment holds a "world record" of 240 days for deployment of OBS at sea.[201] During a 6-week period in the summer of 1979, several hundred earthquakes were recorded. Preliminary investigations suggest that the hypocenters of the detected events are concentrated along the inclinded (Benioff) zone associated with the subduction of the Pacific Plate beneath Kodiak Island. Some shallow earthquakes extend into the island ridge, thus delineating currently active faults.[180]

Teleseismic P-waves were recorded by four ocean-bottom seismographs located in the Vema fracture zone.[149] Large differences between J-B travel time residuals at different OBS are the result of structural variations near the receivers. To explain the results, low velocities in the upper mantle are required. These low velocities may be related to the extension into the fracture zone of a narrow zone of partial melting from beneath the ridge axis.

High frequency Pn, Sn phases from teleseismic events at distances from 7° to 30° were recorded in the Pacific by several OBS.[154-157] Spectrograms contain frequencies as high as 10 Hz and 7 Hz and confirm earlier evidence for remarkable efficiencies of Pn and Sn propagation, with Sn often more efficient (higher apparent Q) than Pn.

Another example of earthquake detection at teleseismic distances is a recording of T waves (seismic water waves) by OBS deployed in the western Pacific Basin. The sources of these T waves were located at the bottom of deep-sea trenches. Their intensity implies that the Q value along the descending lithosphere is more than 4000. The arrivals indicate a stratified structure of the upper mantle with a thin high velocity layer (about 8.6 km/sec) underneath the 8.2 km/sec lithosphere.[150]

Hyndman and Rogers[179] have detected nearly 100 events off Vancouver Island in a period of 16 days. They were able to locate them with an accuracy better than ±10 km, compared with 25 to 100 km errors in older determinations of epicenters. They found that over the magnitude range of 1 to 3 the magnified frequency of occurrence relations are in good agreement with 75 years of land station data. They also found P wave velocity in the uppermost mantle of 7.6 km/sec and the mean Vp/Vs ratio of 1.71.

Two large cooperative experiments (LADLE[32] and ROSE[182]) were carried out recently with a specific objective to augment controlled-source seismology with recordings of earthquakes. Both experiments were successful but the final interpretations are not yet available.

The original objectives to use OBSs for the detection of underground nuclear explosions has not been pursued. Walker[161] has observed high frequency mantle arrivals on hydrophones from Soviet nuclear explosions at distance ranges of 60° to 90°, but failed to detect the tests conducted at the Nevada Test Site at similar distances. Further research on this topic is in abeyance as long as the ban on underground nuclear explosions holds.

Microearthquakes

Aftershocks have been studied as a means of better understanding the source mechanism and the total energy release in an earthquake.[107] For example, 2 years after the 1968 Tokachi-Oki earthquake (M = 7.9) off Sanriku, Japan,[85-87]microearthquake activity was 2 to 12 shocks per day. The larger number occurred on the oceanic side of Japan Trench. One explanation for these observations is that the tectonic deformation of the continental slope caused by the large earthquake gradually migrated into the ocean basin cross the deep trench.[101,113]

Some of the earliest observations of microearthquake activity associated with a fracture zone were made in the vicinity of the Mendocino Fracture Zone.[90,105] This OBS detected five times as many events from the fracture zone as did an equidistant land station. This observation illustrated the attenuation of seismic waves crossing an ocean-continent boundary. The observation further confirmed that the Mendocino Fracture Zone is a true transform fault since the microearthquakes were not observed beyond the junction of the fracture zone with the spreading axes.

During a period of 40 hr following Gorda ridge earthquake of June 26, 1968, about 50 earthquakes were recorded by an OBS located 220 km south of Mendocino Fracture Zone.[104] The magnitude vs. the length of the fault for this sequence is approximately similar to that of other recent sequences in California, but there is an indication that the sequence was rather rich in large-magnitude earthquakes. The composite fault-plane solution for all of the aftershocks was similar to that of the main shock and gave a fault trend of N41°E.

In the more stable regions, microearthquakes have been observed on the Hawaiian Ridge, on the Fiji Plateau[102] and off the Makran coast of Iran.

During an 8-month period an OBS off Oahu, Hawaii, detected 359 events, 40% of which were not detected by land stations and that probably occurred in the oceanic lithosphere. The significant earthquake activity that exists along the Hawaiian Ridge presumably is a result of slumping, fracture zone activity, and possible submarine volcanism.[91]

Three OBS were moored about 170 km south of Makran coast in the Gulf of Oman for 2 weeks. During this period 40 events were identified as aftershocks of the destructive Bandar-Abbas earthquake of March 21, 1977. In comparison with Zagros frontal fold, no seismicity associated with the postulated subduction of the Arabian plate beneath the Eurasian lithosphere could be detected.[103] (Three 180-kg calibration shots fired during this experiment gave a mantle velocity of 8.0 km/sec at a depth of 20 km overlain by a 6.7-km thick layer of velocity 6.7 km/sec).

The microearthquakes on spreading centers or the accretion plate boundaries have been studied on a number of spreading centers: on Reykjanes Ridge,[99] on Mid-Atlantic Ridge near 45°N,[92,97,98] near 37°N,[94,183] near 23°N,[125] and near the intersection with St. Paul Fracture Zone.[93] Other studies were conducted in the Indian Ocean near the triple junction[108] and in the Pacific near 9°S,[100] near the intersection with Rivera Fracture Zone,[107,130] and on Juan de Fuca[123] and Explorer[124] Ridges.

The objective of most of these studies was to detect the largest number of small earthquakes occurring near the spreading axis, to locate them accurately in time and space, and to identify as much as possible the nature of the source mechanism. The key element in these studies is the accurate determination of hypocenters.[83] Small arrays of three or more OBS can locate hypocenter with an accuracy of 1 to 2 km.[84] The observations suggest that the activity is confined to a zone less than 5 km wide and a depth of 10 km or less. The time distribution of microearthquakes is not uniform. Swarm-like sequences suggest that these events are associated with magma movements.[97,98] A preliminary conclusion from these studies is that the character and frequency of microearthquake activity near spreading centers greatly varies along any ridge segment, between different ridges, and at different intersections between fracture zones and spreading axes. A systematic study of these differences is yet to be undertaken.

The studies of microearthquakes near the intersection of fracture zones and spreading centers conclude that the source mechanism is the strike-slip along faults in the direction of plate motions. Of 50 microearthquakes detected by Project ROSE[182] during a 15 day period, half were aligned along a narrow fault with hypocentral depths between 0 and 17 km. In the case of a slow spreading center (MAR near 37°N), the seismicity is concentrated on the corner of the plate between the spreading center and the transform fault.[183] This study concluded that diffuse seismicity (20 to 30 km) probably reflects internal deformations of the corner by secondary faults away from the plate boundary.

One OBS operated for 13 days near the Indian Ocean triple junction at 25°S, 70°E.[108] Epicenters could not be determined with a single instrument but from the S-P time differences it was concluded that at least 38 microearthquakes were local events (again giving an activity rate of 3 microearthquakes per day). S-wave corner frequencies of 15 to 25 Hz were measured, and by relating these to source mechanism, a fault length of 30 to 45 m was inferred. All the events occurred on the African-Indian plate boundary. A conclusion was reached that triple junction earthquakes have moments and fault lengths which scale as if the events occurred on a normal spreading center.

The observations to date can be summarized by stating that microearthquakes occur with a frequency that varies from 0 to 100 events per day.[106] On the slow spreading centers, the hypocenters are less than 10 km deep and possibly reflect movements in the magma chamber. Other microearthquakes near spreading centers reflect strike-slip movements on transform faults with depths up to 17 km. The direction of slip is consistent with the direction of movement of tectonic plates.

Microseismicity and Seabed Noise

Considerable research effort has been expended on acquiring a better understanding of seismic noise on the seabed.[22-25,29,165] Two types of noise are recognized: microseismicity due to ocean storms and attendant pressure fluctuations, and the general ambient noise due to other natural and manmade phenomena.

An ocean basin is a noisy place with a variety of sound sources: marine mammals, especially whales; passage of ships; offshore drilling and exploration; ocean storms and pressure fluctuations; and precipitation. The net result is a noise level often higher than that at quiet continental sites. In the band of 0.1 to 9 Hz spectra exhibit peaks corresponding to the organ-pipe modes in the water column.[17-21] Noise measurements in deep water, 200 km off the California coast[22] show microseisms moving in from further out to sea and being correlated with storm activity. Central Pacific noise levels range from 1 μm of ground displacement at 1 Hz to 10 μm at the microseismic peak between 0.1 and 0.2 Hz.[31]

More recent investigations have found fairly quiet sites in the NE Pacific. These studies confirmed that, around 1 Hz, noise levels fall rapidly with increasing frequency, and above about 4 Hz reach the 1 nm level. Similar results above 1 Hz were obtained by Russian researchers in the Black Sea.[26]

Controlled Source Seismology

Continental Margins and Adjacent Ocean Basins

Despite many good reasons in favor of OBS experiments on continental shelves, surprisingly little work has been done so far. Working on shelves is closer to shore bases, and therefore logistics of the operation should be easier. The experiment can be augmented with shore-based stations in so-called onshore/offshore experiments. The areas are of great practical importance and have a considerable scientific interest, which should have provided an added incentive. But, a combination of impediments has slowed down this work.

Shallow water sites are usually noisy due to wind and wave action, precipitation, and propeller noise from passing vessels. There is a possible loss of instruments due to fishermen dragging the bottom. Thick sedimentary wedge attenuates acoustic energy, thus requiring very large charges (400 kg or larger) if a full crustal cross section is to be penetrated. Large charges are difficult to handle and could be destructive in shallow water. There is a considerable reverberation in the water column due to the wave-guide propagation between the surface and sea bottom. Finally, these programs face a multitude of environmental protection regulations as they are perceived to be in conflict with the fishing interests.

Some observations of local interest were made around Japan[95,110] and as a part of the Lopez Island experiment,[173] but only two shallow water OBS crustal refraction lines have been reported in the literature. A combined sonobuoy and OBS experiment was carried out on the Texas shelf.[109] This experiment showed that signal-to-noise ratio for the OBS is about three to four times better than that of the sonobuoy. A velocity of 7.7 km/sec was encountered at a depth of 14.5 km and was interpreted as a lower crust or an anomalous mantle.

An OBS experiment on the Nova Scotian shelf provided a convincing picture of the crustal cross section.[184] The crust beneath the outer shelf includes 9 to 16 km of sediments. A layer with a compressional wave velocity of 6.3 km/sec comprises the main crustal layer, below which the mantle is estimated to lie at total depths of 26 to 30 km. A seismic stratigraphy for the sediments can be deduced by comparing the refraction results with sonic log velocities from nearby deep exploration wells. The total thickness of crustal rocks of prerift age is between 14 and 21 km, which is significantly thinner than the 35 to 38 km thickness measured beneath mainland Nova Scotia in earlier studies. Both the thin crust and the high rate of subsidence during the Lower Jurassic age are consistent with extension of the lithosphere during the initial rifting of this margin in late Triassic time. Beneath the upper continental rise of Nova Scotia about 7 km of sediments lie on top of the oceanic layer 2, with a velocity of 5.3 km/sec. Mantle velocities of 8.3 to 8.48 km/sec are encountered at depths between 12 and 18 km.

On the east side of the Atlantic, a similar experiment near the ocean-continent boundary was recently completed.[185] An interesting study of a rifted margin and a thinned and subsided continental crust was carried out northeast of Newfoundland in the Orphan Basin.[186] The results show that these outer regions of the margin are underlain by thinned continental crust with a total depth to the M discontinuity of about 22 km. There are two main crustal layers with P wave velocities of about 6.1 and 7.0 km/sec that appear to be homogeneous and separated by sharp interfaces. These are overlain by a layer with a velocity of 5.5 km/sec, interpreted to be Precambrian or Paleozoic basement. Mesozoic and Cenozoic sediments above this basement are 4 km thick.

A similar transition crust was observed around the Gulf of Mexico.[187] Beneath the 5- to 9-km thick sedimentary layer is a 10- to 12-km thick crustal layer with velocities of 6.4 to 7.0 km/sec. This layer is only 3 to 6 km thick in the central part of the Gulf and is designated oceanic layer 3.

The data confirm that most of the deep Gulf was formed by rapid ocean floor spreading and that the transition crust presumably formed by the initial stretching and subsidence is confined to the rims of the Gulf.

A rather unique study of the continent-ocean transition was carried out in the Arctic Ocean, north of Beaufort Sea.[188] An OBS was lowered through a hole in the ice to the bottom on 3000 m of Keflar rope. A 12-channel hydrophone array, that could be steered through digital processing for the best phase velocity discrimination, was deployed just below the ice surface. Explosive charges carried by a helicopter were dropped at seven locations up to 73 km from the receivers. A total crustal thickness (to a 7.9 km/sec layer) of 13 to 16 km is comprised of 4.3 km of sediments (2.7 km/sec), an assumed oceanic layer 2 (4.9 km/sec) 4.4 to 5.6 km thick, and a layer 3 (6.8 km/sec) of less than 2.2 km thickness, if present at all. It was concluded that the crust was oceanic, perhaps 100 to 130 million years old. A possibility that this is a transitional (stretched continental) crust cannot be ruled out as the nearby measurements during Project AIDJEX had identified another layer with a velocity of 8.1 to 8.5 km/sec at a depth of 20 to 24 km.

The renewed interest in the nature and extent of the transition crust and the considerable potential for hydrocarbon reserves in this region will see a major expansion of OBS research into the transition zone between continents and oceans.

Subduction Zones

Subduction zones occur where two plates collide. Deep-sea trenches are a manifestation of these collisions in the oceanic regions.

All the deep-sea trenches studied with OBSs are on the rim of the Pacific Ocean: off Vancouver Island,[195,203] Middle America Trench,[137,138] off Colombia,[190] Peru,[141] New Hebrides,[139,189] off Japan,[110,113,142,143,147] and Kurile Trench.[140,145]

In the case of the Middle America Trench, the oceanic mantle of 8.1 km/sec velocity was at a depth of 13 to 17 km near the trench axis. The overlying 6.5 km/sec layer could be traced 30 km landward from the trench axis. Similar results were obtained further south off Colombia:[189] layer 2 with a velocity of 5.1 km/sec; layer 3 with a velocity of 6.8 km/sec; and a mantle velocity of 8.2 km/sec at a depth of less than 20 km below the trench axis and dipping at an angle of about 15° in the subduction under the continent.

Off Peru, at 18°S, a low mantle velocity of 7.9 km/sec increases rapidly to a value of 8.5 km/sec. It is not clear whether this low velocity is related to petrologic effects or to subduction effects (stress).[141] A surprising result of this experiment was a low velocity of 6.5 km/sec at the base of the crust. The thickness of this zone, based on extremal inversion is between 0.43 and 5.5 km and is attributed to a partially serpentinized periodotite. The thickening of the crust was also observed under the other trenches studied. It appears that a combination of controlled source and earthquake seismology holds the greatest promise for future studies of the subduction zones.

Deep Ocean Basins

The objective of seismic refraction measurements in the deep ocean basins is to describe the "typical" or average structure of the oceanic crust. On the assumption that such a typical crust exists and remains homogeneous over considerable distances, several very long (about 1000 km) profiles have been completed, four in the Pacific and one in the Atlantic Ocean.

A detailed investigation of a small area of 40-to 50-million-year-old crust in the North Atlantic was carried out by firing closely spaced charges around a circle of about 50 km diameter with an OBS in the center.[121] Significant lateral variations in the seismic structure of layer 2 was recorded by delay time variations of up to 0.1 sec, probably caused by lateral changes in the structure of the uppermost few hundred meters of the volcanic basement. The upper portion of layer 2 (several hundred meters thick) has a low velocity of 3.9 km/sec. It is underlain by a step in velocity to about 4.5 km/sec and a uniform velocity gradient of about 0.7 km/sec/km to the uppermost basement level at a depth of about 7.5 km. A number of profiles in the North Atlantic show a similar positive velocity gradient in layer 2 of 0.85 to 1.35 km/sec/km. An explanation for this gradient is that it is due to the reduction downwards of bulk porosity, particularly due to the closure of cracks.[122]

Inside the magnetic quiet zone off Morocco the structure of Concepcion Bank was interpreted as due to a volcanic origin unrelated to sea floor spreading.[120] The anomalous mantle of 7.6 km/sec velocity at a depth of less than 13 km is overlain by a volcanic layer with a P wave velocity of 5.60 km/sec.

A comparison of crust of equal age (80 million years bp) on both sides of the Mid-Atlantic Ridge was carried out as a part of the site surveys for the Deep Sea Drilling Project.[117] On both sides, a typical crust was found: a 0.6-km thick layer 2B with a velocity of 4.70 km/sec, overlying a 1.5-km thick layer 2C with a velocity of 6.25 km/sec, overlying a layer 3 with a velocity of 6.75 km/sec (the base of layer 3 was not observed). Also observed was a consistent increase in layer 2 velocity with age from the ridge crest.

The question of seismic anisotropy of the upper mantle has been of great interest to geophysicists ever since the concept of ocean floor spreading gained acceptance. One OBS experiment to investigate this anisotropy was carried out 360 km southeast of Reykjanes Ridge.[196] There were only about 30 mantle arrivals that could be used in this study, and complicated statistical arguments led to the conclusion that it cannot be said with certainty that there was any anisotropy in this area. By comparison, in other areas of the Atlantic and the Pacific, where anisotropy was found, 443, and 1143 mantle arrivals were available for the analysis. The apparent absence of anisotropy may be related to the slow spreading rate of the Reykjanes Ridge, though this experiment was not conclusive and needs to be repeated.

Anisotropy at a deeper level in the lithosphere was detected in the western Pacific.[118,119,151,152] Two "longshot" profiles,[82,198] perpendicular to each other and over 1000 km in length, and a third profile in the Marianas Basin,[114] suggest that there is anisotropy in the lithosphere both regionally and also as a function of azimuth. A layer of high velocity (8.4 to 8.7 km/sec) was detected below the normal mantle velocity of 8.2 km/sec. A very thin lower velocity layer (8.4 km/sec) was detected at the base of the lithosphere, which was estimated to be 80 km thick.

A 600-km long line along the 70 million year isochrone off the east Pacific Rise found considerable velocity variations in the upper lithosphere.[115,116] There is a general tendency for velocity to decrease with depth except at about 27 and 52 km depth where there are rapid velocity increases, perhaps associated with phase or compositional changes.

In spite of great operational complexity and considerable cost, long profiles shot with large charges offer a powerful tool for the exploration of the lithosphere to depths of 100 km or more.

Spreading Centers

Currently active spreading centers, or mid-ocean ridges, have attracted the largest number of OBS experiments. Measurements have been made and results reported in the literature for Dellwood Knolls,[131] Explorer Ridge,[124] Juan de Fuca,[123] East Pacific Rise at 23°N,[127] at 20°N,[130] on Cocos plate near 14°N,[112,197] and at 8°N,[129] in Shikoku Basin,[128] and on Mid-Atlantic Ridge at 23°N,[125] 37°N,[133,134,191,192] 45°N,[193] north of the Azores,[132] and Reykjanes Ridge.[96,99,194]

The general feature of the refraction profiles near Pacific spreading centers are (1) narrow zone of lower velocities and thin crust near the axis of spreading centers, and (2) thickening crust and more normal oceanic crust and upper mantle velocities away from the crest.

On the Explorer Ridge (at 50°N)[124] the low-velocity mantle (7.3 km/sec) is overlain by unusually low-velocity layer 3 (6.1 km/sec). Away from the crest, the velocity structure is significantly different with a near normal layer 3 velocity of 7.1 km/sec. Immediately to the south on the Juan de Fuca Ridge (at 49°N)[123] a low velocity lower crust and upper mantle exist while under the young flanks layer velocities of 5.2, 6.1, 6.4, and 7.1 km/sec are found above the mantle with a velocity of 8.3 km/sec.

Off the mouth of the Gulf of California, at 23°N, a zone of low velocity in the crust was found to be less than 10 km wide.[127] Thin crust of only 2.5 km thickness is associated with the axial valley. Below it, P-wave velocities may be as low as 4.2 km/sec. Away from the crest, the mantle has a normal velocity of 8.1 km/sec and dips up towards the crest at an angle of 11°.

A pronounced low velocity was found on the East Pacific Rise near Siqueiros Fracture Zone at 8°N.[129] P-wave velocities increase quite rapidly from about 5 km/sec at the lip of the low velocity zone. Mantle velocity under the crest (7.7 km/sec) remains anomalously low (7.6 km/sec) under crust 2.9 million years old, but at 5 million years, distinct layering with velocity gradients is observed above a normal mantle with 8 km/sec velocity.

The most systematic study of the change of oceanic crust with distance from the East Pacific Rise was carried out on the Cocos plate at 14°N.[103,197] This study concluded that:

1. The crust thickens systematically with increasing age. This conclusion is independent of the interpretation of a low velocity zone at the base of the crust.
2. The thickening is due to the development of a zone in the lower crust which has a velocity somewhat lower than that in the middle crust.
3. The lower oceanic crust may be laterally heterogeneous, in some places having a sharp crust-mantle boundary and in some places being gradational.
4. The degree of upper mantle anisotropy appears to decrease by a factor of 2 away from the rise axis, having a value of about 0.6 km/sec at zero age, and about 0.3 km/sec beyond about 2 million years.
5. The data strongly suggest that the velocity-depth function in the upper crust is better represented by a linear gradient than constant velocity layers.

No OBS seismic refraction lines have been reported from either the South Pacific or South Atlantic. In the North Atlantic two areas have been studied intensively: Mid-Atlantic Ridge at 37°N (Project FAMOUS)[133,134,191,192] and Reykjanes Ridge.[96,99,194]

At the ridge crest at 37°N most seismic arrivals were from a main refractor of an apparent velocity of 5.4 to 6.3 km/sec. This main layer corresponds to a low velocity layer 3, and overlays a low velocity mantle of 7.2 km/sec. The crust is less than 3 km thick at the crest and doubles in thickness in less than 10 km, where a normal upper mantle velocity of 8.1 km/sec exists. A study of time residuals along two profiles across the median valley indicates the presence of a 2- to 3-km wide low velocity zone (about 3.2 km/sec) beneath the median valley floor. This zone extends over the upper 2.5 km of the crust and is believed to represent a zone of intrusion through which magma passes on its way to the sea floor.[134] A further study showed that shear waves propagate across the axis, thus precluding the existence of any sizeable magma chamber at shallow depth.[192]

A study of aging of the crust away from the axis, similar to the Cocos plate study, was carried out over Reykjanes Ridge[194] at 0, 30, and 90 km from the ridge axis. The observations show the following trends with increasing age:

1. The velocity of the sea floor refractor increases from 2.2 km/sec at the crest, to

3.8 km/sec for crust 9 million years old. The layer has a relatively constant thickness of 400 m.

2. The 4.6 to 4.7 km/sec layer becomes thinner (from 1.3 to 0.8 km) but its velocity remains constant.
3. The 5.4 to 6.2 km/sec layers become shallower but their combined thickness and mean velocity remain relatively constant. (The velocity structure of these layers is not well defined, as neither the first arrivals nor the synthetic seismograms analysis are well controlled for this velocity or depth.)
4. The 6.4 to 6.6 km/sec layer also becomes shallower though maintaining its velocity and thickness (0.8 km).
5. The 6.6 to 7.2 km/sec layer increases its thickness and mean velocity with age.
6. The maximum velocity measured increased with age. The transition to this velocity is a gradient rather than a first order discontinuity. The maximum velocity measured at the Ridge axis was 7.1 km/sec, corresponding to the lower limit of anomalous mantle velocities recorded. At 9 million years, a typical Moho velocity of 8.2 km/sec was encountered.

A set of experiments was carried out over the Arctic Mid-Ocean Ridge by combining a near-surface hydrophone array with an OBS lowered to the bottom on a line through a hole in the ice.[12] The Mid-Arctic Ridge is spreading at a very slow rate of 5mm per million years, the crust is thin (2 to 3 km) and associated magnetic anomalies are of low amplitude. The study concluded that a generally thin crust is probably due to basaltic productivity of the ridge crest, as a result of increased cooling at the very slow spreading rate. In the follow-up experiments in the Pole Abyssal Plain a total crustal thickness of 4 to 7 km was detected over magnetic anomaly 22 again indicating correlation between thin crust and slow spreading rate.[202]

These experiments have confirmed that: (1) there are substantial differences between slow (MAR) and fast (EPR) spreading ridges, and (2) there are significant differences in structure with increasing age away from the ridge axis. On the slow-spreading ridges a substantial faulting and fracturing takes place,[126] probably associated with magma chambers.[193] Fast spreading ridges also have a narrow zone of partial melting and lower mantle velocity under the axis, and more uniform thickening of the lithosphere. A low velocity zone has been observed underlying a fast-spreading rise crest,[148] while a zone of P-wave attenuation was observed beneath the crest of Mid-Atlantic Ridge.[132,153] This attenuation was related to hydration (probably including serpentinization) of the upper periodotite layers.

Fracture Zones

Fracture zones or transform faults are tectonic features connecting offset portions of moving plate boundaries.[144] Typical fracture zones, such as Kane and Oceanographer in the Atlantic Ocean, connect portions of the Mid-Atlantic Ridge spreading axis offset by 100 to 200 km. The zones are sites of considerable seismic activity and so have attracted OBS research, in particular studies of microearthquake activity.

A split refraction profile 74 km long was conducted along the sediment filled axis of the Oceanographer fault zone.[136] A layer 2 km thick with P-wave velocity of 4.4 km/sec overlies a layer with velocity 6.5 km/sec. The first layer was interpreted as either a broken and fractured pile of basaltic rock of a layer of serpentinite. The thickness of the 6.5 km/sec layer was not precisely determined but was estimated to be about 4 km thick. It is either a layer of metagabbro or partially serpentinized ultramafic rock.

Kane Fracture Zone (at 24°N, 44°W in the North Atlantic) was also studied in a detailed seismic refraction experiment.[135] The shooting lines and receivers formed a "T" configuration across the fracture zone, with two receivers located about 50 km apart in the fracture zone trough and the remaining six receivers positioned 25 to 30 km apart on either side of the fracture zone. The crustal thickness and velocities observed at the receivers located north and south of the Kane Fracture Zone fall within the range of those typically observed for normal oceanic crust. There is no convincing evidence for significantly different crustal thicknesses or upper mantle velocities on either side of the fracture zone despite a 10 million year age difference. Anomalously thin crust is present beneath the Kane Fracture Zone trough with a total crustal thickness of only 2 to 3 km — about half the thickness of normal crust. This crust is also characterized seismically by low P-wave velocities (about 4.0 km/sec) at shallow depths, and the absence of a normal layer 3 refractor. Results from other geological and geophysical studies of fracture zones suggest that this type of crustal structure may be typical of many Atlantic fracture zones.

Other Studies

The technique of firing explosive charges on the deep sea floor has not been perfected though it could provide means for detailed studies of the nature and properties of the upper sedimentary layers. For example, propagation of Stonley[158,174] and Rayleigh[159] waves is the best method for the study of shear velocities in sediments. Density and attenuation have been studied using multibounce phases.[160] All these properties are important in deducing reflectivity and scattering properties of the ocean floor, which in turn are important for the target resolution and identification in sonar detection.

FUTURE PROSPECTS

Certain developments can be forecast without attempting to predict the order in which they might occur, or the time scale. The cost of information per function performed will continue to decrease following further advances in microelectronics. This will lead to sophisticated digital instruments with computational power similar to some main frame computers of a generation ago. Advances in solid state data storage devices (bubble memories and similar) will decrease the overall size of instruments and increase their reliability. Decrease in size will decrease the overall cost, though considering the cost of pressure housing, buoyancy, and recovery aids, there is clearly a bottom cost which is not far below the cheapest instruments constructed today.

Communication with the instrument through an acoustic link will be improved and the channel capacity will be increased to allow for an examination of portions of data records. This will be feasible when random access mass storage devices (bubble memories) are available. More sophisticated on-board data processing will lead to better signal quality, fidelity, response, and signal-to-noise ratio. Particularly important will be absolute amplitude calibration in terms of true ground motion. This in turn will require further understanding of coupling between the instrument and the bottom medium on which it is resting.

Availability of large numbers of instruments may lead to cooperative, multi-institution experiments employing sophisticated two-dimensional arrays. In terms of application, the most important may be earthquake monitoring, for example off Japan, Alaska, west coast of North America, etc. Studies of local seismicity and terrain stability may be important in terms of offshore development related to hydrocarbon exploration and production.

Seismic refraction methodology is an averaging geophysical technique with a limited resolution. Methods today are thus not directly applicable to oil and gas exploration. It is a powerful method, however, and it is likely that future advances in processing of array data, and new approaches to data analysis and interpretation, will contribute directly to man's quest for resources and protection of the environment.

APPENDIX 1
LIST OF OBS RESEARCH GROUPS ACTIVE IN EARLY 1981

Canada

1. Atlantic Geoscience Centre
   Geological Survey of Canada
   Bedford Institute of Oceanography
   Dartmouth, N.S. B2Y 4A2
2. Earth Physics Branch
   Pacific Geoscience Centre
   Patricia Bay Institute of Ocean Sciences
   9860 West Saanich Road
   Sidney, B.C., V8L 4B2

France

3. Centre Océanologique de Bretagne (COB)
   CNEXO
   B.P. 337
   29273 Brest, Cédex
4. Laboratoire d'Étude Géophysique
   Institute de Physique du Globe
   Tour 14-24, 4e Étage
   4, Place Jussieu
   Paris

Germany

5. Institut für Geophysik
   Bundesstrasse 55
   D2 Hamburg, 13

Japan

6. Earthquake Research Institute
   University of Tokyo
   Bunkyo-Ku
   Tokyo 113
7. Department of Geophysics
   Hokkaido University
   Sapporo 060

U.K.

8. Institute of Oceanographic Sciences
   Wormley near Godalming
   Surrey, GU8 5UB
9. Institute of Oceanographic Sciences
   Blacknest
   Brimpton near Reading
   Berkshire, RG7 4RS
10. Bullard Geophysical Laboratory
    University of Cambridge
    Madingley Road
    Cambridge, CB3 OEZ

U.S.

11. Department of Geology and Geophysics,
    Woods Hole Oceanographic Institution
    Woods Hole, MA 02543
12. Lamont-Doherty Geological Observatory
    Columbia University
    Palisades, NY 10964
13. Geological Research Division
    Scripps Institution of Oceanography
    University of California San Diego
    La Jolla, CA 92093
14. Hawaii Institute of Geophysics
    University of Hawaii
    Honolulu, HI 96822
15. Department of Oceanography
    University of Washington
    Seattle, WA 98195
16. School of Oceanography
    Oregon State University
    Corvallis, OR 97331
17. Marine Science Institute
    University of Texas
    Austin, TX 78712
18. Department of Earth and Planetary Sciences
    Massachusetts Institute of Technology
    Cambridge, MA 02139
19. Department of Geological Sciences
    University of California Santa Barbara
    Santa Barbara, CA 93106

U.S.S.R.

20. Faculty of Physics
    Moscow State University
    Institute of Oceanology
    Academy of Science
    Moscow, 117234

REFERENCES

1. **Ewing, M., Crary, A. P., and Rutherford, H. M.,** Geophysical investigations in the emerged and submerged Atlantic Coastal Plain, Part I: Methods and results, *Geol. Soc. Am. Bull.,* 48, 753, 1937.
2. **Ewing, M. and Vine, A. C.,** Deep-sea measurements without wires or cables, *Trans. Am. Geophys. Un.,* 19th Annu. Meet., 1938.
3. **Ewing, M., Woolard, G. P., Vine, A. C., and Worzel, J. L.,** Recent results in submarine geophysics, *Geol. Soc. Am. Bull.,* 57, 909, 1946.
4. **Bullard, E. C. and Gaskell, T. F.,** Submarine seismic investigations, *Proc. R. Soc., London, Ser. A.,* No. 971, 177, 476, 1941.
5. **de Launay, L.,** The geology of the bottom of the sea, *Smithson. Inst. Annu. Rep.,* 2334, 329, 1915.
6. **Anon.,** Collection and analysis of Pacific ocean-bottom seismic data, *Geophysics,* 29(5), 745, 1964.
7. **Arnett, R. A. and Newhouse, T. W.,** Ocean-bottom seismograph, *IEEE Proc.,* 53(12), 1899, 1965.
8. **Frosh, R. A., Anderson, V. C., and Bradner, H.,** Instrumenting the sea floor, *IEEE Spectrum,* 1, 101, 1964.
9. **Schneider, W. A. and Backus, M. M.,** Ocean-bottom seismic measurements off the California coast, *J. Geophys. Res.,* Vol. 69(6), 1135, 1964.
10. **Schneider, W. A., Farrell, P. J., and Brannian, R. E.,** Collection and analysis of Pacific ocean bottom seismic data, *Geophysics,* 29, 745, 1964.
11. **Thomson, J. T. and Schneider, W.,** An automatic marine seismic monitoring and recording device, *Proc. Inst. Radio Eng.,* 50, 2209, 1962.
12. **Jackson, H. R., Reid, I., and Falconer, R. K. H.,** Crustal structure near the arctic mid-ocean ridge, *J. Geophys. Res.,* 85(B3), 1773, 1982.
13. **Sutton, G. H., McDonald, W. G., Prentiss, D. D., Thanos, S. N.,** Ocean-bottom seismic observatories, *IEEE Proc.,* 53(12), 1909, 1965.
14. **Ewing, J. I. and Ewing, W. M.,** A telemetering ocean-bottom seismography, *J. Geophys. Res.,* 66(11), 3863, 1961.
15. **Bradner, H.,** Seismic measurements on the ocean bottom, *Science,* 146, 208, 1964.
16. **Bradner, H.,** Geophysical measurements at ocean bottom, *IEEE Spectrum,* 1(11), 108, 1964.
17. **Bradner, H. and Dodds, J. G.,** Comparative seismic noise on the ocean bottom and on land, *J. Geophys. Res.,* 68, 4339, 1964.
18. **Bradner, H., Dodds, J. G., and Foulks, R. E.,** Investigation of microeism sources with ocean-bottom seismometers, *Geophysics,* 30(4), 511, 1965.
19. **Bradner, H. and Latham, G. V.,** Prograde Rayleigh waves and microseism sources, *J. Geophys. Res.,* 77, 6422, 1972.
20. **Bradner, H., Dodds, J., and Foulks, R.,** Coherence measurements with time sampling ocean-bottom seismometers, *IEEE Proc.,* 53(12), 1906, 1965.
21. **Foulks, R. E.,** Investigation of microseism sources with ocean-bottom seismometers, *Geophysics,* 30(4), 511, 1965.
22. **Latham, G. V. and Nowroozi, A. A.,** Waves, weather, and ocean-bottom microseisms, *J. Geophys. Res.,* 73, 3945, 1968.
23. **Monakhov, F. I.,** Microseisms at the bottom of the Baltic Sea and in the northern part of the Atlantic Ocean, *Izv. Acad. Sci. U.S.S.R.,* Geophys. Ser., 573, 1962.
24. **Monakhov, F. I.,** Microseisms at the bottom of the Black Sea, *Izv. Acad. Sci. U.S.S.R.,* Geophys. Ser., 461, 1961.
25. **Prentiss, D. D. and Ewing, J.,** The seismic motion of the deep ocean floor, *Bull. Seismol. Soc. Am.,* 53, 765, 1963.
26. **Rykunov, L. N. and Sedov, V. V.,** Seismic noise in the 2 to 15 cps frequency range on the bottom of the Black Sea, *Izv. Phys. Solid Earth,* 7, 443, 1965.
27. **Whitmarsh, R. B. and Lilwall, R. C.,** Ocean-bottom seismographs, in "Structure and Development of the Greenland-Scotland Ridge — New Methods and Concepts", *NATO Conf. Ser. IV: Marine Sciences,* Bott, M. H. P., Saxov, S., Talwani, M., and Thiede, J., Eds., Plenum Press, New York, 1981.
28. **Bolt, B. A.,** Ocean bottom seismometry; a new dimension to seismology, *Boll. Geofis. Teor. Appl.,* 20, 107, 1977.
29. **Suwa, A., Yamakawa, N., and Iinuma, T.,** Research and development of permanent ocean-bottom seismograph observation system off the Pacific coast to central Honshu, Japan. Wind and seismic effects, Lew, H. S., Ed., *U.S. Spec. Publ.,* 523, III.30, 1978.
30. **Ezell, G. D.,** Historical review of the development of a deep sea ocean bottom seismograph system, in *Trans. 3rd Annu. MTS Conf. Exhibit,* Marine Technology Society, Washington, D.C., 1967, 147.
31. **Latham, G. V. and Sutton, G. H.,** Seismic measurements on the ocean floor, *J. Geophys. Res.,* 71, 2545, 1966.

32. **Whitmarsh, R. B., Lilwall, R., Donegan, M., Keen, C. E., Loncarevic, B. D., and Steinmetz, L.,** Progress report on LADLE, *Geophys. J. R. Astron. Soc.*, 65, 263, 1981.
33. **Avedik, F., Renard, V., Busine, D., and Cornic, J. Y.,** Ocean bottom refraction seismograph (OBRS), *Mar. Geophys. Res.*, 3(4), 357, 1978.
34. **Asada, T., Matsumoto, H.A.,** Sensor for an ocean bottom seismometer, *Zisin (Seismol. Soc. Jpn. J.)*, 27(1), 83, 1974.
35. **Barrett, D. L., Heffler, D. E., and Keen, C. E.,** Ocean bottom seismometer developments, *Can. Geol. Surv.*, 77-1B, 129, 1977.
36. **Blackington, J. G. and Odegard, M. E.,** An ocean bottom seismograph using digital telemetry and floating-point conversion, *IEEE Trans. Geosci. Electr.*, GE-15, 74, 1977.
37. **Bookbinder, R. G., Hubbard, A. C., McDonald, W. G., and Pomeroy, P. W.,** Design of an ocean bottom seismometer with response from 25 Hz to 100 sec, *Oceans*, 178, 510, 1978.
38. **Carmichael, D., Carpenter, G., Hubbard, A., McCamy, K., and McDonald, W.,** A recording ocean bottom seismograph, *J. Geophys. Res.*, 78(35), 8748, 1973.
39. **Carmichael, D., Ludwig, W. J., Ewing, J. I., Carpenter, G., and Hubbard, A. C.,** A pop-up ocean bottom hydrophone, *Mar. Geophys. Res.*, 3, 225, 1977.
40. **Cranford, M. D., Johnson, S. H., Bowers, J. E., McAlister, R. A., and Brown, B. T.,** A direct-recording ocean-bottom seismograph, *Seismol. Soc. Am. Bull.*, 66(2), 607, 1976.
41. **Duschenes, J. D., Barash, T. W., Mattaboni, P. J., and Solomon, S. C.,** On the use of an externally deployed geophone package on an ocean bottom seismometer, *Mar. Geophys. Res.*, 4, 437, 1981.
42. **Francis, T. J. G., Porter, I. T., Lane, R. D., Osborne, P. J., Pooley, J. E., and Tomkins, P. K.,** Ocean bottom seismograph, *Mar. Geophys. Res.*, 2(3), 195, 1975.
43. **Francis, T. J. G.,** Ocean bottom seismographs, *Earthquake Inf. Bull.*, 9(5), 9, 1977.
44. **Francis, T. J. G. and Porter, I. T.,** Experience gained with the Blacknest ocean bottom seismograph, *Mar. Geophys. Res.*, 3(1), 143, 1977.
45. **Heffler, D. E. and Barrett, D. L.,** OBS development at Bedford Institute of Oceanography, *Mar. Geophys. Res.*, 4(2), 227, 1979.
46. **Herber, R.,** Ocean-bottom seismograph of the Institut fuer Geophysik, Hamburg, *Mar. Geophys. Res.*, 4(2), 247, 1979.
47. **Johnson, R. V., II, Lister, C. R. B., and Lewis, B. T. R.,** A direct recording ocean bottom seismometer, *Mar. Geophys. Res.*, 3(1), 65, 1977.
48. **Johnson, S. H., Cranford, M. D., Brown, B. T., et al.,** A free-fall direct-recording ocean bottom seismograph, *Mar. Geophys. Res.*, 3(1), 103, 1977.
49. **Kasahara, J., Koresawa, S., Ouchi, T., and Nagumo, S.,** Free-fall pop-up ocean bottom seismograph system, *Spec. Bull. Earth. Res. Inst.*, 13, 1, 1974.
50. **Kishinouye, F., Yamazaki, Y., Kobayashi, H., and Koresawa, S.,** A submarine seismograph, the first paper, *Bull. Earth Res. Inst.*, 41, 819, 1963.
51. **Koelsch, D. E. and Purdy, G. M.,** An ocean bottom hydrophone instrument for seismic refraction experiments in the deep ocean, *Mar. Geophys. Res.*, 4, 115, 1979.
52. **Koresawa, S.,** Construction of ocean bottom seismograph, *Earthquake Res. Inst. Bull., Tokyo*, 43(4), 671, 1966.
53. **Langford, J. J. and Whitmarsh, R. B.,** Pop-up bottom seismic recorder (PUBS) of the Institute of Oceanographic Sciences, U. K., *Mar. Geophys. Res.*, 3(1), 43, 1977.
54. **Latham, G. V., Donoho, P. L., Griffiths, K. H., Roberts, A., and Ibrahim, A. B. K.,** The Texas ocean-bottom seismograph, *Offshore Tech. Conf., Proc.*, 10(2), 1467, 1978.
55. **Lister, C. R. B. and Lewis, B. T. R.,** An ocean-bottom seismometer suitable for arrays, *Deep-Sea Res.*, 23(1), 113, 1976.
56. **Mattaboni, P. J. and Solomon, S. C.,** MITOBS: A seismometer system for ocean-bottom earthquake studies, *Mar. Geophys. Res.*, 3(1), 87, 1977.
57. **Moore, R. D., Dorman, L. M., Huang, C-Y., and Berliner, D. L.,** An ocean bottom, microprocessor-based seismometer, *Mar. Geophys. Res.*, 4, 451, 1981.
58. **McDonald, W. G., Hubbard, A. C., Bookbinder, R. G., et al.,** Design and shipboard operation of a multipurpose ocean bottom seismograph, *Mar. Geophys. Res.*, 3(2), 179, 1977.
59. **Nagumo, S., Kobayashi, H., Koresawa, S.,** Construction of ocean bottom seismograph, *Earthquake Res. Inst. Bull., Tokyo*, 43(4), 671, 1965.
60. **Neprochnov, Yu. P., Sedov, V. V., Rykunov, L. N., and Pokryshkin, A. A.,** Automatization of observations and preliminary processing of recordings of ocean-bottom seismographs during DSS in the sea, *Phys. Solid Earth*, 13(2), 127, 1977.
61. **Prothero, W. A., Jr.,** An ocean-bottom seismometer capsule, *Seismol. Soc. Am. Bull.*, 64(4), 1251, 1974.
62. **Prothero, W. A., Jr.,** An operationally optimized ocean-bottom seismometer capsule, *Phys. Earth Planet. Inter.*, 18(2), 71, 1979.

63. **Prothero, W. A., Jr.**, A digital event-recording ocean bottom seismometer capsule, *Mar. Geophys. Res.*, 3(1), 119, 1977.
64. **Sutton, G. H., Kasahara, J., Ichinose, W. N., and Byrne, D. A.**, Ocean bottom seismograph development at Hawaii Institute of Geophysics, *Mar. Geophys. Res.*, 3(2), 153, 1977.
65. **Whitmarsh, R. B.**, An ocean-bottom pop-up seismic recorder, *Mar. Geophys. Res.*, 1, 91, 1970.
66. **Berger, J., Agnew, D. C., Parker, R. L., and Farrel, W. A.**, Seismic system calibration: 2 cross-spectral calibration using random binary signals, *Seismol. Soc. Am. Bull.*, 69, 271, 1979.
67. **Gilbert, R. L. G.**, Data-collecting systems in oceanography, *Boll. Geofis. Teor. Appl.*, 8(32), 286, 1966.
68. **Nagumo, S., Kobayashi, H., and Koresawa, S.**, Improvements of ocean-bottom seismograph construction of a long-life magnetic tape recorder (with Japanese ABS.), *Earthquake Res. Inst. Bull. Tokyo*, 46(4), 861, 1968.
69. **Shimamura, H. and Asada, T.**, A cassette recorder for an ocean bottom seismometer, *Bull. Hokkaido Univ.*, 32, 17, 1974.
70. **Kasahara, J., Nagumo, S., Koresawa, S., and Ouchi, T.**, Low power amplifier and tape recorder for the ocean bottom seismograph, *Zisin (Seismol. Soc. Jpn. J.)*, 27(1), 75, 1974.
71. **Hasegawa, S. and Nagumo, S.**, Construction of a long-life magnetic tape recorder and some features of ocean-bottom seismograms, *Earthquake Res. Inst. Bull., Tokyo*, 48(5), 967, 1970.
72. **Davis, R. E. and Ambuter, B. P.**, An oceanographic data recorder, *Geophysics*, 46(2), 203, 1981.
73. **Ambuter, B. P. and Solomon, S. C.**, An event-recording system for monitoring small earthquakes, *Seismol. Soc. Am. Bull.*, 64, 1181, 1974.
74. **Heffler, D. E. and Locke, D. R.**, A deep ocean release mechanism, *Mar. Geophys. Res.*, 3(2), 229, 1977.
75. **Thanos, S. N. and Hubbard, A. C.**, Two-way hydroacoustic communications link for an ocean-bottom seismograph, *IEEE Trans. Geosci. Electron.*, GE-4(1), 17, 1966.
76. **Jacob, A. W. B.**, Dispersed shots at optimum depth — an efficient seismic source for lithospheric studies, *J. Geophys.*, 41, 63, 1975.
77. **White, R. S. and Bunch, A. W. H.**, Explosive sink rates, *Geophys. J. R. Astr. Soc.*, 56, 473, 1979.
78. **Nagumo, S., Kobayashi, H., Koresawa, S., and Takeyama, T.**, Pressure vessels for ocean-bottom seismograph, *Earthquake Res. Inst. Bull. Tokyo*, 48(5), 955, 1970.
79. **Steinmetz, R. L., Donoho, P. L., Murff, J. D., and Latham, G. V.**, Soil coupling of a strong motion, ocean bottom seismometer, *Offshore Tech. Conf., Proc.*, 11(4), 2235, 1979.
80. **Kennett, B. L. N.**, Towards a more detailed seismic picture of the oceanic crust and mantle, *Mar. Geophys. Res.*, 3(1), 7, 1977.
81. **Spudich, P. K. P. and Helmberger, D. V.**, Synthetic seismograms from model ocean bottoms, *J. Geophys. Res.*, 84(B1), 189, 1979.
82. **Asada, T. and Shimamura, H.**, Observations of earthquakes and explosions at the bottom of the western Pacific; structure of oceanic lithosphere revealed by longshot experiment, in The Geophysics of the Pacific Ocean Basin and Its Margin, *American Geophysical Union of Geophysists Monograph*, 19, 135, 1974.
83. **Ichikawa, M.**, Redeterminations of hypocenters of earthquakes occurring in northern Japan by use of ocean-bottom seismographic data, and some related problems, *Geophys. Mag. (Tokyo)*, 34(3), 333, 1969.
84. **Lilwall, R. C. and Francis, T. J. G.**, Hypocentral resolution of small ocean bottom seismic networks, *Geophys. J. R. Astron. Soc.*, 54(3), 721, 1978.
85. **Nagumo, S.**, Ocean-bottom seismographic observation of the Tokachi-Oki earthquake of 1968, in *General Report on the Tokachi-Oki Earthquake of 1968*, Earthquake Research Institute, Tokyo, 1971, 225.
86. **Nagumo, S., Hasegawa, S., Koresawa, S., and Kobayashi, H.**, Ocean-bottom seismographic observation off Sanriku; aftershock activity of the 1968 Tokachi-Oki earthquake and its relation to the ocean-continent boundary fault, *Earthquake Res. Inst. Bull., Tokyo*, 48(5), 793, 1970.
87. **Nagumo, S., Kobayashi, H., and Koresawa, S.**, Foreshock phenomena of the 1968 Tokachi-Oki earthquake observed by ocean-bottom seismographs of Sanriku, *Earthquake Res. Inst. Bull., Tokyo*, 46(6B), 1355, 1968.
88. **Prothero, W. A.**, Earthquake signal processing and logging with a battery-powered microcomputer, *Seismol. Soc. Am. Bull.*, 70, 2275, 1980.
89. **Walker, D. A. and Sutton, G. H.**, Oceanic mantle phases recorded on hydrophones in the NW Pacific at distances between 9° and 40°, *Seism. Soc. Am. Bull.*, 61, 65, 1971.
90. **Auld, B., Latham, G., Nowroozi, A., and Seeber, L.**, Seismicity off the coast of northern California determined from ocean bottom seismic measurements, *Seismol. Soc. Am. Bull.*, 59(5), 2001, 1969.
91. **Estill, R. E. and Odegard, M. E.**, Hawaiian ridge seismicity from the Makapuu ocean bottom seismograph, *Geophys. Res. Lett.*, 5(6), 485, 1978.

92. **Francis, T. J. G. and Porter, I. T.**, Microearthquake survey of the Mid-Atlantic Ridge, *Nature (London)*, 290, 547, 1972.
93. **Francis, T. J. G., Porter, I. T., and Lilwall, R. C.**, Microearthquakes near the eastern end of St. Paul's fracture zone, *Geophys. J. R. Astr. Soc.*, 53, 201, 1978.
94. **Francis, T. J. G., Porter, I. T., and McGrath, J. R.**, Ocean-bottom seismograph observations on the Mid-Atlantic Ridge near lat. 37 degrees N., *Geol. Soc. Am. Bull.*, 88(5), 664, 1977.
95. **Kasahara, J., Nagumo, S., and Koresawa, S.**, Seismic activity in Sagami Bay and off the Boso Peninsula observed by ocean bottom seismographs in 1971, *J. Phys. Earth (Tokyo)*, 26(2), 199, 1978.
96. **Goldflam, P., Weigel, W., and Loncarevic, B. D.**, Seismic structure along RRISP — Profile I on the southeast flank of the Reykjanes Ridge, *J. Geophys.*, 47, 250, 1980.
97. **Lilwall, R. C., Francis, T. J. G., and Porter, I. T.**, Ocean-bottom seismograph observations on the Mid-Atlantic Ridge near 45°N; further results, *Geophys. J. R. Astron. Soc.*, 55(1), 255, 1978.
98. **Lilwall, R. C., Francis, T. J. G., and Porter, I. T.**, Ocean-bottom seismograph observations on the Mid-Atlantic Ridge near 45°N, *Geophys. J. R. Astron. Soc.*, 51(2), 357, 1977.
99. **Lilwall, R. C., Francis, T. J. G., and Porter, I. T.**, Some ocean-bottom seismograph observations on the Reykjanes Ridge at 59°N, *Geophys. J. R. Astron. Soc.*, 62, 321, 1977.
100. **Lilwall, R. C., Francis, T. J. G., and Porter, I. T.**, A microearthquake survey at the junction of the east Pacific rise and the Wilkes (9°S) fracture zone, *Geophys. J. R. Astron. Soc.*, 66, 1981.
101. **Nagumo, S., Kasahara, J., and Koresawa, S.**, Structure of microearthquake activity around Japan Trench, off Sanriku, obtained by ocean-bottom seismographic network observation, *J. Phys. Earth (Tokyo)*, 24(3), 215, 1976.
102. **Nagumo, S., Kasahara, J., and Ouchi, T.**, Active seismicity in the Fiji Plateau observed by ocean-bottom seismograph, *J. Phys. Earth (Tokyo)*, 23(3), 279, 1975.
103. **Niazi, M., Shimamura, H., and Matsu'ura, M.**, Microearthquakes and crustal structure off the Makran coast of Iran, *Geophys. Res. Lett.*, 7(5), 297, 1980.
104. **Nowroozi, Ali. A.**, Seismicity of the Mendocino Escarpment and the aftershock sequence of June 26, 1968; ocean bottom seismic measurements, *Seismol. Soc. Am. Bull.*, 63(2), 441, 1973.
105. **Seeber, L., Barzangi, M., and Nowroozi, A.**, Microearthquake seismicity and tectonics of coastal California, *Seism. Soc. Am. Bull.*, 60, 1669, 1970.
106. **Shimamura, H., Asada, T., and Takano, K.**, Measurements of microearthquakes at sea bottom, *Bull. Soc. Fr. Jpn. Oceanogr.*, 8, 1970.
107. **Reid, I. and Prothero, W. A.**, An earthquake sequence studied with ocean-bottom seismographs, *Geophys. J. R. Astron. Soc.*, 64, 381, 1981.
108. **Solomon, S. C., Mattaboni, P. J., and Hester, R. L.**, Microseismicity near the Indian Ocean triple junction, *Geophys. Res. Lett.*, 4(12), 597, 1977.
109. **Ibrahim, A. B. K. and Latham, G. V.**, A comparison between sonobuoy and ocean bottom seismograph data and crustal structure of the Texas shelf zone, *Geophysics*, 43(3), 514, 1978.
110. **Kasahara, J., Nagumo, S., Koresawa, S., and Kobayashi, K.**, Ocean bottom seismographic observation at Suruga Bay utilizing oil filled battery, *Zisin (Seismol. Soc. Jpn. J)*, 26(2), 171, 1973.
111. **Latham, G. V. and Dorman, J.**, Progress report on ocean bottom seismology in the Gulf of Alaska, 1979, *Alaska Sci. Conf., Proc.*, 30, 70, 1979.
112. **Lewis, B. T. R. and Snydsman, W. E.**, Fine structure of the lower oceanic crust on the Cocos Plate, *Tectonophysics*, 55(1-2), 87, 1979.
113. **Nagumo, S. and Kasahara, J.**, Ocean-bottom seismograph study of the western margin of the Pacific, *Am. Geophys. Union Geophys. Monogr.*, 19, 155, 1976.
114. **Nagumo, S., Ouchi, T., Kasahara, J., Koresawa, S., Tomoda, Y., Kobayashi, K., Furomoto, A. S., Odegard, M. E., and Sutton, G. H.**, Sub-Moho seismic profile in the Marian Basin — ocean bottom seismograph long-range explosion experiment, *Earth Planet. Sci. Lett.*, 53, 93, 1981.
115. **Orcutt, J. A. and Dorman, L. M.**, Results of Scripps long-range profiles in the Pacific, *Tectonophysics*, 56(1-2), 83, 1979.
116. **Orcutt, J. and Dorman, L. M.**, An oceanic long-range explosion experiment; a preliminary report, *J. Geophys.*, 43(1-2), 257, 1977.
117. **Rabinowitz, P. D. and Ludwig, W. J.**, Geophysical measurements at candidate drill sites along an east-west flow line in the central Atlantic Ocean, *Mar. Geol.*, 35(1-3), 243, 1980.
118. **Shimamura, H. and Asada, T.**, OBS studies on P and S velocities beneath the northwestern Pacific basin, in *Science Council of Japan, Tokyo*, 1978, 146.
119. **Shimamura, H., Tomoda, Y., and Asada, T.**, Seismographic observation at the bottom of the central basin fault of the Philippine Sea, *Nature (London)*, 253, 177, 1975.
120. **Weigel, W., Goldflam, P., and Hinz, K.**, The crustal structure of the Conception Bank, *Mar. Geophys. Res.*, 3, 381, 1978.
121. **White, R. S. and Matthews, D. H.**, Variations in oceanic upper crustal structure in a small area of the northeastern Atlantic, *Geophys. J. R. Astron. Soc.*, 61, 401, 1980.

122. **Whitmarsh, R. B.**, Seismic refraction studies of the upper igneous crust in the North Atlantic and porosity estimates for layer 2, *Earth Planet. Sci. Lett.*, 37, 451, 1978.

123. **Davis, E. E., Lister, C. R. B., and Lewis, B. T. R.**, Seismic structure of the Juan de Fuca Ridge; ocean bottom seismometer results from the median valley, *J. Geophys. Res.*, 81(20), 3541, 1976.

124. **Hyndman, R. D., Rogers, G. C., Bone, M. N., Lister, C. R. B., Wade, U. S., Barrett, D. L., Davis, E. E., Lewis, T., Lynch, S., and Seemann, D.**, Geophysical measurements in the region of the Explorer Ridge off western Canada, *Can. J. Earth Sci.*, 15(9), 1508, 1978.

125. **Kasahara, J., Hussong, D. M., and Sutton, G. H.**, A seismic study using ocean-bottom seismometers near the Mid-Atlantic Ridge at 23°N, *Mar. Geol.*, 35(1-3), 199, 1980.

126. **Lilwall, R. C.**, Fault mechanisms and subcrustal seismic velocities on the Mid-Atlantic Ridge, *Geophys. J. R. Astron. Soc.*, 60, 245, 1980.

127. **McClain, J. S. and Lewis, B. T. R.**, A seismic experiment at the axis of the East Pacific Rise, *Mar. Geol.*, 35(1-3), 147, 1980.

128. **Nagumo, S., Ouchi, T., and Koresawa, S.**, Seismic velocity structure near the extinct spreading center in the Shikoku Basin, North Philippine Sea, *Mar. Geol.*, 35(1-3), 135, 1980.

129. **Orcutt, J., Kennett, B. L. N., and Dorman, L. M.**, Structure of the East Pacific Rise from an ocean bottom seismometer survey, *Geophys. J. R. Astron. Soc.*, 45(2), 305, 1976.

130. **Prothero, W., Reid, I., Reichle, M. S., and Brune, J. N.**, Ocean bottom seismic measurement on the East Pacific Rise and Rivera fracture zone, *Nature (London)*, 262(5564), 121, 1976.

131. **Riddihough, R. P., Currie, R. G., and Hyndman, R. D.**, The Dellwood Knolls and their role in triple junction tectonics off northern Vancouver Island, *Can. J. Earth Sci.*, 17(5), 577, 1980.

132. **Steinmetz, L., Whitmarsh, R. B., and Moreira, V.**, Upper mantle structure beneath the Mid-Atlantic Ridge north of the Azores based on observations of compressional waves, *Geophys. J. R. Astr. Soc.*, 50, 353, 1977.

133. **Whitmarsh, R. B.**, Median valley refraction line, Mid-Atlantic Ridge at 37°N, *Nature (London)*, 246, 297, 1973.

134. **Whitmarsh, R. B.**, Axial intrusion zone beneath the median valley of the Mid-Atlantic Ridge at 37°N detected by explosion seismology, *Geophys. J. R. Astron. Soc.*, 42, 189, 1975.

135. **Detrick, R. S. and Purdy, G. M.**, The crustal structure of the Kane Fracture zone from seismic refraction studies, *J. Geophys. Res.*, 85, 3759, 1980.

136. **Fox, P. J., Schreiber, E., Rowlett, H., and McCamy, K.**, The geology of the Oceanographer Fracture zone: a model for fracture zones, *J. Geophys. Res.*, 81, 4117, 1976.

137. **van Huene, R., Aubouin, J., Azema, J., Blackinton, G., Coulbourn, W. T., Cowan, D. S., Curiale, J. A., Dengo, C. A., Faas, R. W., Harrison, W., Hesse, R., Hussong, D. M., Laad, J. W., Muzylov, N., Shiki, T., Thompson, P. R., Westberg, J., and Carter, J. A.**, Leg 67; the deep sea drilling project Mid-America Trench transect off Guatemala, *Geol. Soc. Am. Bull.*, 91(7), 421, 1980.

138. **Ibrahim, A. B. K., Latham, G. V., and Ladd, J.**, Seismic refraction and reflection measurements in the Mid-America Trench offshore Guatemala, *J. Geophys. Res.*, 84(B10), 5643, 1979.

139. **Ibrahim, A. K., Pontoise, B., Latham, G., Larue, M., Chen, T., Isacks, B., Recy, J., and Louat, R.**, Structure of the New Hebrides arc-trench system, *J.Geophys. Res.*, 85(B1), 253, 1980.

140. **Kasakhara, D. and Kharvi, R. R.**, Ocean bottom seismograph study of the Kuril Trench area, *Akad. Nauk. S.S.S.R. Sib. Otd. Sakhalin. Nauchno-Issled. Inst., Tr.*, 54, 17, 1977.

141. **Meeder, C. A., Lewis, B. T. R., and McClain, J.**, The structure of the oceanic crust off southern Peru determined from an ocean bottom seismometer, *Earth Planet. Sci. Lett.*, 37(1), 13, 1977.

142. **Nagumo, S., Hasegawa, S., Koresawa, S., and Kobayashi, H.**, Ocean-bottom seismographic observation at the offside of Japan trench near the Erimo seamount; seismic activity of the oceanic lithosphere and velocity structure around the geophysical "ocean-continent boundary", *Earthquake Res. Inst. Bull., Tokyo*, 48(5), 769, 1970.

143. **Shimamura, H. and Asada, T.**, Ocean bottom seismograph as a tool to study the arc-trench system, in *Island Arc and Marginal Sea, Proc. 1st Jpn-U.S.S.R. Symp. Solid Earth Sci.*, Asano, Shuzo and Udintsev, G. B., Eds., Tokai University Press, Tokyo, 1971.

144. **Whitmarsh, R. B.**, The Owen Basin off the south-east margin of Arabia and the evolution of the Owen Fracture zone, *Geophys. J. R. Astr. Soc.*, 58, 441, 1979.

145. **Kasahara, J. and Harvey, R. R.**, Seismological evidence for the high-velocity zone in the Kuril Trench area from ocean bottom seismometer observations, *J. Geophys. Res.*, 82(26), 3805, 1977.

146. **Lewis, B. T. R. and McClain, J.**, Converted shear waves as seen by ocean bottom seismometers and surface buoys, *Seismol. Soc. Am. Bull.*, 67(5), 1291, 1977.

147. **Nagumo, S., Kasahara, J., and Koresawa, S.**, Large Poisson's ratio and low S-wave velocity within the Japan Trench inner wall toe, *Mar. Geol.*, 35(1-3), 129, 1980.

148. **Orcutt, J., Kenneth, B., Dorman, L., and Prothero, W.**, A low velocity zone underlying a fast-spreading rise crest, *Nature (London)*, 256(5517), 475, 1975.

149. **Rowlett, H. and Forsyth, D.**, Teleseismic P-wave delay times in a major oceanic fracture zone, *Geophys. Res. Lett.*, 6, 273, 1979.

150. **Shimamura, H. and Asada, T.,** T waves from deep earthquakes generated exactly at the bottom of deep trenches, *Earth Planet. Sci. Lett.*, 27, 137, 1975.
151. **Shimamura, H. and Asada, T.,** Apparent velocity measurements on an oceanic lithosphere, *Phys. Earth Planet. Inter.*, 13(1), P15, 1976.
152. **Shiamura, H., Asada, T., and Kimazawa, M.,** High sheer velocity layer in the upper mantle of the western Pacific, *Nature (London)*, 269(5630), 680, 1977.
153. **Steinmetz, L., Hirn, A., Sapin, M., Whitmarsh, R. B., and Moreira, V.,** Zones of P wave attenuation beneath the crest of Mid-Atlantic Ridge, *Bull. Soc. Geol. Fr., 7ᵉ Ser.*, 18, 931, 1976.
154. **McCreery, C. S. and Sutton, G. H.,** Wave train characteristics of long-range, high-frequency Pn, Sn crossing an ocean bottom hydrophone array, *Seismol. Soc. Am. Bull.*, 70(2), 437, 1980.
155. **Sutton, G. H., McCreery, C. S., Duennebier, F. K., and Walker, D. A.,** Spectral analyses of high-frequency Pn, Sn phases recorded on ocean bottom seismographs, *Geophys. Res. Lett.*, 5(9), 745, 1978.
156. **Walker, D. A.,** High frequency Pn phases observed in the Pacific at great distances, *Science*, 197, 257, 1977.
157. **Walker, D. A., McCreery, C. S., Sutton, G. H., and Duennebier, F. K.,** Spectral analysis of high-frequency Pn and Sn phases observed at great distances in the western Pacific, *Science*, 199, 1333, 1978.
158. **Bucker, H. P., Whitney, J. A., and Kerr, D. L.,** Use of Stonley waves to determine the shear velocity in ocean sediments, *J. Acoust. Soc. Am.*, 36, 1595, 1979.
159. **Anderson, R. S. and Latham, G. V.,** Determination of sediment properties from first shear-mode Rayleigh waves recorded on the ocean bottom, *J. Geophys. Res.*, 74, 2747, 1969.
160. **Helmberger, D. V., Engen, G., and Scott, P.,** A note on velocity, density, and attenuation models for marine sediments determined from multibounce phases, *J. Geophys. Res.*, 84(B2), 667, 1979.
161. **Walker, D. A.,** Hydrophone recordings of underground nuclear explosions, *Geophys. Res. Lett.*, 7, 465, 1980.
162. **Wielandt, E.,** Generation of seismic waves by underwater explosions, *Geophys. J. R. Astron. Soc.*, 40, 421, 1975.
163. **Kebe, H-W.,** Optimierung verankerter akustischer Aufnehmersysteme fuer Refraktionsseismische Messungen auf see unter Beruecksichtigung der einwirkungen des Seeganges und des Anstroemens, (Dissertation, University of Hamburg), *Hamburger Geopgysikalische Einzelschriften*, 36, 152, 1978.
164. **Barnhard, P.,** Signal strength of marine seismograph explosives, *Geophysics*, 32, 827, 1967.
165. **Duennebier, F. K., Blackington, G., and Sutton, G. H.,** Current-generated noise recorded on ocean bottom seismometers, *Mar. Geophys. Res.*, 5, 109, 1981.
166. **Kasahara, J., Nagumo, S., Koresawa, S., and Daikubara, T.,** Experimental results of vortex generation around ocean bottom seismograph due to bottom current, *Earthquake Res. Inst. Bull., Tokyo*, 58, 1980.
167. **Buskirk, R. E., Frohlich, C., Latham, G. V., Chen, A. T., and Lawton, J.,** Evidence that biological activity affects ocean bottom seismograph recordings, *Mar. Geophys. Res.*, 5, 117, 1981.
168. **Sutton, G. H., Lewis, B. T. R., Ewing, J., Duennebier, F. K., Iwatake, B., and Tuthill, J. D.,** An overview and general results of the Lopez Island OBS experiment, *Mar. Geophys. Res.*, 5, 3, 1981.
169. **Zelikovitz, S. J. and Prothero, W. A.,** The vertical response of an ocean bottom seismometer: analysis of the Lopez Island vertical transient tests, *Mar. Geophys. Res.*, 5, 53, 1981.
170. **Sutton, G. H., Duennebier, F. K., and Iwatake, B.,** Coupling of ocean bottom seismometers to soft bottoms, *Mar. Geophys. Res.*, 5, 35, 1981.
171. **Tréhu, A. M. and Solomon, S. C.,** Coupling parameters of the MIT OBS at two nearshore sites, *Mar. Geophys. Res.*, 5, 69, 1981.
172. **Lewis, B. T. R. and Tuthill, J. D.,** Instrumental waveform distortion on ocean bottom seismometers, *Mar. Geophys. Res.*, 5, 79, 1981.
173. **Johnson, S. H. and McAlister, R. E.,** Bottom seismometer observation of airgun signals at Lopez Island, *Mar. Geophys. Res.*, 5, 87, 1981.
174. **Tuthill, J. D., Lewis, B. T. R., and Garmany, J. D.,** Stonley waves, Lopez Island noise and deep sea noise from 1 to 5 Hz, *Mar. Geophys. Res.*, 5, 95, 1981.
175. **Sinha, M. C., Owen, T. R. E., and Mason, M.,** An ocean-bottom hydrophone recorder for seismic refraction experiments, *Mar. Geophys. Res.*, 5, 130, 1981.
176. **Kebe, H-W,** Self-noise measurements using a moored sonobuoy with a suspended hydrophone, *Mar. Geophys. Res.*, 5, 150, 1981.
177. **Keen, C. E., Loncarevic, B. D., and Nichols, B.,** AGC report on LADLE data analysis and interpretation, Bedford Institute of Oceanography, Dartmouth, Nova Scotia, 1981.
178. **Hyndman, R. D. and Ellis, R. M.,** Queen Charlotte fault zone: microearthquakes from a temporary array of land stations and ocean bottom seismographs, *Can. J. Earth Sci.*, 18, 776, 1981.

179. **Hyndman, R. D. and Rogers, G. C.**, Seismicity surveys with ocean bottom seismographs off western Canada, *J. Geophys. Res.*, 86, 3867, 1981.
180. **Steinmetz, R. L., Murff, J. D., Latham, G., Roberts, A., Donoho, P., Babb, L., and Eichel, T.**, Seismic instrumentation of the Kodiak shelf, *Mar. Geotech.*, 4, 193, 1981.
181. **Lawton, J., Frohlich, C., Pulpan, H., and Latham, G.**, Earthquake activity at the Kodiak continental shelf, Alaska, determined by land and ocean bottom seismograph networks, *Seism. Soc. Am. Bull.*, 72(1), 107, 1982.
182. Project Rose Scientists, Microearthquake activity on the Orozco Fracture zone: preliminary results from Project Rose, *J. Geophys. Res.*, 86, 3783, 1981.
183. **Rowlett, H.**, Seismicity at intersections of spreading centers and transform faults, *J. Geophys. Res.*, 86, 3815, 1981.
184. **Keen, C. E. and Cordsen, A.**, Crustal structure, seismic stratigraphy and rift processes of the continental margin off eastern Canada: ocean-bottom seismic refraction results off Nova Scotia, *Can. J. Earth Sci.*, 18, 1523, 1981.
185. **Avedik, F., Camus, A. L., Ginzburg, A., Montadert, L., Roberts, D. G., and Whitmarsh, R. B.**, A seismic refraction and reflection study of the continent-ocean transition beneath the North Biscay margin, in *The Evolution of Sedimentary Basins*, Kent, P., Bott, M. H. P., McKenzie, D. P., and Williams, C. A., Eds., *Philos. Trans. R. Soc. London*, Ser. A, 305, 5, 1982.
186. **Keen, C. E. and Barrett, D. L.**, Thinned and subsided continental crust on the rifted margin of eastern Canada: crustal structure, thermal evolution and subsidence history, *Geophys. J. R. Astron. Soc.*, 65, 443, 1981.
187. **Ibrahim, A. K., Carye, J., Latham, G., and Buffler, R. T.**, Crustal structure in Gulf of Mexico from OBS refraction and multichannel reflection data, *AAPG Bull.*, 65, 1207, 1981.
188. **Baggeroer, A. B. and Falconer, R.**, Array refraction profiles and crustal models of the Canada Basin, *J. Geophys. Res.*, (in press), 1981.
189. **Coudert, E., Isacks, B. L., Barazangi, M., Louat, R., Cardwell, R., Chen, A., Dubois, J., Lathma, G., and Pontoise, B.**, Spatial distribution and mechanism of earthquakes in the southern New Hebrides arc from a temportary land and ocean-bottom seismic network and from worldwide observations, *J. Geophys. Res.*, 86, 5905, 1981.
190. **Meyer, R. P., Mooney, W. D., Hales, A. L., Helsley, C. E., Woollard, G. P., Hussong, D. M., Kroenke, L. W., and Ramirez, J. E.**, Project Narino III: refraction observation across a leading edge, Malpelo Island to the Colombian Cordillera Occidental, in *Am. Geophys. Union, Geophys. Monogr.*, 19, 105, 1976.
191. **Fowler, C. M. R. and Matthews, D. H.**, Seismic refraction experiment on the Mid-Atlantic Ridge in the Famous area, *Nature (London)*, 249, 752, 1974.
192. **Fowler, C. M. R.**, Crustal structure of the Mid-Atlantic Ridge crest at 37°N, *Geophys. J. R. Astron. Soc.*, 47, 459, 1976.
193. **Fowler, C. M. R.**, The Mid-Atlantic Ridge: structure at 45°N, *Geophys. J. R. Astron. Soc.*, 54, 167, 1978.
194. **Bunch, A. W. H. and Kennett, B. L. N.**, The crustal structure of the Reykjanes Ridge at 59°30'N, *Geophys. J. R. Astron. Soc.*, 61, 141, 1980.
195. **Au, D. and Clowes, R. M.**, Crustal structure from an OBS survey of the Nootka fault zone off western Canada, *Geophys. J. R. Astron. Soc.*, 65, XX-XX, 1981.
196. **Whitmarsh, R. B.**, Seismic anisotropy of the uppermost mantle absent beneath the east flank of the Reykjanes Ridge, *Seismol. Soc. Am. Bull.*, 61, 1351, 1971.
197. **Lewis, B. T. R. and Snydsman, W. E.**, Evidence for a low velocity layer at the base of the oceanic crust, *Nature (London)*, 266(5600), 340, 1977.
198. **Asada, T. and Shimamura, H.**, Long-range refraction experiments in deep ocean, *Tectonophysics*, 56, 67, 1979.
199. **Spudich, P. and Orcutt, J.**, A new look at the seismic velocity structure of the oceanic crust, *Rev. Geophys. Space Phys.*, 18, 627, 1980.
200. **Spudich, P. and Orcutt, J.**, Petrology and porosity of an oceanic crustal site: results from wave form modeling of seismic refraction data, *J. Geophys. Res.*, 85, 1409, 1980.
201. **Frolich, C.**, Personal communication.
202. **Duckworth, G. L., Baggeroer, A. B., and Jackson, H. R.**, Crustal structure measurements near Fram 2 in the Pole abyssal plain, *Tectonophysics*, in press, 1982.
203. **Cheung, H. P. Y. and Clowes, R. M.**, Crustal structure from PS-wave analysis: ocean bottom seismometer results in the northeast Pacific, *Geophys. J. R. Astron. Soc.*, 65, 47, 1981.

INSTRUMENTATION SYSTEMS

A DEEP-TOWED SOUND SOURCE AND HYDROPHONE ARRAY SYSTEM — PERFORMANCE, ANALYSIS, AND DESIGN

Norman H. Gholson and Martin G. Fagot

INTRODUCTION

The need to improve fleet system performance prediction has increased with the advent of more sophisticated Navy sonar systems. Definitive models of the ocean floor and subbottom as transmission media that refract, diffract, and dissipate, as well as reflect acoustic energy are critical to this need. A multichannel hydrophone array and low frequency sound source system, as depicted in Figure 1, towed near the bottom in the deep ocean, can provide data to generate the high resolution geoacoustic parameter inputs required for these models.[1]

The system concept, towed array (streamer) and low frequency sound source, is similar to that employed by the marine oil exploration community, with one primary and distinct exception. The deep-towed system is required to operate a few hundred meters off the bottom in ocean depths up to 6000 m; by contrast, oil exploration systems operate near the ocean surface in relatively shallow water. In addition, the deep-towed system is required to penetrate only the upper 500 m of subbottom structure to meet the geoacoustic modeling needs. The oil exploration systems are designed to penetrate many thousands of meters, at the expense of resolution in the upper sediment section.

The fundamental advantage of a deep-towed system, as opposed to a surface-towed system operating in a deep ocean environment, is simply that the measurement system is much closer to the geophysical features being investigated. Clearly, the signal-to-noise ratio is much better for the deep-towed system, since the signal has a much shorter travel path. More subtle advantages of the deep-towed system include a measurable "moveout" that is more sensitive to the geophysical parameters being estimated. Benefits of the fundamental advantage will become evident in the discussions on layer sound velocity estimation.

Performance analysis presented in this chapter derives error statistics of estimated layer Root Mean Square (RMS) sound velocity as a function of system characteristics and subbottom geometry. The analysis begins with a definition of RMS sound velocity associated with a laterally homogeneous layer. The effect of array altitude (towing altitude) array length, travel time "picking errors", etc., are investigated. Most of the analysis addresses estimating RMS velocity, but some attention is given to estimating gradients. While all analysis presented is for reflection data, one should realize that the deep-towed system is an excellent tool for investigating other modes of propagation.

Critical for successful implementation of the system concept is a hardware suite that can operate at deep ocean depths. The chapter will conclude with a review of the major hardware areas: sound source, hydrophone array, and telemetry system to meet the geophysical data acquisition requirements. This review will identify technical considerations, design approaches, and recommend a hardware implementation solution in each of these major areas.

ESTIMATING LAYER RMS SOUND VELOCITY

An important objective of the deep-towed system is to accurately estimate layer RMS sound velocity. Analysis in the following sections will show how accuracy of the RMS

FIGURE 1. Deep-towed system configuration.

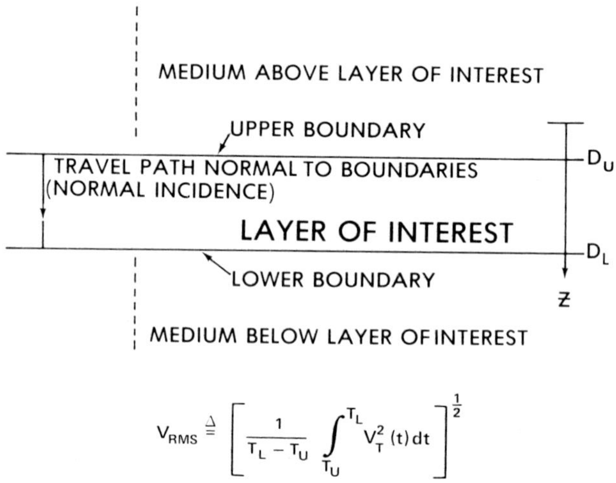

FIGURE 2. Defining RMS sound velocity.

velocity estimates is enhanced by the fundamental advantage of deep tow. Appropriately, the presentation will begin with a definition of RMS velocity.

Root Mean Square (RMS) Velocity

An RMS sound velocity is associated with a laterally homogeneous layer bounded above and below by acoustic reflectors and is defined as the time root-mean-square (RMS) velocity of sound traversing the layer in a direction normal to the reflecting boundaries. Mathematically, RMS velocity is defined below with the aid of Figure 2.

$$V_{RMS} \triangleq \left[\frac{1}{T_L - T_u}\int_{T_u}^{T_L} V_T^2(t)\,dt\right]^{1/2} \quad (1)$$

where

$V_{RMS} \triangleq$ RMS sound velocity of layer of interest
$V_T(t) \triangleq$ sound velocity as a function of time (t) for sound traveling normal to layer boundaries
T_u = time when sound energy is at the upper boundary
T_L = time when sound energy is at the lower boundary

RMS velocity, as defined by Equation 1, can also be expressed by the following (see Gholson and Fagot[2] for derivation):

$$V_{RMS} = \left[\frac{\int_{D_u}^{D_L} V_Z(z)dz}{\int_{D_u}^{D_L} \frac{dz}{V_Z(z)}} \right]^{1/2} \tag{2}$$

where

V_{RMS} = RMS velocity of layer of interest
D_u = depth of upper boundary (see Figure 2)
D_L = depth of lower boundary (see Figure 2)
$V_z(z)$ = sound speed as a function of depth

Equation 2 is significant, not only because it is equivalent to defining Equation 1, but also because it is usually much easier to work with. The second feature results from the fact that sound velocity functions are usually defined in terms of depth as opposed to time.

Several examples will serve to clarify the meaning of RMS velocity and confirm the equivalence of Equation 2 and defining Equation 1.

Example 1. Layer of Constant Sound Velocity. Constant sound velocity implies that sound velocity is not a function of depth and, therefore, not a function of time. Hence, $V_T(t)$ and $V_Z(Z)$ become C, where C is the sound velocity (and is constant). Exercising defining Equation 1 results in:

$$V_{RMS} = \left[\left(\frac{1}{T_L - T_u} \right) C^2 (T_L - T_u) \right]^{1/2} = C$$

Exercising Equation 2 results in:

$$V_{RMS} = \left[\frac{C(D_L - D_u)}{\frac{1}{C}(D_L - D_u)} \right]^{1/2} = C$$

This example illustrates the fact that RMS sound velocity of a laterally homogeneous layer with constant sound velocity is simply the sound velocity in that layer.

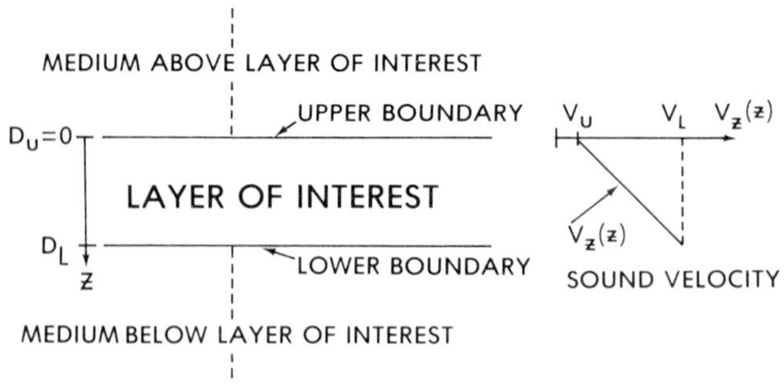

FIGURE 3. Layer with constant sound velocity gradient.

Example 2. Layer of constant sound velocity gradient. Constant sound velocity gradient implies the derivative of sound velocity with respect to depth $dV_Z(Z)/dZ$ is constant. Figure 3 serves to illustrate an example of this condition. Exercising Equation 2 to determine RMS velocity yields

$$V_{RMS}^2 = \frac{\int_0^{D_L} (V_u+gz)dz}{\int_0^{D_L} \frac{dz}{V_u+gz}} = \frac{\frac{1}{2g}(V_u+gz)^2 \Big|_0^{D_L}}{\frac{1}{g}\text{Ln}(V_u+gz)\Big|_0^{D_L}}$$

where g is the sound velocity gradient. Evaluating, and substituting $V_u + gD_L = V_L$ results in the following expression for RMS velocity:

$$V_{RMS} = \left[\frac{1}{2}\,\frac{(V_L^2 - V_u^2)}{\text{Ln}(V_L/V_u)}\right]^{1/2} \qquad (3)$$

where Ln(·) denotes natural logarithm.

In order to employ the defining Equation 1, we must first determine the sound velocity as a function time, $V_T(t)$, from the sound velocity as a function of depth $V_Z(Z)$. This task can be accomplished by first expressing the travel time (t) to a depth Z as a function of the sound velocity vs. depth relationship $V_Z(Z)$.

$$t(z) = \int_0^z \frac{dx}{V_z(x)} = \int_0^z \frac{dx}{V_u+gx} = \frac{1}{g}\,\text{Ln}\left(\frac{V_u+gz}{V_u}\right) \qquad (4)$$

Solving Equation 4 for depth (Z) in terms time (t) results in

$$z(t) = \frac{V_u}{g}\left[\epsilon^{gt} - 1\right] \qquad (5)$$

Differentiating Equation 5 with respect to time yields the following expression for sound velocity as a function of time:

$$V_T(t) \triangleq \frac{dz(t)}{dt} = V_u \epsilon^{gt} \qquad (6)$$

Equation 4 can be used to find T_L (travel time to lower boundary) and T_u (travel time to upper boundary).

$$T_L = T(z{=}D_L) = \frac{1}{g} \mathrm{Ln}(V_L/V_u)$$

$$T_u = T(z{=}0) = \frac{1}{g} \mathrm{Ln}(V_u/V_u) = 0 \qquad (7)$$

Substituting Equation 6 and the results of Equation 7 into the defining Equation 1 results in

$$V_{RMS} = \left[\frac{1}{\frac{1}{g} \mathrm{Ln}(V_L/V_u)} \int_0^{\frac{1}{g}\mathrm{Ln}(V_L/V_u)} (V_u \epsilon^{gt})^2 \, dt \right]^{1/2}$$

Evaluating the above integral results in the following expression for RMS velocity, which is identical to that obtained earlier.

$$V_{RMS} = \left[\frac{1}{2} \frac{(V_L^2 - V_u^2)}{\mathrm{Ln}(V_L/V_u)} \right]^{1/2}$$

This example derives RMS sound velocity for a constant gradient layer, but perhaps more importantly, illustrates the equivalence of Equation 2 and defining Equation 1.

The term "RMS Velocity" is sometimes used synonymously with "Normal Moveout (NMO) Velocity" and "Interval Velocity". The three velocities are *not* identical. The following brief discussion is an attempt to clarify the differences. Root Mean Square (RMS) velocity, at least in the context of this chapter, is clearly defined by Equation 1. The term "NMO Velocity" is related to the slope (derivative with respect to X^2) of the reflection path travel time squared as a function of source to receiver horizontal offset squared T^2 (X^2). The term "Interval Velocity" is particularly difficult to define and perhaps is best described by the following quotation, taken directly from Hubral and Kray,[3] and dedicated to interval velocity measurements: "However, as we see here, if the presumption of constant velocity within the interval is not fulfilled, interval velocities are quantities that depend on the method by which they are calculated."

There is a case, however, where all three velocities are not only clearly defined, but also the same. Such a case is illustrated by Figure 4, where an acoustic source and an array of acoustic receivers are observing a single laterally homogeneous layer bounded below by an acoustic reflector. In this case, the RMS velocity, NMO velocity, and Interval velocity all converge to the sound velocity in the observed layer. Such a case, in an absolute sense, is of academic interest only. In many practical cases, however, such as nearly constant sound velocity and small source to receiver horizontal offsets, RMS, NMO, and Interval velocity are nearly the same.

Geoacoustic Model

The model used for performance prediction as described in this chapter is a simple "flat layer-cake" model. More specifically, the model assumes a laterally homogene-

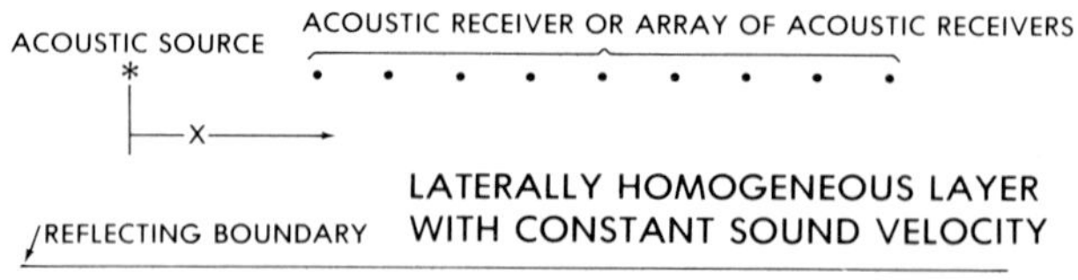

FIGURE 4. Case where NMO, RMS, and Interval Velocities are the same.

ous medium with nonsloping boundaries. Justification for choosing this model is given below.

- Anticipated areas of operation are reasonably characterized by the simple model
- Lack of detailed modeling information
- Simple model allows for tractable mathematical analysis

Further, a model was chosen in an effort to address U.S. Navy interests; and primarily to provide a tool for comparing performance of candidate measurement devices.

Dix Equation

The Dix equation, which appears formally in Reference 4, is an algorithm for estimating layer sound velocities from reflection data. The algorithm was derived assuming small source-to-receiver horizontal offsets and constant sound velocity layers. Later analysis (Gholson and Fagot[2]), however, shows the Dix equation to be valid for nonconstant sound velocity layers and large source-to-receiver horizontal offsets providing strict requirements and definitions are placed on input and output parameters. Features of the Dix equation include simplicity, low computational burden, and remarkable robustness with respect to errors in modeling and input data.

The Dix equation was derived to handle the problem of determining sound velocity of a laterally homogeneous layer bounded above and below by acoustic reflectors. Such a problem is illustrated by Figure 5, where an acoustic source and array of receivers are used to observe a series of constant sound velocity laterally homogeneous layers. The Dix equation incorporates acoustic reflection data obtained from the "upper" and "lower" reflecting boundaries of the layer of interest (see Figure 5). Specifically, the important data are the reflection path travel time as a function of horizontal source to receiver offset, T(x). Figure 6 illustrates a sample T(x) for the upper and lower reflection paths of Figure 5. Notice T_L and T_u of Figure 6 are the normal incidence (x = 0) reflection path travel times for the lower and upper boundaries,respectively. Also notice that the approximate equations of Figure 6 implicitly define the "moveout" velocities V_L and V_u as well as normal incidence arrival times T_L and T_u. For many practical cases, the hyperbolic approximations of Figure 6 are very accurate. The Dix equation uses these parameters just defined to obtain an estimate of the sound velocity in the layer of interest, viz.:

$$\hat{V} = \left[\frac{V_L^2 T_L - V_u^2 T_u}{T_L - T_u}\right]^{1/2} \tag{8}$$

where the $\hat{}$ denotes "estimate". Although the Dix equation was derived for constant sound velocity layers, the equation can be used for cases involving nonconstant sound velocity layers. In such cases, the estimate is an estimate of the RMS sound velocity

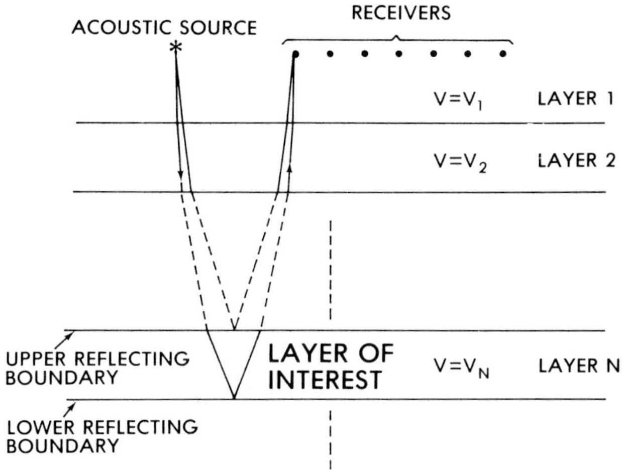

FIGURE 5. Series of laterally homogeneous layers with constant sound velocity.

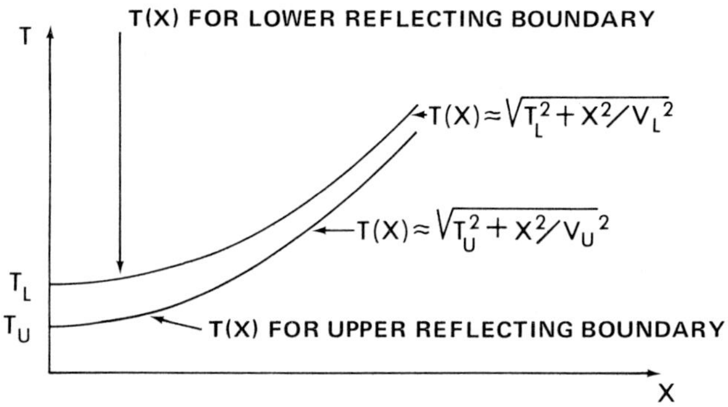

FIGURE 6. Reflection path travel times as a function of source-to-receiver horizontal offset.

of the layer of interest. *Therefore, the Dix equation can be defined in a more general sense as an estimator of RMS sound velocity; constant sound velocity layers being a simple case of the more general problem.*

Errors in Estimating RMS Velocity

Errors in RMS sound velocity estimates obtained from the Dix equation are primarily a result of (1) nonhyperbolic T(x), and (2) errors in measuring the required input parameters T_L, T_u, V_L, V_u. In many practical cases, errors due to nonhyperbolic T(x) are negligible compared to errors due to measurement errors. For this reason, the nonhyperbolic nature of T(x), will be ignored until estimation of *gradients* is discussed. The effect of errors in measured T_L, T_u, V_L, V_u, can be seen by substituting into Equation 8 to obtain

$$\hat{V}_{RMS} = \left[\frac{(V_L+\epsilon_{V_L})^2\,(T_L+\epsilon_{T_L}) - (V_u+\epsilon_{V_u})^2\,(T_u+\epsilon_{T_u})}{(T_L+\epsilon_{T_L}) - (T_u+\epsilon_{T_u})} \right]^{1/2} \quad (9)$$

where

$E_{V\,L}$ = error in measured V_L
$E_{V\,M}$ = error in measured V_u
$E_{T\,L}$ = error in measured T_L
$E_{T\,u}$ = error in measured T_u

Analyzing Equation 9 to determine an "error" i.e., mapping all effects to the real line, is difficult. One approach is to determine a maximum $\hat{V}$, $\hat{V}_{max}$ and a minimum $\hat{V}$, $\hat{V}_{min}$, and define the difference $\hat{V}_{max} - \hat{V}_{min}$ as an error bound. Unfortunately, $\hat{V}_{max}$ and $\hat{V}_{min}$ are difficult to determine. As an example, to determine $\hat{V}_{max}$, one would intuitively maximize the numerator of Equation 9 while minimizing the denominator. A close look at Equation 9 reveals that the max-min procedure generates a conflict regarding how errors in T_u and T_L are treated. An assumption for errors in T_u and T_L which maximizes the numerator, simultaneously maximizes the denominator. Therefore, determining $\hat{V}_{max}$ and $\hat{V}_{min}$ requires an exhaustive search over possible combinations of errors in T_L and T_u. Further, finding an absolute error bound provides no insight into the probability of such an error occurring. Deficiencies of the deterministic approach, such as those presented here, motivated development of a statistical error analysis approach.

The most straightforward approach for statistical error analysis is the Monte Carlo approach, which consists of conducting large numbers of Monte Carlo "trials" to determine error statistics. Problems with the Monte Carlo approach are that it is computationally cumbersome and inherently does not provide insight concerning the sensitivity output errors with respect to input errors or other parameters of interest.

The statistical analysis used here is based on a second order Taylor series expansion of Equation 8. For brevity, only the first order derivation is presented here. Empirical results have indicated the first order expansion to be sufficiently accurate for most all practical applications.

The approach is derived easily by first defining the following function.

$$F(V_L, V_u, T_L, T_u) \triangleq \left[\frac{V_L^2 T_L - V_u^2 T_u}{T_L - T_u}\right]^{1/2} \tag{10}$$

The RMS velocity estimate $\hat{V}_{RMS}$ can then be approximated by

$$\hat{V}_{RMS} = F(V_L+\epsilon_{V_L}, V_u+\epsilon_{V_u}, T_L+\epsilon_{T_L}, T_u+\epsilon_{T_L}) \approx F(V_L, V_u, T_L, T_u)$$

$$+ \frac{\partial F}{\partial V_L}\epsilon_{V_L} + \frac{\partial F}{\partial V_u}\epsilon_{V_u} + \frac{\partial F}{\partial T_L}\epsilon_{T_L} + \frac{\partial F}{\partial T_u}\epsilon_{T_u} \tag{11}$$

where the partial derivatives are evaluated at V_L, V_u, T_L, T_u. Recognizing $F(V_L, V_u, T_L, T_u)$ to be identically V_{RMS} * and rearranging, allows the *error* in $\hat{V}_{RMS}$ to be approximated by

$$\epsilon_{\hat{V}_{RMS}} = \hat{V}_{RMS} - V_{RMS} \approx \frac{\partial F}{\partial V_L}\epsilon_{V_L} + \frac{\partial F}{\partial V_u}\epsilon_{V_u} + \frac{\partial F}{\partial T_L}\epsilon_{T_L} + \frac{\partial F}{\partial T_u}\epsilon_{T_u} \tag{12}$$

* Errors due to honhyperbolic moveout are ignored.

Defining the vectors $\overline{\mathrm{P}}$ and $\overline{\mathrm{E}}$ as follows,

$$\overline{P} \triangleq \left[\frac{\partial F}{\partial V_L} \; \frac{\partial F}{\partial V_u} \; \frac{\partial F}{\partial T_L} \; \frac{\partial F}{\partial T_u} \right] \quad \text{and}$$

$$\overline{\epsilon} \triangleq \left[\epsilon_{V_L} \; \epsilon_{V_u} \; \epsilon_{T_L} \; \epsilon_{T_u} \right]^T$$

where "T" denotes transpose, allows approximation (Equation 12) to be expressed in a more compact form by

$$\hat{V}_{RMS} - V_{RMS} \approx \overline{P}\,\overline{\epsilon} \tag{13}$$

Error variance of $\hat{V}_{RMS}$ can be now expressed quite easily, using Equation 13, to obtain

$$* \; \sigma^2_{\epsilon_{\hat{V}_{RMS}}} \triangleq E\left[(\hat{V}_{RMS} - V_{RMS})^2 \right] \approx \overline{P}\; E\left[\overline{\epsilon}\,\overline{\epsilon}^T \right] \overline{P}^T \tag{14}$$

where E(•) denotes statistical expected value. Taking the square root of Equation 14 yields an expression for the parameter of interest (standard deviation $\hat{V}_{RMS}$ error)

$$\sigma_{\epsilon_{\hat{V}_{RMS}}} \approx \left[\overline{P}\; E\left[\overline{\epsilon}\,\overline{\epsilon}^T \right] \overline{P}^T \right]^{1/2} \tag{15}$$

The expression $E[\overline{E}\overline{E}^T]$ is the error *covariance matrix* of the input errors. In general, this matrix consists of 16 nonzero elements describing the variances and cross variances of the errors in measured V_L, V_u, T_L, T_u. The analysis in this chapter simplifies the matrix by assuming errors in measured V_L, V_u, T_L, T_u to be zero mean and statistically independent. Therefore, the off diagonal terms of $E[\overline{E}\overline{E}^T]$ are zero and the covariance matrix is diagonal as shown below.

$$E\left[\overline{\epsilon}\,\overline{\epsilon}^T \right] = \begin{bmatrix} \sigma^2_{\epsilon_{V_L}} & 0 & 0 & 0 \\ 0 & \sigma^2_{\epsilon_{V_u}} & 0 & 0 \\ 0 & 0 & \sigma^2_{\epsilon_{T_L}} & 0 \\ 0 & 0 & 0 & \sigma^2_{\epsilon_{T_u}} \end{bmatrix} \tag{16}$$

Substituting Equation 16 into Equation 15 results in a very simple expression for determining the standard deviation of RMS velocity estimate.

* Assuming zero mean errors.

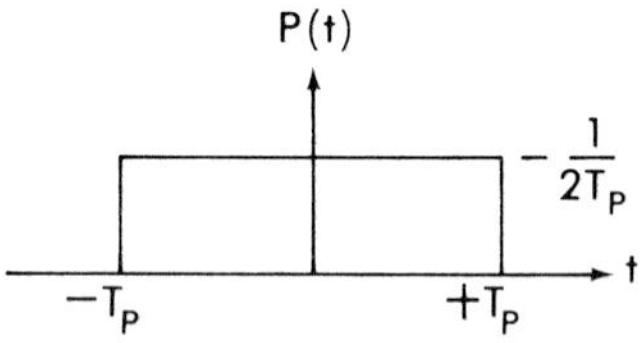

FIGURE 7. Uniform probability— density function for errors in measured T_L and T_u.

$$\sigma_{\epsilon_{\hat{V}_{RMS}}} \approx \left[\left(\frac{\partial F}{\partial V_L}\right)^2 \sigma^2_{\epsilon_{V_L}} + \left(\frac{\partial F}{\partial V_u}\right)^2 \sigma^2_{\epsilon_{V_u}} + \left(\frac{\partial F}{\partial T_L}\right)^2 \sigma^2_{\epsilon_{T_L}} + \left(\frac{\partial F}{\partial T_u}\right)^2 \sigma^2_{\epsilon_{T_u}}\right]^{1/2} \tag{17}$$

Implementing approximation (Equation 17) requires knowledge of measurement error variances of T_L, V_L, etc. Errors in the measured T_L and T_u can be reasonably modeled by a uniform probability density between a maximum picking error bounds ;$\pm T_p$ as shown in Figure 7. Error variance for the uniform model is given by

$$\sigma^2_{\epsilon_T} = \int_{-\infty}^{+\infty} t^2 P(t)dt = \frac{T_P^2}{3}$$

Deriving error statistics for measured V_L and V_u is more difficult than for T_L and T_u. The technique presented here determines a maximum and minimum V for both V_L and V_u and then assigns uniform probability density functions between the extremes. The maximum and minimum V's are determined by examining the extremes of T(x) (moveout) curves which could fit within $\pm T_p$ error bounds of the true* hyperbolic curve for a maximum effective X offset of L_e.

Such curves are intended to represent the extreme cases encountered in processing the acoustic reflection data to determine "moveout velocities" V. The timing (picking) error is modeled to account for such unavoidable factors as finite source band width, poor signal-to-noise ratios, and ambiguous receiver array geometry. Figure 8 illustrates the extreme T(x) curves which implicitly define the extreme V's. Solving for the extremes results in

$$V_{max} = \frac{L_e}{([[T_o^2 + L_e^2/V^2]^{1/2} - T_p]^2 - (T_o + T_p)^2)^{1/2}} \tag{18a}$$

$$V_{min} = \frac{L_e}{([[T_o^2 + L_e^2/V^2]^{1/2} + T_p]^2 - (T_o - T_p)^2)^{1/2}} \tag{18b}$$

where

$V_{min} \triangleq$ velocity for extreme minimum (see Figure 8)

* Errors due to nonhyperbolic T(x) are ignored.

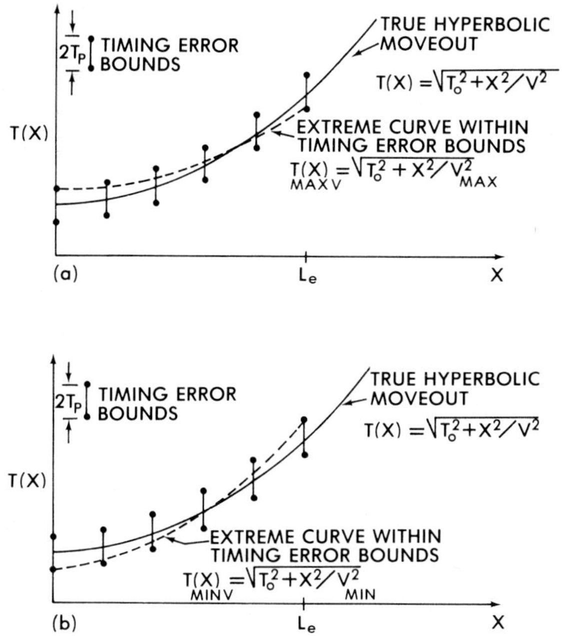

FIGURE 8. Determining (a) maximum V, (b) minimum V.

V_{max} $\triangleq$ velocity for extreme maximum (see Figure 8)
T_o $\triangleq$ normal incidence arrival time $T(x = 0)$
V $\triangleq$ true "moveout" velocity
T_p $\triangleq$ error bound on timing errors
L_e $\triangleq$ maximum *effective* source-to-receiver horizontal offset

From Equations 18a, 18b, and Figure 8, it is easily seen that V_{max} and V_{min} approach V (the true quantity) as L_e becomes larger. This implies that error in measured V_u and V_L can be made arbitrarily small by increasing the maximum effective source-to-receiver horizontal offset, L_e. Unfortunately, the apparent benefit of very large offsets is nullified by the mechanical complexity of very long arrays of acoustic receivers and the potential nonhyperbolic nature of reflection data for large offsets.* Also, one must be careful when relating L_e of Equations 18a and 18b to the measurement system. Intuitively, L_e is identically the horizontal distance from the acoustic source to the most distant receiver, which is simply the length of the acoustic array (streamer). Stoffa[5] points out, however, that due to problems such as poor signal-to-noise ratio, etc., it is more realistic to expect an effective array length approximately *½ the physical array length.* Therefore, L_e of Equations 18a and 18b will be set to ½ the physical acoustic array length.

Assuming a uniform probability density function with bounds V_{max}, V_{min} for V_u and V_L errors, results in the following expression for error variance.

* Further complications include the transition from reflection to other modes of source-to-receiver propagation.

$$\sigma^2_{\epsilon_V} = \frac{(V_{max} - V_{min})^2}{12} \tag{19}$$

where V_{max} and V_{min} are given by Equations 18a and 18b.

The following is a collection of results just derived for calculating errors in RMS sound velocity estimates as a function of measured travel time errors and overall system parameters.

$$\sigma_{\hat{V}_{RMS}} \approx \left[\left(\frac{\partial F}{\partial V_L}\right)^2 \sigma^2_{\epsilon_{V_L}} + \left(\frac{\partial F}{\partial V_u}\right)^2 \sigma^2_{\epsilon_{V_u}} + \left(\frac{\partial F}{\partial T_L}\right)^2 \sigma^2_{\epsilon_{T_L}} + \left(\frac{\partial F}{\partial T_u}\right)^2 \sigma^2_{\epsilon_{T_u}}\right]^{1/2} \tag{20}$$

where

$$F \triangleq \left[\frac{V_L^2 T_L - V_u^2 T_u}{T_L - T_u}\right]^{1/2}$$

$$\sigma^2_{T_L} = T_p^2/3$$

$$\sigma^2_{T_u} = T_p^2/3$$

T_p $\triangleq$ error bounds of measured travel times

$$\sigma^2_{\epsilon_{V_L}} = \frac{(V_{max} - V_{min})^2}{12} \quad \text{evaluated for ``lower'' reflector}$$

$$\sigma^2_{\epsilon_{V_u}} = \frac{(V_{max} - V_{min})^2}{12} \quad \text{evaluated for ``upper'' reflector}$$

$$V_{max} = \left[\frac{L_e^2}{[(T_o^2 + L_e^2/V^2)^{1/2} - T_p]^2 - (T_o + T_p)^2}\right]^{1/2}$$

$$V_{min} = \left[\frac{L_e^2}{[T_o^2 + L_e^2/V^2)^{1/2} + T_p]^2 - (T_o - T_p)^2}\right]^{1/2}$$

L_e = 1/2 physical length of receiver array (see text)

T_o $\triangleq$ normal incidence travel time ($T(x) = 0$) for reflecting boundary of interest

V $\triangleq$ RMS sound velocity of normal incidence path from source to receiver (zero offset) via the reflecting boundary of interest

partial derivatives are evaluated at V_u, V_L, T_u, T_L.

Performance Analysis

The purpose of this section is to illustrate expected performance of the deep-towed system as a function of several design parameters of interest. Tools for performing this analysis are essentially those just developed, except for some cases where a second

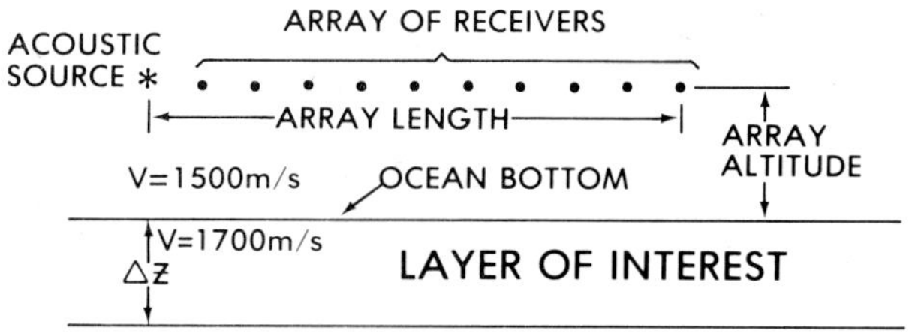

FIGURE 9. Simple two-layer model.

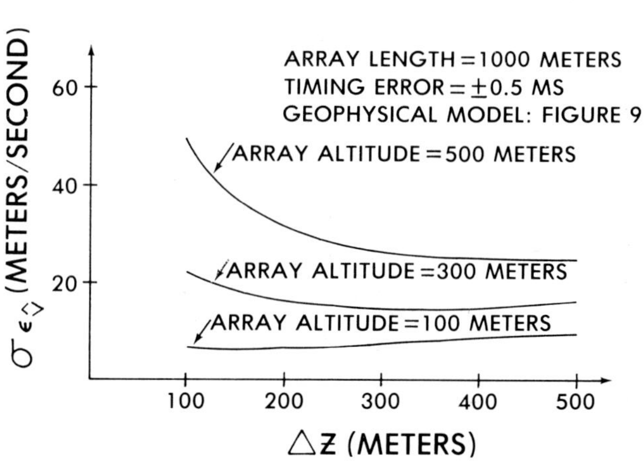

FIGURE 10. Error in velocity estimates as a function of array altitude (towing altitude) and layer thickness.

order expansion is used as opposed to the first order expansion derived in the text. The analysis illustrates the effect of array length, array altitude (towing altitude) and system geometry, etc., on the accuracy of estimating layer sound velocity. Laterally homogeneous layers are assumed and errors due to nonhyperbolic "moveout" are ignored.* Nonhyperbolic T(x) will be explored later in the chapter.

Effect of Towing Altitude

The effect of towing altitude will be explored by using the simple model of Figure 9. While the model is quite restrictive in a geophysical sense, it serves quite well to illustrate the effect of towing altitude. Figure 10 is a graph showing standard deviation error of sound velocity estimate $\sigma_{\epsilon_{\hat{v}}}$ as a function of layer thickness, ΔZ, for several different altitudes. Array length is 1000 m and travel time measurement error bounds are ± 0.5 msec. Notice that errors in estimated sound speed increases rapidly with increasing array altitude. This demonstration establishes a foundation for comparing a surface system with the deep-towed system. Figure 11 compares a long (3600 m) surface system with the deep-towed system. The surface system is 4000 m above the bottom whereas the deep system is 100 m above the bottom. Timing errors for both systems are assumed ± 0.5 msec. Improvement of the deep-towed system over the surface system is pronounced for thin layers, and is consistent for layers to 500 m thick.

* The nonhyperbolic nature of T(x) is very nearly negligible for the examples presented here.

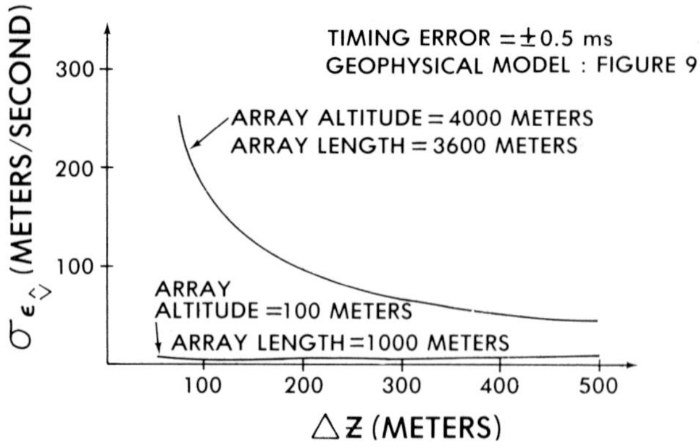

FIGURE 11. Comparison of long shallow-towed system with deep-towed system.

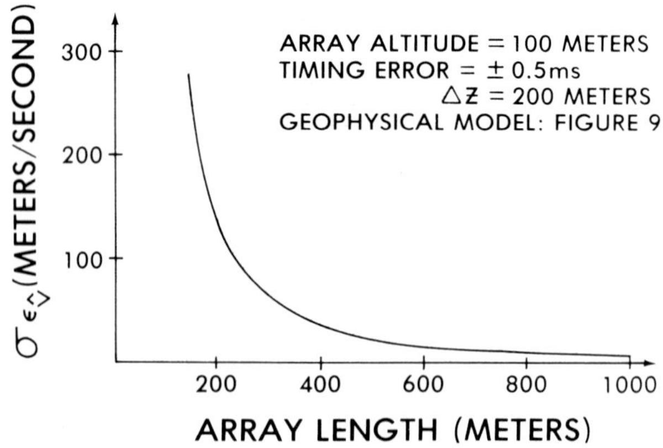

FIGURE 12. Standard deviation velocity error as a function of array length.

Effect of Array Length

The effect of array length will be illustrated by plotting standard deviation velocity estimate error vs. array length for the model shown in Figure 9. Figure 12 plots standard deviation of errors as a function array length. Estimation error decreases with increasing array length due to better estimates of the true "moveout velocity". The extremely rapid degradation for short array lengths is partially due to the nonlinearity of the Dix equation.

Effect of Errors in Travel Time Measurements

Errors in measured reflection path travel times simultaneously affect measured "moveout" velocities V_u, V_L, and normal incidence travel time measurements, T_u, T_L, of Equation 8. Figure 13 displays standard deviation velocity errors, $\sigma_{\epsilon\hat{v}}$ as a function of layer thickness and travel time measurement error bounds for the model shown in Figure 9. The analysis assumes an array length of 1000 meters and an array altitude of 100 m.

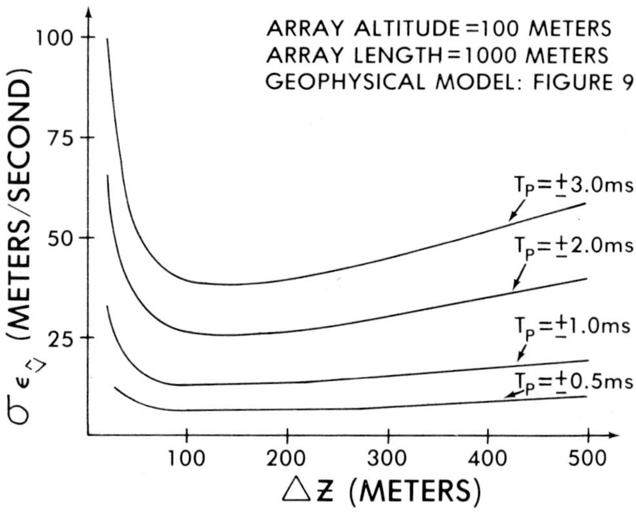

FIGURE 13. Standard deviation velocity error as a function of timing error bounds.

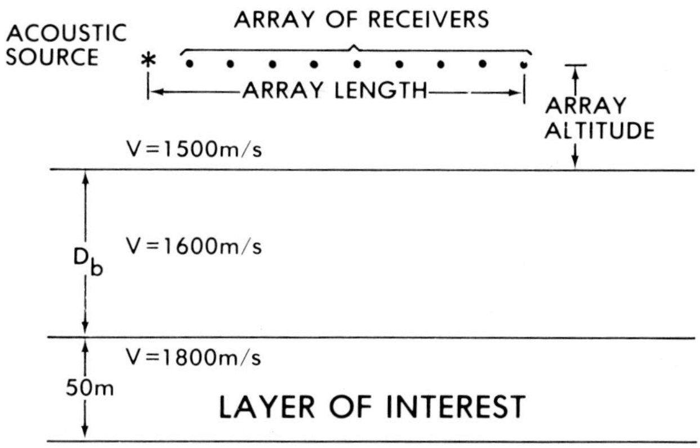

FIGURE 14. Buried layer of interest.

Deeply Buried Layer

Errors in estimated sound velocity for a layer buried below a relatively thick layer, as shown in Figure 14, are described by Figure 15. Notice that estimation errors increase as the depth of burial increases. This trend is in agreement with the fundamental concept of a deep-towed system, i.e., estimation errors generally increase as the distance increases between the measurement device and layer of interest.

Hybrid Configuration

An analysis was performed to compare performance of the deep-towed system with a hybrid system. The hybrid system consists of deep array with a *surface source.* Figure 16 represents geometry of the assumed configuration. Timing error bounds were assumed ± 0.5 msec for both systems. Also, an array length of 1000 m was assumed for both systems. Figure 17 displays the predicted performance of both systems as a function of layer thickness.

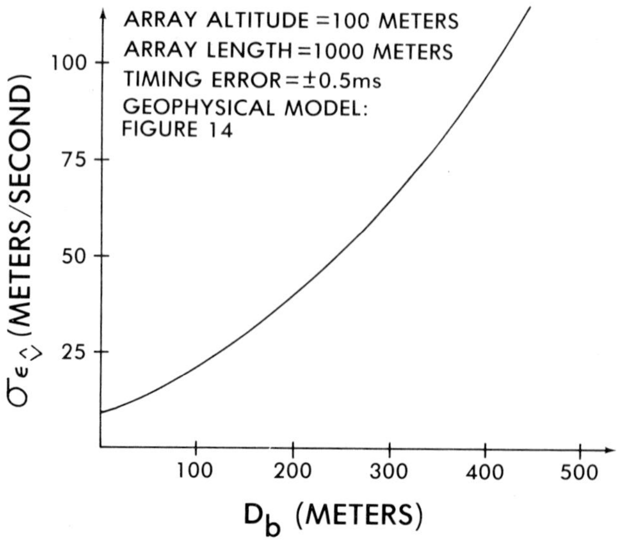

FIGURE 15. Standard deviation velocity error as a function of layer depth.

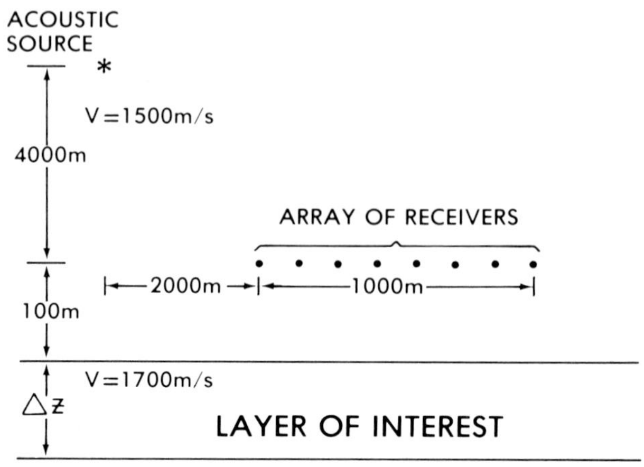

FIGURE 16. Measurement configuration for hybrid system.

ESTIMATING LINEAR SOUND SPEED VS. DEPTH RELATIONSHIP

The fundamental advantage of a deep-towed system allows the subbottom to be characterized in a more general fashion than with a surface towed system. In particular, it may very well be practical to estimate sound velocity *gradients* of the subbottom layers as opposed to only estimating RMS sound velocity. A previous investigation by Gibson et al.[6] shows that estimating gradients from the surface* using reflection data is not practical. The analysis described here is in agreement with that of Gibson et al. but shows that perhaps gradients *can* be estimated from deep-towed reflection data.

* In deep ocean.

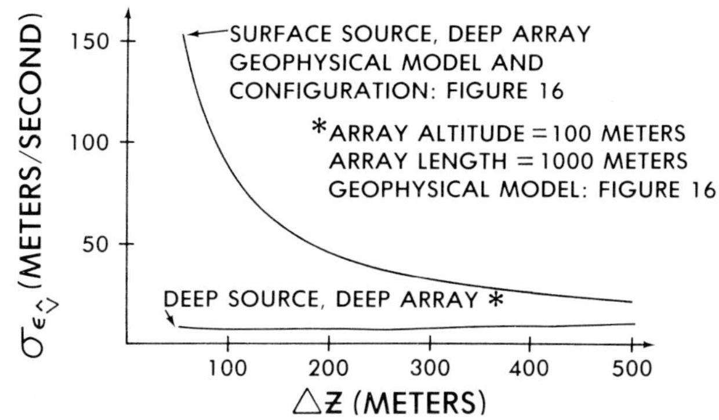

FIGURE 17. Comparing hybrid system with deep-towed system.

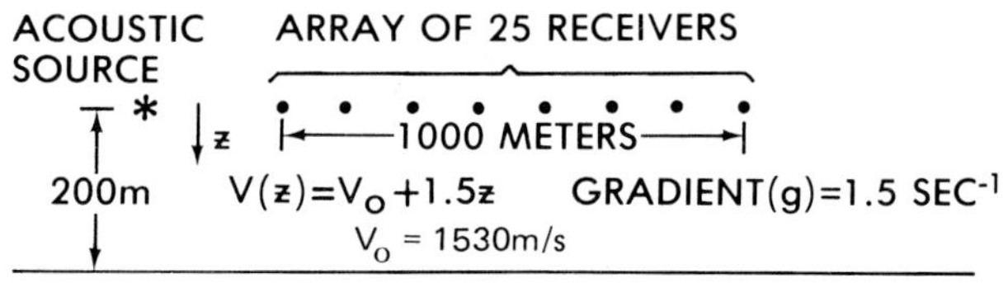

FIGURE 18. Single layer with sound velocity gradient.

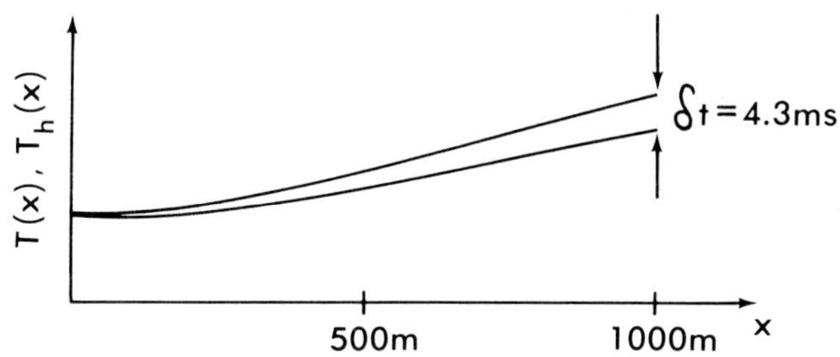

FIGURE 19. Hyperbolic moveout and true moveout.

An experiment was performed using the model of Figure 18 which consists of a *single* 200-m thick layer with a sound velocity gradient of 1.5/sec. Figure 19 illustrates the *nonhyperbolic* nature of T(x) by plotting the hyperbolic* T_h (x) and the true T(x). The nonhyperbolic nature is approximately 4.3 msec at 1000 m offset for the deep-towed system.

It is precisely the nonhyperbolic nature of T(x) that provides information concerning the sound velocity gradient. For a surface system, the nonhyperbolic portion of T(x) would be much less (< 1msec) i.e., the surface system is much less sensitive to the parameter of interest. Figure 20 illustrates predicted performance of the deep-towed system for measuring the gradient (for the example of Figure 18) by plotting standard deviation gradient errors as a function of timing error bounds. The data displayed in

* Hyperbolic curve was generated using $T_h(x) = (T_o^2 + X^2/V^2)^{1/2}$ where V is RMS sound velocity of the layer.

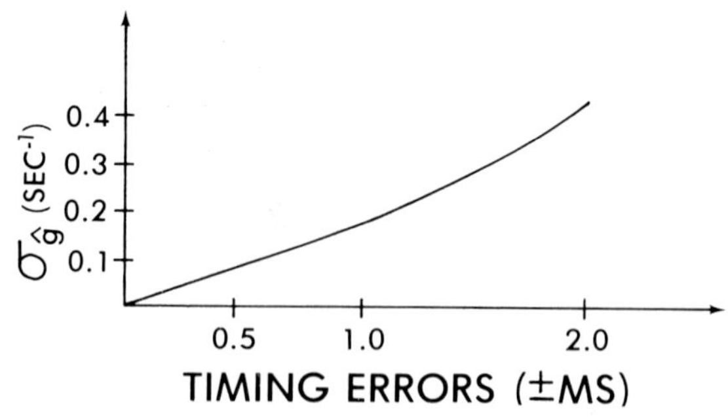

FIGURE 20. Error in gradient estimate.

Figure 20 were obtained by Monte Carlo trials using the simple moments approach of Gholson and Fagot[2] to perform the seismic inversion of reflection data.

HARDWARE DESCRIPTION

Critical for successful implementation of the system concept is a low frequency sound source and hydrophone array that can operate in a towed configuration at deep ocean depths. As was noted earlier, a hybrid configuration such as deep array and surface-towed source will not provide the data quality achievable with both major system components being towed at depth. In addition, the raw data obtained from the deep-towed system must be compatible with high data transfer rates, high power levels, and bidirectional signal paths over a single coaxial steel tow cable.

The component designs must be optimized to provide data for detailed characterization of the first 500 m of subbottom structure while still maintaining a hardware suite that can be easily deployed and operated. These requirements impose strict constraints on the system engineering design. Each major hardware area, sound source, hydrophone array, and telemetry system will be reviewed, identifying technical considerations, design approaches, and recommended hardware implementation solutions for a deep-towed system.

Sound Source

The sound source system for a deep-towed application is a compromise between a set of conflicting requirements. The acoustic performance requirements to meet the 500 m penetration depth imposes the need to operate at a relative low frequency and high source levels to overcome subbottom attenuation. In addition, the requirement is imposed for large source bandwidth to achieve high layer resolution. Conventional sources to meet these requirements and operate at deep depths can result in physically large systems due to their low efficiency and, in many designs, need for depth pressure compensation.

The hardware solutions to these sets of requirements results from a series of tradeoffs between the geophysical data needs and a practical sound source implementation. The following tasks were performed to establish a sound source implementation solution for a deep-towed, high powered, low frequency, and large bandwidth sound source system.

- Implement the sonar equation for typical operating areas and projected system noise to establish acoustic performance requirements.[7]
- Assess acoustic requirements against sound generation techniques and existing

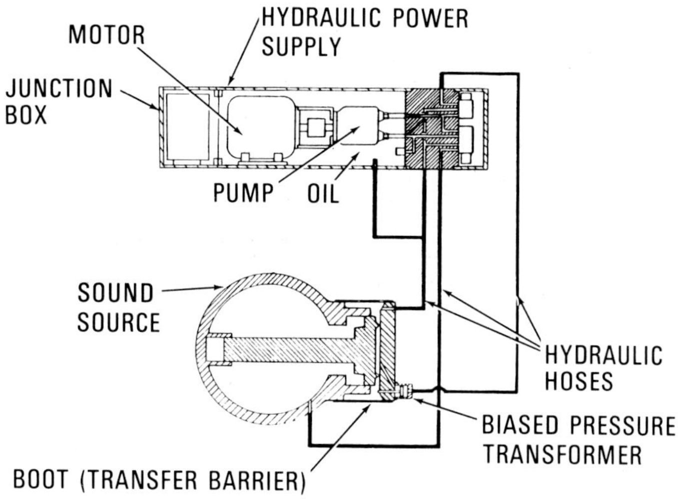

FIGURE 21. Impulse sound source system components.

hardware systems considering the practicality of implementing and operating in a deep-towed configuration to establish technical approach.[8] This assessment established two primary candidate approaches: a Helmholtz resonator transducer and a hydroacoustic impulse source. Although the Helmholtz resonator transducer is a viable technical approach, the hydroacoustic impulse technique is more acoustically optimum for the geophysical application.

- Assess hydroacoustic impulse technique to establish baseline system acoustic parameters.[9]
- Tradeoffs between acoustic parameters and engineering design to establish system hardware characteristics.[10]

These studies have resulted in an implementation solution employing a hydroacoustic impulse sound source system optimized for the geophysical data needs and the practical operations requirements for deep towing.

Hydroacoustic Impulse Sound Source System Mechanical Description

The impulse source consists of two major components, a hydraulic power supply and the oil-backed source (depicted in Figure 21).

The hydraulic power supply is housed in a pressure-equalized vessel and is oil-filled. The supply uses a constant power pump driven by a single phase AC motor. Attached to the back of the supply is a junction box at atmospheric pressure that contains motor start capacitors and control electronics.

A more detailed view of the oil-backed source is given in Figure 22. The source consists of an oil-filled sphere, radiating piston, and a biased pressure transformer. The basic operation of the system is as follows: the radiating piston is moved into the cocked position (as depicted in Figure 22) by increasing the pressure in the oil-filled sphere until the latch seal comes in contact with the latch plate. At this point, the pressure in the oil-filled sphere and latch cavity is reduced with the biased pressure transformer maintaining the pressure in the latch cavity slightly below the pressure in the oil-filled sphere, thus keeping the piston in the cocked position. The pressure is further reduced to approximately 5000 psi below the outside ambient pressure, completing the cocking sequence. The source is fired by suddenly exposing the latch cavity

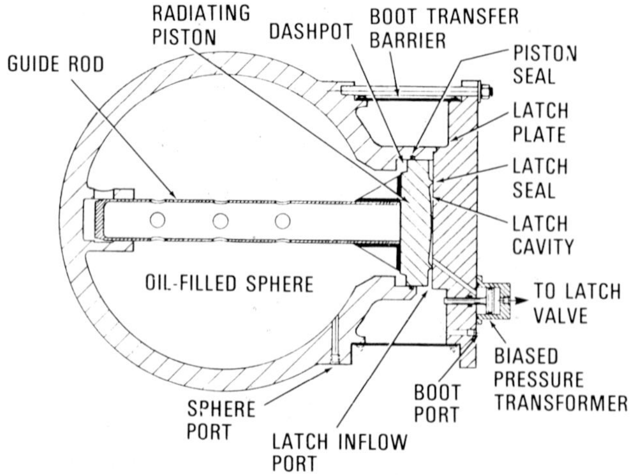

FIGURE 22. Oil-backed impulse sound source.

Table 1
HYDROACOUSTIC IMPULSE SOUND SOURCE MECHANICAL PARAMETERS

Source sphere		
ID (m)		0.65
OD (m)		0.75
Radiating piston diameter (m)		0.30
Radiating piston stroke (m)		0.038
Motor size (hp)		10
Input power (kw)		12
System weight (kg)	Air	Water
Source	423	182
Hydraulic power supply	282	227
Junction box	136	114
Total	841	523
System total size envelope (m)		
Height		0.94
Width		0.84
Length		2.34

to ambient pressure, thus allowing the radiating piston to rapidly retract, and finally, the sudden deceleration of this piston by the dashpot converts the kinetic energy of the piston into a positive high level acoustic impulse.

A summary of the mechanical parameters are given in Table 1. Although the source is relatively large, it is still a physical size that is practical for towing.

Hydroacoustic Impulse Sound Source System Acoustic Description

The acoustic parameters are summarized in Table 2. A projected time signature (idealized) of the acoustic output pulse is given in Figure 23 and its frequency spectrum is shown in Figure 24. Although the prime acoustic band (−3 dB points) is from 140 Hz to 525 Hz, substantial energy exists over even a broader band, covering the range

Table 2
HYDROACOUSTIC IMPULSE SOUND SOURCE ACOUSTIC PARAMETERS

Peak source level (dB re 1 μPa @ 1 m)	226.6
Spectral density level crest (dB re 1μ Pa^2·sec/Hz @ 1 m)	167.0
Bandwidth @ − 3 dB (Hz)	140—525
Depth for full output (m)	4000—8000
Repetition period — minimum (sec)	13

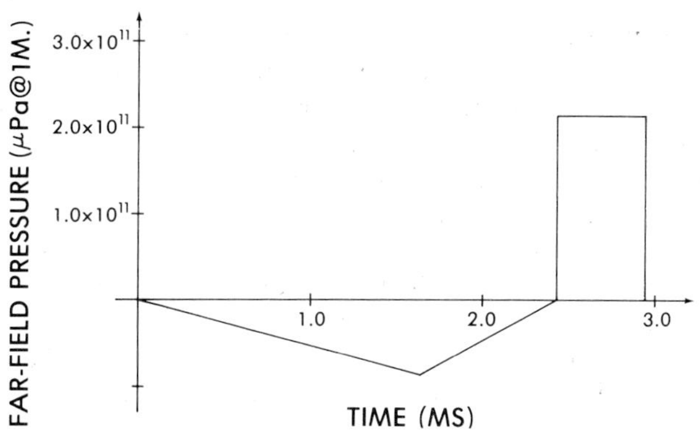

FIGURE 23. Impulse sound source acoustic output time signature (idealized).

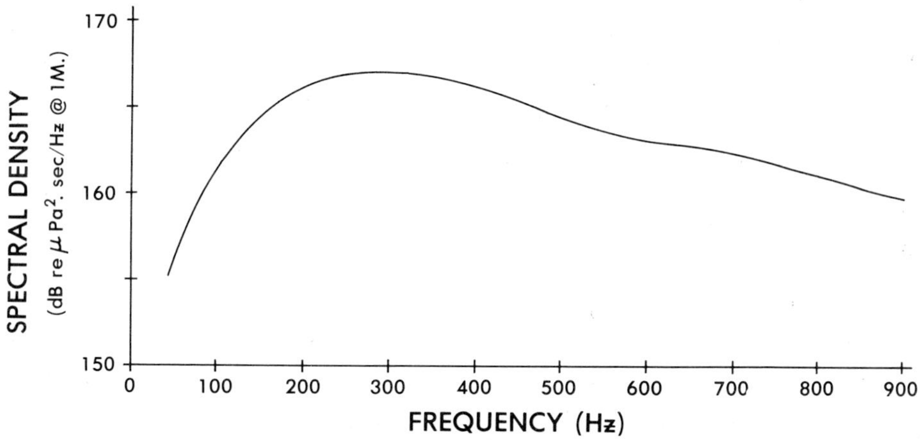

FIGURE 24. Impulse sound source acoustic output frequency spectrum (see Figure 23).

Hz to 525 Hz, substantial energy exists over even a broader band, covering the range from 95 Hz to 810 Hz (−6 dB points). It is worth noting that a surface source operating in 4000 m of water would require an additional 32 dB of source level to match the deep source operating at an altitude of 100 m, due to spreading loss.

Since the source is a differential pressure device, there is only full output power when the device reaches a depth of approximately 4000 m. In shallower operating depths the source level will be reduced, but substantial energy still exists; for example, at a 2000 m operating depth the level drops by about 6 dB.

In summary, the hydroacoustic impulse sound source system projected acoustic parameters will meet the geophysical data acquisition requirements and the hardware mechanical parameters are compatible with a deep-towed operational scenario.

Array

The array system design for a deep-towed configuration requires addressing many unique problems. As was depicted in Figure 1, the array will be attached as a tail to a deep-towed instrument vehicle (fish) containing the sound source. In addition to the obvious requirement of withstanding the high pressure experienced at these tow depths, the array will be sensitive to motion imparted by the combination forcing function of the tow ship, tow cable, and fish. System measurement performance will degrade if this motion and its resulting relative displacement of each hydrophone location with respect to each other and their relation to the sound source, are not known with high precision. Therefore, major technical considerations are designs to minimize array motion and determining techniques and instrument systems capable of measuring any residual deformation of the array-sound source system that may exist. This is a statics correction problem as termed by the oil exploration community.

Typical tow speeds for a deep-towed system range from 0.5 to 1.5 m/sec (1 to 3 kn).[11] Motion sensitivity analysis at these low tow speeds show that the fish altitude (or tow depth) is controlled by the drag on the tow cable and, therefore, is a function of the tow-ship speed.[12,13] This imposes the desired fundamental system operational requirement to maintain a constant tow speed, a requirement that is not always easy to obtain with environmental variables such as wind, waves, and ocean currents. Coupled with this relative long-term altitude variation are short-term variations imposed by the ship tow point motion, such as heave and surge, transmitted to the array-fish system through the tow cable.

The construction of the array itself, as related to motion sensitivity, poses additional problems. The candidate array design for the deep-towed system will employ an oil-filled approach similar to that used by the oil exploration community, but with the array components (hydrophones, engineering sensor, and interconnecting cabling) compatible with the deep operating depth. A requirement will be to ballast the array such that it will tow horizontal. The oil exploration community maintains a horizontal tow by using active depth controllers (birds), an approach difficult (but not impossible) to implement for deep-operating depths. The effect of a slight misballasting of the array as a function of tow speed is revealed by solving an equilibrium equation for a typical array design. For example, an array 1000 m long and 6.35 cm diameter with a 1% uniform trim error (equally distributed along the array) being towed at 1.0 msec (2 kn) results in a tail-up tow angle of about 5°.[13] In addition, the tow angle is sensitive to tow speed. For the same array parameters given above, but at tow speeds of 0.5 msec (1 kn) and 1.5 msec (3 kn), the tow angle is 9.8° and 3.2°, respectively.

Therefore, the array system design will be an optimization to minimize motion effects and development of array motion measurement techniques and hardware measurement subsystems to measure residual deformation. Some engineering and operational deployment options available to minimize the long- and short-term motion effects include:

- Maintaining constant tow altitude by operating in low sea states where good ship handling exists; providing the ship pilot with visual read-out of system altitude to allow varying ship speeds; providing the tow cable winch operator with a visual read-out of system altitude to allow varying tow cable length.

- Reducing short-term motion by: rendering the fish neutrally buoyant through addition of floatation, and attaching a "clump" weight near the fish tow-point junction to absorb the heave-surge motion; installing heave motion accumulators in the tow cable line at the tow ship and/or near the fish.
- Reducing array tilt angle changes due to tow speed variation by: attaching a dead array section and drogue parachute at the tail of the array;[11,13] implementing an active control system similar to the "bird" concept.

There are certainly alternatives and more subtle areas that could be addressed, but these provide the reader with an insight into some of the practical considerations and technical options available to minimize system motion. Implementing some, or all, of these approaches does not eliminate the need to instrument the deep-towed system to provide a high precision measure of the hydrophone/sound-source locations.

A timing error of ± 0.5 msec for a medium sound speed of 1500 msec would require knowing the relative sensor (hydrophones/source) positions within ± 0.75 m. This assumes, for simplicity, that the acoustic raypaths are normal to the array longitudinal axis and in the hydrophone/source vertical plane. It should be noted that the normal raypath assumption would only be true for deep and nondipping layers. This high measurement precision poses stringent requirements on the position measurement system. High resolution depth sensors along the array and at the fish, but not necessarily with precise absolute depth, present one possible approach. Calibration studies[14] for a sensitive quartz-type pressure transducer designed for full ocean depths, but operating over a limited depth range as in a deep-towed application, revealed that measurement uncertainty of about 1.0 dbar (approximately 1.0 m) can be expected. Although this uncertainty is slightly greater than relative sensor position requirement, the results indicate the feasibility of achieving high resolution measurements.

An alternative to the depth sensor approach, with potentially higher precision, is a differential pressure device. This approach would rely on measuring small pressure change about a reference depth. Depth sensors operating over this limited depth range (± 200 m for example) can achieve higher resolution than those requiring a full ocean depth (6000 m) capability. The reference depth would be preset to the nominal deep-towed operating depth. Depth changes around this reference depth would then be used to correct sensor position to the horizontal datum.

An additional approach for measuring vertical motion is an acoustic technique through triangulation of the water break (direct) and water surface reflected arrivals to each hydrophone.[13,14] Since the travel path from the source to sea surface to array results in nearly a hyperbolic moveout, and knowing the water column sound speed, the moveout for a horizontal array can be accurately predicted. Deviation of arrival times from the predicted moveout due to a nonhorizontal array can be determined and used to correct for array deformation. The precision of this technique will require assessing the effects of wave height variations (sea state) and S/N ratios on measuring arrival times. In addition, the lag time between the subbottom arrival and the deformation measurement (up to 6 sec for 4500 m tow depths) will need to be considered. An advantage of this approach is knowing the relative position at each hydrophone location as compared to sparsely spaced depth sensors. Undoubtedly, the final approach for measuring residual deformation will employ a hybrid technique, hydrostatic and acoustic, to reduce the position error by employing independent measurements.

The array will also contain magnetic heading sensors to provide data on the horizontal configuration (snaking). Horizontal displacements couple into vertical arrival time errors, but with less magnitude than vertical displacements. Under the condition of strong crosscurrents, a condition not generally encountered at deep ocean depth, precise measurements of horizontal displacements may be necessary.

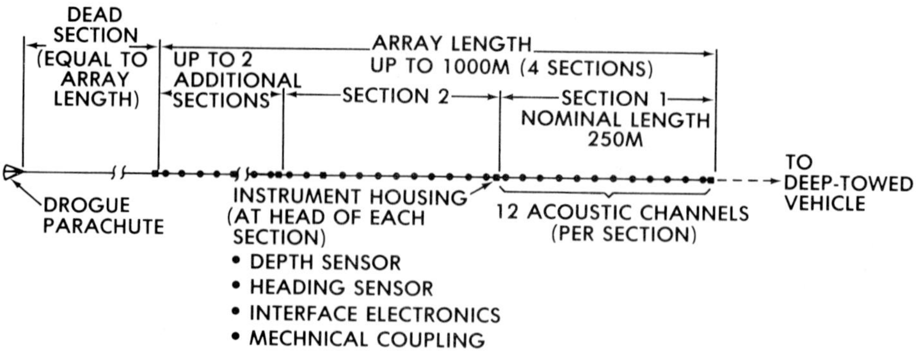

FIGURE 25. Candidate array design configuration.

The hydrophones used in the deep array do not pose any severe problems. There are units manufactured for deep ocean applications. If more than one hydrophone per group (a channel) is to be used, a group response analysis must be performed.[16] This analysis defines the operating and environmental parameters needed to minimize destructive interference between hydrophones in each group. The analysis will include such variables as operating frequency, layer velocity structure, and typical layer dip angles.

A candidate array design configuration for the deep-towed application is depicted in Figure 25. A multisection oil-filled design is envisioned. Each section will have a nominal length of 250 m, and up to 4 sections will be used, giving a total length of 1000 m. The active sections, including the instrument housings containing a heading sensor, depth sensor, and interface electronics, will be interchangeable. This will allow varying the active array length and provide ease in handling and maintenance. A dead section equal to the active length and a drogue parachute will be attached as a tail for motion stability.

The telemetry system, as discussed in the next section, will have the capability to handle up to 25 acoustic channels. One channel will be used for a near field source monitor at the fish, and the remaining 24 channels will be used for the array. Each array section will contain 12 acoustic channels with a minimum of 2 sections used, thereby employing the 24 acoustic channels. When the array length is increased beyond 500 m, only selected acoustic channels in each section will be used to maintain the 24 total channels.

The array system design encompasses many unique technical considerations for the deep-towed application. Some of the major considerations were reviewed and design approaches presented. A candidate design configuration has been formulated to provide maximum flexibility for both acoustic and operational performance.

Telemetry

Telemetry for the deep-towed system is complex due to the required high data rate over a long transmission line. Further, the telemetry must be bidirectional to allow for a control line while simultaneously sharing the line with higher power distribution for the acoustic source.

Uplink (array-to-ship) telemetry utilizes a 16-bit floating point format to achieve an instantaneous dynamic range greater than 100 dB. Pseudo dynamic range is even greater (> 150 dB) via remote control of preamplifier gains. The high dynamic range allows measurement flexibility and also minimizes predeployment setup and calibration time.

Signals from the acoustic channels are low pass filtered to prevent aliasing in the analog to discrete conversion process. The low pass filters are a 6 pole Bessel realization in order to minimize phase distortion of the received signals and perhaps eliminate some early "dephasing" stages of signal processing.

The features just described are gained at the expense of requiring a high data rate (≈ 2 meg b/sec) over a long (10 km) transmission line. Further, it is desired that the transmission line be free of repeaters. Such performance, under the given constraints, is very difficult and requires careful attention to designing equalizers to overcome the nonlinear phase and nonconstant attenuation characteristics of the transmission line.

SUMMARY

The U.S. Navy's need for a more definitive model of the ocean floor and subbottom has motivated the development of a deep-towed geophysical system. This chapter reviews a statistical performance prediction analysis technique for the deep-towed system to estimate error in measuring one of the more critical geoacoustic parameters; layer sound velocity. Although the analysis technique focuses on the deep-towed configuration, it is applicable to more conventional systems and will provide an analysis method for predicting performance of other specific system configurations.

Velocity error estimate results are presented comparing the deep-towed configuration with more conventional and hybrid configurations. These results predict substantial performance improvements for the deep-towed system.

Velocity estimates as a function of system parameters (towing altitude and array length), timing error bounds, and layer depth of burial are presented. The potential of the deep-towed system with its unique data set to estimate layer sound velocity gradient for the case of a linear sound velocity vs. depth relationship is illustrated.

The chapter concludes by identifying hardware approaches in the areas of the sound source, hydrophone array, and telemetry system to meet the geophysical data needs while providing a system compatible with operating in the deep ocean environment.

REFERENCES

1. **Fagot, M. G., Gholson, N. H., Moss, G. J., and Milburn, D. A.,** Deep-Towed Geophysical Array System — Development Program Review, NORDA Tech. Note 75, Naval Ocean Research and Development Activity, NSTL Station, Miss., 1980.
2. **Gholson, N. H. and Fagot, M. G.,** A Moments Approach for Analyzing Geophysical Reflection Data, NORDA Rep. 30, Naval Ocean Research and Development Activity, NSTL Station, Miss. 1980.
3. **Hubral, P. and Krey, T.,** *Interval Velocities from Seismic Reflection Time Measurements,* Society of Exploration Geophysicists, Tulsa, 1980.
4. **Dix, C. H.,** Seismic velocities from surface measurements, *Geophysics,* 20(1), 68, 1955.
5. **Stoffa, P.,** *Design Criteria for a Deep Towed Source and Multi-Channel Array,* Chute Stoffa, Inc., Nyack, N.Y., 1979.
6. **Gibson, B., Odegard, M., and Sutton, G.,** Nonlinear least-squares inversion of traveltime data for a linear velocity depth relationship, *Geophysics,* 44(2), 185, 1979.
7. **Fagot, M. G. and Eckstein, B. E.,** Deep-Towed Geophysical Array Development Program Prog. Rep. (FY 1978), NORDA Tech. Note 41, Naval Ocean Research and Development Activity, NSTL Station, Miss., 1979.

8. **Groves, E. D.,** Deep tow low freqency sound source technology assessment, in *Workshops on Deep-Towed, Low Frequency Acoustic Sources and Receivers,* Lewis, B.T.R., Ed., Office of Naval Research, Washington, D.C., 1979, 5.
9. **Technical Staff,** Design Study of Hydroacoustic Impulse Source for Deep Towing, Report No. HA 132-79, Hydroacoustic, Inc., Rochester, N.Y., 1979.
10. **Technical Staff,** Phase II Design Study of Hydroacoustic Impulse Source for Deep Towing, Report No. HA 115-81, Hydroacoustics, Inc., Rochester, N.Y., 1981.
11. **Milburn, D. A. and Fagot, M. G.,** A Steady-State Analysis of Candidate Towing Cables for a Deep-Towed Geophysical Array System, NORDA Tech. Note 40, Naval Ocean Research and Development Activity, NSTL Station, Miss., 1979.
12. **Mudie, J. D. and Ivers, W. D.,** Simulation studies of the response of a deeply towed vehicle to various towing ship maneuvers, in *Ocean Engineering,* Pergamon Press, Oxford, 1975, 3.
13. **Fagot, M. G., Moss, G. J., Milburn, D. A., and Gholson, N. H.,** Deep-towed seismic system design for operation at depths up to 6000 m, in, *Proc. 1981 Offshore Technology Conference,* Offshore Technology Conference, Houston, 1981, 3.
14. NOAA, Report of Calibration of Deep Sea Pressure Transducers, Ocean Technology and Engineering Services, Engineering Support Office, Washington, D.C., 1981.
15. **Koenigs, P. D and Martin, R. L.,** Measurement of Large Aperture Array Shape, Tech. Memo. No. 781025, Naval Underwater Systems Center New London Laboratory, New London, Conn., 1978.
16. **Savit, C. H.,** A 500-channel streamer system, in, *Proc. 1977 Offshore Technology Conference,* Houston, 1977, 2.

HYDROPHONE CABLES FOR SEISMIC EXPLORATION

Albert J. Berni

INTRODUCTION

Hydrocarbon exploration is performed offshore by the seismic survey method. Seismic waves are generated just astern of the ship. The most common source is rapid injection of compressed air. The seismic waves propagate downward into the earth and reflect from the boundaries between rock layers. The reflected waves return to the surface and are detected by pressure sensitive transducers known as hydrophones. The hydrophones are housed in a cable towed behind the survey ship. Figure 1 shows the overall configuration. The arrival times and reflection amplitudes are analyzed to determine the best drilling sites for hydrocarbon discovery.

The conventional cable consists of a series of 100 m sections. The total length is variable but 3000 m is typical. Figure 2 shows the cable cross section and a side view of the junction between sections. Cylindrical plastic spacers are attached to the steel ropes in order to maintain the cable shape, see Figure 3. Hydrophones are mounted at approximately 1 m intervals. Electrical wire bundles pass through holes in the spacers and connect the hydrophones. Covering this structure is a thin-walled plastic hose of approximately 6 cm diameter. The hose is filled with kerosene or another oil to provide buoyancy. Lead weights are strapped to the outside of the cable for buoyancy trimming if necessary.

There are several operational methods. In water depths greater than 10 m the cable is pulled through the water at 4 to 7 kn. The depth is controlled by the attitude of winged surfaces attached to the cable as shown in Figure 4. These depth controllers are attached at approximately 200 m intervals.

A bottom reference cable is used for shallow water operations. Chains are attached every 30 m to a normally buoyant cable. As the chain settles, weight is released until a condition of neutral buoyancy occurs. The cable then remains a meter above the bottom. A cable weighted by chains can be dragged at approximately 2 kn.

Several hydrophones are connected together to form a detector array. A dedicated pair of wires connects each array to the shipboard recording system. A typical "cable" consists of 100 "arrays". The array signal is actually a weighted summation of individual hydrophone signals. The weighting function is chosen so as to obtain some desirable frequency-wavenumber sensitivity pattern. In this way certain seismic waves are accepted while others are discriminated against. The theory and design of such arrays are the subjects of this chapter. We begin with a discussion of the basic array design problem.

ARRAY DESIGN PROBLEM

We approach the array design as a problem in linear filtering theory. Below, we describe the array as an anti-alias filter, that removes energy at the higher wavenumbers prior to spatial sampling. For the offshore problem, the wavefield is only sampled along one spatial dimension.

Figure 5 illustrates the effect of spatial sampling. At upper left is a representation of a sampled wavefield at some instant of time. These spatial samples represent individual hydrophone signals in our application. The corresponding figure on the right represents the Fourier transform amplitude spectrum of the sampled wavefield. In this example, the spectrum is assumed to have no significant energy above the Nyquist

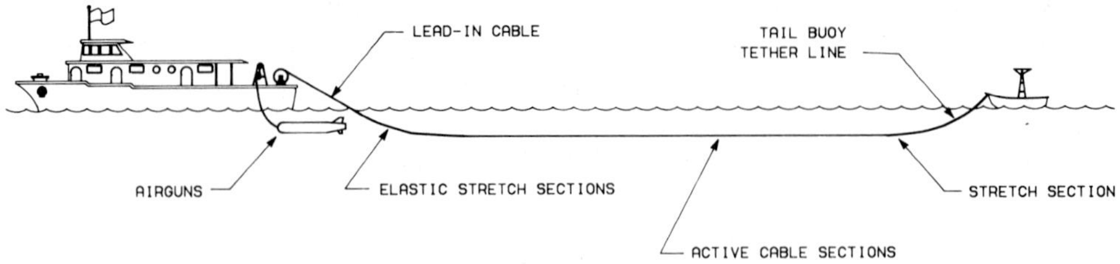

FIGURE 1. Overall configuration of exploration ship and hydrophone cable.

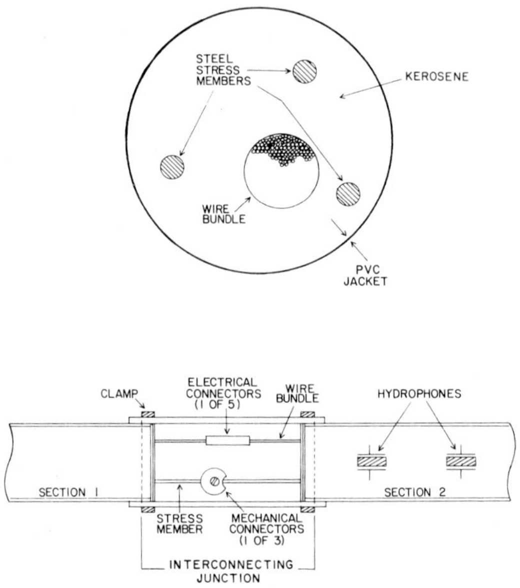

FIGURE 2. Diagram of conventional hydrophone cable.

wavenumber (0.5/Δx). Therefore, the wavefield is accurately represented by the sampled values. An array signal is formed by making a weighted summation of these samples. A set of weights is shown in Figure 5B along with the corresponding wavenumber spectrum. In order to cast the array design as a problem in digital filtering, we must consider the conceptual signal that is the convolution of the sampled wavefield and the sequence of weighting coefficients. This conceptual signal does not physically exist at all increments of Δx.

In the wavenumber domain, the sampled wavefield is multiplied by the array function to form the conceptual signal in Figure 5C. Only at certain intervals, shown in Figure 5D, does the conceptual signal correspond to an actual array output. Formation of the array output can be viewed as a spatial sampling of the conceptual signal. The spatial interval for this second level of sampling is D. The corresponding spectra are repeated at 1/D wavenumber increments. In order to avoid aliasing due to this second, wider sampling, the array response must adequately attenuate energy at wavenumbers above 0.5/D.

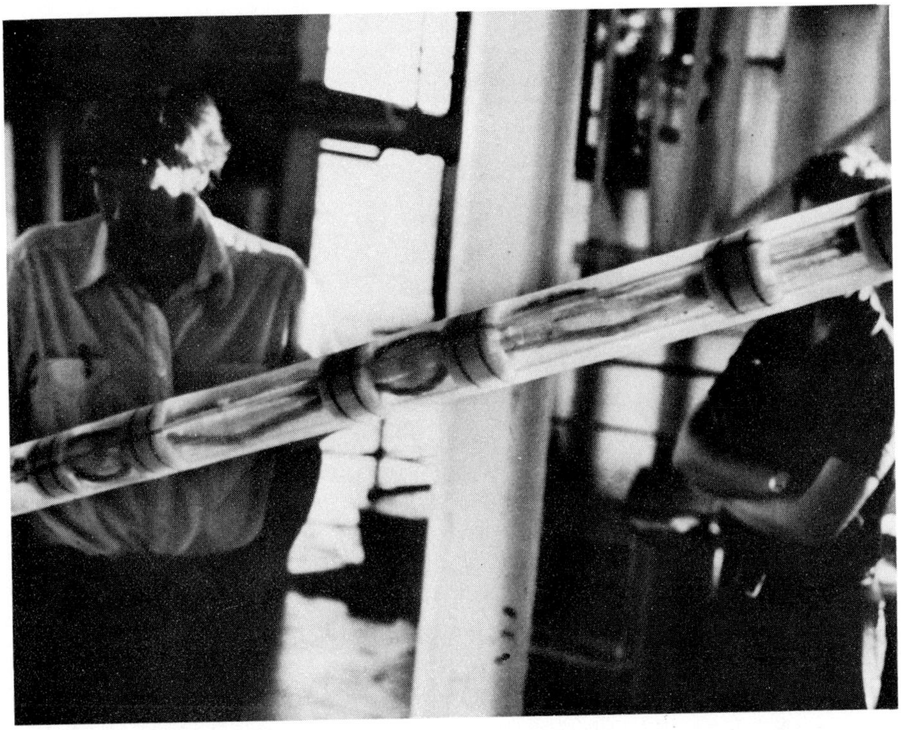

FIGURE 3. Photograph of hydrophone cable on ship.

FIGURE 4. Hydrophone cable in water. Courtesy of Seismic Engineering Company, a subsidiary of Whitehall Corporation.

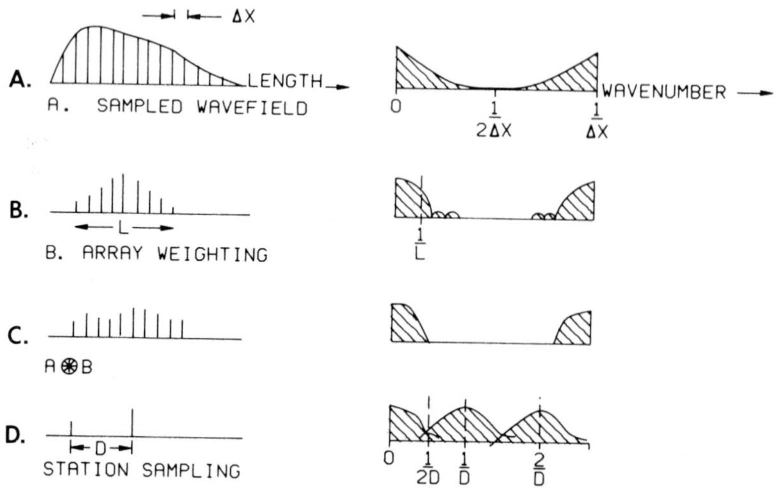

FIGURE 5. Array design problem. On the left are samples along the length axis. On the right are the corresponding wavenumber spectra.

The anti-alias criterion for array design differs significantly from the more conventional one of maximizing the output signal-to-noise ratio (SNR). The latter is not a reasonable criterion for the marine seismic cable problem. For the cable, the numerous array output signals are digitized and subsequently processed by computer. Many array output signals are combined to enhance the SNR for event detection and parameter estimation purposes. Much finer wavenumber resolution is available at this stage. The job of the basic hardware array, then, is not to enhance SNR by wavenumber selectivity, but instead, to ensure that the process of spatial sampling does not cause an irreversible degradation of SNR in the frequency-wavenumber bands of interest. Intelligent choices must be made for the sampling intervals, Δx and D, and the weighting coefficients. These choices will depend in part on the desired signal and noise characteristics.

We are usually dealing with array lengths that are much shorter than the transmission path of the seismic waves. The plane wave approximation can therefore be used. Plane waves are transformed into spectral lines in the frequency-wavenumber domain, as illustrated in Figure 6 (see Appendix). The slope of the spectral line is V_a, the apparent velocity in the x direction. As frequency increases, so does wavenumber. The requirement of sharp attenuation at the Nyquist wavenumber can be met by limiting the frequency content of the signal. Alternately, we say that the spatial sampling interval D must be short enough to prevent aliasing at the highest frequencies of interest. To assess the impact of this requirement, we must evaluate the apparent velocities in the x direction for both seismic waves and propagating noises.

Of particular importance is the determination of the signal with the slowest velocity. For a given frequency, the wavenumber is inversely proportional to the velocity. The slowest velocity will determine the sampling requirements.

SEISMIC SIGNAL AND NOISE

Reflection Signal Characteristics

Figure 7(a) is a raypath diagram for the simplest earth reflection model, a single flat boundary. The apparent velocity, V_a, in the horizontal direction is $V_p \cdot \sin^{-1} \phi_i$ where V_p is the propagation velocity in the layer and ϕ_i is the incidence angle at the receiver.

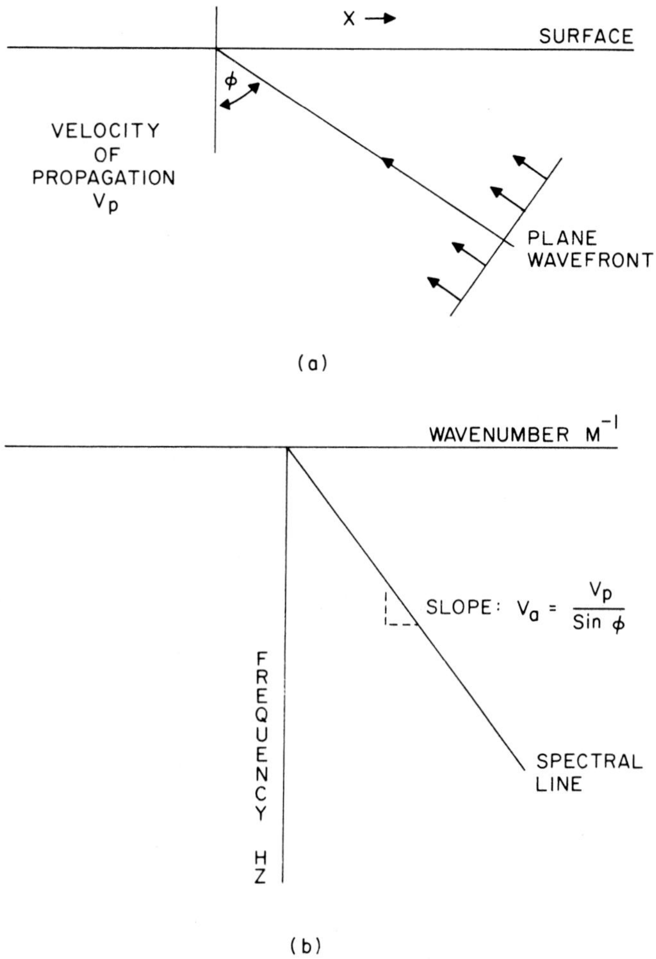

FIGURE 6. (A) Plane wavefront impinging on surface. (B) Sketch of frequency-wavenumber domain showing the spectra associated with sampling the plane wavefront along the horizontal axis.

$$V_a = \frac{V_p\,(X^2 + 4Z^2)^{0.5}}{X} \tag{1}$$

The raypath length is $(X^2 + 4Z^2)^{.5}$ and equals the propagation velocity times the arrival time T.

←Substituting this relation in Equation 1

$$V_a = \frac{V_p^2 \cdot T}{X} \tag{2}$$

Figure 7(b) shows the case of a reflector dipping with angle Θ. The following relationships can be obtained by geometric analysis,

$$(L_1 + L_2)^2 = (X + 2\ Z_S \cos\theta\ \sin\theta)^2 + (2\ Z_S \cos^2\theta)^2$$

$$X \tan\theta = Z_R - Z_S \tag{3}$$

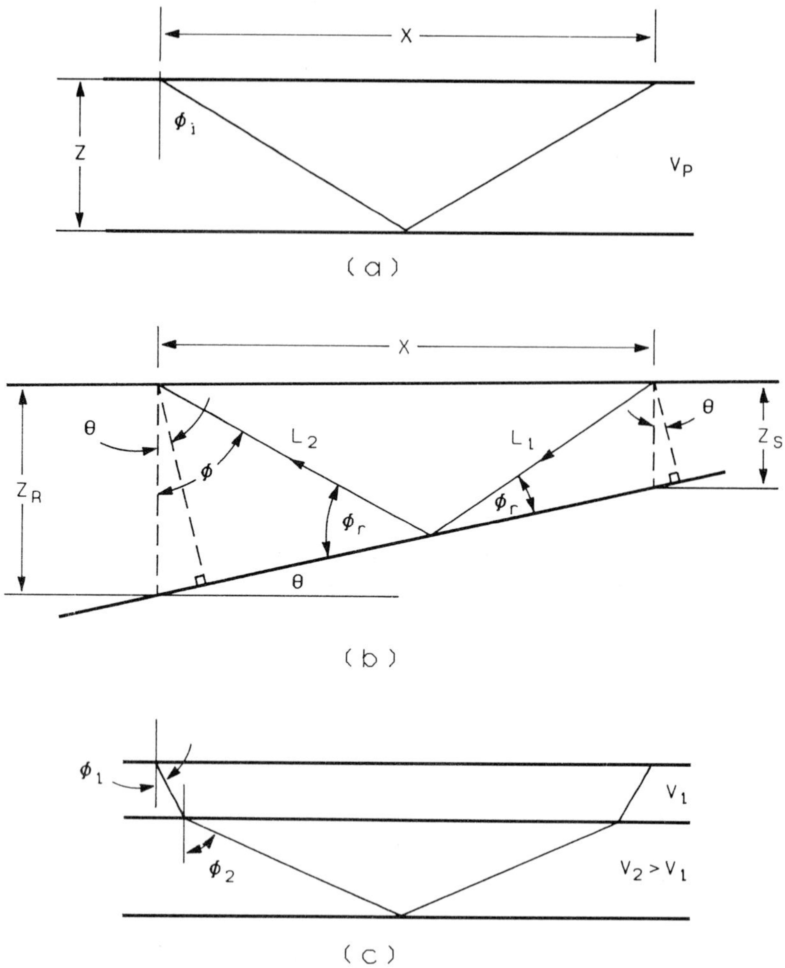

FIGURE 7. Reflection models, (A) single flat reflector; (B) single dipping reflector; (C) two subsurface reflectors.

where

$L_1 + L_2$ is the raypath length

Z_R is the depth of the reflector under the receiver

Z_S is the depth of the reflector under the shotpoint

Θ is the dip angle

X is the shotpoint to receiver distance

The apparent velocity can be expressed in terms of these quantities,

$$V_a = \frac{V_p}{\sin \theta}$$

$$V_a = \frac{V_p(L_1 + L_2)}{X + 2\ Z_S \cos \theta \sin \theta} \tag{4}$$

for the case $X = Z_S$

Table 1

Θ	Ratio
0	1
10	0.74
20	0.57
30	0.46

$$\left.\frac{V_a}{V_p}\right|_{\text{Dip Angle } \theta} = \frac{(1 + 4 \sin\theta \cos\theta + 4 \cos^2\theta)^{0.5}}{1 + 2 \cos\theta \sin\theta} \quad (5)$$

We want to compare this result to the case of no dip. For comparison, consider a flat reflector at depth $Z = 0.5\,(Z_R + Z_S)$. From (3) and (1),

$$V_a = \frac{V_p\,(X^2\,(1 + \tan^2\theta) + 4\,Z_S^2 + 4\,X\,Z_S \tan\theta)^{0.5}}{X} \quad (6)$$

for $X = Z_S$

$$\left.\frac{V_a}{V_p}\right|_{\text{Flat Reflector at } Z=.5(Z_R+Z_S)} = (5 + 4 \tan\theta + \tan^2\theta)^{0.5} \quad (7)$$

Table 1 shows the ratio of apparent velocities (dipping case over flat) for various angles. Even for large dip angles, the apparent velocity is almost half the no dip case. The apparent velocity is greater for $X < Z$. The alternate case of shotpoint down dip from receiver yields apparent velocities greater than the no dip case.

Figure 7(c) shows a two-layer case with the usual situation of increasing velocity with depth. It is clear from this diagram that the arrival angle (incidence angle ϕ_1 at surface) is smaller than for a single layer. The apparent velocity is therefore increased. By Snell's law, the apparent velocity at the top of the second layer is the same as at the surface.

$$\frac{V_1}{\sin\phi_1} = \frac{V_2}{\sin\phi_2} \quad (8)$$

This argument can be extended to the case of multiple layers.

In a typical marine case, V_1 is water velocity, 1500 m/sec, V_2 is 4 km/sec or greater and the water layer is much shallower than the rock layer. The model degenerates to the single layer case with velocity V_2 and an apparent velocity (from Equation 1) of 8.9 km/sec for $X = Z$ and no dip of the reflector. For a 20° dip angle, the apparent velocity (updip shooting) is greater than 5 km/sec.

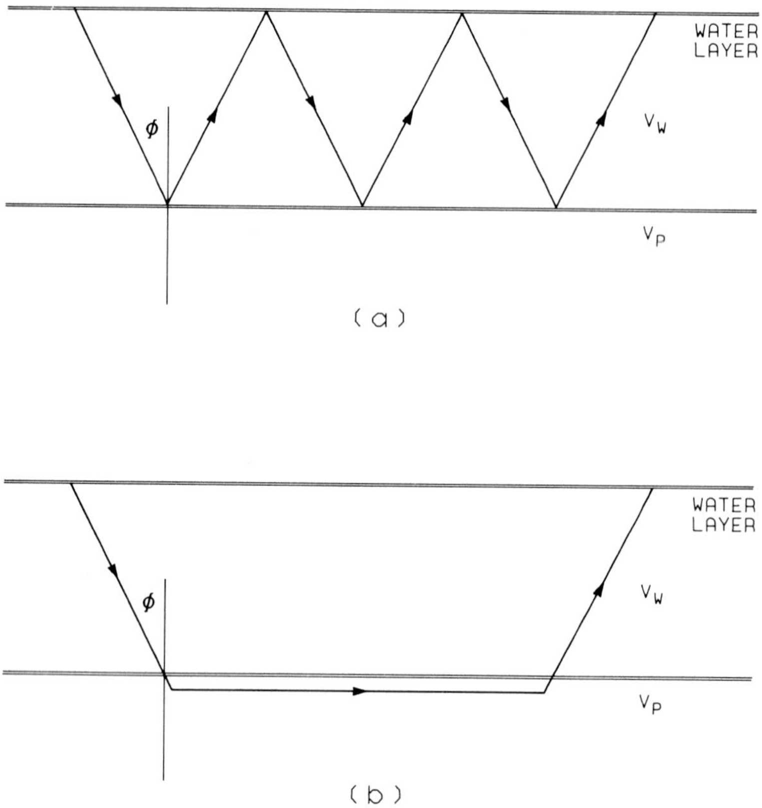

FIGURE 8. Shot-generated noise, (A) the water layer as a waveguide; (B) a subbottom refraction.

Propagating Noise

Shot-Generated Noise

The seismic source generates energy that is trapped in the near-surface layers. This energy is usually not useful in the search for hydrocarbons. It has been termed shot-generated noise. There are several known mechanisms for the propagation of this energy. In marine seismology, the dominant propagation paths are in the water and the sea bottom. The water layer acts as a waveguide for waves whose incidence angle at the sea bottom is greater than the critical angle. Figure 8(a) shows a raypath for a wave whose incidence angle just equals the critical angle. Figure 8(b) shows the refracted arrival. Both rays arrive with angle ϕ. The apparent velocity for both cases, by Snell's law, equals V_p, the velocity of propagation in the subbottom.

Clearly waves with angles of incidence ϕ greater than the critical angle also propagate in the water layer. For incidence angles of 70° or more, the apparent velocity is no more than 10% higher than V_w. Rays with incidence angles below critical will be attenuated because of the reflection loss. Indeed there is some reflection loss even for the high incidence angle rays because of compressional to transverse mode conversion at the sea bottom. It follows that those rays which have fewer bottom bounces are relatively less attenuated than others. For further discussion of shot generated noise, see References 1, 2, and 3.

Scattering

Shot-generated noise can be scattered by near-subsurface irregularities such as rough

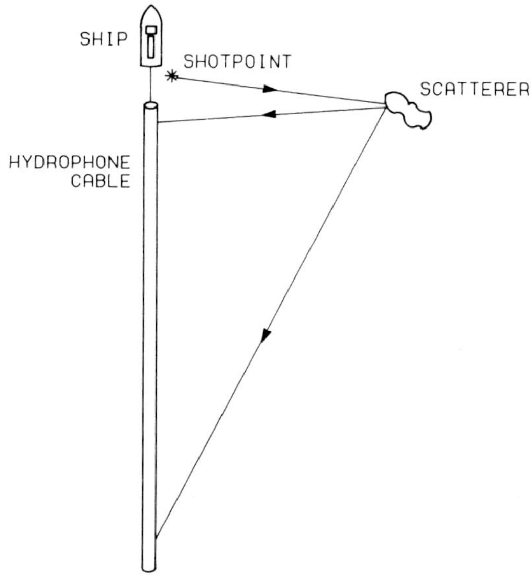

FIGURE 9. Top view of a near-subsurface scatterer.

sea bottom features. Figure 9 shows the case of a single scatterer and two return ray paths in the azimuthal plane. Apparent velocities for scattered energy range from V_w to infinite. Both negative as well as positive velocities must be considered.

Towing Noise

There are several noise waves that propagate by vibration of the cable components. These waves are generated and travel along the cable when it is towed. There are several propagation mechanisms because the cable consists of a combination of different materials such as steel stress members, hard plastic spacers, enclosed fluid, soft plastic hose jacket, etc. In addition to these internal noise waves, there is also the noise caused by pressure fluctuations outside the jacket. This phenomenon is associated with turbulent water flow and has been termed flow noise. All of these noise sources taken together have been called towing noise, since they occur when the cable is moved, and they are not shot generated.

Figure 10 shows the noise recordings from individual hydrophones in a special section built for the study of cable-borne noise. We see low velocity (≅ 30 m/sec) waves emanating from the ends of the section. These are tube waves propagating by motion of the jacket. They are generated by motion of solid internal components, such as spacers. Strong waves are generated at section terminations and at oil blocking spacers where the enclosed fluid cannot flow around the moving object. The action is analogous to a piston moving in a cylindrical cavity with elastic walls.

Vibrations propagating along the stress members at high velocity (≅ 1300 m/sec) are converted to tube waves at the points of attachment of spacers and terminating bulkheads. Hydrophones should be mounted equidistant from adjacent spacers. This facilitates cancellation of the nearly out-of-phase waves emanating from the two spacers. A longitudinal vibration of the stress member causes one spacer to move toward the hydrophone (pressure increase) while the other moves away (pressure decrease). Similarly, arrays of hydrophones should be centered between oil blocking spacers to facilitate cancellation of the tube waves. Unfortunately, perfect cancellation cannot be achieved because excitation is not simultaneous, due to the finite vibration velocity on the stress members.

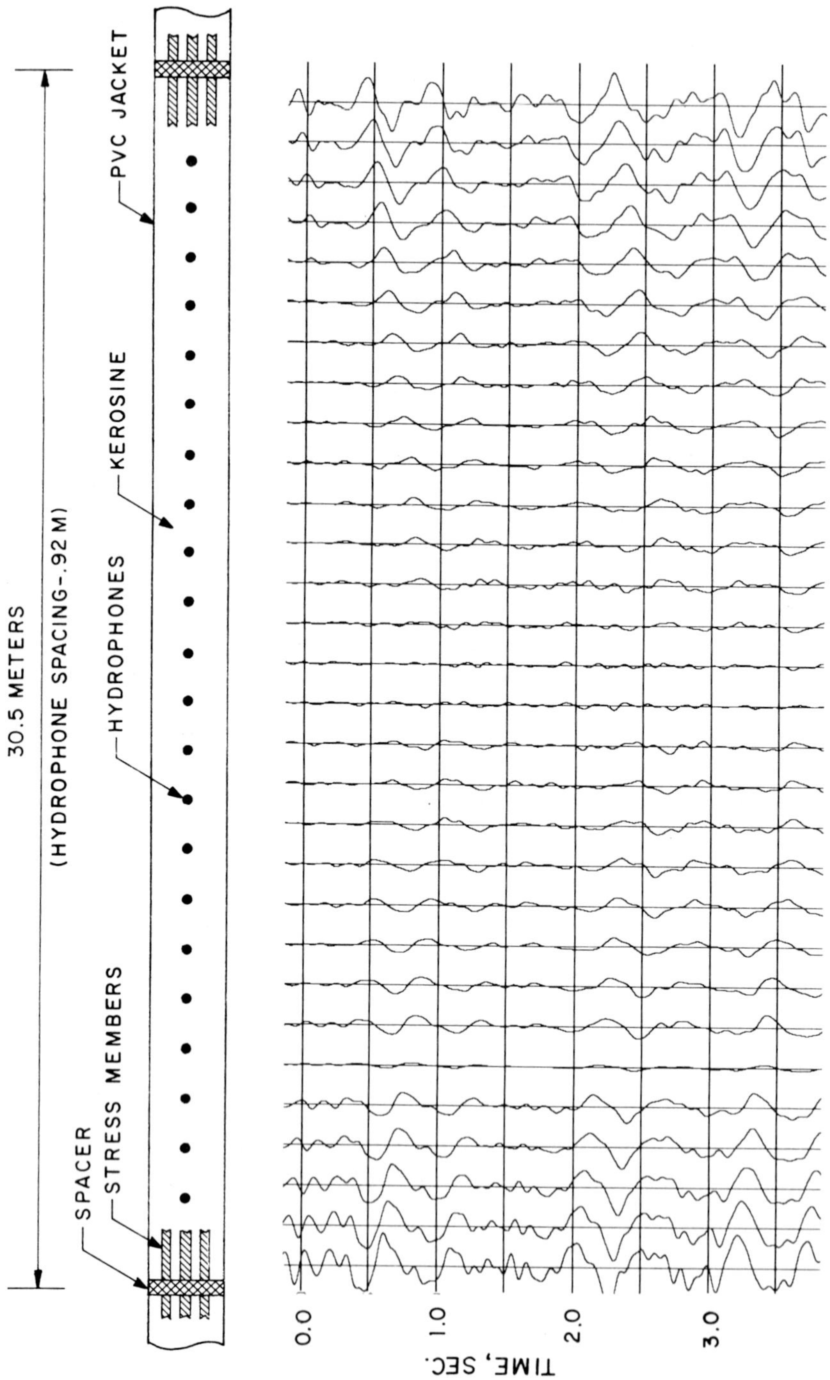

FIGURE 10. Noise recordings of 30 hydrophones in a seismic cable.

Studies have been performed[4,5] on the turbulent flow of fluid past smooth cylinders and other solid objects. From this work, we will infer conditions that should roughly be true for a hydrophone cable streaming through the water. Over short distances, flow noise has an effective velocity slightly less than the ship speed. The propagation is always in the direction of flow, i.e., toward the tail buoy. The noise becomes uncorrelated (no identifiable velocity) at longitudinal intervals greater than a wavelength. Ship speed is approximately 3 m/sec. For frequencies from 10 to 100 Hz, the wavelength of flow noise ranges from 30 cm down to 3 cm. Hydrophones in seismic cables have almost always been mounted at wider intervals. Therefore, flow noise appears white over the wavenumber spectrum. The flow noise contribution can be minimized by increasing the number of hydrophones in the array or by significantly increasing the pressure sensing area of an individual hydrophone. See References 6, 7, 8 for further discussion of towing noise on seismic cables.

DESIGN SPECIFICATION EXAMPLE

Hydrophone Spacing

As stated previously, the choice of hydrophone spacing (Δx in Figure 5) constitutes a choice of spatial sampling interval. If the sampling interval is too large, then aliasing of high wavenumber (short wavelength) noise into the desired signal spectrum occurs and the SNR is degraded. The highest wavenumber noise determines the hydrophone spacing. Towing noise components such as cable-borne vibration exhibit high wavenumber energy.

These phenomena have been studied through the use of specially constructed cables with very fine spatial sampling. Figure 11 is a frequency wavenumber spectrum computed from the noise recordings of a noise test cable section. Individually recorded hydrophones were spaced 15 cm apart. We see two distinct high noise regions. The low velocity region (10 to 30 m/sec) is diffuse. Note that the low velocity noise does not decay into the background ''white noise'' level until wavelengths are less than 0.6 m. The low velocity region is strongest at frequencies up to 30 Hz. At higher frequencies, we see dominance by high velocity noise. This noise is limited to the cone bounded by the 1.2 km/sec lines. The high velocity noise is probably a combination of vibration traveling along the stress members and some water-borne arrivals from the ship. Ships are a source of seismic noise even when there is no shot because of machinery vibration and propeller cavitation.

Array signals can be formed by summing the individual hydrophone signals. The effect of aliasing can be quantified by successively forming arrays of the same length but with different hydrophone intervals. This has been done over a 5 m length for hydrophones spaced respectively at 0.3, 0.6, 0.9 and 1.5 m intervals. The output signals are divided by the number of hydrophones in the particular array. In this way, a signal common to all detectors (such as a broadside desired signal) has the same strength in the array output, independent of the number of elements or spacing. The resulting spectra are shown in Figure 12. As expected, the only effect is at the lower frequencies. From 0.3 to 1.5 m there is approximately a 5 dB degradation. There is very little degradation in going from 0.3 to 0.6 m. Typically, hydrophones are spaced at 1 m intervals in cables used for exploration.

Array Wavenumber Selectivity Function

The Passband-Stopband Philosophy

Figure 13 is a frequency-wavenumber diagram showing the track of two propagating waves. The 4.5 km/sec line represents the slowest reflection signal of interest in this example. Typically, desired signal energy resides in the region between the frequency

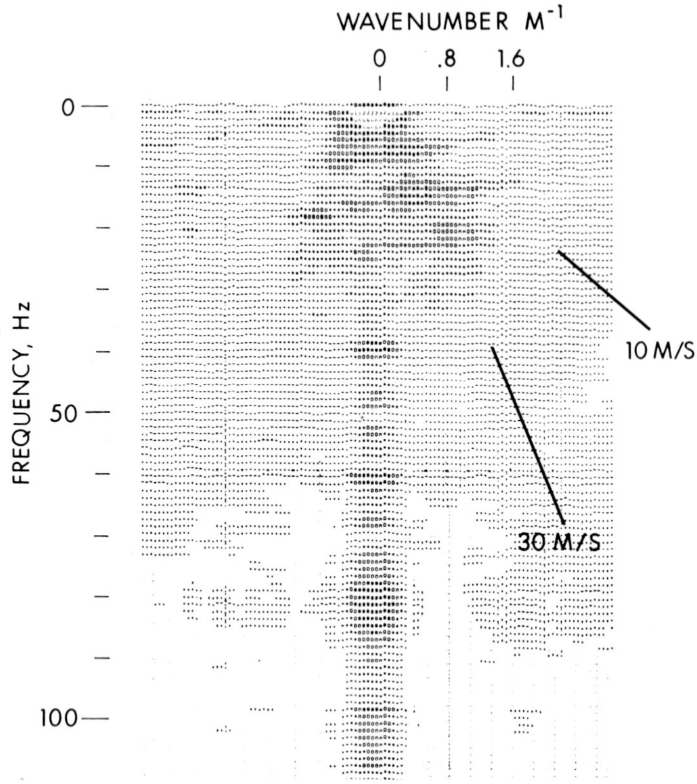

FIGURE 11. A frequency-wavenumber (reciprocal wavelength) diagram calculated from the tracings of individually recorded hydrophones. Hydrophones were spaced at 15 cm intervals in this study. Noise recordings.

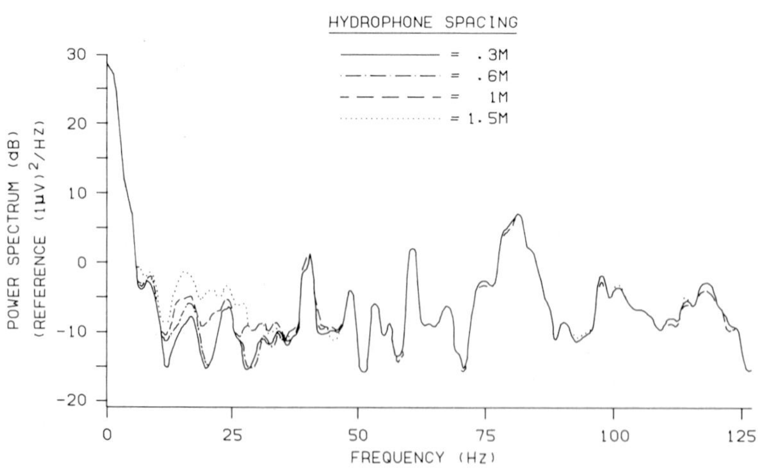

FIGURE 12. Power spectra of 5 m length array successively measured with different hydrophone spacing. Noise recordings.

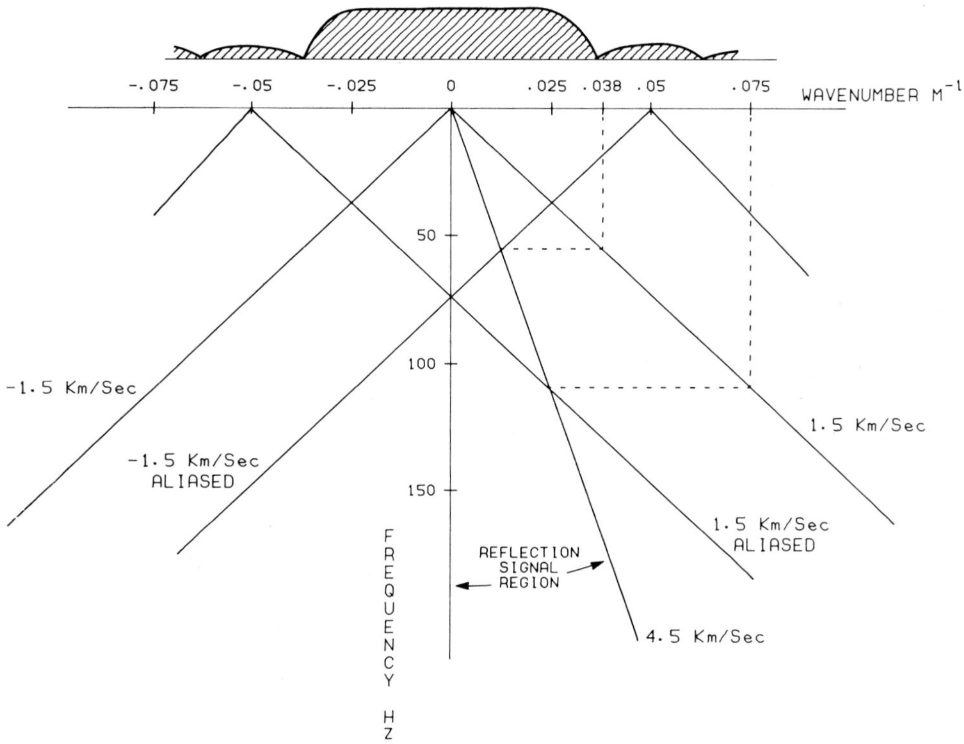

FIGURE 13. Sketch of frequency-wavenumber diagram showing effect of aliasing of the seismic reflection signal and noise cones.

axis (infinite velocity line) and the 4.5 km/sec line. The other wave, propagating at water velocity, is considered shot-generated noise. This noise energy resides in the cone bounded by the positive and negative water velocity lines. The regions of strongest noise tightly surrounded the 1.5 km/sec water velocity lines.

Aliasing of the propagating noise is also shown in Figure 13. In this example, we consider an array midpoint spacing (interval D in Figure 5) of 20 m. The spectra therefore repeat at wavenumber intervals of 0.05/m. For purposes of clarity, the aliased reflection signal is not illustrated. Also not seen is the effect of hydrophone spacing (Δx in Figure 5) because this causes spectrum repetition at far wider intervals than shown in Figure 13.

At the top of the diagram is shown an array wavenumber response function. As discussed previously, the array wavenumber response function attenuates the reflection signals and noise prior to aliasing. The traditional anti-alias design calls for sharp attenuation at and beyond the Nyquist point, in this example 0.025/m. Our problem does not need such a severe requirement since there is no serious degradation of SNR until the aliased noise crosses the desired signal cone. In our example, this occurs at a wavenumber of 0.012/m or when the aliased spectra is 0.038/ away from twice Nyquist. This means that sharp attenuation of the spectra by the array need not begin until 0.038/m. A further requirement on the array design is that the desired signal is not significantly attenuated. This means that wavenumbers from 0 to 0.018/m be unattenuated in order to obtain frequencies up to 100 Hz.

These considerations have led to the passband-stopband philosophy of design specification. In this approach, one specifies a passband up to some wavenumber, then a stopband in another wavenumber interval. In this example, the crossing of the desired signal cone by the noise requires a critical stopband from 0.038 to 0.075/m. The passband extends from 0 to 0.018/m Between 0.025 and 0.038/m is a transition region which is neither passband nor stopband. The array length and the choice of weighting coefficients are the means for accomplishing the design specifications.

In summary, the design specifications are set by the signal and propagating noise characteristics as well as the array midpoint spacing (interval D). The fineness of the sampling interval D is usually limited by hardware constraints such as the number of shipboard recording channels or the number of wire pairs in the cable.

Stopband Level

The next section will describe methods for the weighted summation of hydrophone signals in order to obtain desirable array wavenumber selectivity functions. One of the factors to be considered there is the required stopband level. The stopband level cannot be zero over an arbitrary wavenumber interval because the array length is finite. For a given array length, tapering the weights can yield better stopband rejection at the expense of a broader passband and transition region. These tradeoffs must be balanced properly in any specific design. In order to scope the problem, some finite stopband level must be specified that is compatible with the other design specifications. It will now be shown that the accuracy of the hydrophone pressure response is a principal factor in determining the attainable stopband level. It is unwise to choose array weights that yield idealized stopband levels far below the level that will actually be obtained because of inaccuracies in the hydrophone response.

Consider an array output signal y that is the weighted summation of hydrophone signals s().,

$$y = \sum_{n=1}^{N} s(n) \cdot w_n \cdot (1 + \epsilon_n) \tag{9}$$

$$s(n) = \sigma_s \cdot e^{i \cdot 2\pi \cdot k \cdot n \cdot \Delta x} \cdot e^{-i \cdot 2\pi \cdot f \cdot t} \tag{10}$$

The signal power is σ_s^2. In this analysis, the hydrophone pressure sensitivity is nominally unity with an error term ε_n. The $\{\varepsilon_n\}$ are zero mean and uncorrelated between hydrophones.

$$\overline{\epsilon_n} = 0 \tag{11}$$

$$\overline{\epsilon_n \, \epsilon_m} = \begin{cases} \sigma_\epsilon^2 & n = m \\ 0 & n \neq m \end{cases} \tag{12}$$

The output can be considered the sum of two arrays. The first has the ideal weights $\{w_n\}$ while the second has weights $\{w_n \cdot \varepsilon_n\}$. The contribution of the "second array" must be considered in the statistical sense. The mean square value of this contribution is independent of the wave number of the incident wave.

$$\overline{|y_\epsilon|^2} = \sum_{n=1}^{N} \sum_{m=1}^{N} s(n) \cdot s^*(m) \cdot w_n \cdot w_m \cdot \overline{\epsilon_n \cdot \epsilon_m} \tag{13}$$

$$= \sigma_s^2 \cdot \sigma_\epsilon^2 \cdot \sum_{n=1}^{N} w_n^2 \tag{14}$$

where * denotes the complex conjugate operator.

Usually hydrophone inaccuracy is small, $\sigma_\epsilon^2 \ll 1$, and the effect is insignificant in the passband where the nominal response is strong. However, in the stopband the same error causes a large percentage difference. The ratio of the nominal peak level to error stopband level is the significant factor. For weighting coefficients that are real, positive, definite, and symmetric about the midpoint, the peak passband level is at k = 0.

$$|y_p|^2 = \sigma_s^2 \left(\sum_{n=1}^{N} w_n \right)^2 \qquad (15)$$

and the peak to stopband ratio follows by taking the ratio of Equation 15 to Equation 14.

$$PSR = \frac{N \cdot F_w}{\sigma_\epsilon^2} \qquad (16)$$

$$F_w = \frac{\left(\sum_{n=1}^{N} w_n \right)^2}{N \sum_{n=1}^{N} w_n^2} \qquad (17)$$

We need to know the function F_w in order to calculate the PSR. The function F_w is less than or equal to 1. Unity occurs for the case of the uniformly weighted array, $w_n = 1$, all n. For moderately tapered arrays (from Reference 9) $F_w = 0.5$. Figure 14 shows the PSR as a function of hydrophone pressure sensitivity inaccuracy for $F_w = 0.5$ and arrays of 20, 80 and 160 hydrophone elements. Uniformly weighted arrays would be 3 dB higher. For a typical seismic cable case, 30 hydrophones would be used with element accuracy of 10%. Therefore, the stopband level due to errors in weighting would approximately be −30 dB.

In summary, finite tolerances of hydrophone pressure sensitivity cause actual array stopband levels to be different than design values. The design specification for stopband level should not be much lower than the level caused by the hydrophone sensitivity errors.

WEIGHTING NETWORK IMPLEMENTATION

Hydrophone Networks

Hydrophones made of piezoceramic have the electrical equivalent circuit shown in Figure 15. This equivalent circuit is valid in the frequency range used for seismic exploration, 5 to 100 Hz. The most popular array is the uniformly weighted one formed by wiring all hydrophones in parallel. This array and the equivalent circuit is shown in Figure 16. Figure 17 shows the wavenumber selectivity function calculated by digital computation. Note the −13 dB level in the stopband specified in the example of the previous section. Furthermore, there is a 7 dB variation across the passband (0 to 0.018/m).

A simple way to obtain tapered arrays is to connect the individual hydrophones together in various serial and parallel arrangements. Figure 18 shows the basic network

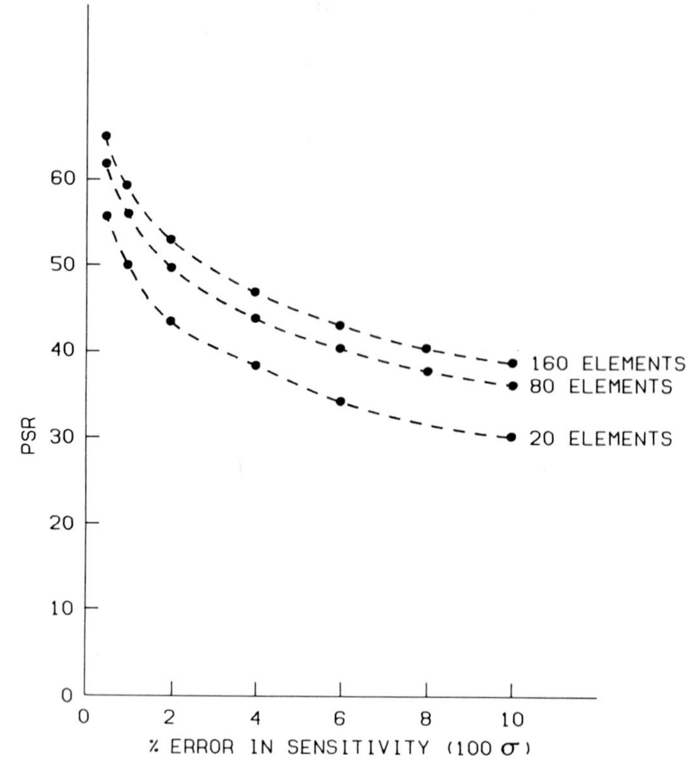

FIGURE 14. Peak signal to weighting error stopband level ratio as a function of hydrophone pressure sensitivity variation.

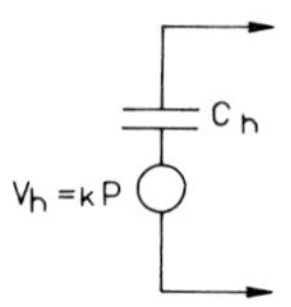

HYDROPHONE TYPE	PRESSURE SENSITIVITY κ (VOLTS/BAR)	CAPACITANCE C (nf)
GEOSPACE MP 8F	13	13
GEOSPACE MP 18	14	17
HOUSTON PRODUCTS WM1-018	28	17

FIGURE 15. Hydrophone electrical equivalent circuit.

transformations that are useful in analyzing circuits of this type. Figure 19 is an example of a reasonably complicated network of six elements to provide a 2, 4, 5, 5, 4, 2 relative weighting. One way to use this six-element scheme in an actual array of 30 hydrophones is shown in Figure 20. Here five individual hydrophones are wired in parallel. The resulting subarray is uniformly weighted and has capacitance 5 C_h. This subarray can be used with identical adjacent subarrays as the elements in the network

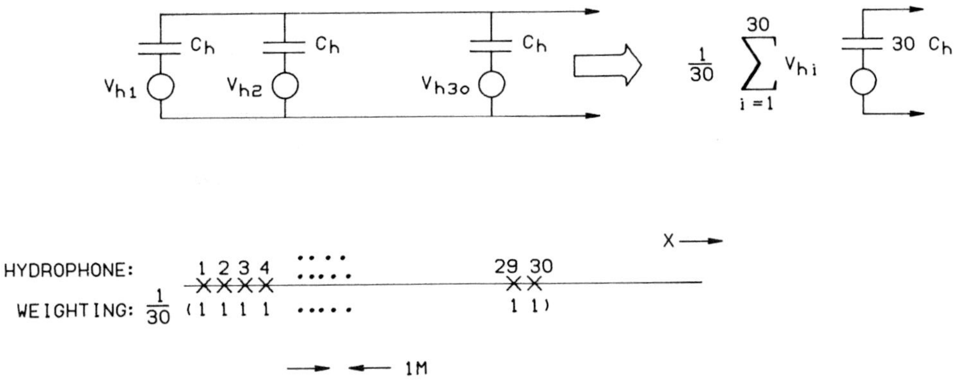

FIGURE 16. Uniformly weighted array of 30 hydrophones.

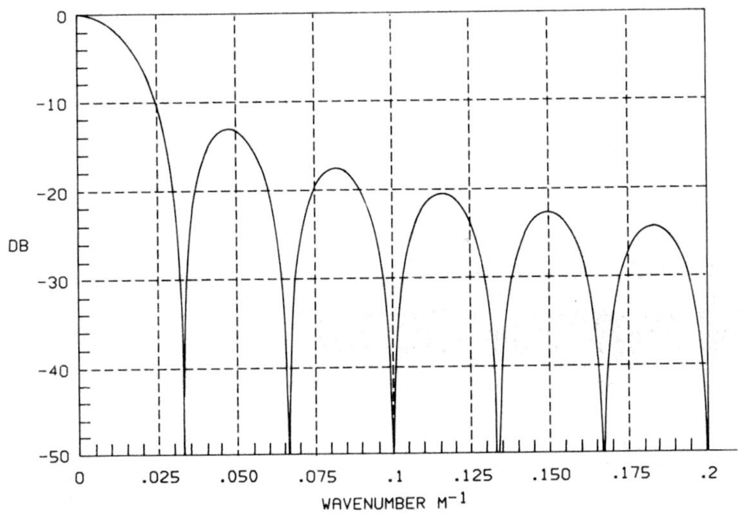

FIGURE 17. Wavenumber selectivity function for a uniformly weighted array.

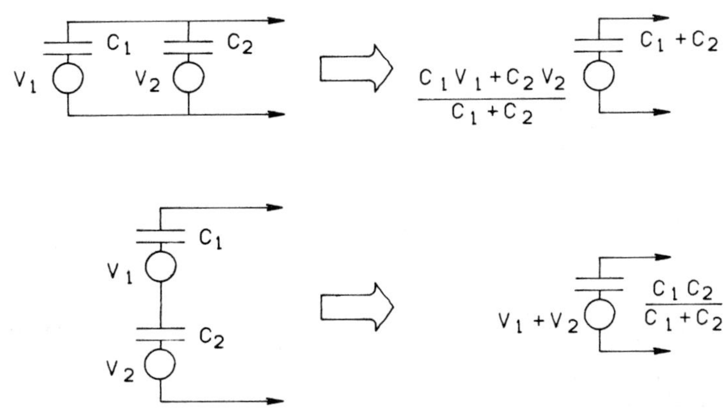

FIGURE 18. Basic network transformations useful in analyzing hydrophone arrays.

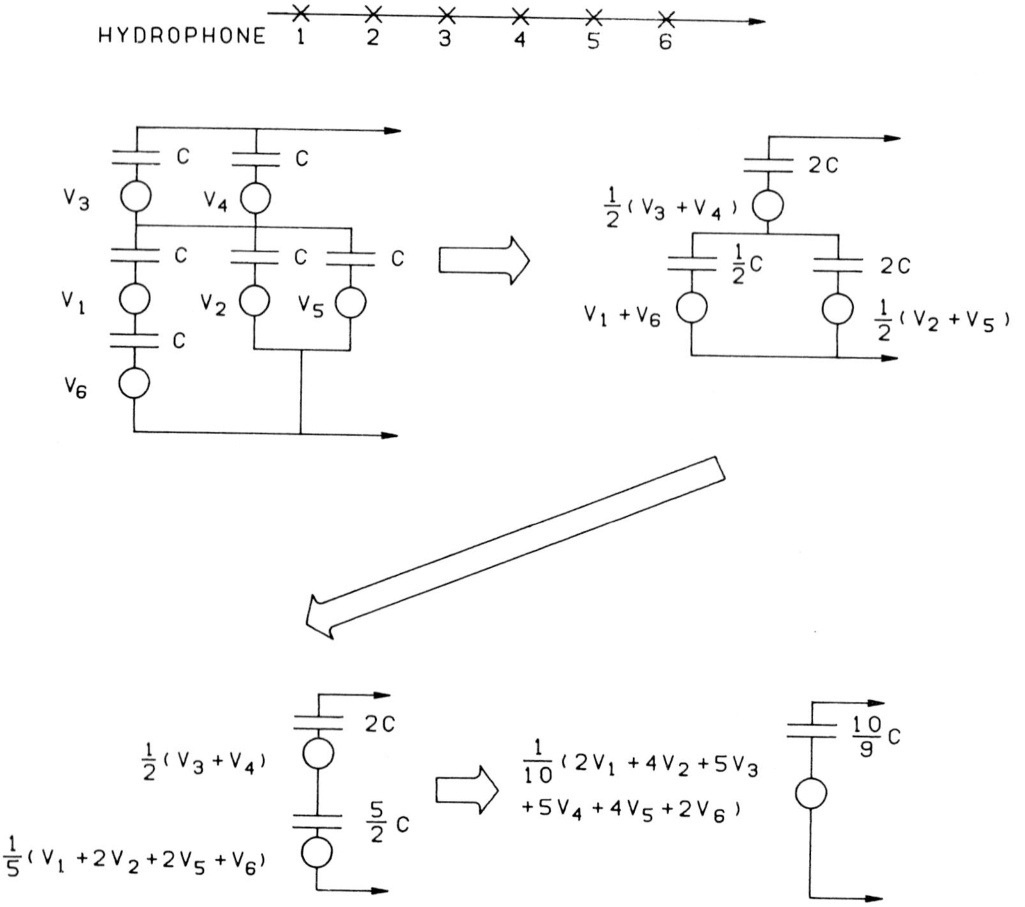

FIGURE 19. An interconnection of elements to obtain a 2, 4, 5, 5, 4, 2 relative weighting.

of Figure 19. For example, let $C = 5\ C_h$, $V_l = 1/5(V_{hl} + V_{h2} + V_{h3} + V_{h4} + V_{h5})$, $V_2 = 1/5\ (V_{h6} + V_{h7} + V_{h8} + V_{h9} + V_{h10})$ etc. The resulting relative weighting over the 30 hydrophones is shown in the lower part of Figure 20. The array capacitance is 5.55 C_h. The wavenumber selectivity function for this case is illustrated in Figure 21. Note that the stopband (0.038 to 0.075/m) level is now 22 dB below the peak. The passband variation is less than 4 dB. More intricate interconnections can provide improved sensitivity functions.

As will be shown below, there are other ways to obtain the same wavenumber selectivity function. One figure of merit that distinguishes the various approaches is the charge sensitivity of the array. This quantity is simply the multiplication of broadside signal pressure sensitivity and array capacitance. The higher charge sensitivities yield greater acoustic signal to electronic noise ratios. The max charge developed by the array in this example is 12.2 k.P.C_h.

Capacitor Coupling Networks

An alternate way to obtain tapered weighting is to interconnect the hydrophones through an external network of capacitors. Figure 22 shows an arrangement in which the *C* represent hydrophone capacitance while the C_t represent external capacitors. Each hydrophone element can actually be a parallel combination of individual hydrophones such as shown previously in Figure 20.

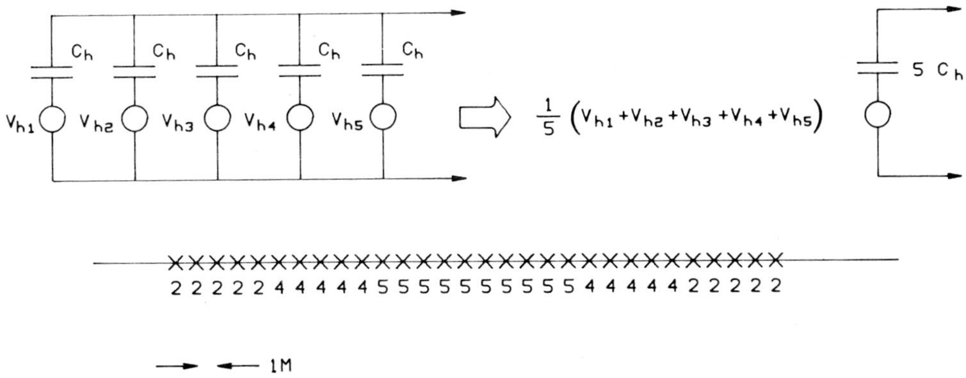

FIGURE 20. (Upper) Five hydrophones wired in parallel to form an element for the network on Figure 19. (Lower) The resulting 30-hydrophone array.

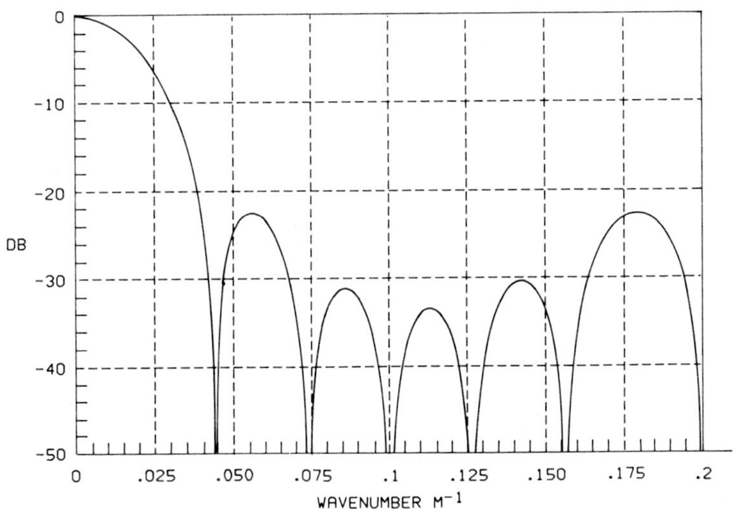

FIGURE 21. Wavenumber selectivity function for a 2, 4, 5, 5, 4, 2 relative weighting spanning 30 m.

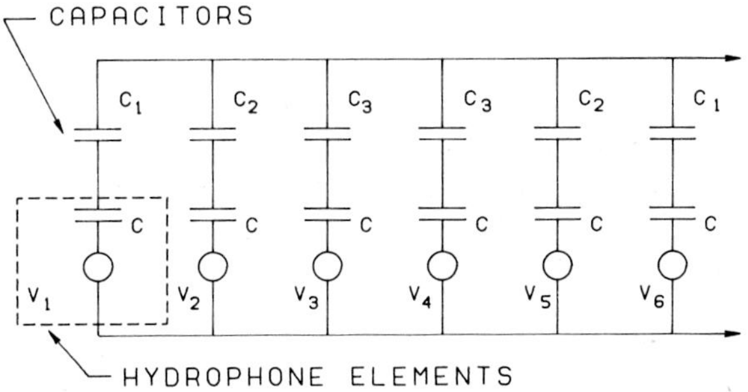

FIGURE 22. Array elements connected through external capacitors.

$$\frac{1}{2C_1+2C_2+2C_3}\begin{pmatrix}C_1V_1+C_2V_2+C_3V_3\\ +C_3V_4+C_2V_5+C_1V_6\end{pmatrix}$$

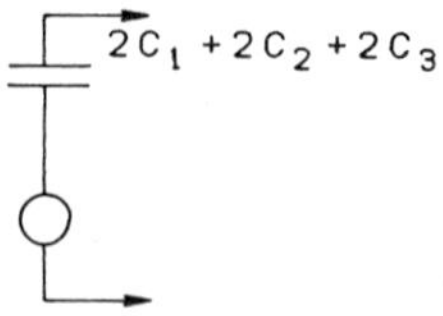

FOR 2.4.5.5.4.2 WEIGHTING

$C_1 = .4C_3$

$C_2 = .8C_3$

$C_3 = .01C$

$C = 5C_h$

FIGURE 23. An electrical equivalent of Figure 22 with values that allow a 2, 4, 5, 5, 4, 2, relative weighting.

Consider that the largest capacitor of the C_i is much smaller than the hydrophone subarray capacitance *C*. The hydrophone capacitance can therefore be ignored for purposes of network analysis since it is in series with a much smaller capacitance. As an example, in Figure 23 the 2, 4, 5, 5, 4, 2, relative weighting is implemented using subarrays of five parallel hydrophones, $C = 5\ C_h$, and a value for the largest coupling capacitor C_3 that is 0.01 C.

The resulting array has the wavenumber selectivity pattern shown previously in Figure 21. The array capacitance is 0.22 C_h and the max charge is 0.22 k.P.C_h. Note that both capacitance and max charge are lower than obtained with the hydrophone network approach. The salient feature of the capacitor network approach is the greater flexibility in choice of weights.

Iterative Networks

In our design example, the array length (30 m) is greater than the spacing (20 m) between adjacent array midpoints (spacing D in Figure 5). This is always the case when one uses the anti-alias design philosophy. The reciprocal of array length is roughly the low pass cutoff point of the wave number selectivity function. The reciprocal of the length D is twice the Nyquist wavenumber. If the array must attenuate energy above Nyquist, its length should be at least 2D. For the reasons stated in Section IV, our requirement is not quite this severe. Nonetheless, considerable overlap of adjacent arrays is necessary to obtain reasonable wavenumber selectivity functions. Both the coupling capacitor and the hydrophone network approaches must duplicate hydrophones in the same length interval to achieve overlap. Duplication can be avoided if the same hydrophone can be shared between two arrays. This can be done with continuous, iterative, connections of hydrophones.

Figure 24 is an example of a continuous network interconnection of all the hydrophone elements in the cable. The amplifier input terminals are attached at the points (A+,A−), (B+,B−), etc. Each amplifier represents an array output point. The distance between amplifier input terminal locations is D, the array midpoint spacing.

Both transformer coupled and operational amplifiers are used in the seismic industry with the trend towards the latter type. The operational amplifier in the so-called

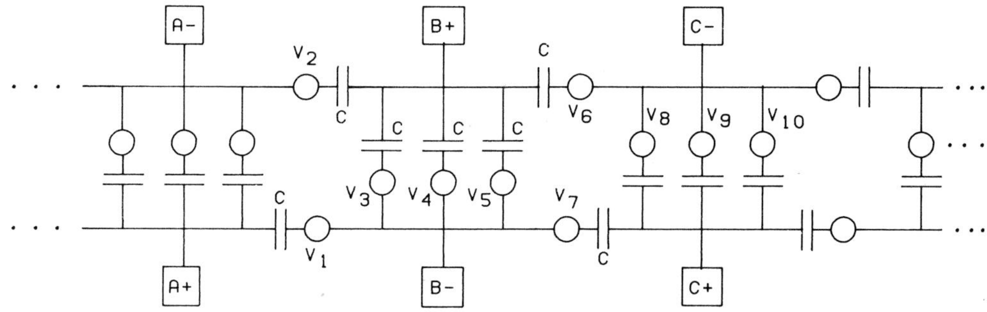

FIGURE 24. An iterative network composed of the interconnections of many elements.

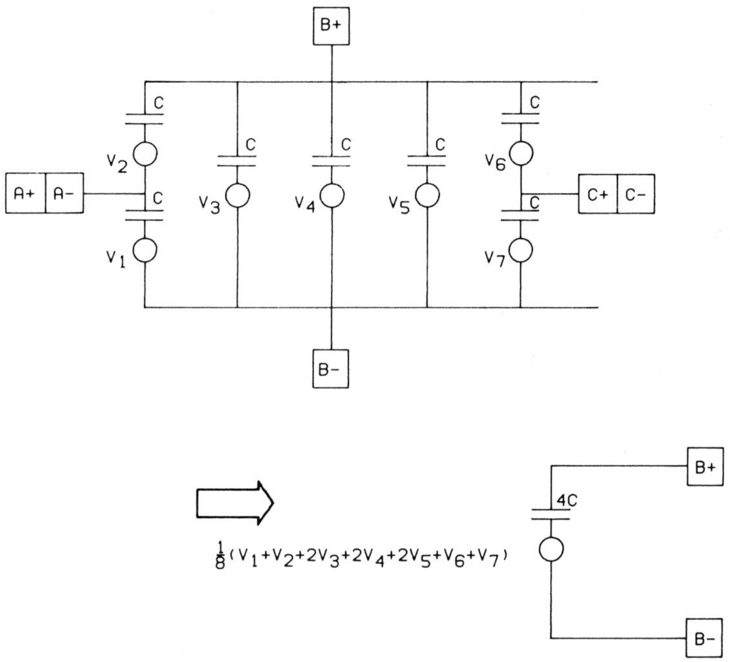

FIGURE 25. Equivalent circuits to the network of Figure 24 valid for analyzing the "B" array when low input impedance amplifiers are used. A 1, 1, 2, 2, 2, 1, 1 relative weighting is obtained.

"charge" amplifier[10] configuration has a very low input impedance. As such, when considering the "B" array we can consider that the other amplifier input terminals present short circuits to the network. The simplified circuit of Figure 25 then results. The equivalent circuit shown in the lower part of Figure 25 illustrates the 1, 1, 2, 2, 2, 1, 1 relative weighting that is obtained. We therefore have seven effective elements in the array.

In order to compare this scheme to the previous ones, we will wire four adjacent hydrophones in parallel and use this as a basic element in the iterative network. The resulting 28-element array spans 29.4 m if hydrophones are spaced at 1.05 m intervals. The spatial weighting is shown in Figure 26. Note that eight hydrophones are shared and the spacing D equals 21 m. The wavenumber selectivity function is shown in Figure

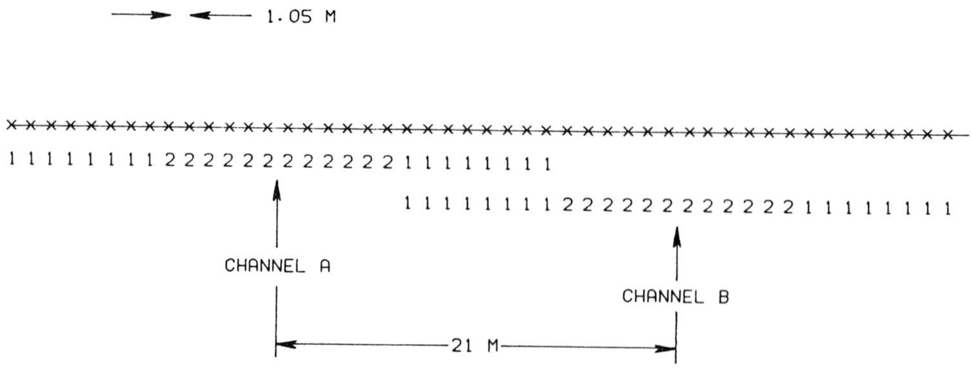

FIGURE 26. An array of 28 hydrophones formed by using the network of Figure 24 where each element is the parallel combination of four hydrophones.

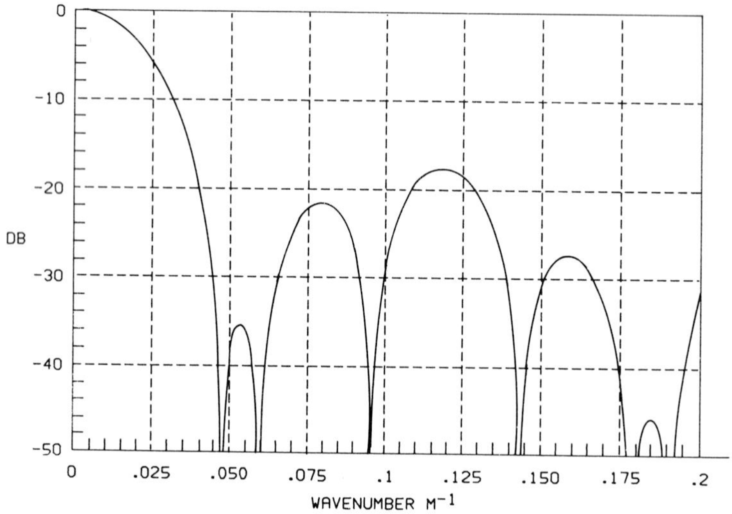

FIGURE 27. Wavenumber selectivity function for a 1, 1, 2, 2, 2, 1, 1 relative weighting spanning 29.4 m.

27. It has passband, transition region, and stopband characteristic similar to the 2, 4, 5, 5, 4, 2, array shown previously in Figure 21. The max charge is 20 $k \cdot P \cdot C_h$ and the array capacitance is 16 C_h.

Iterative networks using external capacitors can also be used with similar results. Lower array capacitance and lower max charge will result as compared to the hydrophone network case illustrated above.

SUMMARY

We have described the design of seismic cable hydrophone arrays as a problem in linear filtering theory. The array is viewed as an anti-alias wavenumber filter. A design example was given and several implementations were described.

APPENDIX

A plane wave propagating with velocity V in the positive x direction can be defined by the following convolution:

$$s(t - x/V) = \int_{-\infty}^{+\infty} p(t - \tau)\, \delta(\tau - x/V)\, d\tau \qquad \text{(A-1)}$$

where

δ() is the dirac delta function, and
p() is the time signature of the wave.

The two-dimensional transform of s() into the frequency-wavenumber domain is particularly easy for a plane wave.

$$S(f,k) = \iint_{-\infty}^{+\infty} s(t - x/V)\, e^{-i2\pi ft}\, e^{-i2\pi kx}\, dt\, dx \qquad \text{(A-2)}$$

$$= \iiint_{-\infty}^{+\infty} p(t - \tau)\, \delta(\tau - x/V)\, e^{-i2\pi ft}\, e^{-i2\pi kx} d\tau\, dt\, dx \qquad \text{(A-3)}$$

$$= \iint_{-\infty}^{+\infty} V\, p(t - \tau)\, e^{-i2\pi ft}\, e^{-i2\pi kV\tau}\, d\tau\, dt \qquad \text{(A-4)}$$

$$= \int_{-\infty}^{+\infty} V\, P(f)\, e^{-i2\pi f\tau}\, e^{-i2\pi kV\tau}\, d\tau \qquad \text{(A-5)}$$

$$= VP(f)\delta(f + kV) \qquad \text{(A-6)}$$

REFERENCES

1. **Grant, F. S. and West, G. F.,** *Interpretation Theory in Applied Geophysics,* McGraw Hill, New York, 1965.
2. **Lofthouse, J. H. and Bennett, G. T.,** Extended arrays for marine seismic acquisition, *Geophysics,* 43(1), February, 1978.
3. **Arnold, M. E.,** Effect of hydrophone arrays on offshore Texas seismic signals, *Geophysics,* 43(6), October, 1978.
4. **Willmarth, W. W. and Yang, C. S.,** Wall pressure fluctuations beneath turbulent boundary layers on a flat plate and a cylinder, *J. Fluid Mech.,* 41, 1970.
5. **Bakewell, H. P.,** Turbulent wall pressure fluctuations on a body of revolution, *J. Acoust. Soc. Am.,* 43, 1968.
6. **Bedenbender, J. W., Johnston R. C., and Neitzel, E. B.,** Electroacoustic characteristics of marine seismic streamers, *Geophysics,* 35(6), December, 1970.
7. **Schoenberger, M. and Mifsud, J. F.,** Hydrophone streamer noise, *Geophysics,* 39(6), December, 1974.
8. **Weichart, H. F.,** Acoustic waves along oil-filled streamer cables, *Geophys. Prospecting,* 21, 1973.
9. **Harris, F. J.,** On the use of windows for harmonic analysis with the discrete Fourier transform, *IEEE Proc.,* 66(1), January, 1978.
10. **Mastin, M.,** U.S. Patent 3,939,468.

THE OCEAN SUBBOTTOM SEISMOMETER

F. K. Duennebier and G. Blackinton

INTRODUCTION

Since the beginning of the Deep Sea Drilling Project (DSDP) more than ten years ago, there has been interest in emplacing instruments in the holes and leaving them to monitor geophysical parameters, especially seismic, after the drilling ship has left the site. Until 1979, however, no serious attempt was made to achieve this goal because of conceptual problems in the emplacement and recovery of data. Briefly, the main problem is how to get the data out of the hole after the drilling ship leaves the site. During 1977, a series of meetings were held at DSDP to attack this problem and develop a system to be emplaced in a deep sea drill hole. Although the system conceived during these meetings was too ambitious for funding agencies to sponsor, many of the ideas bore fruit in the system subsequently developed at the Hawaii Institute of Geophysics.

The drilling schedule at the end of 1977 put the GLOMAR CHALLENGER in the Gulf of California in early 1979. Since a large seismic experiment (ROSE) was to take place then in the same area, an effort was made to have an instrument ready for emplacement at that time. National Science Foundation funding was obtained in April 1978 and the first system installed in February 1979. The system is explained in detail in this paper along with the method of emplacement and data recovery. We hope that the simplicity and versatility of the system will be apparent and that others will want to try their own sensors or recording devices in similar packages.

SYSTEM DESIGN

The ocean subbottom seismometer (OSS) system is divided into three main components: the borehole package, the recording package, and the recovery system (Figure 1). The borehole package obtains all of its power from the recording package and thus can remain in the hole indefinitely (Figure 2). The recording package supplies power to the borehole package and records data. The borehole package is connected to the recording package through 9 km of 7-conductor armored cable. The recording package is attached to the free end of the cable and can be pulled to the surface with the recovery rope to resupply the batteries and tapes. While on the surface, the end of the cable can be connected to a test box for real-time monitoring of the borehole package. All data are telemetered from the borehole package in digital form and seismic data are converted to analog form for recording. The recovery system includes the rope used in recovery of the recording package and the anchor-release system that holds it on the ocean floor.

Borehole Package

The borehole package contains all sensors, signal electronics, multiplexer, A to D converter, telemetry, and power conditioning electronics. The pressure case, which has an inner diameter of 65 mm, is housed in a modified density logging tool similar to those routinely used on the GLOMAR CHALLENGER. The tool was specially manufactured for this experiment by Gearhart-Owen, Inc. The maximum tool diameter (3 7/8 in.) is constrained by the inner diameter of the drill pipe (4 1/8 in.). A command can be sent down the cable to enable a spring to push the borehole package out of the

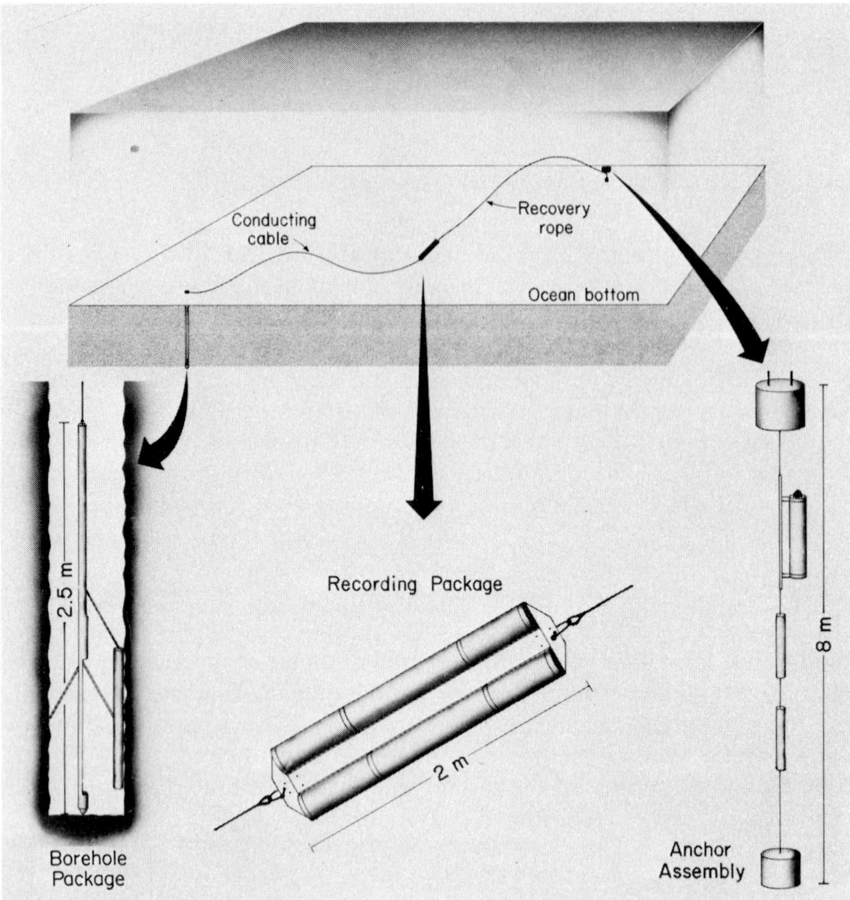

FIGURE 1. The ocean subbottom seismometer system. The borehole package is emplaced in a deep sea drill hole with power transferred to it, and data sent from it, through a conducting cable connected to a recording package. The recording package is attached to a floating rope that is in turn connected to a float-anchor assembly for recovery.

body of the tool and against the side of the hole (Figure 1), insuring good seismic coupling. The electronics in the borehole package are divided into modular units for ease in interchanging sections and for electrical isolation (Figure 3). Although many sensor configurations (up to 16 sensors) are possible, the present configuration has three short-period geophones, two bubble tilt meters, and temperature multiplexed into 16 data channels. The processing of the multiplexed data is discussed in the section on signal handling.

Three 4.5-Hz Geospace model HS-1 high temperature geophones are mounted orthogonally in a single block of aluminum (Figure 4) and secured with set screws. The corners of the two horizontal geophone cases were machined to fit into the cylindrical block. The tilt meters (Fredericks Model 7660) are aligned along the same axes as the horizontal geophones. They are functional over ± 0.2 rad with sensitivities after electronics of 0.3 V/mrad at the absolute tilt output and 30 V/mrad for signals between 1 and 50 sec periods at the differential tilt output. Although these sensitivities are not high enough to resolve tides or reasonable tectonic tilts, they are valuable in deciding the orientation and sensitivities of the geophones and the tilt of the package in the hole. Higher sensitivities are possible with lower-range bubble tilt meters[1] at the cost

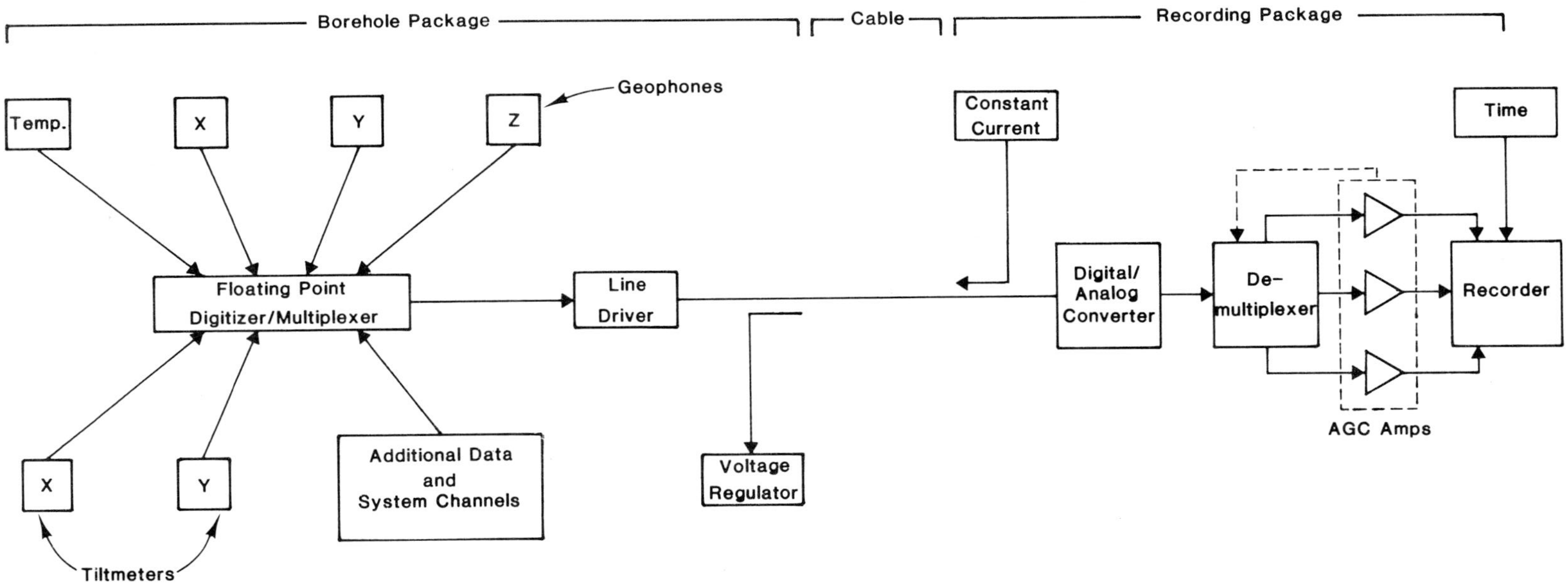

FIGURE 2. Block diagram of ocean subbottom seismometer system.

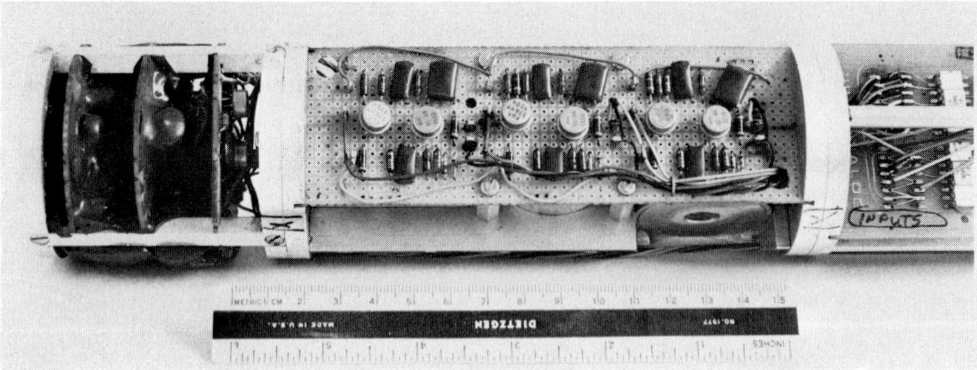

A

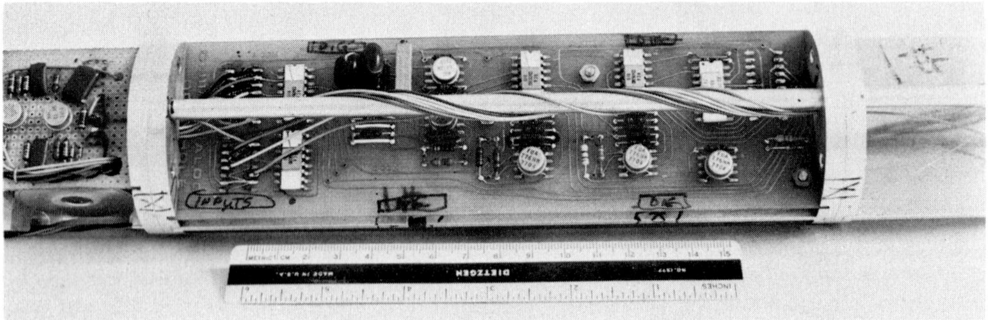

B

FIGURE 3. A and B. Borehole package electronics showing modular design for ease of replacement and modification.

of requiring high-precision mechanical leveling of the bubble. This was not practical in this first instrument although it is certainly possible in future designs. The thermal sensors are Analog Devices AD590, with a sensitivity of 1 m V/°C within the range from −55 to +150°C.

The resolution of the digitized signals from each of the sensors is between 0.8% and 0.2% of the absolute value of the sensor output. This is discussed in more detail in the section on the floating point analog to digital converter.

The package can easily accept other sensors that do not require significant power (less than 30 mW) and that can fit inside the pressure case. The sensors chosen for the package were standard off-the-shelf units and may not be the most sensitive or reliable sensors available. Now that the system is developed, more effort will be put into sensor optimization.

The 9 km of cable that connects the borehole package to the recording package is the single most expensive item in the system. Seven-conductor double armored Vector 7-J46RB cable was chosen for two reasons: DSDP uses this cable for the logging operations aboard the GLOMAR CHALLENGER and has the equipment and experience necessary to handle it and its connectors and splices, and the cable has the strength necessary to survive the wire stripping operation. Standard connectors manufactured by Gearhart-Owen used to connect the cable to the packages were filled and booted with a Devcon® potting compound rather than grease to minimize long-term leakage.

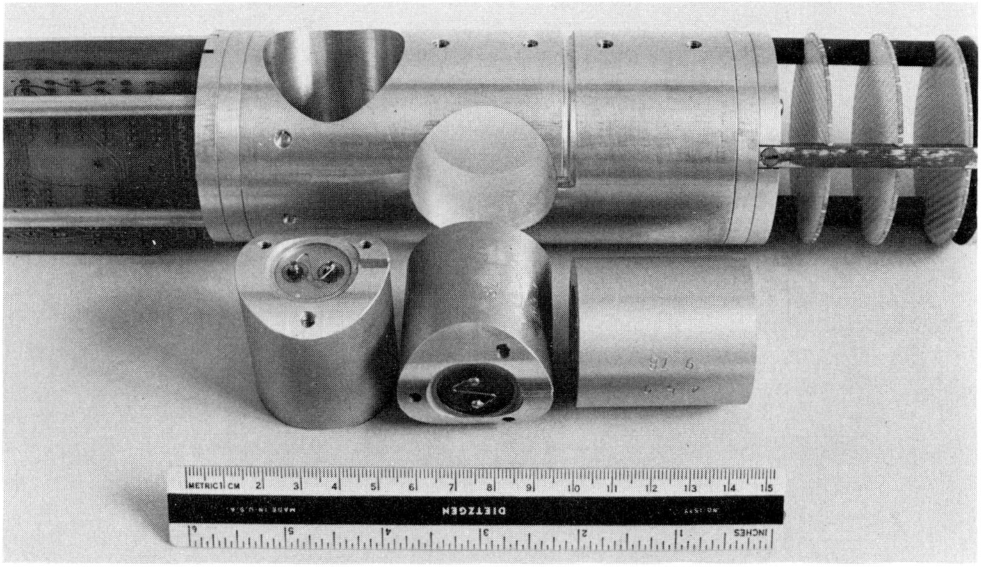

FIGURE 4. Geophones and geophone mounting block in the borehole package. No provision is made for leveling since the geophones will operate up to 5° from vertical and the package tilt can be measured from the tilt meters.

Data Recording/Power Units

Two units are used at the upper end of the cable: a test box to monitor data in real time and the recording package to record data unattended for long periods of time. Other data retrieval methods, such as buoys, digital packages, and satellite links are possible but have not yet been developed.

When the end of the wire is at the ocean surface all data are available in digital form. At this point, the user is free to decimate, compress, or otherwise choose the data he wishes to monitor while recording all data on tape for later analysis. The test box acts as a test set for system check out, a D-to-A unit for real-time monitoring, as a playback unit for playing back tapes of data that were recorded in real time, and as an interface to the computer for data processing. It also is capable of sending the command to extend and retract the borehole package from the tool. The data handling functions of the test box are discussed in the section on data processing.

The recording package pressure case consists of two aluminum tubes approximately 2 m in length (Figure 5). One side contains batteries to power the tool and the other contains electronics, tape recorders, and more batteries. Five cassette recorders (Figure 6) can be preset to start recording at individual times and speeds. Each is capable of recording four tracks of signals up to 20 Hz for 15 days or up to 100 Hz for 3 days. The batteries (standard D-cells) and electronics are amounted in aluminum frames for easy access and removal. Prior to launch a vacuum is pulled in both tubes. The internal pressure of each tube can be monitored through a bulkhead connector to detect pressure leaks.

Recovery System

The recovery rope attaches to the end of wire rope about 5 m above the recording package. This rope, 1¼ in. plaited polypropylene, has a length greater than 1.5 times the depth of water. Its buoyancy is such that the recording package never actually touches the ocean floor. Since the breaking strength of the rope is about 10 tons, the cable (breaking strength 8 tons) should break first if it snags on the bottom, insuring recovery of the recording package. A swivel at each end of the rope limits snarls.

FIGURE 5. Recording package. Two aluminum tubes about 2 m in length contain batteries, electronics, and recorders for recording continuous data from the borehole for up to 75 days unattended.

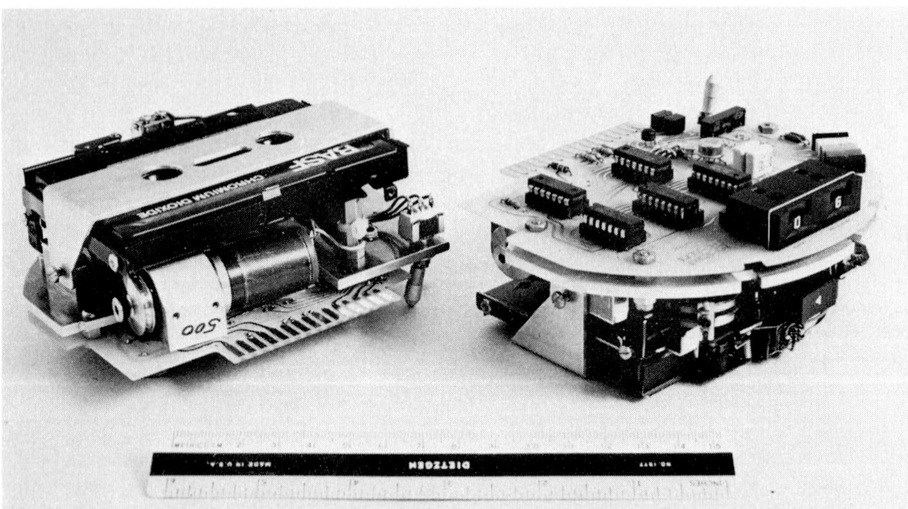

FIGURE 6. Cassette tape recorders. Analog geophone data are recorded on three tracks and time and digital data on the fourth. Each recorder is preset to turn on at a particular time and record continuously for up to 16 days.

The anchor assembly attaches at the free end of the rope (Figure 7). It holds the rope fixed on bottom with a lead anchor weight and a small Danforth® anchor. The rope can be connected to the anchor through three release devices, an acoustic transponding release and, if the recovery time is known, two timed releases, any one of which can release the anchor and allow the rope to float to the surface. The decision

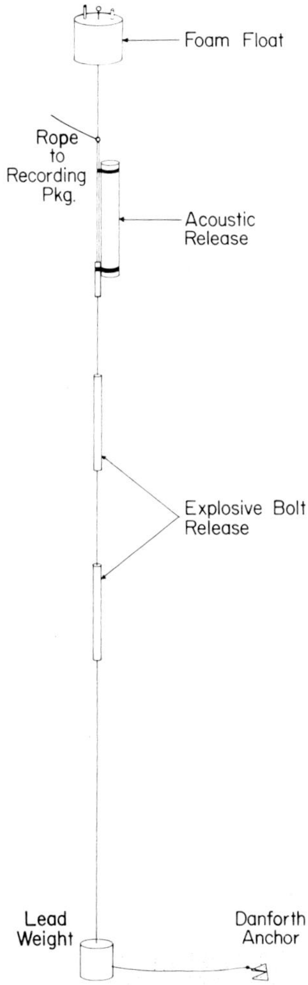

FIGURE 7. Float-anchor assembly. The anchor is released from the float and recovery rope either at a specific time (explosive bolt release) or on command (acoustic release), allowing the float and recovery rope to return to the surface.

to have no equipment on the ocean surface was made because of the risk to (and from) passing ships and susceptibility to damage in poor weather. It was judged considerably easier to design a reliable anchor system than a surface buoy that would survive for the required length of time.

The system was designed to be emplaced in the deep ocean by the GLOMAR CHALLENGER and recovered by an oceanographic research vessel. However, there is nothing to prevent its emplacement in a hole on land, in an offshore exploration well, or possibly in an existing hole in the deep ocean from a research ship.

SIGNAL HANDLING

Multiplexing and Digitizing

The 16 channels of data are sequentially sampled according to the hierarchical

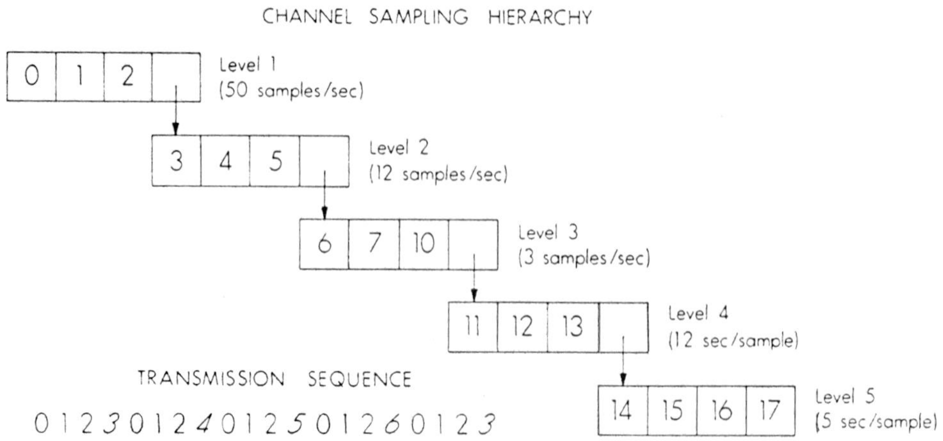

FIGURE 8. Channel sampling hierarchy: 16 channels are sampled at five different rates, from 50 samples/sec to 5 sec/sample.

scheme of Figure 8. The scheme is a modified four-channel scanning sequence with five steps. The fourth channel of each step is one channel of a lower step. The highest step channels are thus sampled 50 or 100 times per sec and those of the lowest step once every 5 or 2.5 sec. The signals that were connected to each channel in the first two instruments are described in Table 1. All signals are conditioned to vary between +2.5 and −2.5 V for digitizing.

As the signal on each channel is connected in turn to the digitizer it is converted into a 12-bit floating point digital number. The number has a 3-bit, base 4 exponent and a 9-bit mantissa (8 bits plus sign).[2] The ranges of the exponent and mantissa allow signals from ±0.6 μV to ±2.5V to be encoded — a dynamic range of 138 dB. This is equivalent to the dynamic range of a 23-bit linear converter. The 9-bit mantissa, however, limits the resolution of each digital representation to 0.2% at best. The base 4 exponent means that the resolution can be as poor as 0.8% of the value of the sample.

For most varying signals with zero average, a resolution better than 1% is sufficient. Few direct tape recordings of the signals will provide much better fidelity. A small signal superimposed on a large steady-state signal, however, may be severely distorted. The 1% resolution of the large signal may represent 10% or even 100% of the small signal. The ratio of signal to digitizing noise throughout the converted dynamic range is summarized in Figure 9.

Telemetry

The digital sample is sent serially to the recording package in the form of an 18-bit delay modulated word.[3] The 18 bit are 1 start bit, 4-bit channel number, 12-bit sample, and 1 stop bit (Figure 10). As each of the high-frequency channels is sampled 50 times per sec, 200 samples or 3200 bit are sent each second. In the recording package, the start and stop bits are sensed and each sample is converted from serial to parallel. A floating point digital-to-analog (FPD/A) converter reproduces the voltage sampled 10 msec earlier in the borehole package. Only three selected channels are demultiplexed and recorded in analog form on tape.

Recording

In the first instruments only the three geophone signals were analogized by the FPD/A. The FPD/A, however, does not reproduce exactly the sampled voltage. Rather,

Table 1
OSS CHANNEL ASSIGNMENTS

Channel (Octal)	Signal
0	Vertical geophone
1	Horizontal geophone (parallel)
2	Horizontal geophone (perpendicular)
3	Differential tilt (parallel)
4	Differential tilt (perpendicular)
5	Differential temperature
6	Absolute tilt (parallel)
7	Absolute tilt (perpendicular)
10	Triangle (system)
11	Absolute temperature
12	Excess current monitor (system)
13	Open (system)
14	+5V Monitor (system)
15	−5V Monitor (system)
16	Common (system)
17	Open (system)

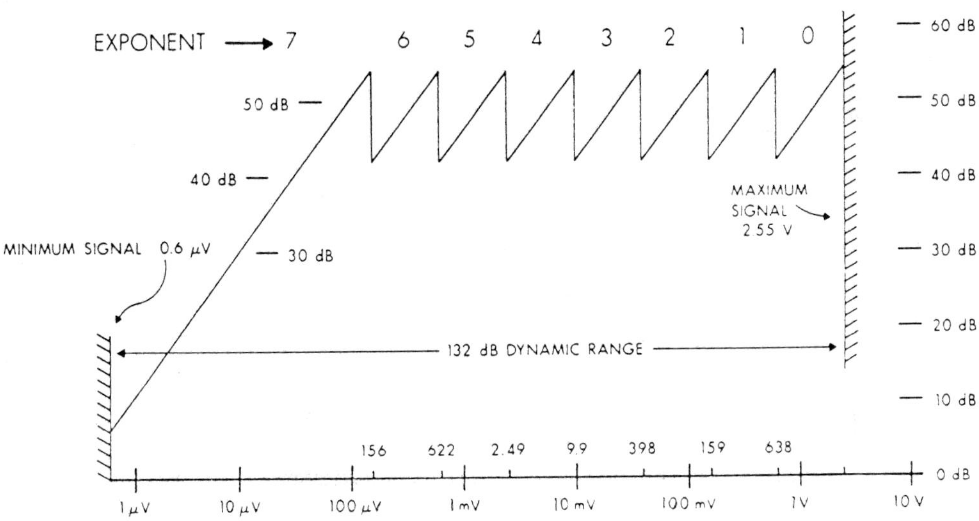

FIGURE 9. Signal amplitude vs. digital signal-to-noise ratio. A 12-bit floating point word with a 3-bit (base 4) exponent and a 9-bit mantissa is used. This is equivalent in total dynamic range to 132 dB with resolution of better than 1% of the signal amplitude.

automatic gain control (AGC) circuits remember the average value of the preceding samples from each channel being recorded. When one of those channels is again sent, the respective AGC instructs the FPD/A to decode the digital word as if its exponent represented a larger signal. The 132-dB dynamic range of the FPA/D process is thus compressed to fit the 40-dB dynamic range of the slow-speed analog tape recorders. The actual level of the signal is continuously recoverable since the AGC command to the FPD/A is allowed to change only at the start of each minute, and the command for each minute for each channel is coded into seconds 50 to 58 of the modified IRIG-H time code (Figure 11) recorded on the fourth data channel. The characteristics of the OSS AGC system are similar to the AGC system used in the HIG ocean bottom seismometers.

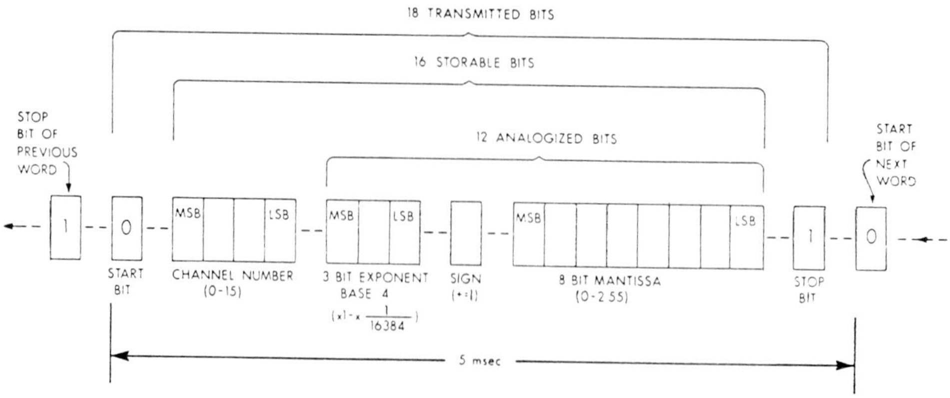

FIGURE 10. Transmission sequence. This sequence is sent up the wire at 200 samples/sec or 3200 bits/sec. The rate is limited by the power available to the system.

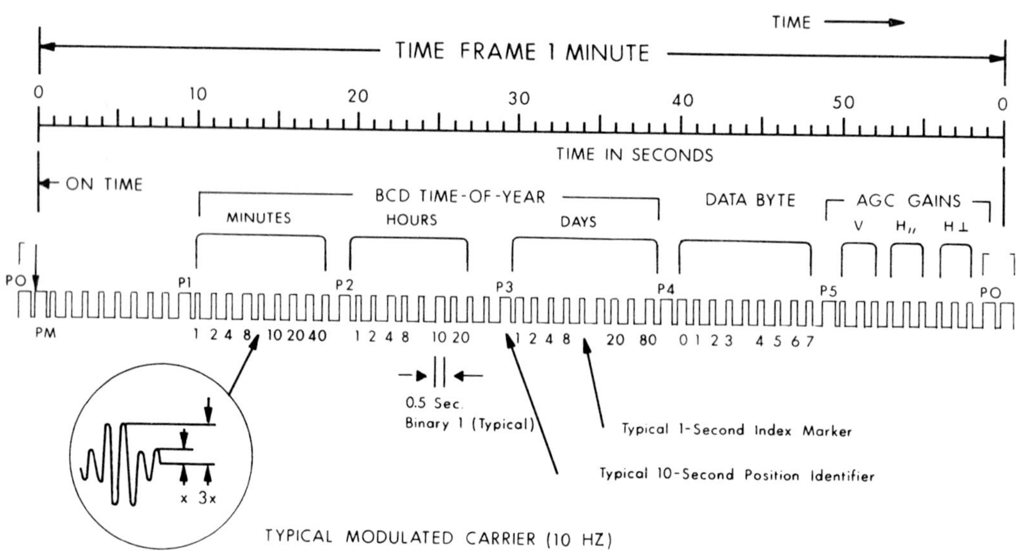

FIGURE 11. Time code. In addition to time, the code contains the recorded geophone gains and the digital data from the borehole package recorded at a rate of one sample/channel/32 min. On playback, the 10 Hz carrier is used in phase lock to control digitization rate.

The AGC system fixes the background level to be within the dynamic range of the tape. If the average level gets above −26dB or below −40 dB the level is allowed to change by 12 dB at the beginning of the next minute. In this way, a background noise signal is always above the tape noise and the smallest signals are always seen. The major disadvantage of this recording technique is that large signals will exceed the dynamic range of the tape and be clipped. For seismic signals, however, the times of the first arrivals are often more important than the amplitude. A sample of a signal showing the time code and a gain change is shown in Figure 12.

The recording package currently in use is designed for long-term unattended recording of continuous data. The recording medium is thus analog slow-speed magnetic

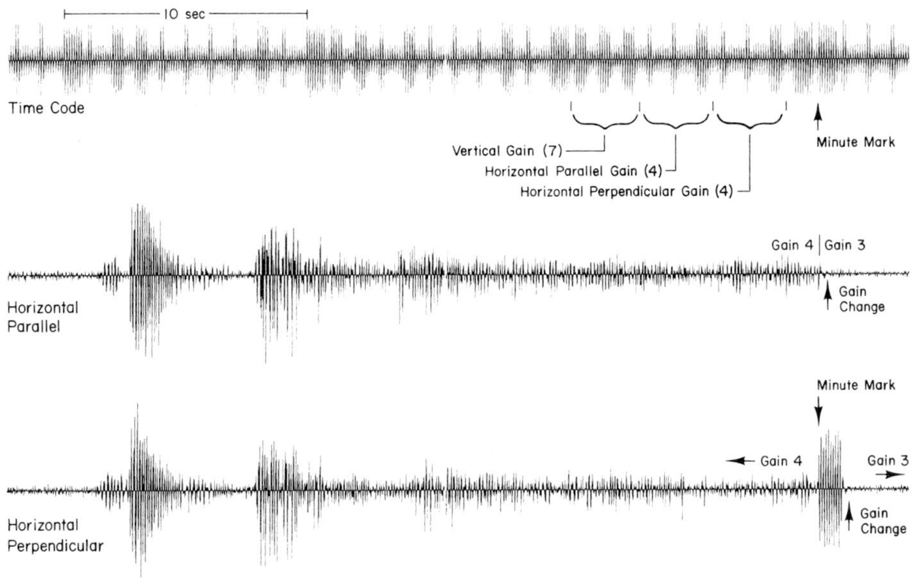

FIGURE 12. Automatic gain ranging. A large change in signal or background level causes the gain to change 12 dB at the next minute mark. The gain is recorded as part of the time code.

tape. Either a digital or an analog format can be used for recording long-term data. Digital recording retains the high quality and dynamic range of the FPA/D system but requires either a large physical volume for data storage or selective recording during specific times. Analog signals can be recorded continuously for long periods of time in small volumes but with reduced dynamic range. For this experiment, with interest in a considerable variety of seismic events from teleseismic earthquakes to small explosions, analog recording was judged more practical than digital recording. If more dynamic range was required, the present analog recording package could be replaced by a digital recording package. Although only the three geophones are recorded on the analog tape, each of the data channels from the borehole package is recorded in digital format on the fourth tape channel in an unused section of the time code. In this way, each channel is sampled once every 32 min. Although this rate is of little use for seismic signals, it has provided an excellent record of the thermal and tilt history of the package as well as other slowly varying parameters.

EMPLACEMENT

The most difficult problem in designing a downhole marine instrument system is the actual installation technique. As the system described here has been successfully emplaced in two DSDP holes, the technique used is discussed in detail.

The first problem is selection of a deployment site. While the site should ideally be chosen on the basis of the information desired, the necessity of using the drill ship restricts possible sites to those scheduled for drilling. In our case this was not a problem, since it is hard to imagine any site without sufficient merit to warrant long-term instrumentation. The site chosen should be known well in advance, as water depth will affect the amount of recovery rope needed, and predicted hole temperature could affect the sensors used.

Once the drilling is complete, the drill bit is dropped to allow normal logging of the hole. Logging of holes to be instrumented is of great value since characteristics of the hole can be important in understanding the OSS signal. When the logging is completed, the OSS borehole package is attached to the logging cable, tested, and lowered to the bottom of the hole with the logging winch. When geophone channels are being monitored, it is obvious when the package reaches the bottom of the hole because package motion is detected by the geophones. The clamping arm is extended and the tilt meters are monitored to insure that the hole angle is not beyond the limit of the horizontal geophones (about 5°). The rate of change of hole temperature also can be monitored to insure that the equilibrium temperature will not exceed the design limits. When testing is complete, the clamping arm is retracted and the package lifted about 10 m from the bottom of the hole in preparation for pipe stripping. The telemetered signal from the OSS borehole package is continuously recorded on magnetic tape during the lowering and testing.

The most important procedure in the installation is removing the drilling pipe from around the logging cable while keeping the borehole package in the hole. To do this, the cable connecting the borehole package to the ship is clamped at the drill floor and tension transferred from the logging winch to the clamp. The cable is cut and terminated with a snap fitting that mates with the overshot on a lifting line. From the time the cable is cut until all pipe is stripped and the cable is rejoined, no data can be obtained from the borehole package.

The sand line on the CHALLENGER was used as the lifting line and connected through a swivel and the overshot to the logging cable. The swivel was found necessary to prevent twist developed by tension on the sand line (not torque balanced) from being passed through the cable to the borehole package. The tension is taken on the sand line and the overshot positioned just above the level of the rig floor as shown in Figure 13A. The pipe is lifted about 90 ft as usual until a joint is level with the overshot (now inside the pipe). The joint is broken, revealing the cable, overshot and mating snap fitting (13B). A fork is inserted in the cable snap fitting and tension is taken on the fork by lowering the sandline until the fork rests on the top of the remaining drill pipe. The overshot joint is broken and the lifted pipe stand is returned to the pipe rack (13C). The sand line is pulled out of the pipe stand, reconnected to the cable snap fitting, lifted, and the fork removed. The process is repeated until all pipe has been removed and only the cable connects the ship to the hole. Lifting pipe in this manner takes 20 to 30% longer than normal. It was feared that a large horizontal motion of the drill pipe as it cleared the mud line could shear the cable or that a gross movement of the ship away from the hole could pull the package from the hole. In the two installations so far, however, neither of these problems has occurred.

After the stripping, the hanging cable is reconnected to the cable remaining on the winch drum with a "torpedo" splice. Following the connection, signals can again be obtained from the borehole package with the test box. The package is lowered to the bottom of the hole while the signals from all 16 channels are carefully monitored. The clamping arm is extended and about 50 m of cable payed out to prevent ship motion from pulling on the clamped package. Again, the telemetered signal is recorded continuously. With the pipe stripped and the borehole package working properly, the ship is allowed to drift away from the hole while the remaining cable is payed out. Care must be taken to pay out cable at a specific rate. Ideally, the cable in the hole should not move. The cable out vs. distance from the hole can be calculated for a given weight in the hole, water depth, and cable weight. The calculated schedule is used during the pay-out operation. If the ship moves too far relative to the amount of the cable out, the package will be pulled up in the hole. If cable is payed out too soon, it will start to pile up in the hole or on the bottom and be more likely to break when pulled during recovery.

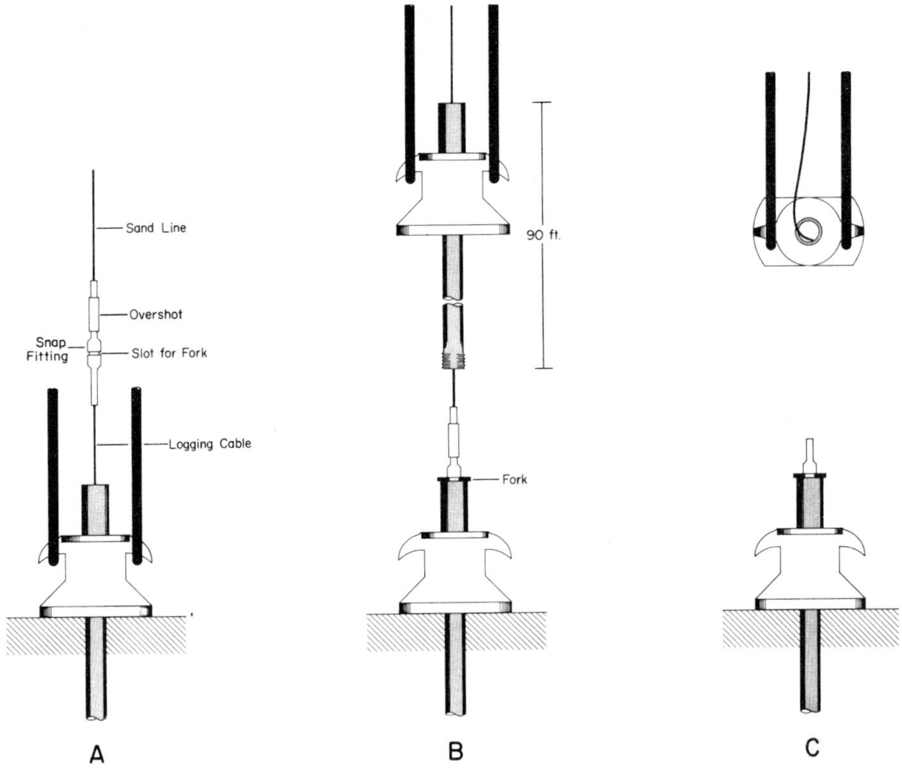

FIGURE 13. Pipe stripping. When the borehole package is in place, the drill pipe is stripped from around the cable as discussed in the text.

Near the end of the cable the ship is stopped. This is an excellent point in the operation for taking real-time data. During the second deployment, the CHALLENGER held this position for 12 hr while a second ship (R/V KANA KEOKI) shot refraction lines to the borehole package and to an array of ocean bottom seismometers. After real-time operations and testing, the wire is again clamped, headed with a new connector, passed to the side of the ship (keelhauled) and the borehole package fully tested. The recording package assembly is tested and connected to the cable. The time code from the recording package is recorded with WWV signals to calibrate the internal chronometer. The time code is checked to be sure that the telemetered signal is being decoded correctly. When the complete system has been checked, the recording package is lifted over the side and tension is taken on the recovery rope.

Once again the ship is allowed to drift and the rope payed out at a speed about equal to the ship speed over the bottom. On the GLOMAR CHALLENGER, rope tension was controlled at a tensioner (Figure 14) before passing to the spool of rope. Tension on the rope decreases as the wire is laid on the ocean bottom and the buoyancy of the rope supports the recording package. The end of the rope is attached to the anchor assembly. The ship moves slowly away to pull the anchor assembly away as far as possible from the recording package (Figure 1), and the anchor assembly is dropped overboard.

RECOVERY

To recover the recording package, the recovery ship locates itself with respect to the

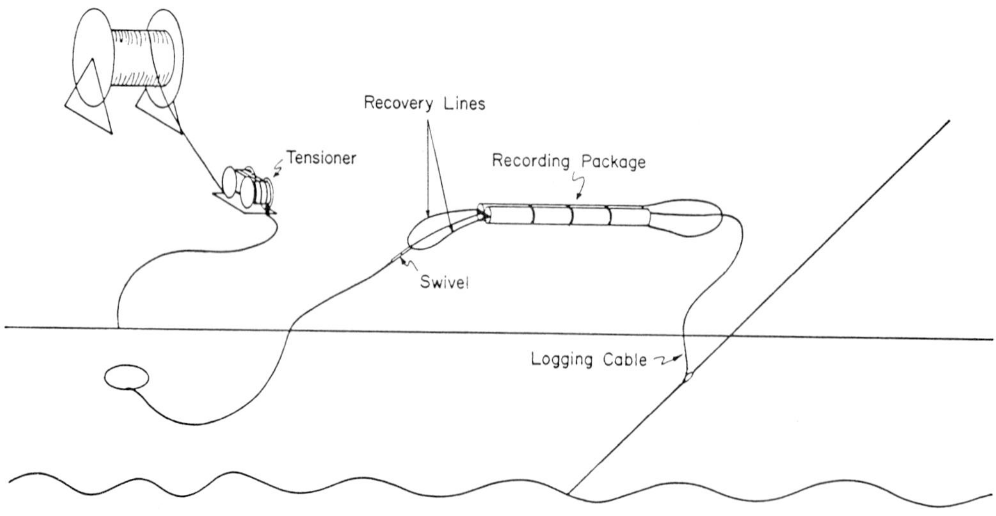

FIGURE 14. Deployment of recording package.

transponders on the anchor assembly and either waits for or commands the rope and float to release from the anchor. The time releases free the buoy and recovery rope from the bottom at a preset time. The transponder release can be commanded to return to the surface when the recovery ship is ready. If neither of the above works, the third recovery technique is grappling for the recovery rope (Figure 1). If the recovery rope cannot be located or is broken, the final resort is to drag for the cable lying between the hole and the recording package. The last two methods are undesirable for reasons of increasing danger to the OSS system and increasing time involved in recovery. Release can be detected by the change in travel time of the acoustic pulses from the recovery rope transponder to the ship. When the recovery buoy arrives at the surface it is located by sight and picked up. The 1¼ in. rope is threaded across a roller on the stern, through a tension measuring device, and around a hauler (Figure 15). The rope is drawn on board with the hauler and stored.

When the recording package is brought up to the ship the time code is immediately recorded for chronometer calibration purposes and the package is removed from the cable.

The borehole package (still in the hole) is tested by connecting the test box to the end of the cable. If the borehole package is working well, the recording package is tested, refurbished, and reinstalled by reversing the procedure outlined above. If the borehole package is not working well, an attempt is made to pull it from the hole and recover it.

CURRENT STATUS AND FUTURE PLANS

The system has been emplaced twice; in DSDP holes 482C and 494A. The first system (Downhole 1) was damaged during the stripping operation (probably because of the lack of a swivel in the sand line) and worked only in the real-time mode, but valuable temperature and seismic noise data were obtained.[5] The Downhole 1 borehole package was pulled from the hole after 42 days. The second system worked as planned except for electronic problems that caused the sensitivity of the seismic sensors to be considerably degraded. Refraction lines were shot to the system and to an OBS array around the hole while the data were recorded in real time. Several more lines were shot

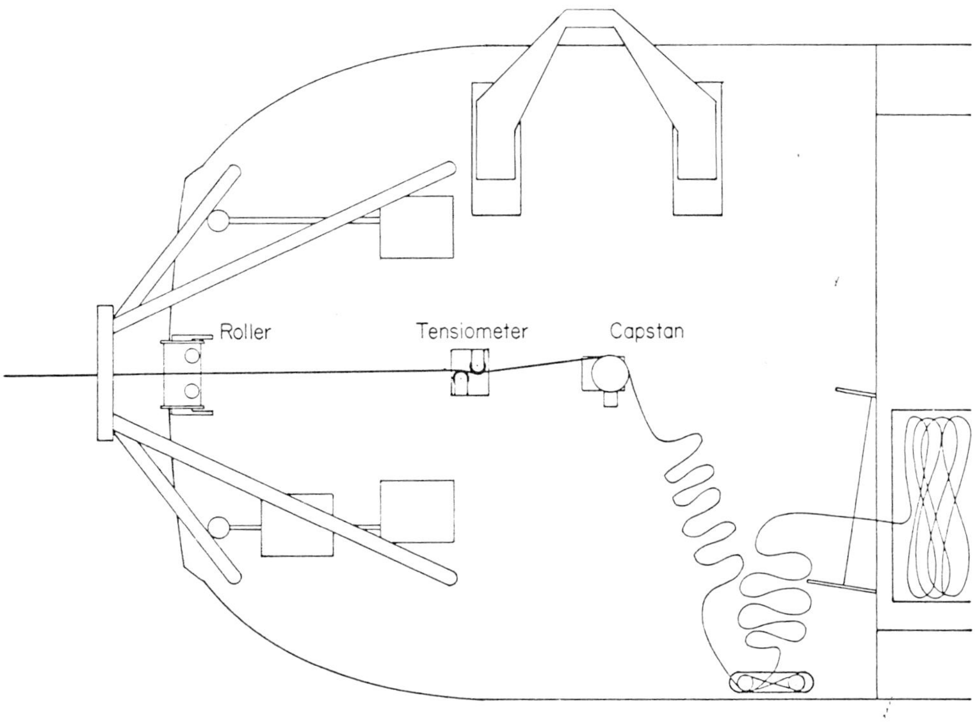

FIGURE 15. Recovery of recording package.

after the recording package was emplaced. The recording package was recovered, refurbished, and redeployed in June 1979, with no degradation noted in the system. Due to lack of suitable ships in the area, the system was not revisited again until May, 1980. By that time the transponding release at the anchor had failed and the recovery rope was grappled. Rather than hooking it, however, the grapple wire cut the rope close to the anchor. The rope floated to the surface and the recording package was recovered. Although the borehole package was still operating, an attempt to pull the borehole package from the hole failed when the wire parted.

Emplacement of the third system is now planned. This system will have considerably more sensitivity and will use 8 of the 16 channels to record and remove the DC levels from the other 8 channels. Nine vertical geophones will be wired in series and two geophones in each horizontal direction to yield more sensitivity. The rate at which data is sent up the wire will be controlled by the amount of power available with 100 samples per sec on the geophones when recording is done in real time and 50 samples per sec when batteries are used. Improvements in the analog tape system have increased the effective dynamic range to about 45 dB and guaranteed constant speed of operation to within a few percent.

CONCLUSIONS

The instrument system described in this paper can be used to obtain large amounts of high quality geophysical data from deep sea drill holes for long periods of time. Although the concept is still in its infancy, the principles of emplacement and recovery have been shown to work and easily fit into the framework of the deep sea drilling program.

ACKNOWLEDGMENTS

This paper has been critically reviewed by George Sutton, Don Hussong, Jerry Carter, and Rita Pujalet. This project would not have occurred without the encouragement and support of Charles Helsley, Director of Hawaii Institute of Geophysics. Several groups have contributed a great deal of help and expertise to this project. Our thanks to the personnel of the HIG Engineering Support Facility, the University of Hawaii Snug Harbor Marine Facility, the Deep Sea Drilling Project, the IPOD Downhole Measurements Panel, and the scientists and crew of the D/V GLOMAR CHALLENGER on Legs 65 and 67 for their aid in making this project a success. The funds for design, manufacture, and emplacement of Downhole 1 were supplied by NSF Grant OCE78-10772. The emplacement of Downhole 2 was funded by the Office of Naval Research. HIG Contribution No. 1138.

REFERENCES

1. **Miller, W. F., Geller, R. J., and Stein, S.,** Use of a bubble tiltmeter as a horizontal seismometer, *Geophys. J. R. Astron. Soc.*, 54, 661, 1978.
2. **Blackinton, J. G. and Odegard, M. E.,** An ocean bottom seismograph using digital telemetry and floating point conversion, *IEEE Trans. Geosci. Elect.*, GE-15(2), 74, 1977.
3. **Bentley, V.,** Squeeze more data onto mag tape, *Electron. Design.*, 23(21), 76
4. **Byrne, D. A., Blackinton, G., and Duennebier, F. K.,** Isolated sensor ocean bottom seismometers, in preparation.
5. **Duennebier, F. and Blackinton, G.,** A manmade hot spring on the ocean floor, *Nature (London),* 284, 338, 1980.

ELECTRICAL RESISTIVITY/CONDUCTIVITY IN SEABED SEDIMENTS

Richard H. Bennett, Douglas N. Lambert, Mathew H. Hulbert, John T. Burns, William B. Sawyer, and George L. Freeland

INTRODUCTION

Resistivity measurements (or its reciprocal, conductivity) have a long history of successful use in geophysical surveying and in exploration for petroleum and other minerals.[1-9] Many such measurements are obtained by using downhole tools and instrumentation that have essentially no applications to the study of surficial sediments. An important use of resistivity by industry is the determination of porosity during downhole logging. The difference between mineral exploration methods and unconsolidated sediment methods is primarily due to the presence of a borehole in the former which allows electrical logging tools to be lowered through the formations without danger of hole collapse. With long sections of sediment to be measured, electrodes may be spaced many feet apart, thus measuring the formation resistivity several feet away from the borehole where there are no effects of disturbance of the rocks themselves, and the interstitial fluids. In unconsolidated sediment, the measuring tool must be pushed into relatively soft sediment, thus disturbing it. Since the depth of penetration is usually small, short spacing between electrodes is necessary and penetration of the induced electrical field into the sediment is limited. Thus, the measurement, in unconsolidated submarine sediments, of electrical conductivity to an accuracy of ±10% with a vertical resolution of approximately 25 cm, presented a challenge.[10-21] Design and testing of instrumentation capable of these accuracies were considered by these authors as being within obtainable goals for prototype systems. Also, such accuracies would provide reasonable data for engineering applications and other practical and scientific purposes. *In situ* testing of conductivity was considered of paramount importance since laboratory measurement of conductivity requires recovery of the sediment in a condition sufficiently undisturbed that the conductivity is the same (within prescribed accuracy) as when the sediment is in place. Measurement with the desired resolution requires that the measuring device itself be emplaced *in situ*. Recovery of sufficiently undisturbed sediment samples is often possible in fine-grained deposits (muds), but recovery of undisturbed samples in coarse sandy material is difficult or virtually impossible with present technology. Thus, the design of an *in situ* system capable of obtaining *in situ* conductivities of surficial sediments would be expected to provide critical data leading to the determination of porosity and wet bulk density of not only cohesive sediments, but also sand deposits (predominantly quartz grains), and coarse-grained carbonates. The determination of the geotechnical properties of undisturbed noncohesive sediments is an area of much-needed research. Advances in technology would add an important dimension to scientific and engineering knowledge and capabilities in the study of the fundamental properties of surficial sediments. Density and porosity data are fundamental information required in the study of sedimentary diagenesis, pore fluid migration, consolidation history, and in numerous engineering applications. Fortunately, the Dutch have made some important advances in the study of the physical properties of sands.[4a] An instrument (similar to a piston corer) for measuring resistivity in thick, very soft deep-sea sediments has been built,[22] but would not be usable within the study area due to the considerably harder nature of the surficial layers.

The purpose of this study was to develop the required instrumentation to meet the above-stated requirements and goals, as no commercial units appeared available. Different approaches to making conductivity measurements offered the potential and capability of yielding satisfactory accuracy and resolution; however, the four-electrode technique was used. Details regarding the choice of the method will be made clear in the following discussion. Simplicity in design, durability, field adaptability, and cost were primary factors in the choice and design of the system.

MEASUREMENT SYSTEMS — TECHNICAL ASPECTS

Review of Measurement Techniques

Measurements of sediment conductivity can be obtained using various techniques. The simplest technique is to apply a known direct current potential (or a known current) to a pair of electrodes having a known separation in the sediment, and to measure the resulting current flow (or the potential) between the electrodes. Application of Ohm's Law yields the apparent resistance from which the apparent resistivity and the apparent conductivity can be calculated (resistivity of a material is the electrical resistance of a cylinder of the material having unit length and cross-sectional area; conductivity is the reciprocal of resistivity). The two-electrode direct current technique has the advantage of requiring a minimum of apparatus and can therefore be made into an operational system relatively cheaply and quickly. It suffers from two severe drawbacks: (1) in addition to the electrical resistance of the sediment itself, there is a contribution to the apparent resistivity due to electrochemical, or polarization, resistance, and (2) the current density in the near vicinity of each of the electrodes is very high so that contact resistance at the electrodes can be a major contributor to the apparent resistivity.

The contribution to the resistance resulting from electrochemical reactions can be minimized by using alternating current rather than direct current. There is a slight increase in cost and complexity. The effect of contact resistance can be minimized by using two pairs of electrodes, one for applying a known current and the other for measuring the resulting electrical potential.[4] In submarine environments, the potentials to be measured for both the two-electrode and four-electrode techniques are quite small; therefore, sensitive electronic instrumentation (most of which is commercially available) is required to detect the signal and discriminate against noise. In addition, emplacement of the electrodes is difficult, especially in sandy sediments, because (1) firm electrode contact with the sediment must be maintained, (2) free communication with the overlying, highly conductive seawater must be avoided, and (3) electrode arrays must be designed with electrode spacing less than 40% of desired resolution and electrode size sufficiently small to minimize sediment disturbance; therefore, they tend to be somewhat fragile.

Emplacement problems can be substantially reduced by using an inductive technique rather than electrodes, but there is a long development time and high device complexity and cost.[9] Inductive techniques rely upon detection of the secondary electromagnetic field established in conductive materials by an applied electromagnetic field. The electromagnetic field to be applied is produced by an alternating current flowing in a loop or coil of wire: no physical contact is required between the material being studied and either the primary (transmitter) or the secondary (receiver) coil. With both contact and inductive techniques, the vertical extent of the sediment contributing to the resistivity measurement can be controlled to a substantial degree by the addition of guard electrodes or coils to which are applied an appropriate electrical waveform, but the equipment is costly and complex.[14]

Choice of Methodology and Design of Systems

Due to mechanical and environmental factors, several major considerations were taken into account before the resistivity probe could be properly developed. They are as follows:

1. The probe must be extremely durable to withstand being driven through semiconsolidated carbonate sediment and reef rock materials by a vibracore with an effective weight of approximately 360 kg and a high rate of vibration.
2. The probe must be shaped such that it will penetrate the substrate with a minimum amount of effort.
3. The electrode tips must be small in diameter to approximate a theoretical point source for the field generated, and must be located far enough ahead of the probe wedge to be out of the main zone of sediment disturbance.
4. The electrodes must be extremely rigid and tough enough so that they will not change geometry by bending or erosion while being driven into the sediment.
5. All materials other than the electrodes must have high volume resistivities.
6. Electrical connections to the electrodes must be waterproof and be able to withstand extreme amounts of vibration.
7. The probe must be easily mounted to the vibracore pipe to minimize assembly and changeover time at sea, and to minimize cost.

The system selected to achieve the measurement goals is based on the four-electrode contact technique. The device took the form of a probe (Figure 1) at the tip of a 6.1 m (20 ft) length of 127 mm (5 in.) diameter vibracore pipe. The probe is wedge-shaped to ease emplacement and minimize sediment disturbance near the electrode tips, and is fabricated of nonconductive plastic (see section entitled Physical System). The electrodes extend beyond the wedge and are spaced so that (1) the total volume of sediment that contributes to the resistivity measurement is large compared with the portion of that volume that has been disturbed, (2) the vertical resolution of the measured resistivity is adequate, and (3) the electrode separation is at least three times the expected grain size of the sediment.[6] The electrodes are equally spaced with the two potential-measuring electrodes on the inside, in the position termed the "Wenner arrangement." The response of systems with electrodes in this arrangement has been derived from theory for conductive materials in a variety of configurations.[3] These theoretical treatments apply strictly only as the area of the electrodes becomes negligible, but the limiting cases are useful in understanding the behavior of real electrode systems.

Physical System

The body of the probe tip (Figure 1) was machined from a 5 in. diameter rod of polycarbonate plastic which has a volume resistivity of 10^{16} ohm/cm, high impact, tensile and compressive strengths, and low water absorption. It also has excellent machining characteristics. From the solid rod, four holes were bored in a straight line the length of the probe (Figure 1). A recess was then cut into the top of the rod to accept a nylon bearing plate which rests on top of the electrode heads.

A 20° angle was then machined on two sides of the polycarbonate rod to form a wedge-shaped leading edge of the probe tip to ease penetration into the sediments. The top of the polycarbonate rod was turned down to make a slip-fit inside a steel coupler. It was held in place inside the coupler by two threaded steel studs.

The electrodes were made from 6.4 mm (¼ in) diameter standard hardened tool-steel hotwork ejector pins. They were selected for their hardness and ability to withstand constant physical punishment. A tube of polycarbonate insulator was solvent-

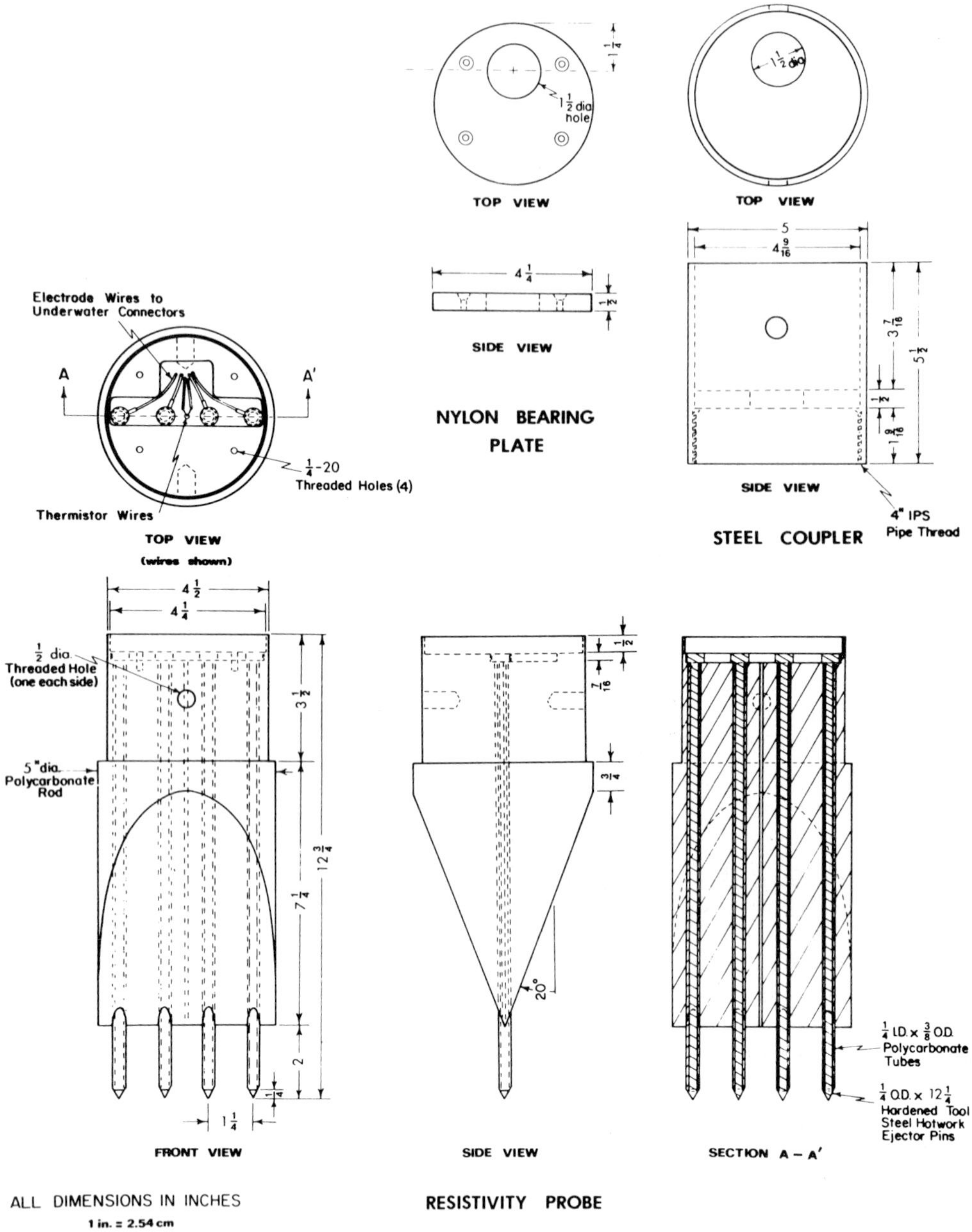

FIGURE 1. Mechanical drawing of resistivity probe and components.

cemented over the entire length of the steel rods. A point was then precision ground on each rod and insulator with an angle of 25° to insure an equal amount of exposed surface area on each electrode tip. Electrical wires were attached to the electrode heads. A waterproof seal was assured by the use of silicone dielectric grease in the space between the electrode insulator and the wedge. This "soft potting" of the system was selected due to the difficulty of "hard potting" several different materials (polycarbonate, nylon, and neoprene wire insulation) and the probability that the large vibrating forces exerted on the materials would break down their waterproof integrity thus

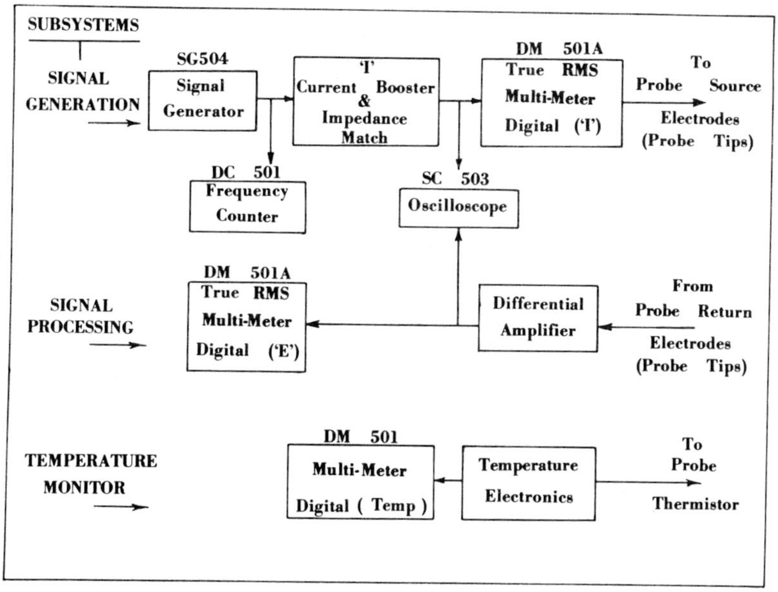

FIGURE 2. Resistivity probe electronics system.

causing electrical shorts across the electrodes. After assembly, all probe tips were pressure tested in a hyperbaric chamber to 0.69 MPa (100 psi) to check for water leakage to the probe electrical components. Additional details of the probe specifications can be found in Hulbert et al.[28]

A miniature probe also was designed and fabricated for use of *in situ* and shore-based laboratory testing of samples. This probe was machined from a 38mm (1½ in.) diameter polycarbonate rod into which were mounted 1.6mm (1/10 in.) diameter steel pins on 9.5mm (3/8 in.) lateral spacing. The response of the miniature probe reflects the contribution of a smaller volume of material than does the larger probe described above; thus it can be used to distinguish conductivity features of smaller scale.

Electronics System

The resistivity probe system electronics are composed of three basic subsystems: signal generation, processing, and temperature monitoring (Figure 2). All three of the subsystems were used with the standard probes, while for the miniature probe, sample temperature could be conveniently determined directly and the temperature monitoring subsystem was not required.

Alternating current is supplied at up to approximately 20 V to the outer electrodes. The inner, potential-measuring electrodes are connected to a unity-gain differential amplifier with a high common-mode rejection ratio and a high input impedance; output of the amplifier as well as the current source are monitored (Figure 3). Temperature at the probe tip is monitored with a thermistor circuit since the sediment resistivity varies 1.5%/°C.[23] All connections to the probe are made with two-conductor shielded cable.

Laboratory tests were conducted to determine an optimal range of frequencies for the source signal. At high or intermediate frequencies a capacitive component of the response interferes, while at very low frequencies polarization resistance interferes and the measurement instrumentation is not certified by the manufacturer. No interference was noted in the range 40 to 115 Hz; in the vicinity of 500 Hz the capacitive component

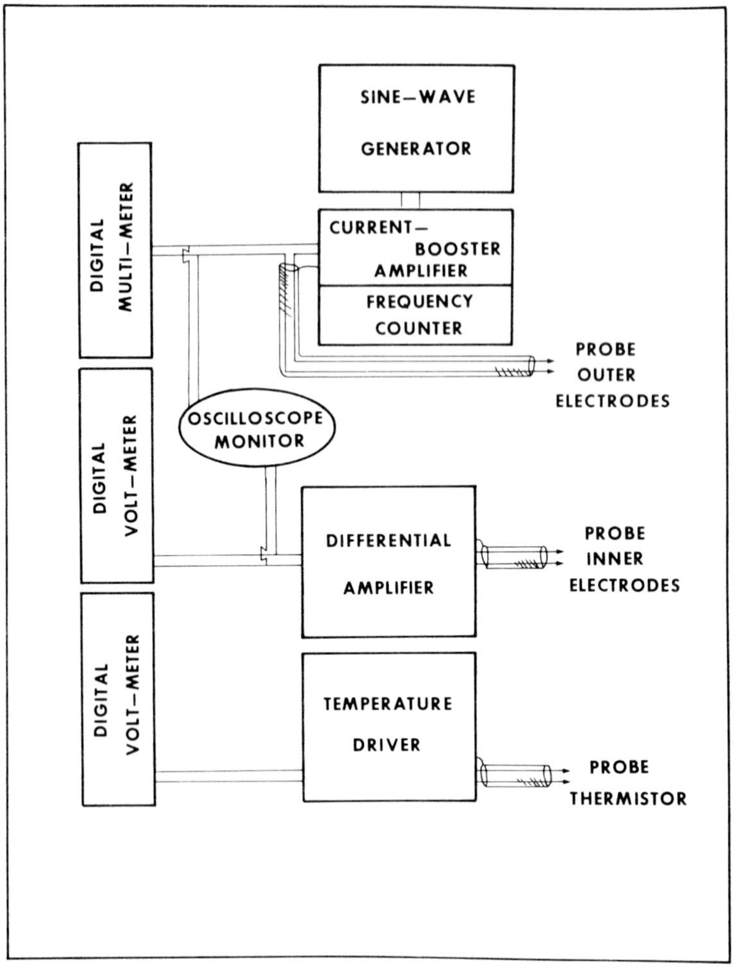

FIGURE 3. Block diagram of resistivity probe electronics subsystem.

was noted to have an effect less than 5% as large as the signal in material of the conductivity range of interest.

A Tektronix® SG504 generator was used to provide a stable 100 Hz signal. A frequency of 100 Hz was chosen to minimize noise problems caused by 60 Hz powerline disturbances and to minimize the polarization resistance components of the measurements.

Since extremely low resistances were expected, a current booster (Figure 4) and impedance-matching system was developed in-house and was designed to keep the loading of the output of the signal generator (SG504) from introducing distortion. The current booster and impedance-matching circuits were required to withstand long periods of almost direct short yet have a very fast response time as the probe encountered higher resistances.

The signal from the current booster was monitored by a Tektronix® true-rms-reading DM501A digital multimeter prior to being sent to the source electrodes (Figure 2). The DM501A digital voltmeter was chosen because of its true-rms-measuring capabilities. This eliminated some of the computer processing time for data reduction.

For the signal processing electronics, a high-input-impedance differential amplifier

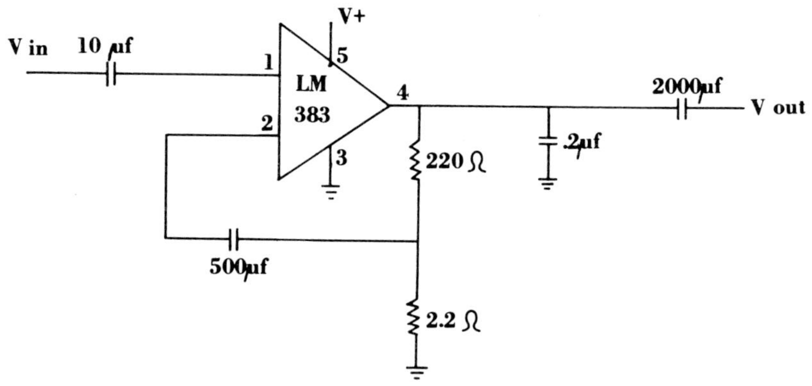

FIGURE 4. Current booster and impedance matching circuitry.

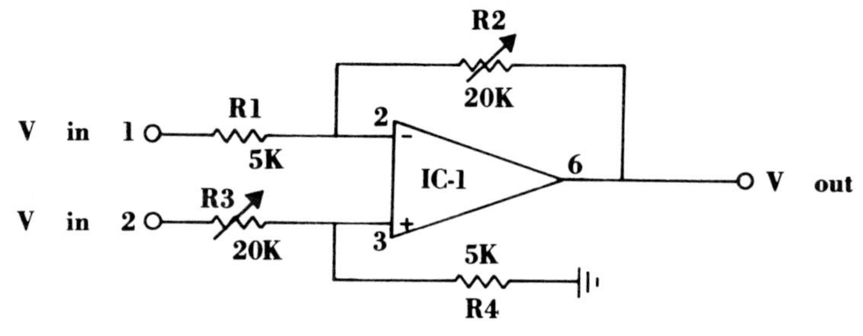

$$V_{out} = -\left(\frac{R2}{R1} \times V_{in\ 1}\right) + \left[\left(\frac{R4}{R3+R4}\right)\left(\frac{R1+R2}{R1}\right) V_{in\ 2}\right]$$

FIGURE 5. Differential amplifier circuitry.

(Figure 5) was developed to process the signal from the potential-measuring electrodes. The equation for the amplifier is

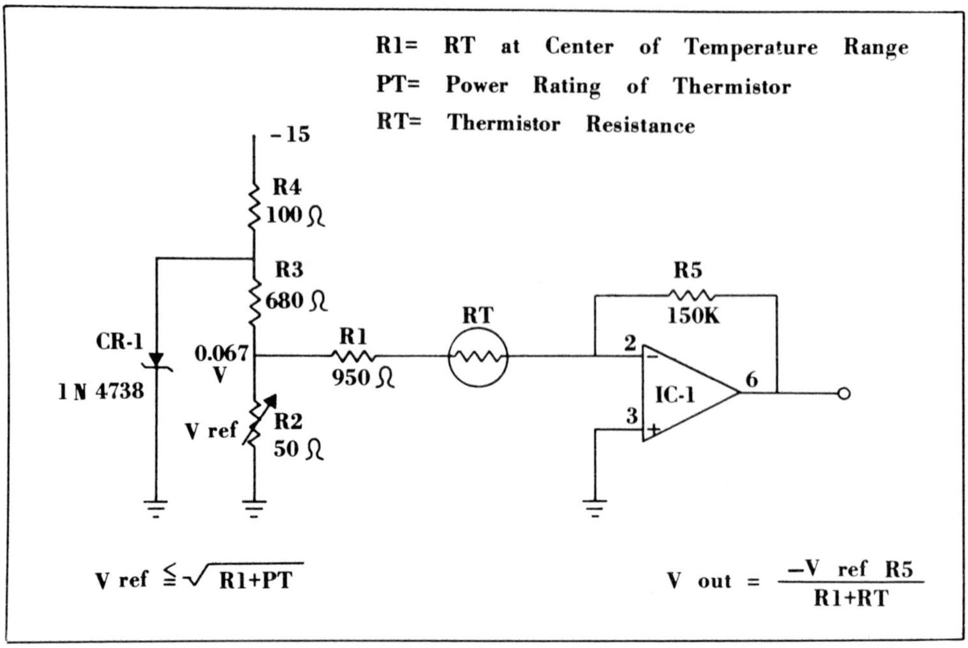

FIGURE 6. Temperature monitor circuitry.

$$V_{out} = -\left(\frac{R2}{R1} \times V_{in1}\right) + \left[\left(\frac{R4}{R3 + R4}\right)\left(\frac{R1 + R2}{R1}\right) V_{in2}\right] \quad (1)$$

The signal out of the differential amplifier is monitored by a Tektronix® true-rms-reading DM501A multimeter (Figure 2).

A temperature measuring system was used to monitor effects of heat caused by vibration of the probe tip as it encountered the stiff bottom material (Figure 6). The output voltage was monitored by a Tektronix® DM501 DC voltmeter (Figure 2).

A Tektronix® DC501 frequency counter and SC503 oscilloscope were used to monitor the output frequency and return signals during initial development of the system. They were not used during *in situ* measuring or calibration in order to minimize loading effects or introduction of errors into the measuring system.

Calibration of the critical components of the electronics system was carried out by the manufacturer's calibration laboratory following documented procedures and using equipment traceable to the National Bureau of Standards, within the limits of the Bureau's calibration services.

CALIBRATIONS

Resistivity Probe Calibrations

Calibration of the resistivity probe tips was accomplished by making admixtures of seawater and freshwater of varying ratios in polyethylene tanks (approximate dimensions 1.2 m [4 ft] in diameter by 1.5 m [5 ft] high), containing approximately 1.51 m^3 (400 gal) of solution. Appropriate tank dimensions were determined in homogeneous electrolyte solutions by monitoring the apparent resistivity as insulating barriers were placed in various geometries around the probe. The tank dimensions used, and the positions of probes within the tank, were well removed from those that had noticeable effects on the apparent resistivity. Ratios of seawater to freshwater used were 1:0, 1:1, 1:3, and 1:8. After thorough mixing, the solutions were allowed to stand overnight to assure consistent salinities and temperatures throughout each tank. A 2-ℓ sample of each tank was then taken at a depth of 0.3m below the surface (equivalent to the depth of exposed electrodes below the water surface) in each of the four calibration solutions for the determination of conductivity by standard techniques.

Conductivity ratios for each calibration solution were determined using a Plessey Environmental Systems, Inc. laboratory salinometer. The instrument was calibrated using standard seawater (I.A.P.S.O. Standard Sea-Water Service, Institute of Oceanographic Services, Wormley, Godalming, Surrey, England) with a chlorinity of 19.376;50‰ or a salinity of 35.0037‰, such that a conductivity ratio of 1.00000 is equivalent to a salinity of 35.000‰. Three determinations were made on each calibration solution and the conductivity ratios of the solutions at the temperature at which they were analyzed were determined by averaging the three conductivity ratios. Maximum variability between each of the three determinations for each solution was 1.0×10^{-4}.

The conductivity ratio of each calibration solution was then corrected for differences between the temperatures at which the ratios were determined in the salinometer and the temperatures of the solutions in the calibration tanks when the conductivity probes were calibrated. This was accomplished by using standard temperature correction factors given in tables compiled by the National Institute of Oceanography of Great Britain and UNESCO (1966).

The true conductivity of each calibration solution was determined by multiplying the corrected conductivity ratio by the specific conductivity of seawater[24] at 35.000 ‰ at the calibration tank temperature when each probe was calibrated.

Each conductivity probe was calibrated by immersion (electrode tips approximately 0.3 m [12 in.] below tank water surface) in each of four calibration solutions during pre-field tests (conductivity ratios of solutions: 0.97756, 0.52581, 0.28239, and 0.14961 based on a conductivity ratio of 1.00000 for standard seawater with a salinity of 35.000‰) and three calibration solutions during post-field tests calibration (conductivity ratios of solutions: 0.98670, 0.28687, and 0.13181 based on 1.00000 for standard 35.000‰ seawater).

The probe and electrodes were wiped dry between measurements in each solution to eliminate contamination from one tank to the next. Each calibration tank was grounded by using a 0.2 m (8 in.) × 0.46 m (18 in.) piece of copper screen connected through a cable to the electronics system. All cables and electronic equipment were the same as used during field testing.

Approximately 0.6 V (measured at the output terminals without the probe attached) at 100 Hz alternating current was applied to each probe while six measurements of resistance were recorded at 30-sec. intervals in each calibration solution. The maximum variability in the resistance readings occurred in the high-conductivity solutions and was on the order of ±0.002 ohm.

The six values of resistance were averaged for each probe in each calibration solution. The reciprocals of these averages were plotted against the true conductivity value of each calibration solution. Best-fit least-squares lines were determined by computer for each probe in both pre- and post-field calibrations (Figures 7 to 12). All correlation coefficients were better than 0.999. Example calculations can be found in Appendix A.

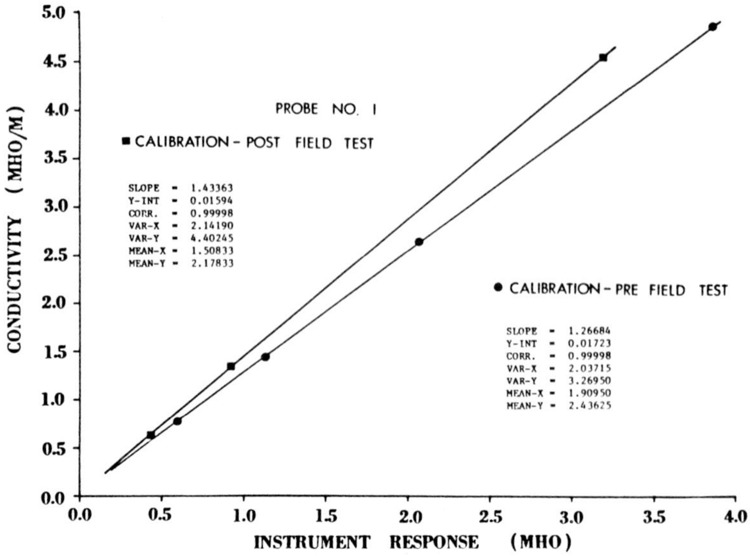

FIGURE 7. Calibration data — pre- and post-field tests, probe 1.

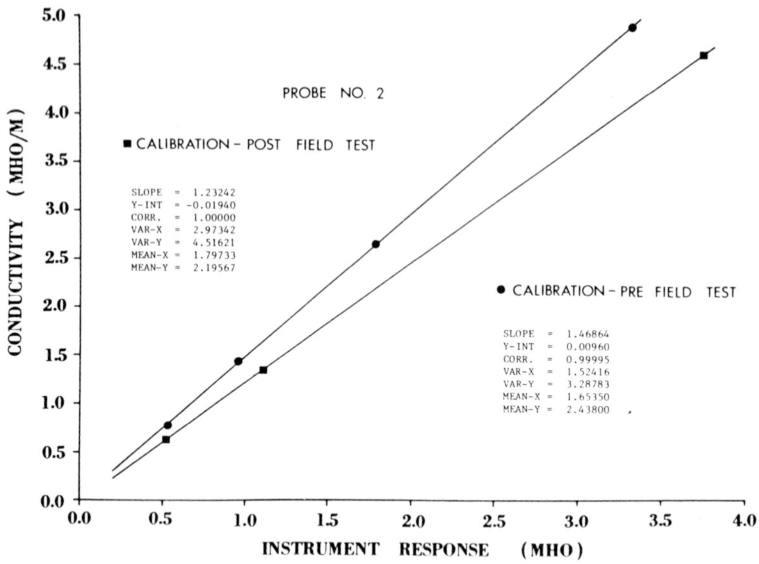

FIGURE 8. Calibration data — pre- and post-field tests, probe 2.

The vertical extent and the volume of material making the primary contribution to conductivity probe response were determined in the tanks of solutions used in the pre-field test calibrations. Insulating materials (most commonly made of polyvinyl chloride) were placed around the probe and the configurations varied to determine those that resulted in a 10% change in the instrument response.

Temperature

Temperatures of the calibration solutions were determined prior to, and immediately

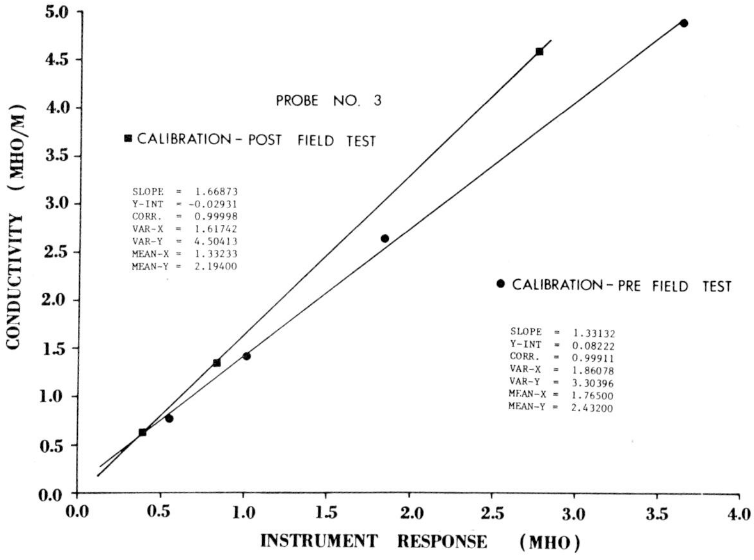

FIGURE 9. Calibration data — pre- and post-field tests, probe 3.

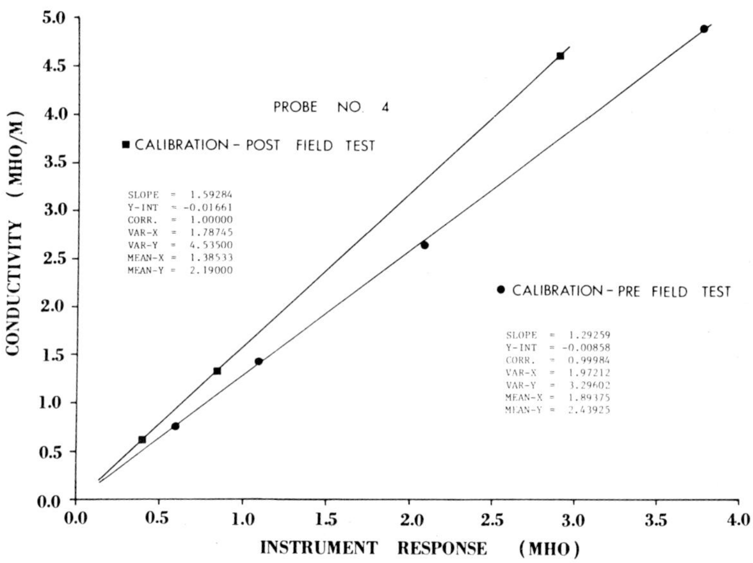

FIGURE 10. Calibration data — pre- and post-field tests, probe 4.

after, each set of resistance measurements. Temperatures above 19°C were measured with precision laboratory thermometer readable to 0.01°C. Temperatures below 19°C were measured using a Yellow Springs Instruments, Inc. (YSI) electronic thermometer Model OT-2. Thermistors mounted within the leading edge of the polycarbonate probe were read during pre-field testing calibrations. Due to several minutes required to equilibrate the polycarbonate to the calibration solution, the temperatures recorded are not thought to be as accurate as those recorded by the laboratory thermometer or the YSI probe.

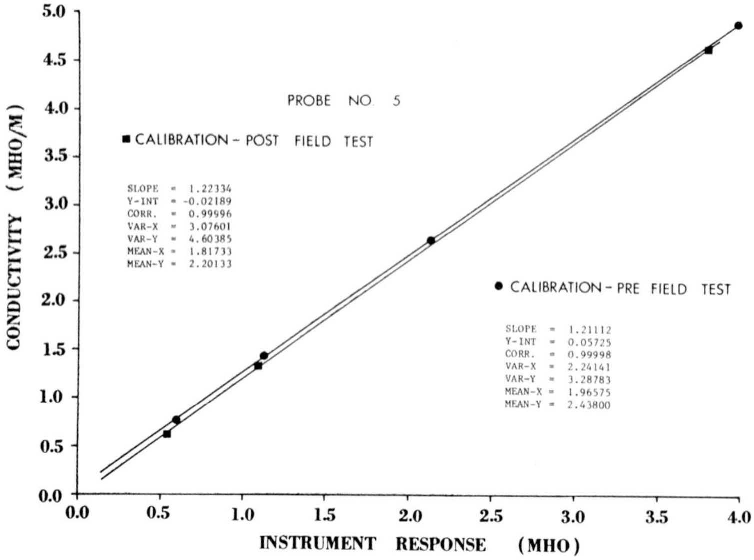

FIGURE 11. Calibration data — pre- and post-field tests, probe 5.

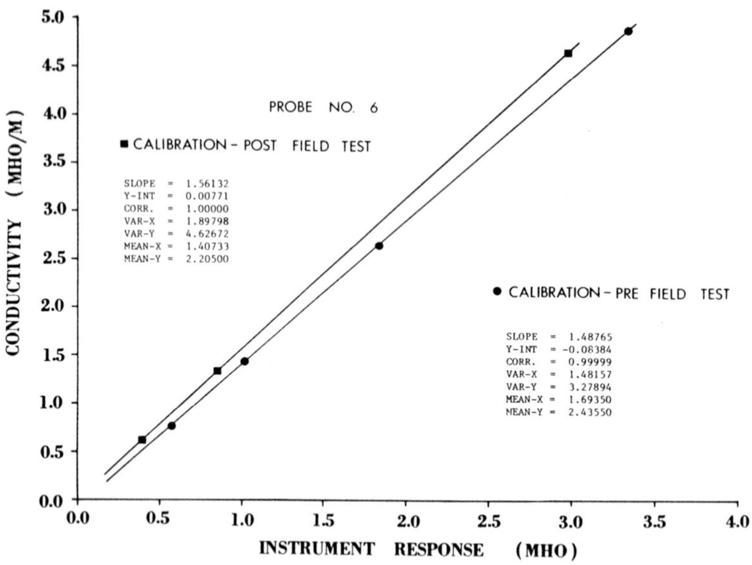

FIGURE 12. Calibration data — pre- and post-field tests, probe 6.

Resistivity measurements taken over a 6-min period with the probe inserted in the sediments were found to be extremely stable. Thus, temperature changes at the electrode tips were considered to have had a minimal effect on the resistivity readings; however, a slight temperature increase of up to 0.75°C was considered when establishing the estimated limits of uncertainty of the reported data.

The thermometers, YSI probes, and resistivity probe thermistors were calibrated prior to field-testing against a Hewlett-Packard Model 2801A quartz thermometer that has been calibrated by the manufacturer against standards traceable to the National

Bureau of Standards. All temperature measuring devices were calibrated to 0.01 °C in a Forma Scientific Model 16-120 calibration temperature bath with tapwater as a solution. All instruments were calibrated between 15 °C and 25 °C in 2 °C increments.

The miniature probe was calibrated in a similar fashion with the probe tips immersed vertically 76mm (3 in.) into calibration solutions in large glass beakers. The effective volume contributing to the measured value was determined by moving the probe in relation to the walls of the beaker. The instrument response (R) is equal to: $R = (0.519 \times \text{conductivity ratio} - 0.00294)$ with the regression coefficient exceeding 0.9999.

Depth of Penetration

Depth of penetration of the resistivity probe was measured using a penetrometer system consisting of a ten-turn potentiometer and a pulley system mounted on the vibracore rig developed by Offshore Scientific Service, Inc. (O.S.S., Inc.). The potentiometer was calibrated by O.S.S., Inc. to make 1 r/2 ft of penetration. Readout of the potentiometer was on a strip chart recorder with 0 to 6.1 m (0 to 20 ft) full-scale. Calibration of the system was accomplished by sliding the vibracore head in 0.61 m (2 ft) increments down the vibracore frame and making sure the recorder registered 2 ft of penetration. Best estimate of the penetrometer system accuracy is considered to be ± 76 mm (3 in.).

GEOLOGY AND MORPHOLOGY OF THE STUDY AREA

The study area is located on the Florida continental shelf in the northern Straits of Florida, through which the Gulf Stream passes northward into the Atlantic Ocean. The narrow Florida shelf (approximately 2.8 km) is bounded by the shoreline to the west and, generally, the 21 m bathymetric contour to the east (Figure 13). The seafloor gradient then steepens down to the broad Miami Terrace in water depths of 220 to 366 m.[25] The shelf in this area is characterized by a series of three reef ridges running roughly north-south parallel to the shoreline. These ridges form highs between a series of steplike linear flats or plateaus. The reefs are composed of hard substrate while the lows are depressions partially filled with white-to-gray unconsolidated calcareous sands and gravels.

The majority of resistivity probe measurements were collected in the outer plateau, which is located between the second and third reef. This trough is approximately 400 m wide in the study area and is filled with 6 to 8 m of carbonate sand and reef rubble, most of which is unconsolidated. Its floor varies in depth from 18 to 21 m with the greater depths near the second reef, giving it a slight shoreward dip. Sediments found in this low are primarily medium carbonate sands with large amounts of gravel-sized to cobble-sized reef debris interspersed. Mean size of the sand fraction is 0.47 mm with a low quartz concentration typically $< 15\%$.[26] Duane and Meisburger[27] have reported that the acid-soluble content of this interreef sediment is generally 80%.

Seismic Survey

A detailed seismic survey was completed in the study area using an EG & G, Inc. "Uniboom" shallow-penetration seismic reflection profiling system. Navigation was accomplished with shore-based twin theodolite towers. The location of the seismic tracklines and the resistivity probe sites are indicated in Figure 14.

The sedimentary structures in the series of linear ridges and plateaus aligned north-south are delineated in the seismic survey. The low at the study site, between the second and third reef, is nearly acoustically transparent except for a strong reflector 0.3 m thick lying 2 to 2.5 m beneath the sediment surface (Figures 15 and 16). This layer

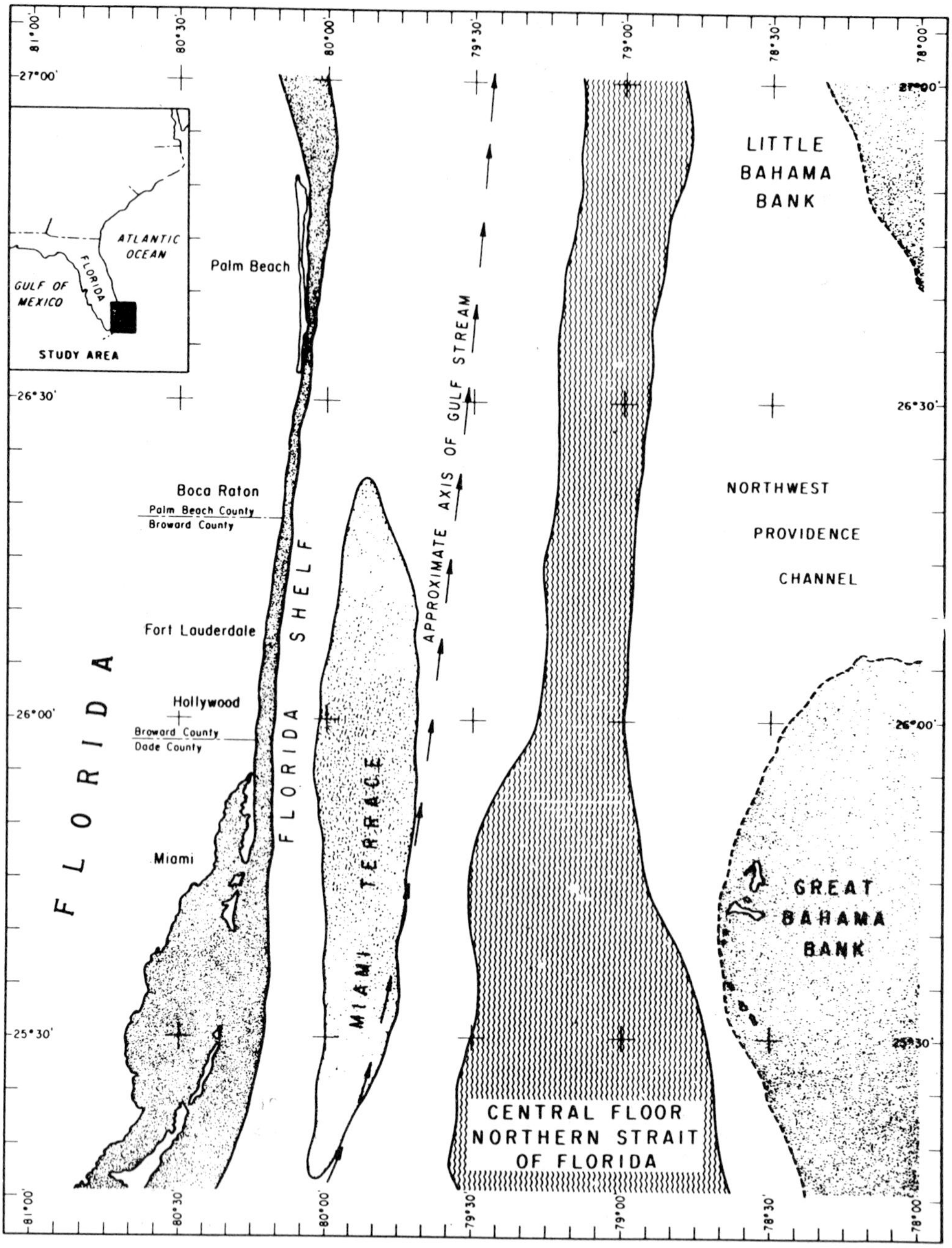

FIGURE 13. Physiographic provinces of the northern Straits of Florida.

consists of larger coral and mollusk fragments packed together with sand, forming a very dense layer and a good acoustic reflector. This reflector can be traced through the second and third reefs and across the second plateau (Figure 15). Depth to the reflector varies only slightly over the whole study area, even over the reefs. It is suggested that this layer may be an unconformity marking the death of the outer barrier reef organisms. Subsequent deposition of large amounts of coral rubble occurred in the depression behind the reef. During this time the second reef began growing coral

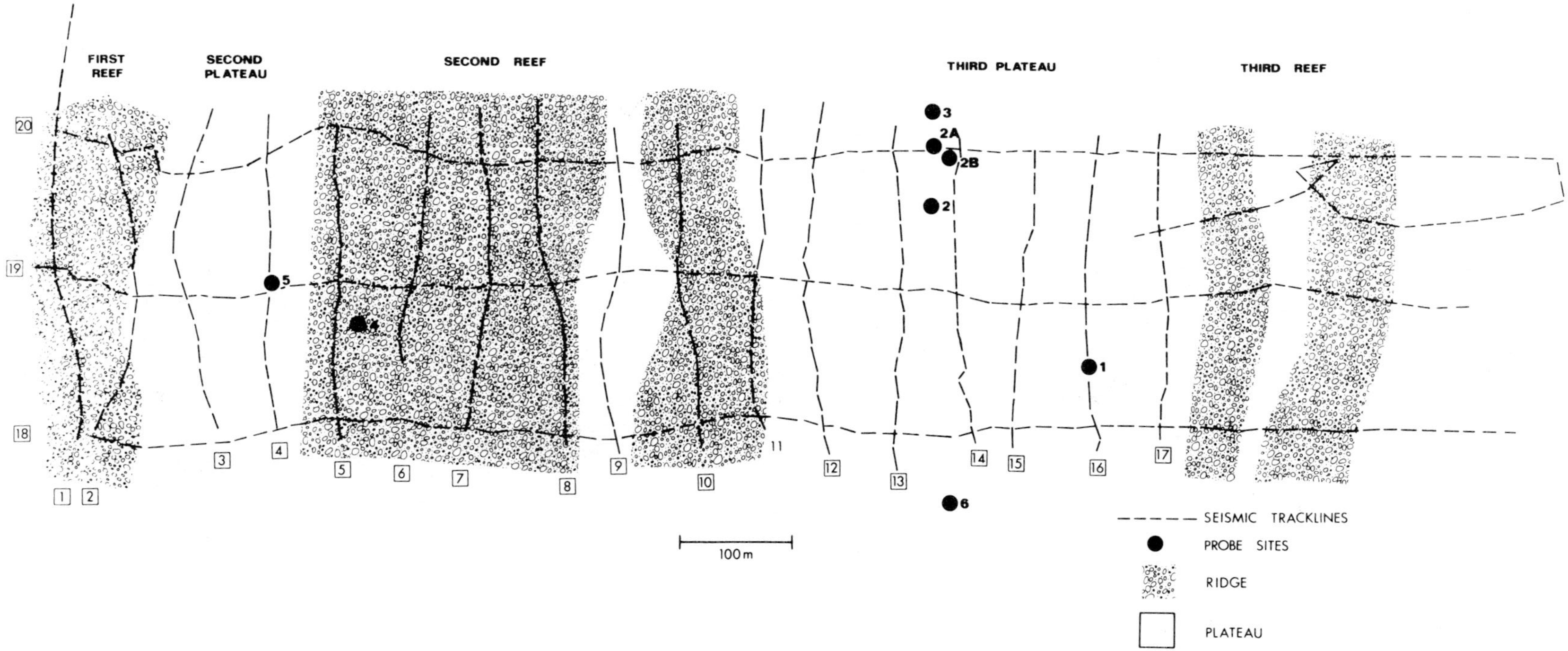

FIGURE 14. Map of the study area showing the location of the seismic tracklines and resistivity probe sites.

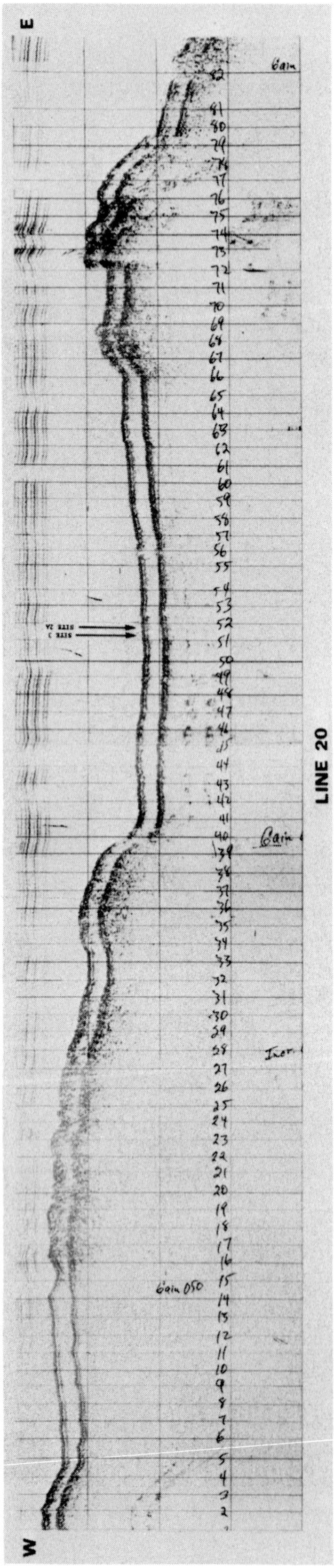
E
W
LINE 20

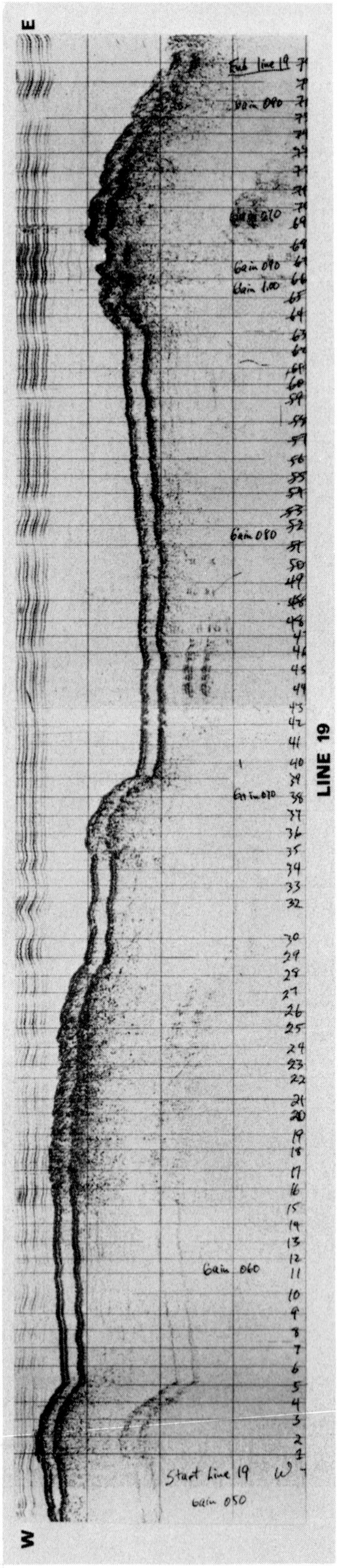
E
W
LINE 19
Start Line 19
Gain 050

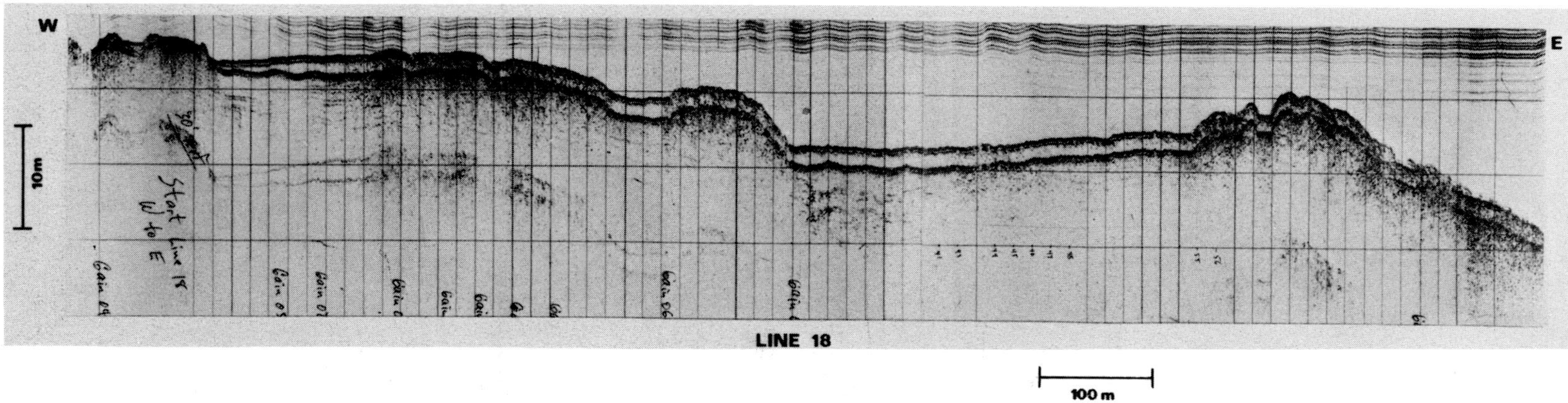

FIGURE 15. Seismic records from the east-west trending tracklines.

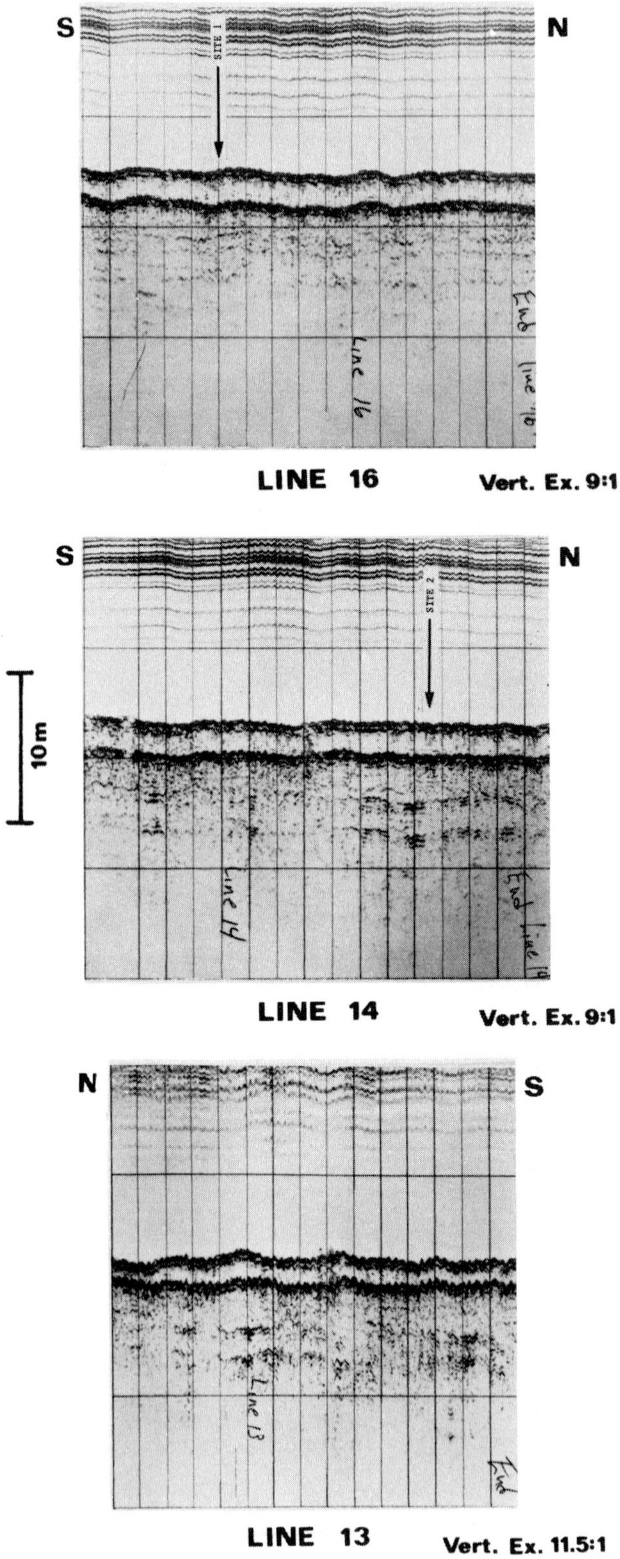

FIGURE 16. Seismic records from the north-south trending tracklines across the resistivity probe sites.

Table 1
DRILLING STATISTICS

Site	Hole no.	Depth (m)	Total blows bottom (0.3 m)
2A	1	2.7—3.0	32
2A	1	4.3—4.6	12
2A	1	5.5—5.8	265
2B	2	3.4—3.7	32
2B	2	6.4—6.7	23
2B	2	7.9—8.2	6
2B	2	9.4—9.7	12
2B	2	11.0—11.3	20
6	3	3.4—3.7	81
6	3	4.9—5.2	71
6	3	6.4—6.7	64
6	3	7.9—8.2	142
6	3	9.4—9.7	71

on the old coastal dune ridge and soilstone crust as sea level transgressed. The seismic profiles over the interreef areas appear nearly transparent (typical of sand deposits) while over the reefs it appears very jumbled, as would be expected due to the general lack of continuous horizons in a reef environment and a back-reef environment.

Coring and Drilling

Vibracoring and rock drilling operations in the third plateau reveal unconsolidated sediment down to subbottom depths of at least 10 m, and to 11.3 m at site 2B (Figure 14). Below the hard layer at approximately 2m, the sediment is a mixture of carbonate sand and mud with occasional larger pieces of coral fragments. The water and mud contents vary with depth and from core to core. Layers 0.1 to 0.2 m thick occur, which are soft and muddy; others are hard, with low water contents, but fine-grained (mud mixed with some sand). This variation in sediment texture and water content can be seen in the seismic records as faint reflectors alternating with acoustically transparent layers.

Core Penetration Resistance

The drilling statistics during coring operations at sites 2A, 2B, and 6 are summarized in Table 1. The last column shows the total number of blows (48.4 kg/m, 350 ft/lb each) required to drive the bottom 0.3 m (1 ft) of the 0.6 m (2 ft) core barrel. Data for the upper 0.3 m (1 ft) were discarded because the material was always loose sand that remained after washing out the casing.

RESULTS

Conductivity measurements were made at six sites, five within the third plateau and one in the second plateau (Figure 14). Hard and soft sedimentary layers were delineated from the Uniboom seismic profiles (Figures 15 and 16). The seismic data were correlated visually with conductivity and rate of penetration data. Seismic stratigraphy is summarized in Table 2.

Conductivity data vs. depth of penetration are plotted for each site (Figures 17 to 22; instrument response and uncorrected resistivity/conductivity field data are given in Appendix B). Best estimates of the actual conductivity are indicated by a box at each level of penetration. Bars in each figure represent the limits of uncertainty in

Table 2
SEISMIC STRATIGRAPHY

Site no.	Depth (m)[a]	Description
1	0—0.3	Soft
	2	Slightly hard reflector
	2.0—2.3	Soft
	2.3—2.4	Very hard reflector
	2.9—3.0	Slightly hard reflector
	4.3—4.6	Slightly hard reflector
	5.2—5.5	Slightly hard reflector
	5.8	Slightly hard (approx. 0.3 m thick) reflector
	7.3	Slightly hard reflector
2, 2A, 3	0—0.46	Soft
	0.46—0.76	Slightly hard reflector
	0.9—2.0	Soft
	2.0—2.3	Very hard reflector
	2.4—2.7	Slightly hard reflector
	4.6—4.9	Slightly hard reflector
	7.3—7.6	Slightly hard reflector
4	0—0.9	Very hard reflector
	1.8—2.4	Hard reflector
	(Seismic data indicate that the entire section from 0 to 6.1 m appears considerably harder than equivalent section at sites in "lows".)	
5	0.76—1.1	Slightly hard reflector
	1.5—1.8	Very hard reflector
	3.0—3.4	Slightly hard reflector
	4.6—4.9	Slightly hard reflector
	5.2—5.5	Slightly hard reflector
	6.1—6.4	Slightly hard reflector
	7.0—7.3	Slightly hard reflector

[a] Depth uncertainty of approximately 0.3 m due to swell.

data. The uncertainty in the data arises from two sources: (1) changes in the geometry of the measured portion of the electrical field due to slight mechanical changes in the conductivity probe during penetration through hard material (electrical insulation was deformed), and (2) calibration uncertainty. Calibration uncertainty is estimated at less than 0.5% of the measured calibration values. The primary contribution to the limits of uncertainty varied from ±2% to ±15%. Changes in electrode geometry due to deformation of the insulation of the probe tip can increase or decrease the response factor of the probe. For example, exposure of a portion of the outer electrodes that is nearest the inner electrode tips would result in a change such as that seen with probe 1 (Figure 7), but the opposite change would occur with partial exposure of the far sides of the outer electrodes (Figure 8). Our best estimates of the conductivity (Table 3) represented by the boxed data (Figures 17 to 22) are based on the assumptions discussed in the next paragraph and are considered to be representative to within approximately ±3% of the estimated values. The estimated values at each data point (depth of penetration) were determined by analysis of the rate of penetration and probe calibration data (calibrations carried out before and after field tests, Figures 7 to 12 and 23 to 28). These figures show the actual measurements of pre- and post-calibrations of probe conductivities and show the least-squares calibration lines.

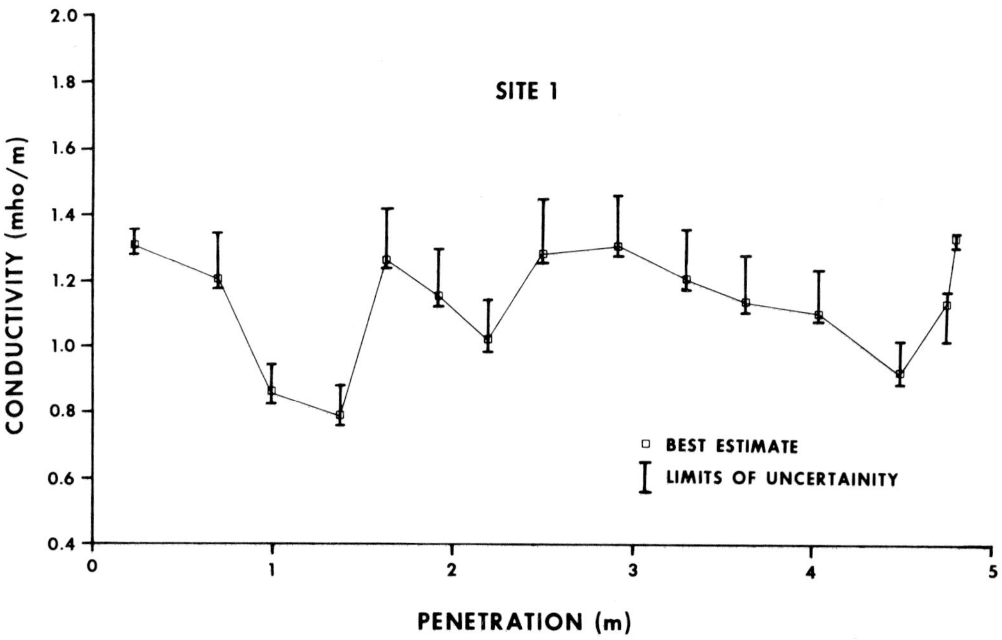

FIGURE 17. Conductivity data vs. depth — site 1.

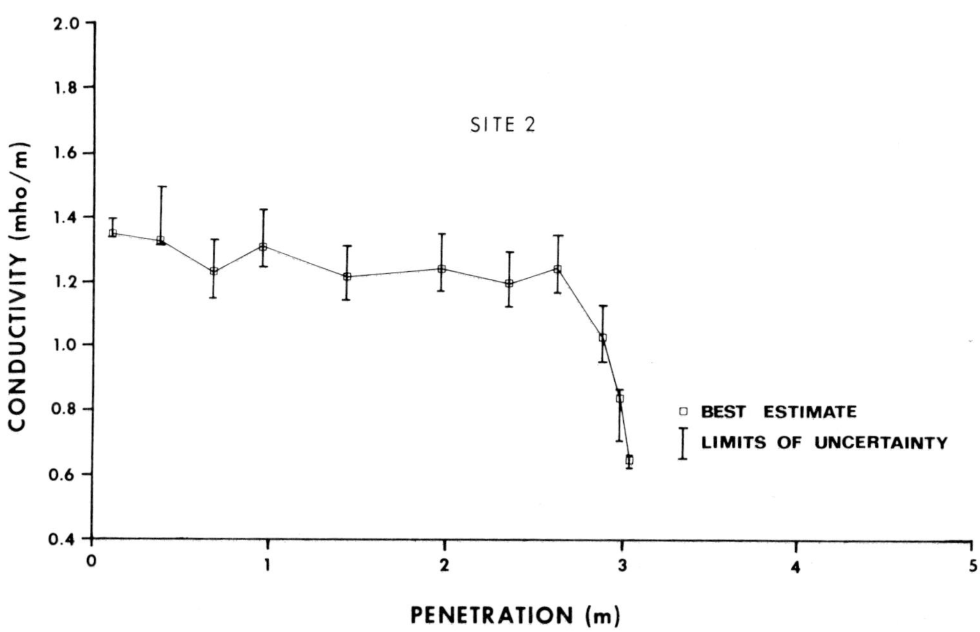

FIGURE 18. Conductivity data vs. depth — site 2.

During the initial sediment penetration of the conductivity probes, negligible deformation of the probe insulation is expected to have occurred. Deformation began upon contact with hard sedimentary layers as indicated by the plots of penetration rate vs.

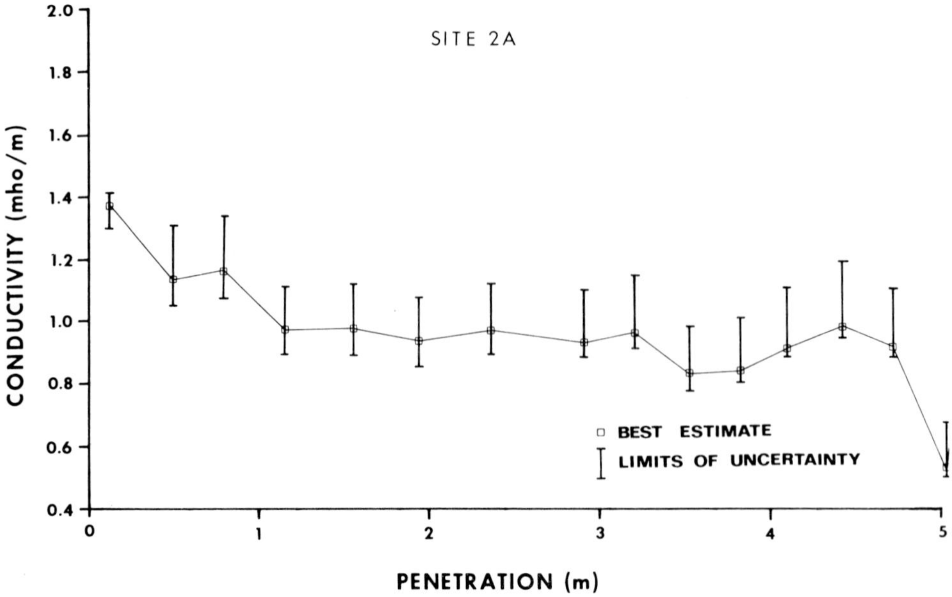

FIGURE 19. Conductivity data vs. depth — site 2A.

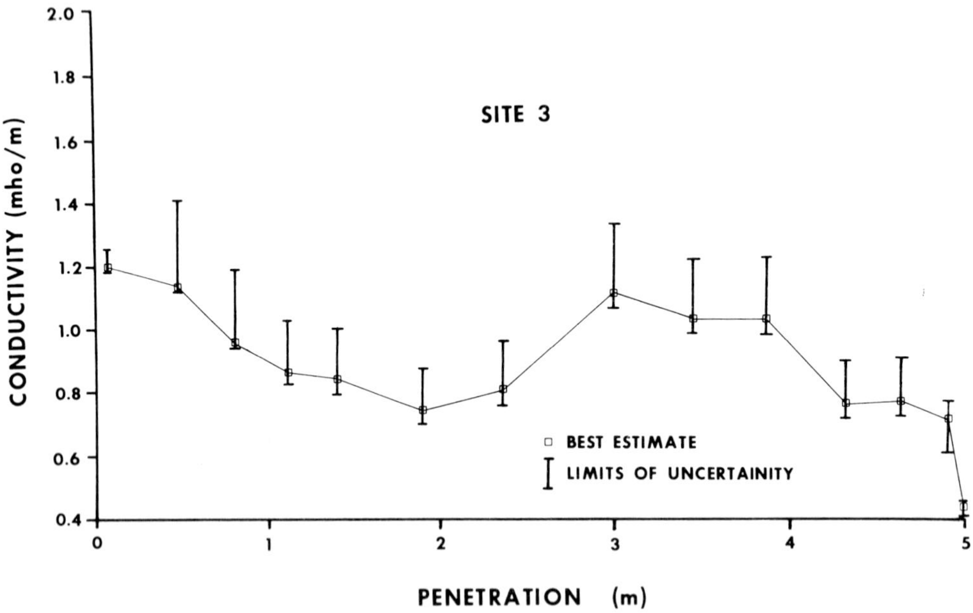

FIGURE 20. Conductivity data vs. depth — site 3.

depth (Figures 23 to 28). Deformation is believed to have occurred progressively as additional hard layers were encountered, with the greatest deformation occurring in the hardest layers. Thus, the appropriate calibration for the initial data point would be that made before the field test, while that for the deepest data point would be the

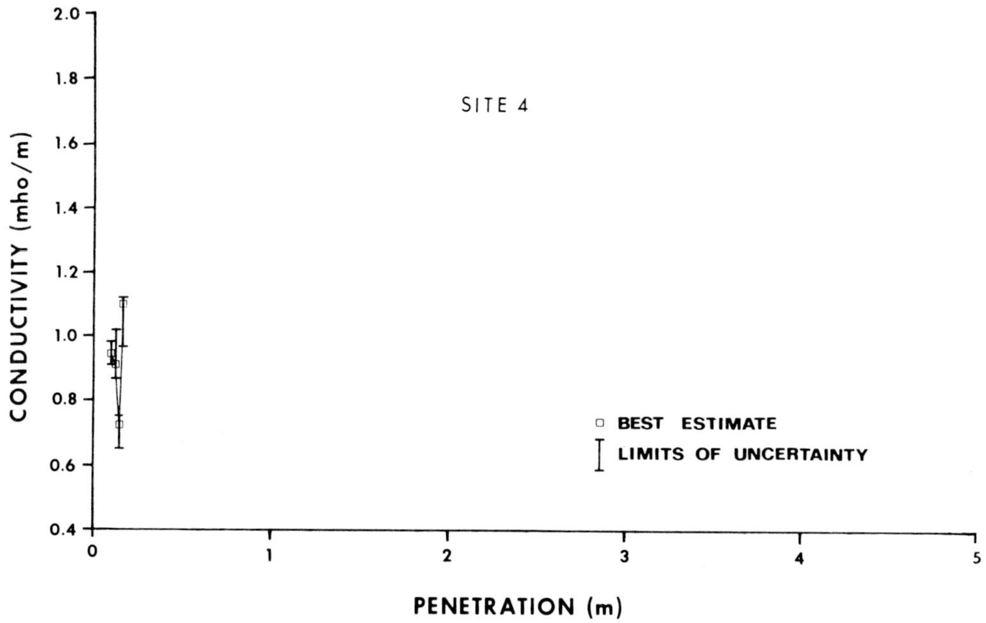

FIGURE 21. Conductivity data vs. depth — site 4.

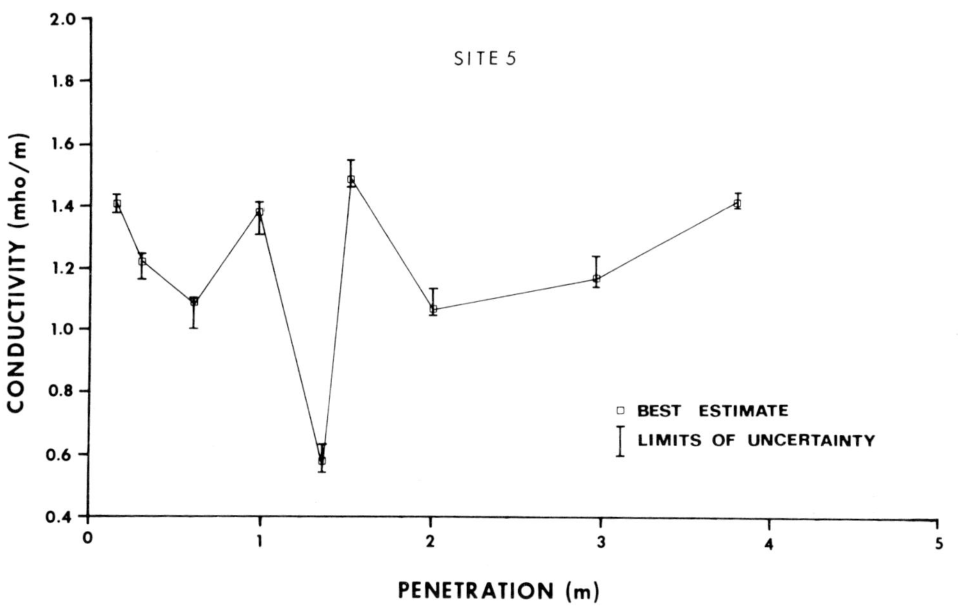

FIGURE 22. Conductivity data vs. depth — site 5.

calibration made after the probe was used in the field. The indicated range of uncertainty in other values measured for conductivity includes the conductivities calculated using both the predeployment calibration and the postdeployment calibration. To assess the maximum effect insulation deformation might have on the conductivity read-

Table 3
BEST ESTIMATE OF CONDUCTIVITY WITH DEPTH AT EACH PROBE SITE

	Depth below sediment/ water interface (m)	Conductivity (mho/m)
Site 1	0.23	1.31
	0.69	1.20
	0.99	0.85
	1.37	0.79
	1.62	1.26
	1.91	1.15
	2.19	1.02
	2.49	1.29
	2.90	1.31
	3.28	1.21
	3.61	1.14
	4.01	1.10
	4.47	0.92
	4.72	1.14
	4.78	1.33
Site 2	0.10	1.35
	0.38	1.33
	0.68	1.23
	0.97	1.32
	1.45	1.22
	1.98	1.25
	2.36	1.20
	2.64	1.25
	2.90	1.04
	3.00	0.84
	3.02	0.65
Site 2A	0.10	1.38
	0.48	1.14
	0.79	1.17
	1.14	0.98
	1.55	0.98
	1.93	0.94
	2.36	0.97
	2.90	0.94
	3.20	0.97
	3.50	0.83
	3.81	0.84
	4.09	0.91
	4.42	0.99
	4.70	0.92
	5.03	0.52
	5.08	0.81
Site 3	0.08	1.20
	0.48	1.13
	0.81	0.96
	1.12	0.86
	1.40	0.84
	1.91	0.74
	2.36	0.80
	3.00	1.11
	3.45	1.03

Table 3 (continued)
BEST ESTIMATE OF CONDUCTIVITY WITH DEPTH AT EACH PROBE SITE

	Depth below sediment/ water interface (m)	Conductivity (mho/m)
	3.86	1.03
	4.32	0.76
	4.62	0.77
	4.90	0.71
	5.00	0.43
Site 4	0.10	0.95
	0.13	0.91
	0.15	0.72
	0.16	1.10
Site 5	0.15	1.42
	0.30	1.23
	0.61	1.09
	0.99	1.40
	1.37	0.56
	1.52	1.50
	2.01	1.06
	2.97	1.17
	3.81	1.43

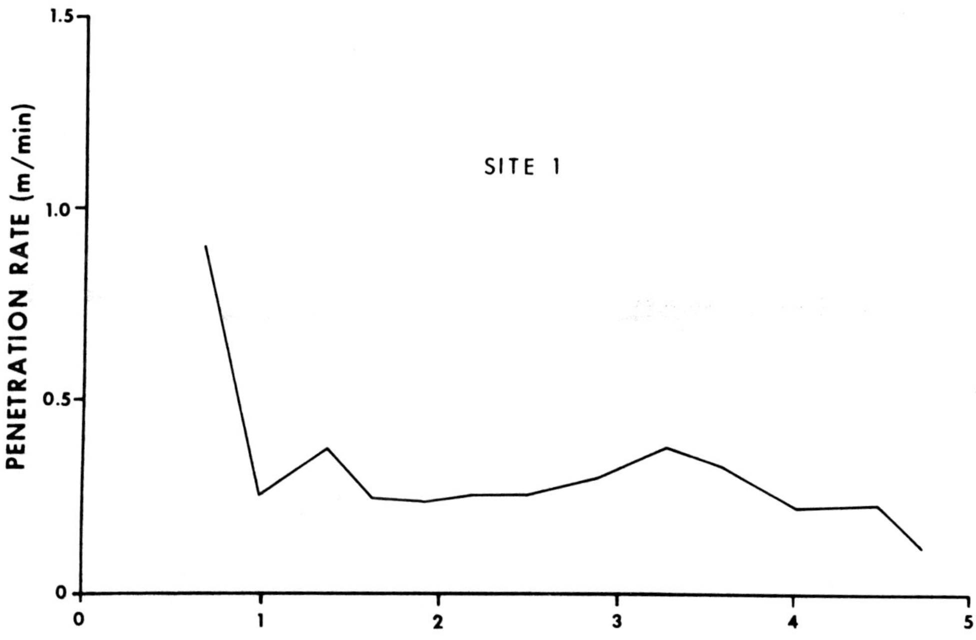

FIGURE 23. Probe penetration rate vs. depth — site 1.

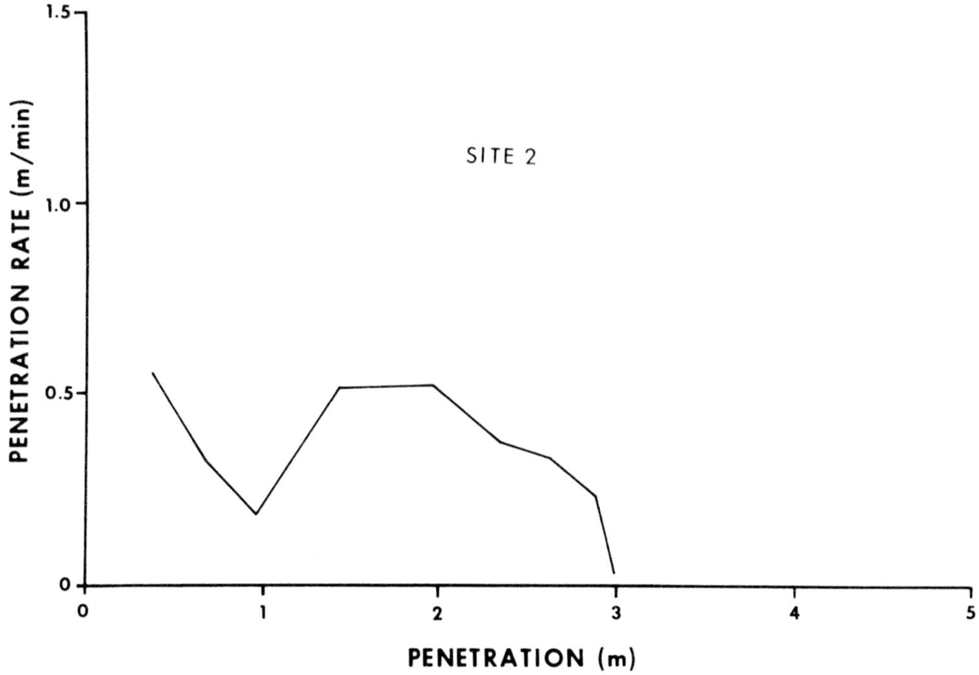

FIGURE 24. Probe penetration rate vs. depth — site 2.

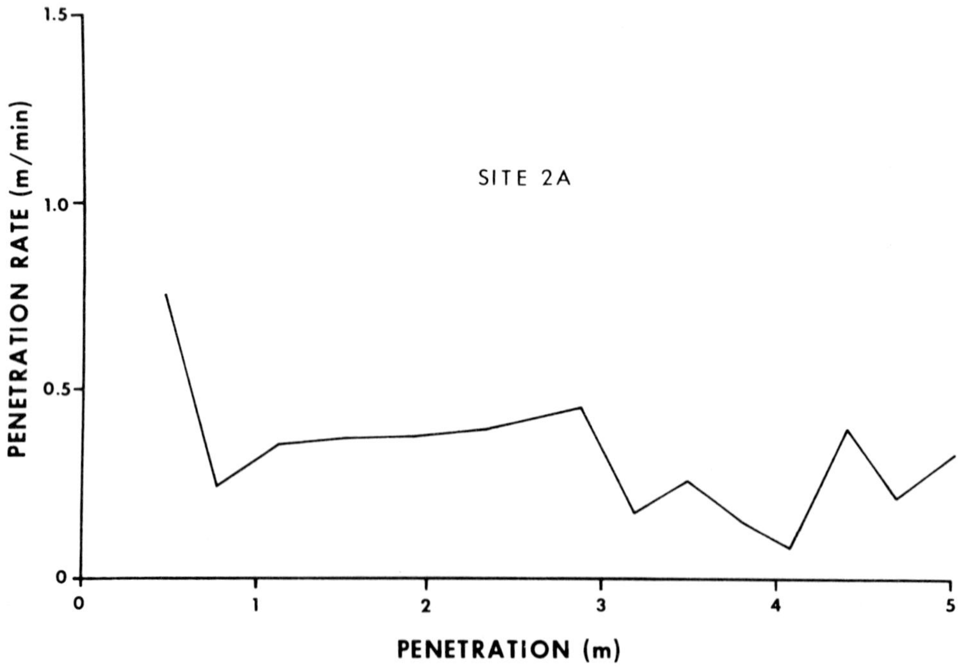

FIGURE 25. Probe penetration rate vs. depth — site 2A.

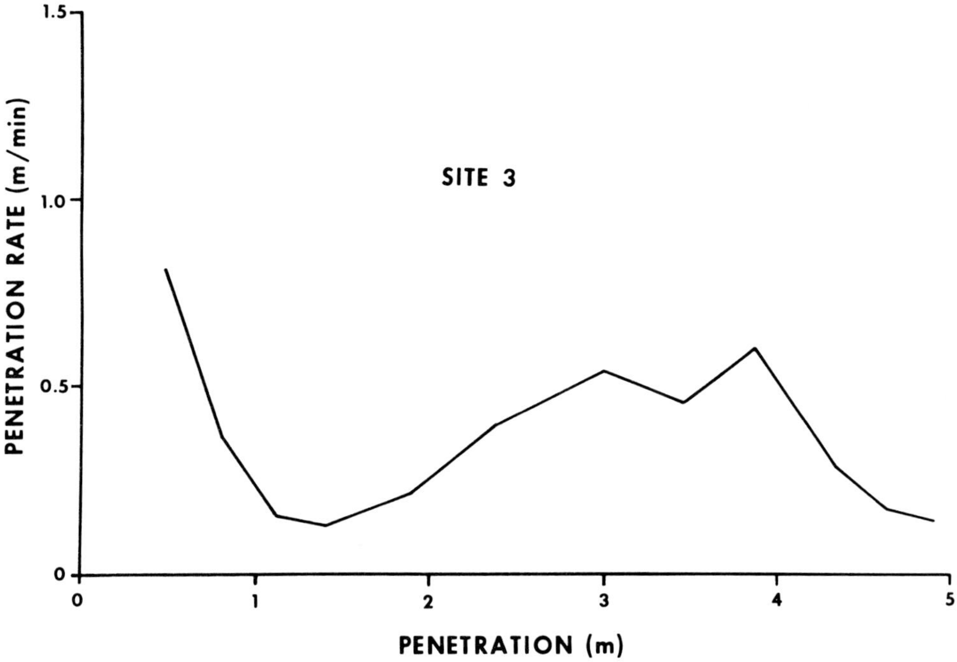

FIGURE 26. Probe penetration rate vs. depth — site 3.

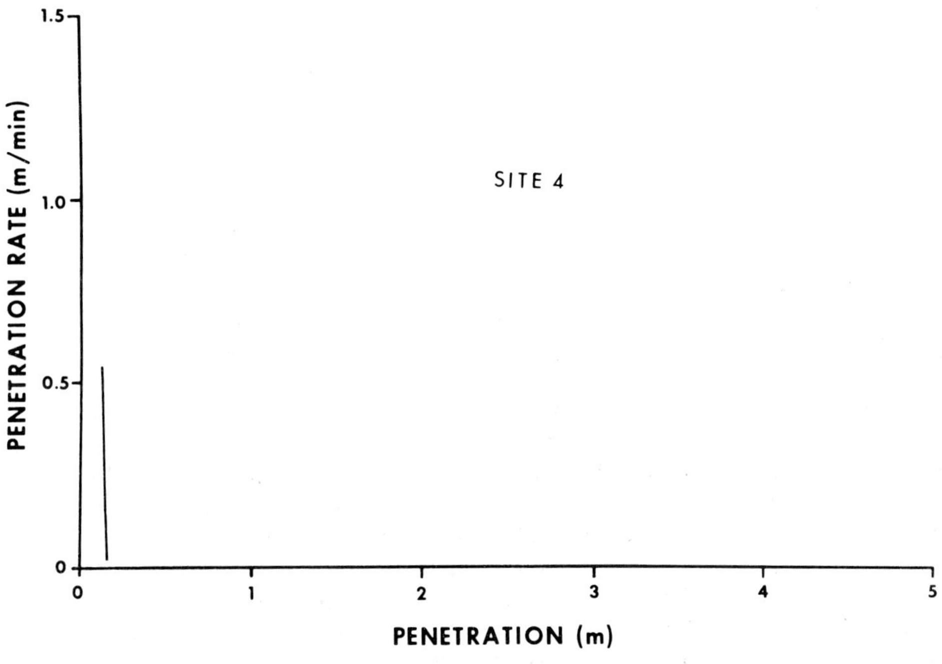

FIGURE 27. Probe penetration rate vs. depth — site 4.

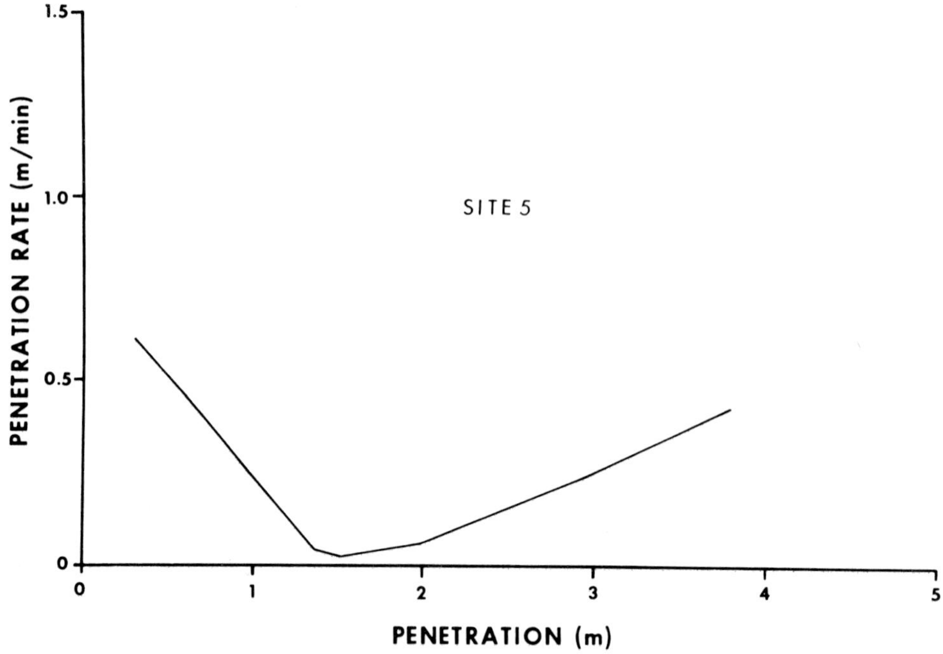

FIGURE 28. Probe penetration rate vs. depth — site 5.

ings, the electrodes of one probe were completely stripped of insulation after recalibration. The results of the calibrations of this probe, and data obtained without any insulation on the electrodes, are shown in Figure 29. Note that even gross changes in the geometry of the exposed electrodes given only moderate changes in the calibration line, particularly in the range of conductivities of interest.

It should be noted that the final data points at all sites, except site 5, were made in material that resulted in penetration refusal for the probe. These data points at some sites gave exceptionally low values of conductivity. This is due to real refusal; the hard material is presumed to be relatively thin sedimentary layers. The authors believe that below these hard layers the sediments become relatively softer and that the sediments have somewhat higher conductivities below the refusal depths (compare Figures 18 and 19 for sites 2 and 2A). Table 4 indicates which probes (No. 1 to 6) were used at each field site and the type of test carried out at each site.

Best estimates of the conductivity between the depth of refusal and depths of approximately 9.1 m (30 ft) are depicted in Figures 30 to 34. Regression analysis was carried out on data from sites 1, 2, 2A, 3, and 5 for a composite of all useful data (low conductivity values obtained at some of the sites at refusal depths were not used in the regression analysis since these values represented a small lithologic unit not representative of the total sediment column) for further interpretation and obtaining an average analysis for all sites with the exception of site 4 (Figure 35). It is important to note that conductivity of sediments will change with temperature (approximately 1.5% increase/C° increase). The reported conductivities were made when the bottom temperatures varied from 21.7 to 22.3°C. Bottom temperatures at the time of field operations are summarized in Table 5.

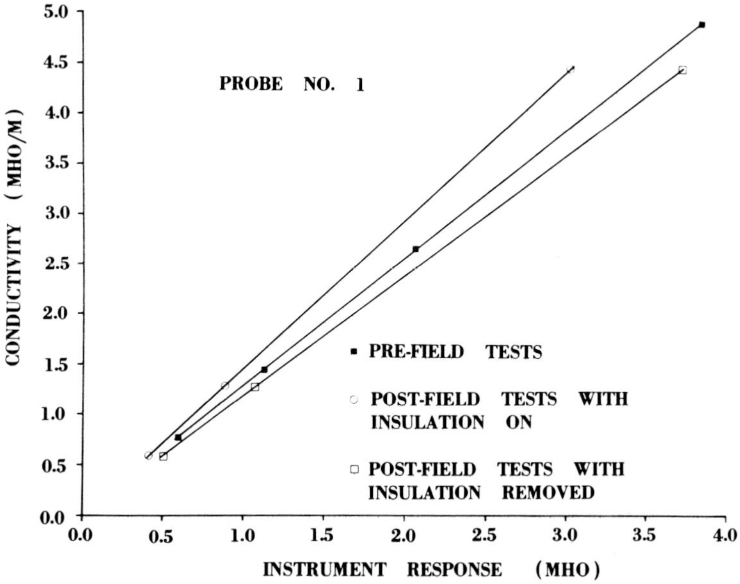

FIGURE 29. Difference in calibration data of Probe No. 1 with and without electrode insulation.

Table 4
NUMBER OF PROBE
USED AT EACH SITE

Site No.	Probe No.
1	1
2	6
2A	2
3	4
4	3
5	5

FIELD OPERATIONS TAKEN
AT EACH SITE

Vibracore	1, 2, 3, 4, 5, & 2A
Probe	1, 2, 3, 4, 5, & 2A
Drill core	2A, 2B, & 6

DISCUSSION

Field Tests

Laboratory tests indicated that in homogenous material of approximately the conductivity of the sediments actually investigated, 90% of the volume of material contributing to the conductivity measurements was within the zone extending vertically 56 mm (2.2 in.) above and below the probe tips. Thus, each measured value of the conductivity reported here represents an average over a vertical range of 110 mm (4.4 in.; see section on calibration). As the conductivity of a material decreases, the vertical extent of the zone containing any fixed fraction of the material contributing to the

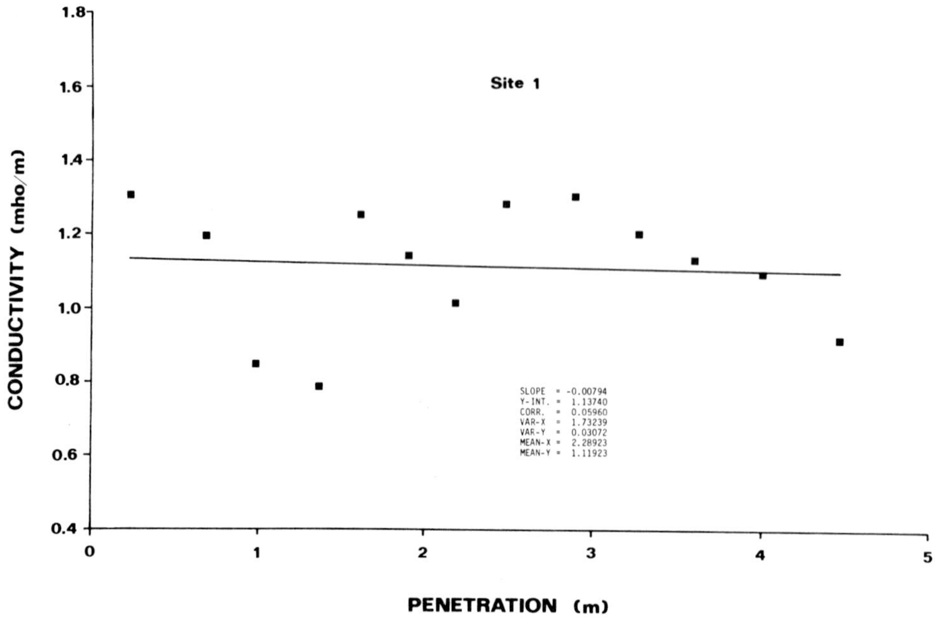

FIGURE 30. Conductivity regression analysis — site 1.

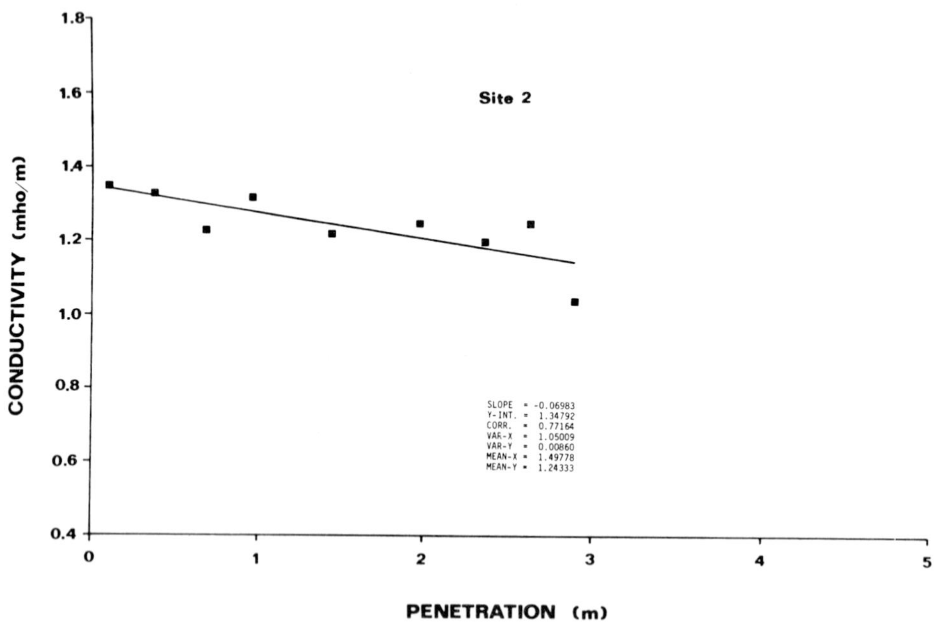

FIGURE 31. Conductivity regression analysis — site 2.

measurement increases. For the conductivity materials encountered in this study, the maximum vertical range containing 90% of the volume contributing to the measurements is estimated as 0.2 m (8.6 in.). In nonhomogeneous materials the current tends to flow preferentially in those portions having the highest conductivity.[4]

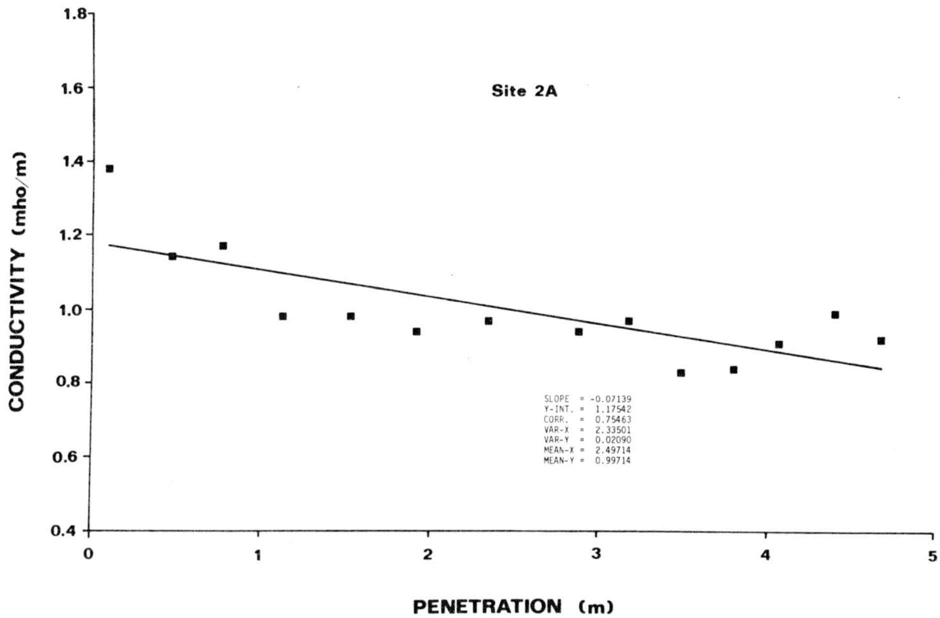

FIGURE 32. Conductivity regression analysis — site 2A.

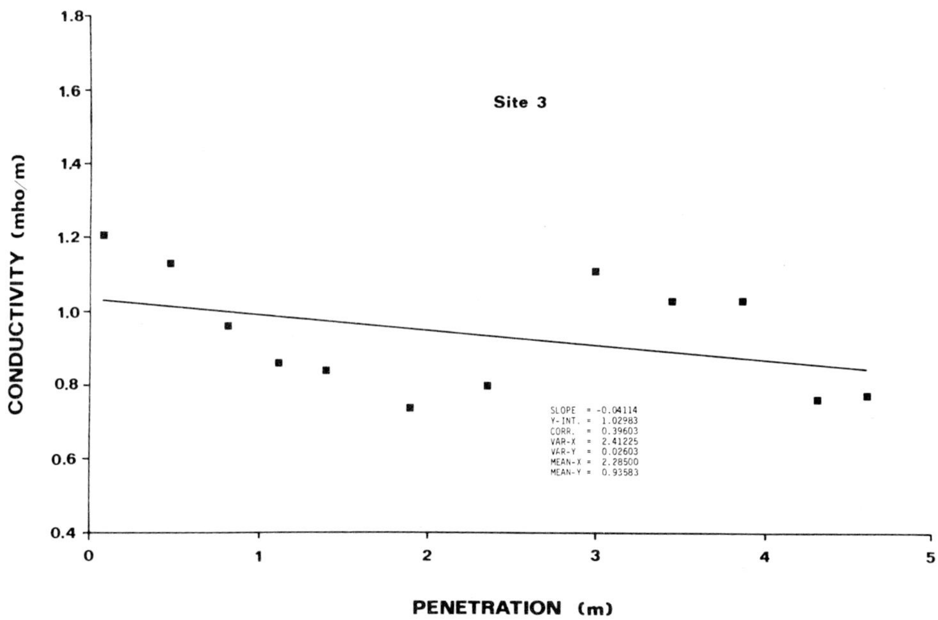

FIGURE 33. Conductivity regression analysis — site 3.

Data comparisons among the observed conductivity values as a function of depth, the rates of probe and core penetration, the seismic reflection records, and the field observations of recovered seabed material, indicate the following:

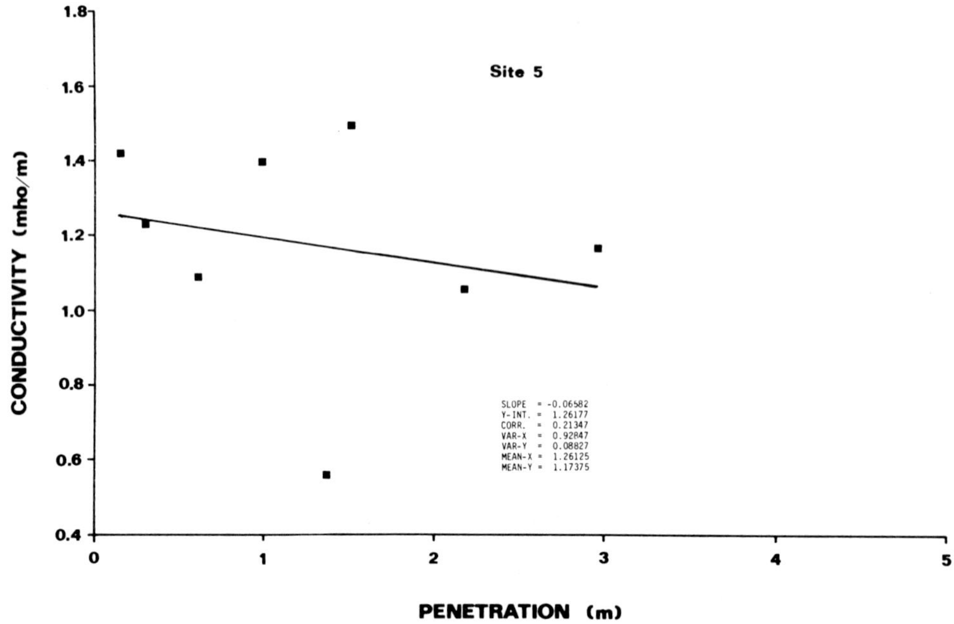

FIGURE 34. Conductivity regression analysis — site 5.

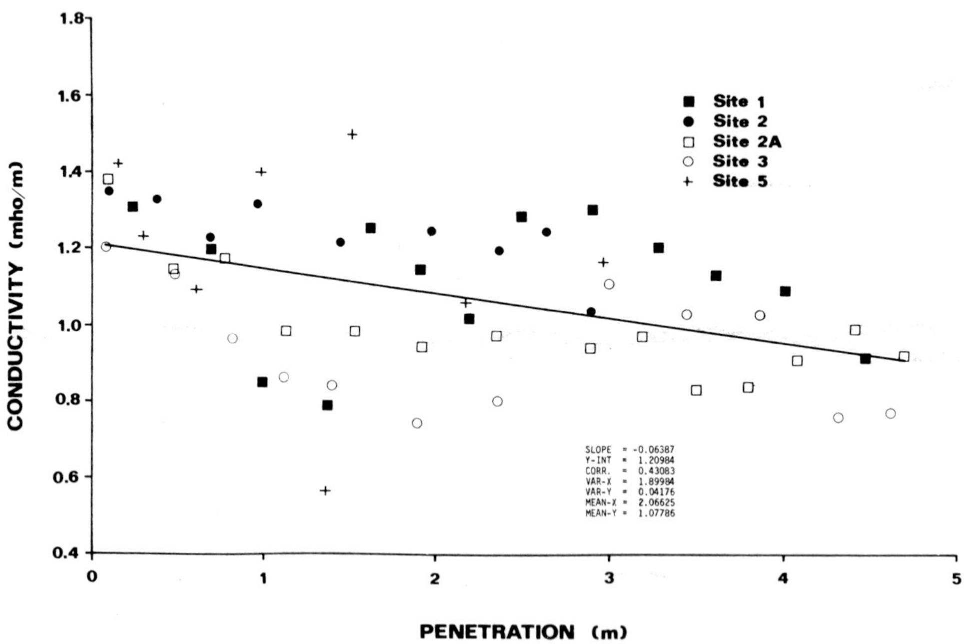

FIGURE 35. A composite conductivity regression analysis for five resistivity probe sites in the study area.

Table 5
BOTTOM WATER TEMPERATURES AND SALINITIES AT CONDUCTIVITY PROBE SITES

Site no.	Bottom water temperature (°C)[a]	Bottom water salinities (‰)[b]
1	22.2	36.06
2	21.7	36.07
2A	21.7	36.05
3	22.0	36.12
4	22.3	35.93
5	22.2	36.08

[a] Temperature determined using a Yellow Springs Instruments electronic thermometer calibrated against a Hewlett-Packard Model 2801 quartz thermometer traceable to the National Bureau of Standards.

[b] Salinity determined using a Model 6230 laboratory salinometer manufactured by Plessey Environmental Systems. All salinities corrected to site bottom temperatures.

1. The upper portion of the seabed is composed of layers of material, distinguishable by the strength of the seismic reflection signal, of differing conductivity.
2. The hard reflectors are typically about 0.3 m (1 ft) thick.
3. Conductivity of the layers giving more intense reflections is approximately 0.5 mho/m, ranging from 0.4 to 0.8 mho/m.
4. Conductivity of the layers giving weak refelctions is typically slightly greater than 1.0 mho/m, ranging from 0.8 to 1.4 mho/m and decreasing slightly with increasing depth of burial.

The layers giving more intense seismic reflections seemed to have greater resistance to penetration by the conductivity probe, probably because of larger coral fragments in these layers. At site 2 and site 3, the probe apparently met penetration refusal with the actual probe tips imbedded in the hard reflector while at site 2A the tips had apparently just broken through the hard reflector at refusal. The conductivity of the sediment in the low area seems to vary within relatively small limits. The average for the area in the upper 4.9 m (16 ft) is about 1.0 mho/m. Regression analysis of the data suggests an average decrease of 0.061 mho/m/m below the sea bottom (Figure 35). The low area including site 5 has similar characteristics (see Figures 30 to 35).

Measurements on the ridge area at site 4 indicate the lithified material has a somewhat lower conductivity than the consolidated sediment. The conductivity of the lithified material is approximately the same as that of the layers giving intense reflection. Because the probe current will preferentially flow through the medium of least resistance, in this case the seawater a few centimeters above the probe tips, the conductivities noted at site 4 represent the combined effect of the seawater and the rock. The relatively high final value of conductivity at site 4 was probably due to the continued operation of the vibrating apparatus resulting in a relatively open current path to the overlying seawater. Rock conductivities at site 4 are believed to be less than 0.7 mho/m, perhaps as low as 0.4 mho/m. The sediment beneath the 2.4 m (8 ft) or so of rock is expected to be less than 1.1 mho/m in uncemented material, and somewhat less in

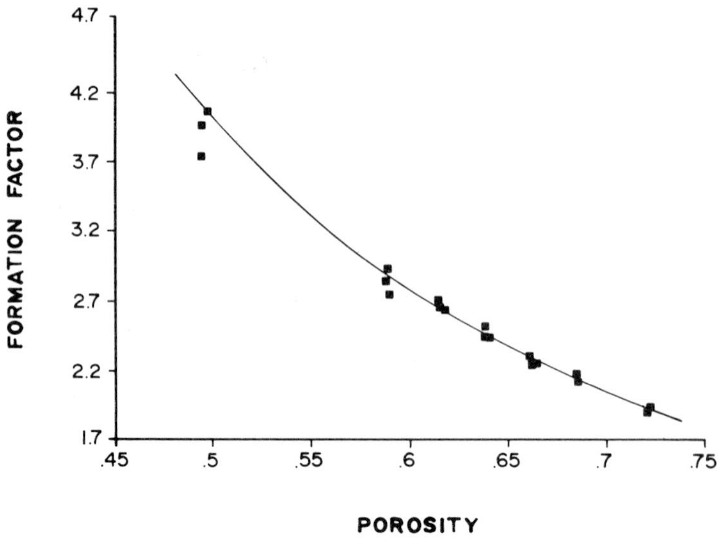

FIGURE 36. Comparison of miniature conductivity probe data with Archies' Law (solid line). Independent measurements of porosity and resistivity were determined on several laboratory sediment samples.

cemented material, on the basis of measurements in the immediate vicinity and measurements in quartz sands.

Sources of Conductivity Anomalies

The porosity of a marine sediment may be expected to be closely related to its electrical conductivity because the electrical current flows through the highly conductive seawater filling the interstitial voids rather than through the nonconducting sediment particles.[4] An experimental confirmation of this expectation resulted from measurements using the miniature probe described above. Submarine sediment clay samples were thoroughly mixed with seawater and triplicate measurements of resistivity determined. Interstitial water salinity and porosity were measured independently for each sample. The ratio of the resistivity of the laboratory tested sediment to the resistivity of the pore water fluid (ratio termed the formation factor), is plotted against the actual porosities (Figure 36). Included in this figure is the curve representing the formulation of the empirical relationship known as Archies' Law:

$$\text{Formation Factor} = (\text{Porosity})^{-2.0}$$

This relationship was initially developed for consolidated, porous rock, and the exponent assumed dependent upon the degree of cementation of the grains; the exponent, -2.0, is reported appropriate for clay-rich sediments.[20] Note that adoption of this exponent leaves no adjustable constants to improve the fit of the curve to the experimental data. However, this simple relationship may be disturbed by a number of factors. In general, these deviations from a simple relationship between porosity and conductivity may be understood as reflections of divergence of the conditions in the sediment from those assumed. A simple example of a source of such divergence is the unusual case of significant discrepancy between the conductivity of seawater at the seafloor and the conductivity of the interstitial pore water in sediments nearby below — usually the two conductivities are assumed equal. A discrepancy of this sort may result from the presence of offshore freshwater springs reducing conductivity of the pore water or from near-surface salt deposits increasing the conductivity. In addition, chemical and biological processes may change the composition of the pore water. These changes have important consequences for many sediment properties, but often change the conductivity minimally. For example, a change in the hydrogen ion concentration by a factor of 1000 may cause important alterations in the biological, chemical, and geological microenvironment, but would typically result in negligible alteration of seawater conductivity.

Hollow particles, such as the tests of a variety of marine organisms, may lead to sediments of very substantial porosity but limited conductivity when the void space within particles is partially or entirely cut off from that external to the particles, and thus cannot make a full contribution to the sediment conductivity. A similar situation may arise in cemented sediments in which there is preferential precipitation in the vicinity of particle contacts rather than in voids. An additional complication that may be expected to arise in situations such as these is the apparent conductivity will depend upon how the measurement is made: the enclosed (partially), conductive material will make a much more significant contribution to conductivities measured by the inductive method than to those measured with classical techniques. The amount of this contribution is expected to be a function of the conductivities of the enclosed material and the pore water, of the size distribution of the enclosed spaces, and the electrical frequency used in the inductive measurement.

A whole class of apparent anomalies in conductivity arise from the fact that while sediments are commonly considered isotropic, conductivity responds to the anisotropy that is common in sediments.[29] Sources of this anisotropy include the differential packing of sediment particles under gravity, the vertical distinctions that exist in bedded sediments, and the filling of interparticle channels by fine particles transported hydraulically. Bioturbation, which may result in sediment sorting and the formation of channels of strongly preferential orientations, may also contribute to anisotropy.

In studies of the conductivity of sediments lying beneath the seabed, it is often possible to make simplifying assumptions valid within the limits of experimental accuracy. Typical assumptions might include: that the conductivity of the interstitial water is equal to the adjacent seawater; that all the electrical current flows in the interstitial water; that the entire pore volume is accessible to the electrical current and is water-filled; that the temperature in the sediment is equal to the temperature of the overlying water; that the sediment is isotropic. As the discussion makes clear, although any or all of these assumptions may be sufficiently accurate for a given investigation in a particular sediment, it is quite possible that they may not. In addition to the problems cited above, there exists the possibility that part of the interstitial volume may be filled with nonconductive gas or that conductive sediment particles may be present. Thus, considerable care is required in the interpretation of conductivity data so that unwarranted assumptions do not invalidate the stated results.

SUMMARY AND CONCLUSIONS

The sediments in the immediate vicinity of sites 1, 2A, 2B, and 3 do not show great variability in electrical conductivity. Average conductivity is 1.0 mho/m and the range is from 0.4 to 1.4 mho/m with lower values in materials showing stronger seismic reflection. This study has demonstrated the feasibility of making *in situ* conductivity measurements in submarine sediments using a four-point contact resistivity technique. The study, combined with laboratory analyses, also indicates that the *in situ* techniques may prove to be a valuable method of obtaining *in situ* soil properties such as porosity and wet bulk density, provided accurate measurements of interstitial water salinity can be obtained during field operations.

Applications of the resistivity probe as a means to provide accurate and rapid measurements of seafloor porosity and wet bulk density are numerous. With an appropriate driving mechanism mounted on a platform or a research submersible, the *in situ* resistivity probe provides the ability to produce a continuous profile of these geotechnical properties. It may be possible to use this technique to measure accurately the sediment-water interface in areas of the seafloor such as the Mississippi Delta where this boundary is often not discernible by conventional techniques. Numerous rapid measurements of porosity and wet bulk density may also be made in sediment cores with a miniaturized resistivity probe in a laboratory or aboard ship.

If a substantial number of high-resolution conductivity measurements were desired, the use of a suitable probe based on the inductive method should be considered. For additional measurements in coarse or fine-grained sediment at penetration depths up to 5.5 m (18 ft), the system designed and used in this project should again prove satisfactory.

APPENDIX 1

OPERATIONS AND CALCULATIONS FOR PROBE CALIBRATIONS

This appendix explains the methods involved in relating the probe/instrument response to actual specific conductivities of the calibration tanks used in this study. From these calibrations, values for the *in situ* conductivities of the sediments were determined.

Step 1. Individual tanks of varying mixtures of fresh and seawater were utilized for both the pre- and post-calibration experiments. The following three operations were carried out:

1. From each tank two 1-ℓ samples were taken for the determination of the conductivities using a precision salinometer.
2. Temperature was measured in each of the calibration tanks.
3. Instrument response was determined by immersion of the probe electrodes in each of the calibration tanks.

Step 2. The water samples were analyzed on the Plessey Laboratory Salinometer to obtain the ratios of their conductivities to the conductivity of Standard Sea-Water (I.A.P.S.O., Institute of Oceanographic Sciences, Wormley, Godalming, Surrey, England). These conductivity ratios required temperature corrections to relate them to the tank temperatures. Standard correction tables are included in the *International Oceanographic Tables,* and generally are used to standardize measurements to either 15°C or 20°C. In this study, tables were used to interpolate temperature corrections such that conductivity ratios were obtained for a given tank mixture at the tank temperature. It should be noted that the temperature corrections for all calibration tanks were very small, but were made to minimize determinate error. Differences in specific conductivities with and without temperature corrections were calculated to be no greater than 0.5% of the observed values.

Example: Post-calibration tank Number 5 at 18.3°C.
Conductivity ratio obtained from the Plessey Laboratory Salinometer = 0.98673 @ 22.8°C (temperature at the time of the salinometer measurement).

From Table IIb in the *International Oceanographic Tables,* the laboratory-measured conductivity ratio was corrected to the tank temperature (see partial reproduction of the table at the end of this Appendix).

Example: 0.98673 (conductivity ratio @ 22.8°C)
−0.00003 (correction for conductivity ratio from 23°C to 18°C)
0.98670

Step 3. From the *Handbook of Marine Science,* a plot was constructed relating temperature to specific conductivity of standard seawater @ 35‰ salinity (conductivity ratio = 1.00000). A linear regression was made with the data and the equation of the relationship obtained (see graph at the end of this Appendix).

C = mT + b
C = specific gravity of $35^{0}/_{00}$ seawater @ temperature "T" (mmho/cm)
m = constant (slope) = 0.99642 mmho/cm°C
T = temperature (°C)
b = Y-intercept = 28.03340 mmho/cm

Step 4. Determine the specific conductivity of $35^{0}/_{00}$ seawater at the tank temperature. This is derived from the equation in Step 3.

Example: Post-calibration tank temperature 18.3°C
Specific conductivity = 0.99642 × 18.3 + 28.03340 of $35^{0}/_{00}$ seawater at tank temperature = 46.268 mmho/cm or 4.6268 mho/m.
Note: conductivity ratio = 1.00000 for $35^{0}/_{00}$ seawater.

Step 5. This step will use the derived data from Steps 2 and 4 to obtain true specific conductivity of the tank mixture at the tank temperature. Since the conductivity ratio is a ratio of specific conductivities based on the conductivity of seawater at $35^{0}/_{00}$, a sample analyzed on the salinometer with a salinity of 35 $^{0}/_{00}$ will yield a conductivity ratio of 1. Therefore, to obtain true specific conductivity of the tank mixture at the tank temperature, the following operation must be performed:

Given: Specific conductivity of $35^{0}/_{00}$ seawater at the tank temperature, 4.6268 mho/m.
Conductivity ratio of the tank number 5 @ 18.3°C = 0.98670. True specific conductivity of the tank Number 5 at calibration temperature = 4.6268 mho/m × 0.98670 = 4.57 mho/m.

The previous steps were followed for each of the calibration tanks to obtain true specific conductivities for all mixtures used in this study.

Additional example using Steps 1 through 5:

Given: Pre-calibration tank Number 2
Tank temperature of 22.5°C
Conductivity ratio of 0.52588 run at 23.0°C.

Calculate specific conductivity of the tank at 22.5°C:

1. Correct conductivity ratio to 22.5°C using the temperature correction tables (see Tables and Step 2):
 0.52588 (conductivity ratio @ 23.0°C)
 −0.00007 (correction interpolated from the tables)
 0.52581 (corrected conductivity ratio @ 22.5°C)
2. Using the equation relating specific conductivity of $35^{0}/_{00}$ salinity to temperature, determine the specific conductivity of $35^{0}/_{00}$ salinity at the tank temperature (22.5°C; Steps 3 and 4):
 Specific conductivity of $35^{0}/_{00}$ seawater at tank temperature (22.5°C) = 0.99642 × 22.5 + 28.03340 = 50.452 mmho/m or 5.0452 mho/m.
3. Calculate the true specific conductivity of the tank by multiplying the results from (a) by the results from (b):
 True specific conductivity of tank number 2 at calibration temperature = 0.52581 × 5.0452 mho/m = 2.653 mho/m.

PRE-FIELD CALIBRATION TABLE

Probe 1, Site 1

Calibration tank number	1	2	3	4
Tank temperature (°C)	22.0	22.5	23.1	23.3
Specific conductivity (mho/m) of 35‰ seawater @ tank temperature	4.9954	5.0452	5.1050	5.1250
Conductivity ratio @ tank temperature	0.97756	0.52581	0.28239	0.14961
True specific conductivity (mho/m) of calibration tank @ tank temperature	4.883	2.653	1.442	0.767
Instrument response V/I (ohms)	0.260	0.484	0.884	1.685
Instrument response I/V (mho)	3.846	2.067	1.131	0.594

POST-FIELD CALIBRATION TABLE

Probe 1, Site 1

Calibration tank number	5	6	7
Tank temperature (°C)	18.3	18.9	19.4
Specific conductivity (mho/m) of 35‰ seawater @ tank temperature	4.6268	4.691	4.7364
Conductivity ratio @ tank temperature	0.98670	0.28687	0.13181
True specific conductivity (mho/m) of calibration tank @ tank temperature	4.565	1.346	0.624
Instrument response V/I (ohms)	0.315	1.091	2.308
Instrument response I/V (mho)	3.175	0.917	0.433

TEMPERATURE CORRECTION TABLE
(International Oceanographic Table IIb, partial reproduction)[a]

Conductivity ratio	18°	19°	20°	21°	22°	23°	24°	25°
0.10	15	8	0	−7	−15	−22	−29	−36
0.15	21	10	0	−10	−20	−30	−40	−49
0.20	25	12	0	−12	−24	−36	−48	−59
0.25	28	14	0	−14	−27	−41	−54	−67
0.30	31	15	0	−15	−30	−44	−58	−72
0.35	32	16	0	−16	−31	−46	−61	−75
0.40	33	16	0	−16	−32	−47	−62	−77
0.45	33	16	0	−16	−32	−47	−62	−76
0.50	32	16	0	−16	−31	−46	−60	−75
0.55	31	15	0	−15	−30	−44	−58	−72
0.60	29	15	0	−14	−28	−42	−55	−68
0.65	27	13	0	−13	−26	−38	−51	−62
0.70	25	12	0	−12	−23	−35	−46	−56
0.95	5	3	0	−3	−5	−7	−10	−12
0.96	4	2	0	−2	−4	−6	−8	−10
0.97	3	2	0	−2	−3	−4	−6	−7
0.98	2	1	0	−1	−2	−3	−4	−5
0.99	1	1	0	−1	−1	−2	−2	−2
1.00	0	0	0	0	0	0	0	0

Note: This table corrects conductivity ratios to 20°C, but in this study we use it to correct conductivity ratios to calibration tank temperatures.

[a] Correction ($\times 10^{-5}$)

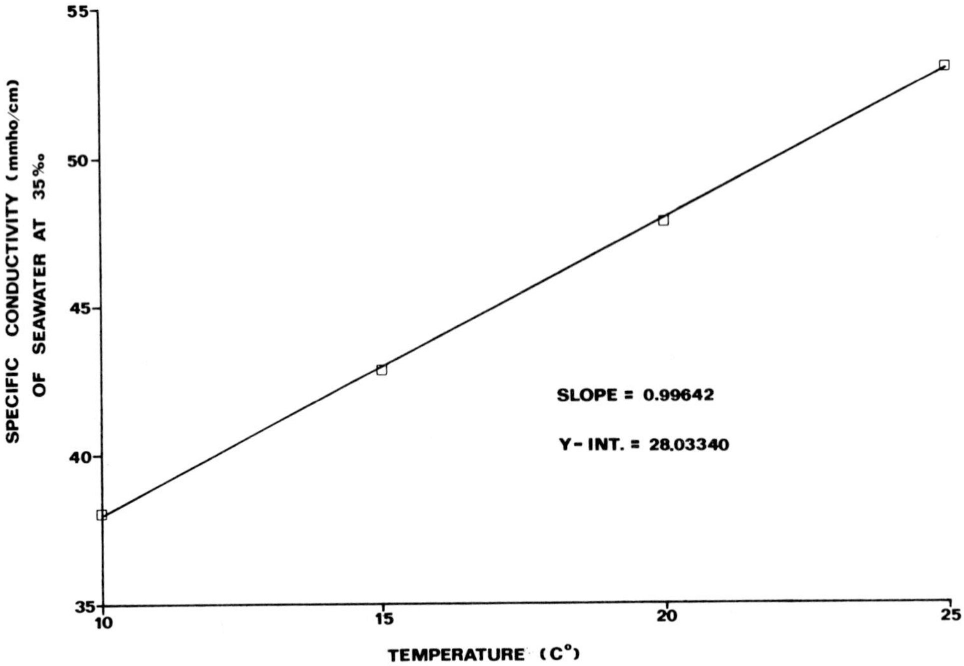
SPECIFIC CONDUCTIVITY (mmho/cm) OF SEAWATER AT 35‰
55
50
45
40
35
10
15
20
25
TEMPERATURE (C°)
SLOPE = 0.99642
Y-INT. = 28.03340

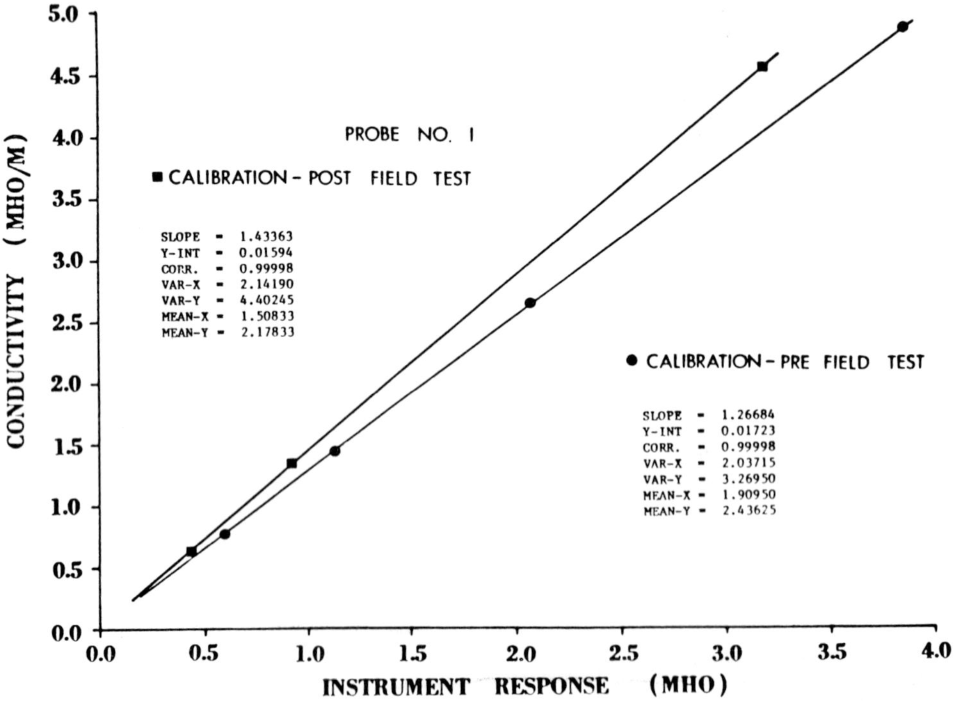
CONDUCTIVITY (MHO/M)
INSTRUMENT RESPONSE (MHO)
PROBE NO. I
■ CALIBRATION - POST FIELD TEST
SLOPE = 1.43363
Y-INT = 0.01594
CORR. = 0.99998
VAR-X = 2.14190
VAR-Y = 4.40245
MEAN-X = 1.50833
MEAN-Y = 2.17833
● CALIBRATION - PRE FIELD TEST
SLOPE = 1.26684
Y-INT = 0.01723
CORR. = 0.99998
VAR-X = 2.03715
VAR-Y = 3.26950
MEAN-X = 1.90950
MEAN-Y = 2.43625

APPENDIX B

RESISTIVITY/CONDUCTIVITY UNCORRECTED FIELD DATA

Site 1, Probe 1

Depth (m)	I[a]	V[a]	V/I (ohm)[b]	I/V (mho)[b]
0.24	22.42	22.04	0.983	1.017
0.67	23.25	25.02	1.076	0.929
0.98	13.54	20.96	1.548	0.646
1.37	21.23	35.57	1.675	0.597
1.62	26.25	26.92	1.026	0.975
1.89	24.67	27.79	1.126	0.888
2.19	22.16	28.48	1.285	0.778
2.50	25.56	25.48	0.996	1.003
2.90	24.97	24.90	0.997	1.003
3.29	25.12	27.03	1.076	0.929
3.60	27.01	30.95	1.146	0.873
4.02	24.60	29.07	1.182	0.846
4.48	22.48	32.44	1.443	0.693
4.72	20.15	25.24	1.253	0.798
4.79	19.26	21.02	1.091	0.916

Site 2, Probe 6

Depth (m)	I[a]	V[a]	V/I (ohm)[b]	I/V (mho)[b]
0.09	17.09	17.58	1.029	0.972
0.37	18.00	18.89	1.049	0.953
0.67	16.74	19.89	1.188	0.842
0.98	16.67	18.38	1.103	0.907
1.46	16.47	19.77	1.200	0.833
1.98	17.63	20.53	1.164	0.859
2.38	16.13	19.65	1.218	0.821
2.65	18.29	21.28	1.163	0.859
2.90	15.39	21.66	1.407	0.711
2.99	13.79	25.15	1.824	0.548
3.02	27.80	65.57	2.359	0.424

Site 2a, Probe 2

Depth (m)	I[a]	V[a]	V/I (ohm)[b]	I/V (mho)[b]
0.09	18.79	20.04	1.067	0.938
0.48	19.94	22.59	1.133	0.883
0.79	12.41	13.55	1.092	0.916
1.16	18.05	24.00	1.330	0.752
1.55	19.83	26.19	1.321	0.757
1.92	18.16	25.08	1.381	0.724
2.38	17.24	22.75	1.320	0.758
2.90	15.86	21.36	1.347	0.742
3.20	17.17	22.18	1.292	0.774
3.51	16.36	24.82	1.517	0.659
3.81	14.57	21.43	1.471	0.680
4.08	15.95	21.35	1.339	0.747
4.42	17.54	21.80	1.243	0.805
4.69	17.16	23.00	1.340	0.746
5.03	14.35	32.51	2.266	0.441
5.09	23.49	34.85	1.484	0.674

Site 3, Probe 4

Depth	I[a]	V[a]	V/I (ohm)[b]	I/V (mho)[b]
0.06	21.13	22.56	1.068	0.937
0.49	20.88	23.41	1.121	0.892
0.82	19.10	25.37	1.328	0.753
1.13	18.71	28.62	1.530	0.654
1.40	14.57	23.00	1.579	0.633
1.89	16.05	28.80	1.794	0.557
2.38	16.16	26.74	1.655	0.604
2.99	18.66	22.06	1.182	0.846
3.44	20.88	26.82	1.284	0.779
3.87	19.75	25.42	1.287	0.776
4.33	18.02	31.51	1.749	0.572
4.91	13.62	27.78	2.040	0.490
5.00	14.97	51.96	3.471	0.288

Site 4, Probe 3

Depth (m)	I[a]	V[a]	V/I (ohm)[b]	I/V (mho)[b]
0.09	9.89	15.13	1.530	0.654
0.12	15.10	24.24	1.605	0.623
0.15	24.06	52.81	2.195	0.456
0.15	27.12	39.40	1.453	0.688

Site 5, Probe 5

Depth (m)	I[a]	V[a]	V/I (ohm)[b]	I/V (mho)[b]
0.15	22.04	19.59	0.888	1.125
0.30	19.79	20.38	1.030	0.971
0.61	20.77	24.34	0.853	1.172
0.98	22.24	20.07	0.902	1.108
1.37	21.06	44.42	2.109	0.474
1.52	28.08	22.58	0.803	1.245
2.01	31.62	35.55	1.124	0.889
2.99	38.76	29.48	1.025	0.976
3.81	32.74	27.58	0.842	1.187

[a] Instrument response.
[b] Uncorrected by calibrations.

REFERENCES

1. **Pirson, S. J.,** *Handbook of Well Log Analysis,* Prentice-Hall, Englewood Cliffs, N.J., 1963, 326.
2. **Wyllie, M. R. J.,** *The Fundamentals of Well Log Interpretation,* 3rd ed., Academic Press, New York, 1963, 238.
3. **Griffiths, G. V. and King, R. F.,** *Applied Geophysics,* Pergamon Press, Oxford, 1965, 223.
4. **Keller, G. V. and Frischknecht, F. C.,** *Electrical Methods in Geophysical Prospecting,* Pergamon Press, New York, 1966, 379.

4a. **Koning, H. L.,** Personal communication, 1980.

5. **Strangeway, D. W.,** *Methods and Techniques in Geophysics,* Vol. 2, Runcorn, S. K., Ed., Interscience, London, 1966, chap. 1.
6. **Parkhomenko, E. I.,** *Electrical Properties of Rocks,* Transl. from Russian by G. V. Keller, Plenum Press, New York, 1967, 314.
7. **Erchul, R. A.,** Ocean engineering applications for electrical resistivity techniques, Offshore Technol. Conf. Pap. 2012, Offshore Technology Conference, 733, 1974.
8. **Ginsburg, A.,** *Resistivity Surveys, Geophysical Surveys,* Vol. 1, 1974, 325.
9. **Dobrin, M. B.,** *Introduction to Geophysical Prospecting,* 3rd ed., McGraw-Hill, New York, 1976.
10. **Boyce, R. E.,** Electrical resistivity of modern marine sediments from the Bering Sea, *J. Geophys. Res.,* 73, 4759, 1968.
11. **Erchul, R. A. and Nacci, V. A.,** The use of electrical resistivity measurements to predict porosity of marine sediments, in Proc. Int. Symp. Eng. Prop. Seafloor Soils and Geophys. Identification, Seattle, 1971, 296.
12. **Erchul, R. A. and Nacci, V. A.,** Electrical resistivity measuring system for porosity determination of marine sediments, *J. Mar. Technol. Soc.,* 6(47), 1972.
13. **Taylor-Smith, D.,** Acoustic and electric techniques for seafloor sediment identification, in Proc. Int. Symp. Eng. Prop. Seafloor Soils and Geophys. Identification, Seattle, 1971, 253.
14. **Sweet, W. E., Jr.,** Electrical Resistivity in Unconsolidated Sediments, Sea Grant Rep. #TAMU-SG-72-205, Texas A & M University, College Station, 1972.
15. **Barnes, B. B., Corwin, R. F., Beyer, J. H., and Hildenbrand, T. G.,** Geologic prediction: developing tools and techniques for the geophysical identification and classification of sea-floor sediments, NOAA Tech. Rep. ERL224-MMT-C2, U.S. Department of Commerce, Washington, D.C., 1972.
16. **Bouma, A. H., Sweet, W. E., Chinelik, F. B., and Heubner, G. L.,** Shipboard and *in situ* electrical resistivity logging of unconsolidated marine sediments, Vol. 1, 3rd Offshore Technol. Conf., Houston, Tex., 1971, 253.
17. **Bouma, A. H., Ferebee, T. W., Chinelik, F. B., and Heubner, G. L.,** Electrical logging systems and results of unconsolidated marine sediments, Vol. 1, 7th Offshore Technol. Conf., Houston, Tex., 1975, 753.
18. **Beckman, H. and Demiray, G.,** The nanolog, a new way to log the resistivity of the seafloor, *Geophysics,* 24, 309, 1976.
19. **Jackson, P. D.,** An electrical resistivity method for evaluating the *in situ* porosity of clean marine sands, *Mar. Geotechnol.,* 1, 91, 1975.
20. **Jackson, P. D., Taylor-Smith, D., and Stanford, R. N.,** Resistivity-porosity-particle shape relationships for marine sands, *Geophysics,* 43, 1250, 1978.
21. **Wang, M. C., Nacci, V. A., and Markert, C. D.,** Electrical resistivity method for marine sediment consolidation study, Pap. No. OTC 2623, 8th Offshore Technol. Conf., Houston, Tex., 1976, 45.
22. **Kermabon, A. J., Gelin, C., and Blavier, P.,** A deep-sea electrical resistivity probe for measuring porosity and density of unconsolidated sediments, *Geophysics,* 34, 554, 1969.
23. **Boyce, R. E.,** Electrical Resistivity of Modern Marine Sediments from the Bering Sea, NUWC Res. Rep. #TP6, National Undersea Warfare Center, New London, Conn., 1976, 106.
24. **Smith, F. C., Ed.,** *Handbook of Marine Science,* Vol. 1, CRC Press, Boca Raton, Fla., 1974.
25. **Siegel, V. B.,** Reconnaissance Survey of the Straits of Florida, Marine Lab., TR 59-3, University of Miami, Coral Gables, Fla., 1959.
26. **Raymond, W. F.,** The marine geology of Broward County, Fla., Masters thesis, Florida State University, Tallahassee, 1972.
27. **Duane, D. and Meisburger, E.,** Geomorphology and sediments of the near-shore continental shelf Miami to Palm Beach, Florida, U.S. Army Corps of Eng., Tech. Memo. 29, U.S. Department of Defense, Washington, D.C., 1969.
28. **Hulbert, M. H., Lambert, D. N., Bennett, R. H., Freeland, G. L., Burns, J. T., and Sawyer, W. B.,** *In situ* electrical resistivity measurements of calcareous sediments, *Am. Soc. Test. Mater. Spec. Tech. Publ.,* in press.
29. **Kutter, B. L., Arulanandan, K., Dafalias, Y. F.,** A comparison of electrical and penetration methods of site investigation, Pap. No. 3488, 11th Offshore Technol. Conf., Houston, Tex., 1979, 1105.

CASE HISTORIES

HIGH-RESOLUTION SEISMIC SOURCES AND MULTIFOLD DATA ACQUISITION

Peter K. Trabant

INTRODUCTION

The term high resolution has acquired a variety of meanings over the past decade within the field of marine geophysics. The term, as used herein, is simply defined as: *The use of seismic sources and acquisition systems that operate at acoustic frequencies greater than 100 Hz.* This chapter shall focus more specifically on those seismic signals contained within a bandwidth of 100 to 1000 Hz. High resolution seismic data obtained at frequencies greater than 1000 Hz are discussed in Chapter 4 of this volume. The physical constraints imposed upon the propagation of acoustic waves through a geologic column during seismic reflection surveys represent somewhat of a dichotomy. Deeper penetration of seismic signals requires lower frequencies which in turn produce less resolution, while higher frequencies, providing greater resolution, are limited in penetration. The cause of this inverse relationship between frequency and penetration of the geologic column is due to the much greater signal decay rate imposed upon higher frequencies. Basically, the earth acts as a low bandpass filter by reducing the higher frequencies with depth of penetration of seismic waves. As acoustic signals transit a geological column the energy loss is proportional to the wavelength.[1] In the exploration for hydrocarbons, seismic data have usually been acquired to two-way times of 6 to 8 sec within a frequency range of 10 to 100 Hz, with emphasis on the lower end of this frequency spectrum.[2] Such parameters have provided for the deeper penetration of sedimentary deposits to accommodate exploration targets.

As a consequence of these objectives, the lack of higher frequencies and consequent resolution has created a transition zone containing little or no data within the uppermost second or two on exploration seismic profiles.

The past decade, however, has seen the refinement and enhancement of singlefold analog data acquisition systems. Sylwester discusses such single channel data acquisition systems in Chapter 3 of this volume. Initially employed by oceanographic institutions, single channel systems have been utilized towards the assessment of shallow (to 1.0 sec two-way time) engineering hazards in the emplacement and construction of offshore drilling platforms.[3] Such systems have presently evolved into small scale versions of the deeper-penetrating seismic exploration systems. Furthermore, refinements to seismic sources, recording instruments, and in data processing have produced a merger of the two types of data acquisition and processing techniques. Current practice in offshore engineering studies involve the utilization of powerful yet high frequency seismic sources, multi-element (6 to 48 trace) hydrophone cables, and floating point digital recording systems. These developments have, in turn, provided enhanced seismic reflection record sections towards resolving shallow to intermediate depth stratigraphic traps to two-way times of 3.0 sec in the exploration for hydrocarbons.[4] Figure 1 depicts a high resolution seismic profile obtained off southern California with a minisleeve exploder (Fairflex) seismic source, a 24-fold hydrophone array, and digitally sampled and processed at 1.0 msec intervals.

HIGH RESOLUTION SEISMIC SOURCES

Following the usage of dynamite until the mid-1960s and widespread use of sparker

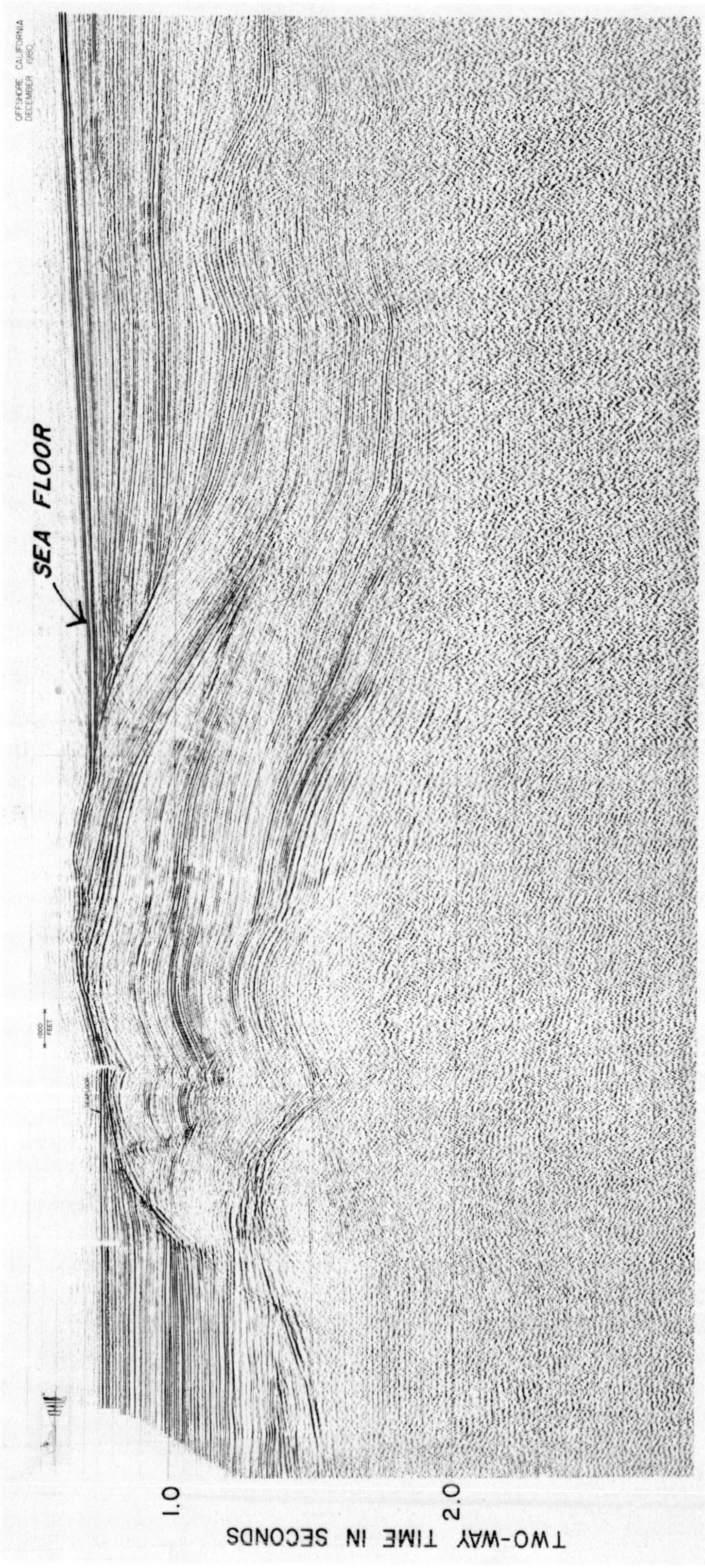

FIGURE 1. Minisleeve exploder (Fairflex) 24-fold seismic profile, Santa Maria Basin, offshore California (Courtesy of Fairfield Industries, Inc.).

systems for the generation of seismic waves at sea, the past decade of the 1970s has seen the development of a number of new and innovative seismic sources for the acquisition of high resolution seismic reflection data. A comparative listing of several of the most used sources is presented in Table 1. Ideally, a seismic source should produce an instantaneous positive shock wave, characterized by a very quick rise in pressure followed by a rapid decay. Dynamite had provided such a source for offshore seismic exploration until the mid-1960s, however the generation of secondary pressure waves (bubble pulse) presented somewhat of a drawback (see Chapter 1). This impediment to a clean seismic pulse has, therefore, received a good deal of attention in the design and development of modern seismic sources.

Sparkers

For the past several decades sparker sources, which operate by discharging a high voltage electrical current between electrodes in the water column, have seen widespread usage in the acquisition of high resolution seismic reflection profiles. Sparkers, operating in the 1000 to 20,000 J power range, have provided a relatively simple and cost effective seismic source. Seismic signatures generated by sparkers, however, leave much to be desired. The combination of variable discharge paths and the generation of bubble pulses tends to produce a relatively "dirty" or complex signature waveform. This is particularly the case when employed for single channel analog applications and when used in shallow (less than 200 ft) water depths.[1] It is not uncommon to generate signatures, including the bubble-pulse train, of 30 to 50 msec in length, depending upon the source depth. The net result has been the masking of seismic reflectors immediately below the sea floor and beneath subsequent reflecting horizons. Thus both resolution and identification of reflectors within 30 to 50 msec below prominent reflectors are wiped out, particularly on single channel data.

On the other hand, the complex sparker source signal becomes an asset when properly recorded in conjunction with wide range dynamic gain in digital format (i.e., floating point). Such near-source recording allows the signature to be correlated or deconvolved with the recorded seismic data during processing, in order to produce a narrow signature or spike through a process of inverse filtering. These techniques, particularly when employed on multifold CDP (Common Depth Point) recordings, permit considerable data enhancement by reducing the effects of reverberation, ghosting, and diffraction signals. A net reduction in bubble size and the resultant ghosting effect may also be obtained by increasing the number and array length of the discharging electrodes. General specifications are listed in Table 1 for several sizes of sparker systems used for high resolution seismic surveys.

Air Guns

While improving the source signature towards a spike when compared to that produced by a sparker, small-chambered (less than 200 in.^{3}) air guns tend to generate a strong bubble-pulse train, with the deleterious effects of both ghosting and reverberation. The problem has been somewhat alleviated by the use of "wave shaping" devices that break up the secondary bubble oscillations. Also, the towing of air guns at very shallow depths, on the order of a few feet, has been helpful towards producing a sharper source pulse.

Large air gun arrays, totaling up to several dozen units, with total chamber volumes on the order of 2000 in.^{3} have served as a major seismic source in exploration geophysics for over a decade. Smaller units and arrays have been used extensively for high resolution seismic surveys. Although receiving considerable use in exploration surveys, high pressure air guns operating at pressures on the order of 5000 lb/in.^{2} have rarely been employed for smaller scale high resolution surveys.

Table 1
COMPARISON OF FREQUENTLY EMPLOYED SEISMIC SOURCES IN THE ACQUISITION OF MARINE HIGH RESOLUTION SEISMIC DATA

		1kJ SPARKER	4.6kJ SPARKER	8kJ SPARKER	15.4kJ SPARKER	FLEXICHOC	SLEEVE EXPLODER	15in³ WATER GUN	40in³ AIR GUN	80in³ WATER GUN
TECHNICAL SPECIFICATIONS	MANUFACTURER	E G & G			FAIRFIELD	I.F.P.	FAIRFIELD	S.S.I.	BOLT	S.S.I.
	MODEL & N°	267A & 231A, 232A	267A & 231A, 232A, 233A	402/7, 231A 2x232A, 2x233A	SS 75	FHC 50	FAIRFLEX	MICRO-MICA "T"	PAR 600	MICA "T"
	SOURCE DIM.	25x25x270cm	25x25x270cm	33x33x275cm	100x60x60cm	65x30x30cm	30x5x5cm each	55x15x15cm	45x10x10cm	90x30x30cm
	SOURCE WT.	8kg	8kg	10kg	25kg	130kg	30kg	15kg	19kg	80kg
	ENERGY SOURCE DIM.	110x80x78cm	161x113x143cm	212x146x208cm	250x110x185cm	170x105x100cm	gas cylinders	200x60x60cm	200x60x60cm	200x200x100cm
	ENERGY SOURCE WT.	157kg	263kg	369kg	600kg	800kg		200kg	200kg	450kg
	TOW SYSTEM	3 ELECTRODE TOW SLED	3 ELECTRODE TOW SLED	9 ELECTRODE TOW SLED	10 ELECTRODE TOW CABLE	CATAMARAN	12 GUN 800kg CATAMARAN	CATAMARAN TOW CABLE	TOW CABLE	TOW CABLE
	NOTES	15kW, 240V supply	15kW, 240V supply	15kW, 240V supply	30kW, 240V supply				WAVE SHAPPING KIT AVAILABLE	
SOURCE PERFORMANCE	ENERGY	ELECTRICAL CAPACITIVE DISCHARGE →	→	→	→	PLATE EXPANDED HYDRAULICS RE-TRACTED by VOID	EXPLOSION within ELASTIC SLEEVE	IMPLOSION of WATER SLUG	IMPLOSION of AIR BUBBLE	IMPLOSION of WATER SLUG
	FIRING RATE	0.75sec	2 sec	2 sec	2 sec	2 sec.	300msec sequentially, 2 sec. otherwise	0.5 sec.	0.25 sec.	8 sec.
	PULSE	200-5,000Hz	100-3,000Hz	70-1,500Hz	40-500Hz 1.0 bar m	30-1,000Hz 3.0 bar m	50-1,000Hz 5.0 bar m	50-1,000Hz 3.8 bar m	1.25 bar m	20-500Hz 7.0 bar m
	PENETRATION	150 m	450 m	1,200 m	3,500 m	1,500 m	1,500 m		1,500m	
	M.T.B.F.	± 100h	± 100h	± 100 h	± 100 h		6 h (maint.)	6 h (maint.)	6 h (maint.)	6 h (maint.)
	NOTES	ELECTRODES	REQUIRE MAINTENANCE EVERY FEW HOURS			REPLACE RUBBER MEMBRANE EVERY 30 HRS.				

In an attempt to reduce drawbacks such as those associated with the sources described, and in order to obtain a ''cleaner'' seismic pulse, the following sources have been developed during the past decade. These have received varying degrees of acceptance in the field of high resolution geophysics.

Water Guns

First developed by SODERA (Societé pour le Développement de la Recherche Applique) in France, water guns employ some of the basic mechanical principles of the air gun, except that instead of evacuating a volume of high pressured air they rapidly expel a fixed volume of water. The outward-moving slug of water, in turn, implodes to form a sharp seismic pulse as a result of cavitation. A major drawback of this system, however, is the generation of a low frequency intermediate amplitude pulse prior to the initial cavitation collapse, due to the initial evacuation of the gun chamber. This portion of the signature makes it difficult to determine the initial time break, which is a very critical element in processing multifold high resolution seismic data. Water guns are operated in arrays of 16 to 20 units of 80 $in.^3$ each in exploration surveys, whereas smaller 15 $in.^3$ units are deployed at fixed depth for high resolution seismic surveys. Additionally, the use of single 80 $in.^3$ units has proven effective as a high frequency seismic source.

Miniflexichoc

Developed by the I.F.P. (Institut Francais des Petroles) the miniflexichoc generates a sharp seismic pulse through the rapid retraction of a flexible metal plate. The unit is subsequently rearmed by separating the plate by the use of a hydraulic jacking system. The ease of operation of hydraulics, the ability to operate at high repetition rates of 0.25 sec, and the clean signature produced make the system very effective as a source for high resolution surveys. Its use has been rather limited but Figure 2 depicts a 12-fold reflection profile off the coast of Venezuela.

Sleeve Exploder

Developed at the EXXON Production Research Co. and scaled down for use as a high resolution seismic source, the system consists of one to two dozen rubber membrane ''sleeves'', each of which is fitted about a mixing/firing chamber mounted under a towable catamaran sled. A mixture of oxygen and propane is fed into the firing chambers and triggered by a spark to produce an explosion contained within each of the rubber sleeve membranes. Upon collapse of the initial explosion, the porous elastic membranes retract allowing the generated gases to diffuse into the water column, thus eliminating the generation of a bubble pulse. In addition, the system operates at the shallow depth of a few feet, virtually eliminating the generation of ghosting. In spite of its size, weight, and use of explosive gases, the minisleeve exploder has gained widespread use in seismic engineering surveys for the detection of hazards to offshore drilling operations, such as faulting and high pressured gas zones. The seismic profile of Figure 3 illustrates a 24-fold seismic section obtained in the Gulf of Mexico.

These last three systems (water gun, Flexichoc, and Minisleeve exploder) represent a relatively new generation of seismic sources employed in the field acquisition of high resolution seismic reflection data.

SEISMIC RECEIVER ARRAYS

The basic requirement for seismic hydrophone receivers used in the acquisition of high resolution marine reflection data is that they be receptive to signals within the

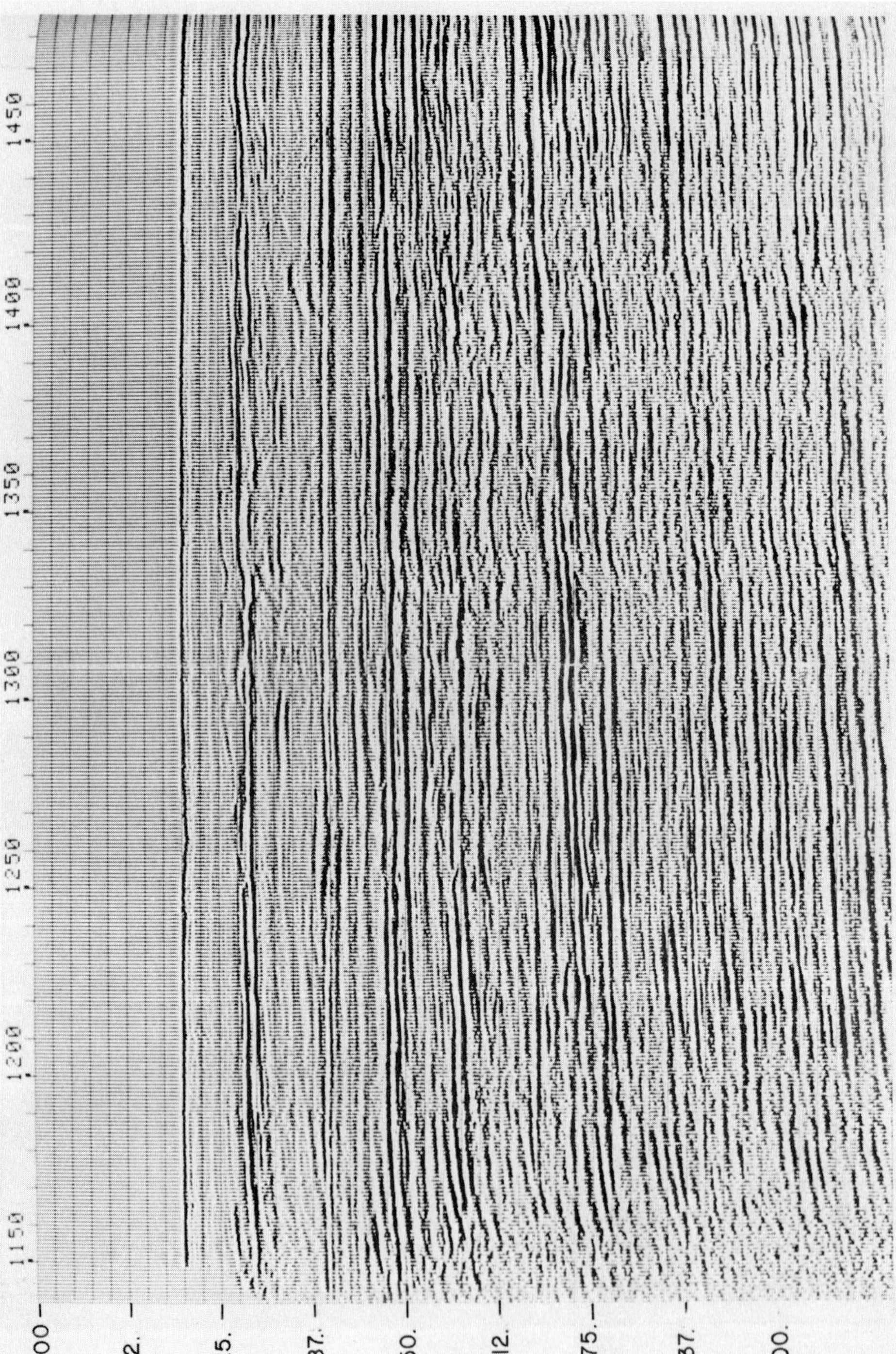

FIGURE 2. MiniFlexichoc 12-fold seismic profile, offshore Venezuela (courtesy of BEICIP).

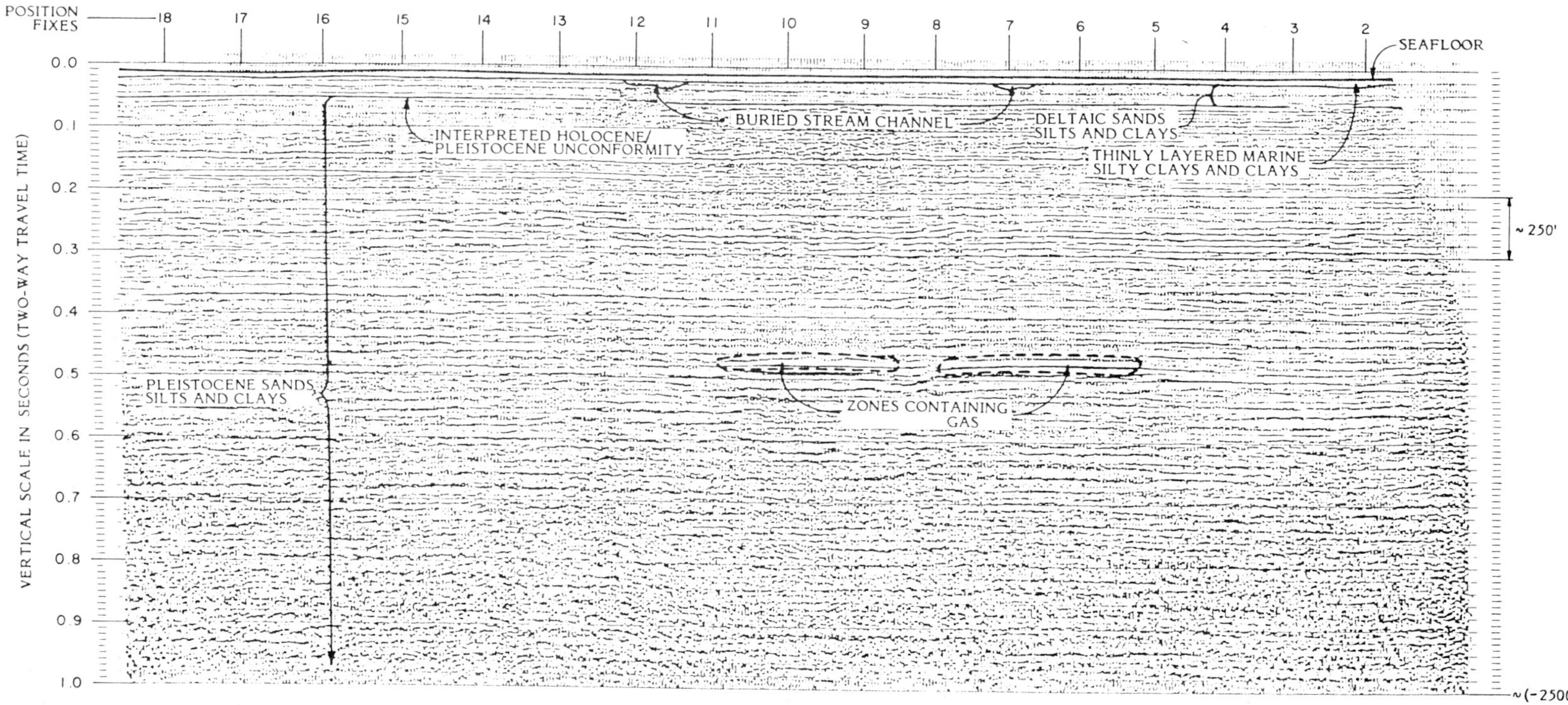

FIGURE 3. Minisleeve exploder (Stratatrac) 24-fold profile, Gulf of Mexico (Courtesy of Odom Offshore Surveys, Inc.).

desired high frequency spectrum, while maintaining a high signal-to-noise ratio. Curves of frequency vs. pressure sensitivity for such hydrophones should display a flat portion indicative of equal receptivity within the range of frequencies from 50 to 1000 Hz. Multifold hydrophone cables usually consist of a number of sections or groups, each containing a number of individual pressure-sensitive hydrophone elements, the arrangement and number of which may vary but are usually linearly separated (equidistant). The hydrophone elements, associated signal conductor wires, and stress cables are usually set within flexible plastic tubing filled with a non-conductive fluid such as kerosene. The number of individual receiver groups has tended to increase over the past decade, with 12 and 24 groups being standard at present compared to only 6 groups at the end of the 1960s. The spacing of individual groups along the hydrophone streamer sbould be adjusted to meet the required objectives of a particular survey, such as depth of penetration and distance between successive shots (firing interval).

For example, if the desired subsurface coverage is in the order of 0.5 sec and within shallow (few hundred feet) water depths, a short group spacing should be adequate, while deeper water depths and subsurface objectives will require longer spacing. The need for such adjustments in receiver geometry is to provide an appropriate offset (normal moveout) in the arrival times of multiples and other unwanted signals, in order to permit their cancellation during the stacking of CDP in data processing. If the spread, or distance between near and far traces, is too small during data acquisition to provide such offsets in the arrival times of multiples, these cannot be removed. This results in reducing the effectiveness of multifold processing for attaining larger signal-to-noise ratios and eliminating undesirable reflection signals.

In addition to the hydrophone array geometry (separation and location of individual groups) critical attention must also be placed upon the depth of the hydrophone streamer. Depth-controlling "birds" are attached to the hydrophone streamer in order to maintain a constant immersion depth according to either preset values or by remote control from the survey vessel. Calibrated depth indicator measuring devices (D.I.s) permit the monitoring of the streamer depth along its length.

The critical element with respect to hydrophone depth is the frequency filtering or aliasing effect generated by ghosting.[1] This effect is due to the reflection of seismic signals off the sea surface, after passing the submerged hydrophone cable, and their recording as a delayed signal. The wave path length between the cable and sea surface may therefore become critical when it is equal to a half signal wavelength. Optimum streamer depth is approximately equal to a quarter wavelength for the seismic frequency spectrum desired. Within the desired frequency bandwidth in high resolution seismic recording, this becomes impossible if all hydrophones are towed at a constant depth. For example, a steamer depth of 12 ft would effectively cancel most incoming signals at a frequency of 200 Hz. Thus, certain frequencies are invariably cut out during the field acquisition of high resolution seismic data.

In an astute attempt to alleviate both the ghosting and filtering problems associated with streamers towed at constant depth, Fairfield Industries, Inc., developed the FLAIR Technique.® This method involves the towing of the hydrophone streamer cable at an incident angle in the water column behind the survey vessel. The near trace group is thus submerged at a shallower depth than the far traces. Through the application of appropriate static corrections to the respective groups during data processing, phase shifts are used to cancel the ghost signals. The process also permits the reception of frequencies along a more normalized curve (reception is practically flat from 40 to 250 Hz with the minisleeve exploder source), eliminating the previously cut-out portions of the frequency spectrum. The method also doubles the fold and increases signal directivity. Figures 4 and 5 each illustrate two detailed comparisons between this tech-

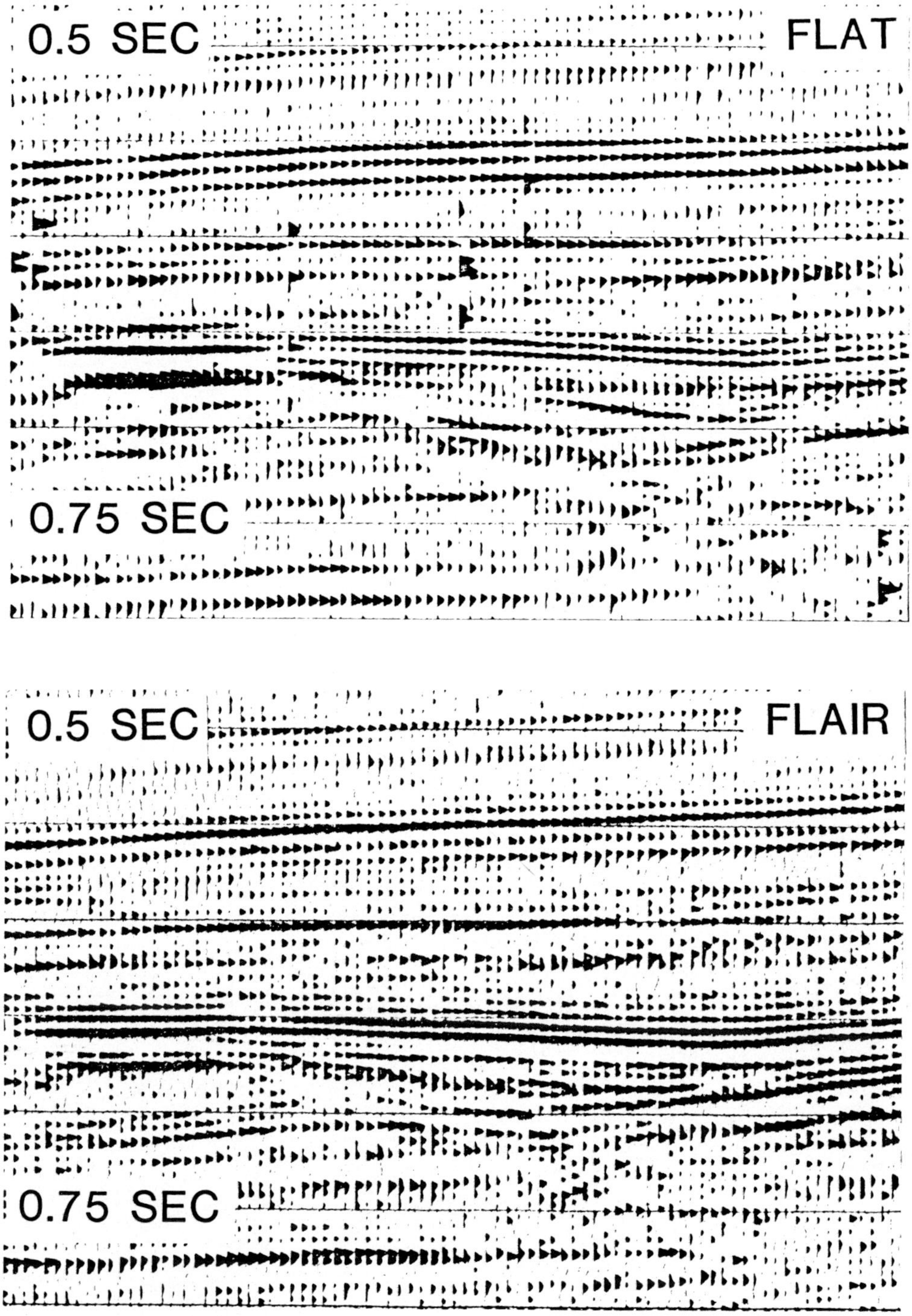

FIGURE 4. Minisleeve exploder (Fairflex) detailed comparison of the Flair® technique vs. standard horizontal (flat) hydrophone cable, from profile of Figure 1, offshore California.

nique and the present standard horizontal (flat) cable method. Additional experimentation along such lines may bring forth the development within the high resolution field of nonlinear and/or closer hydrophone spacing of cable groups in order to further

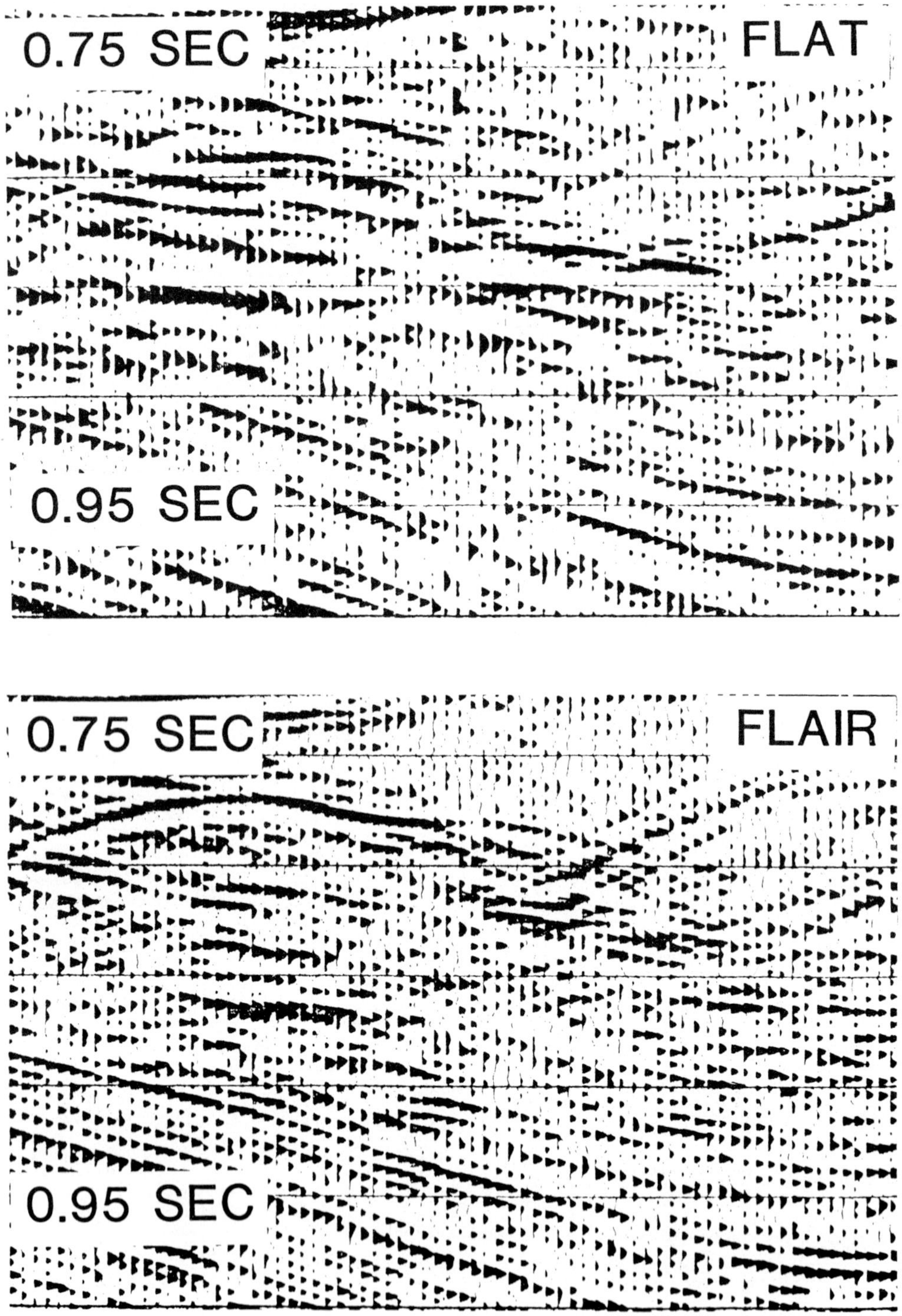

FIGURE 5. Minisleeve exploder (Fairflex) detailed comparison of the Flair® technique vs. standard horizontal (flat) hydrophone cable, from profile of Figure 1, offshore California.

enhance signal-to-noise ratios. Reduction of streamer friction and resultant noise through the application of "nonstick" or water absorbent coatings may further enhance present signal-to-noise ratios and permit greater survey speeds.

SEISMIC PROCESSING

The processing of high resolution multifold seismic reflection data in the early 1980s is identical to that employed for seismic exploration data. The only difference between the two involves the smaller digital sampling rates of from 0.25 to 1.0 msec compared to 2.0 to 4.0 msec for exploration seismic data. Standard recording instruments employed in the acquisition of exploration seismic field data are mostly amenable to the recording of higher frequencies and more frequent sampling rates when employed for shorter recording times of 1 to 2 sec. The smaller sampling intervals employed in the digital recording of high resolution data are required in order to retain the higher frequencies with the necessary fidelity for reproduction. Digital sampling ideally should be on the order of a quarter wave length; if fewer samples are taken aliasing occurs and fidelity of the original signal is lost.[1] Thus a digital sampling rate of 1.0 msec will exclude signals of frequencies greater than 500 Hz and produce a low fidelity recording. In some cases, data are presently sampled at up to 0.25 msec from near-trace phones in order to preserve a very high resolution seismic reflection picture of the immediate subbottom to two-way times of 0.25 sec. An additional factor in the recording of digital high resolution seismic data is the retention of true amplitude.

Single channel analog recordings, to date, essentially have been limited in this respect. The maximum amplitude gain offered by analog recording systems has been limited to approximately 20 dB. Digital seismic recording systems of the late 1970s operating on a fixed point gain technique, however, allow for as much as 80 dB of total gain retention, whereas present-day floating point gain systems allow for a full 100 dB amplitude retention through the digital recording process. This latter value is about the limit in terms of necessary amplitude recovery requirements for high resolution seismic sampling. The use of such gain recovery techniques is particularly helpful for the generation of relative or preserved amplitude seismic sections in data processing. Shallow gas-bearing strata or "bright spots" are easily identifiable on such sections and have provided a powerful tool in the seismic exploration for hydrocarbons and stratigraphic traps over the past decade. Present-day digital field recording systems provide the best of the above parameters as well as the versatility to be modified in the field to change parameters to suit particular survey objectives.

The processing of multifold high resolution seismic reflection data usually follows a sequence of operations or steps including most, but not necessarily all, of the following:

1. Demultiplexing of original field tapes or sorting of the digital field data into individual reflection traces. These were originally split up during recording in the field in order to fit in a continuous flow onto tape, following the analog to digital conversion.
2. Gain recovery of data with respect to time (depth) to boost signal amplitude losses due to spherical spreading of the acoustic signal as it propagates away from the seismic source, and those losses due to filtering effects of the earth.
3. Deconvolution of the data with an inversed filter generated from the seismic source signature to "sharpen" or spike the returned reflection signals. This step has been found to be a prerequisite for the enhancement of high resolution seismic data.
4. Sorting of the seismic reflection traces into CDP (common depth point) gathers for stacking and velocity measurements.
5. Static corrections if the recording hydrophone cable was not maintained at constant depth, as in the use of the FLAIR® technique.

6. Filtering of the data permits additional improvement of the signal-to-noise ratio by the reduction in amplitude of unwanted or noisy portions of the frequency spectrum. Low cut and high cut frequencies are selected, as well as notch filters. Thus, excessive 60 Hz noise from the survey vessel's electrical generating system may be notched out at this processing stage.
7. Velocity analyses provide RMS (root mean square or average) velocities with respect to depth by calculating arrival delay times of select reflectors due to increasing distance along the hydrophone receiver cable. This delay in arrival time is referred to as normal moveout (nmo correction). The number of such velocity measurements are dependent upon the suspected lateral variations in velocity.
8. Stacking of multifold seismic reflection traces according to CDP produces a considerable enhancement of seismic data by increasing the signal-to-noise ratio and canceling multiple reflections. Vertical stacking, which consists of the blending of adjacent seismic reflection traces, is occasionally performed at this stage and is of particular advantage for single trace digital data.
9. Deconvolution and filtering performed after the stacking of multifold data helps cleanup remaining spurious noises prior to the generation of a final display.
10. Migration of the final stacked seismic reflection sections may be performed where strongly dipping reflectors are present. This permits the shifting of reflections to their true spatial location with respect to vertical angles of incidence.

As an illustration of the effects of several of the aforementioned processing steps, a seismic reflection recording obtained offshore Venezuela is presented. Figure 6 displays the effects of a series of filters on the 1200% stacked data; panel A represents a 32 to 64 Hz bandpass filter; panel B a 64 to 120 Hz bandpass filter; panel C a 120 to 240 Hz bandpass, and panel D a 240 to 400 Hz bandpass filter. It becomes evident from such displays that optimum filtering should retain data within a bandpass of 64 to 240 Hz for the 450 msec portion of the displayed record. The seismic reflection profiles of Figures 7 and 8 display the improved data quality obtainable through deconvolution of the signals both before and after stacking, as compared to after stacking only.

CONCLUSIONS

Over the past decade the acquisition and processing of high resolution seismic reflection data obtained at sea has followed a steady course from the simple analog single channel systems to the present-day equivalent of modern exploration seismic systems. Further developments and additional uses of the technique are envisioned for the future. While the major use for high resolution seismic reflection data has been in the determination of engineering hazards to the emplacement of offshore structures, future uses are expected in the assessment of shallow stratigraphic hydrocarbon traps and the detailed geology of ocean basins and continental shelves.

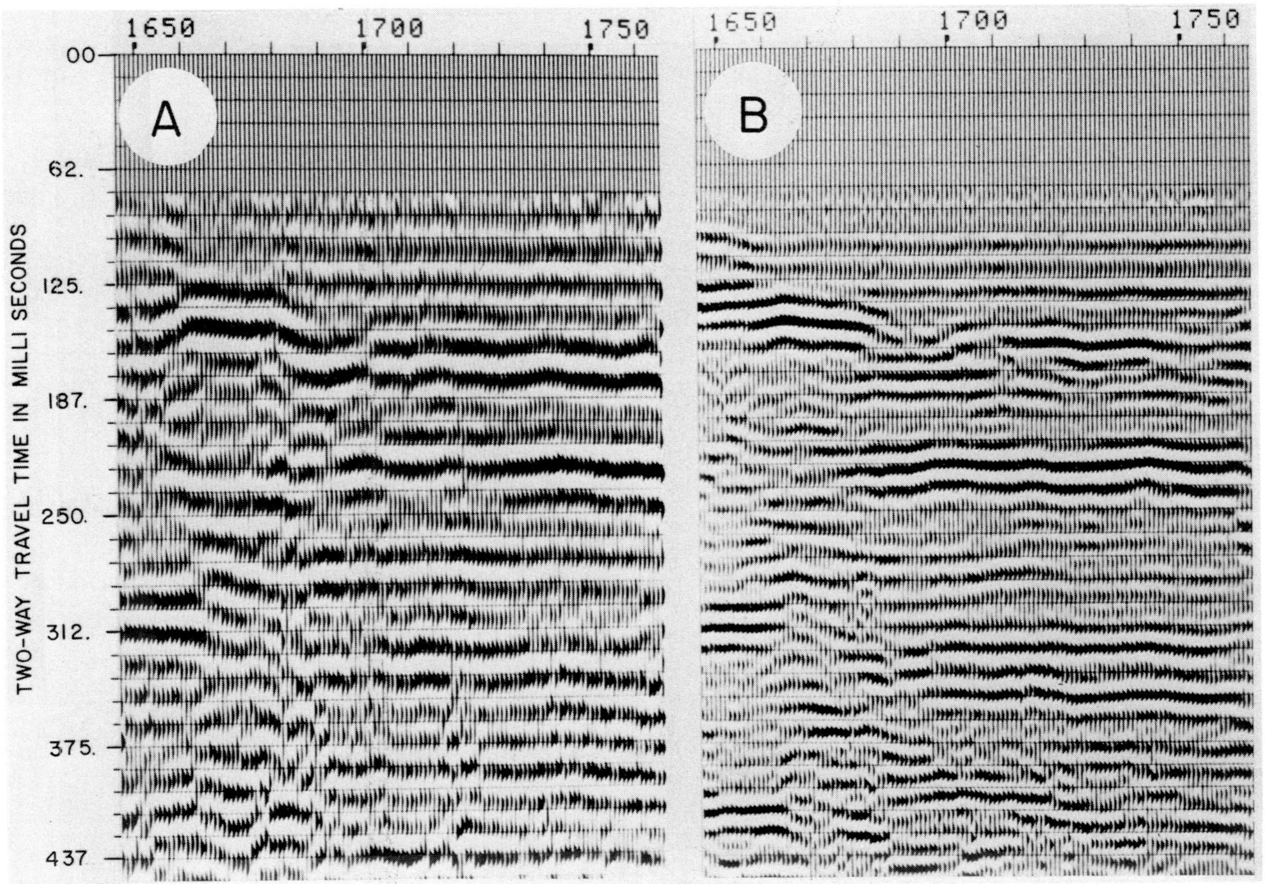

FIGURE 6A. and B. Miniflexichoc 12-fold high resolution seismic reflection section filter panels offshore Venezuela. A: 32 to 64 Hz, B: 64 to 120 Hz.

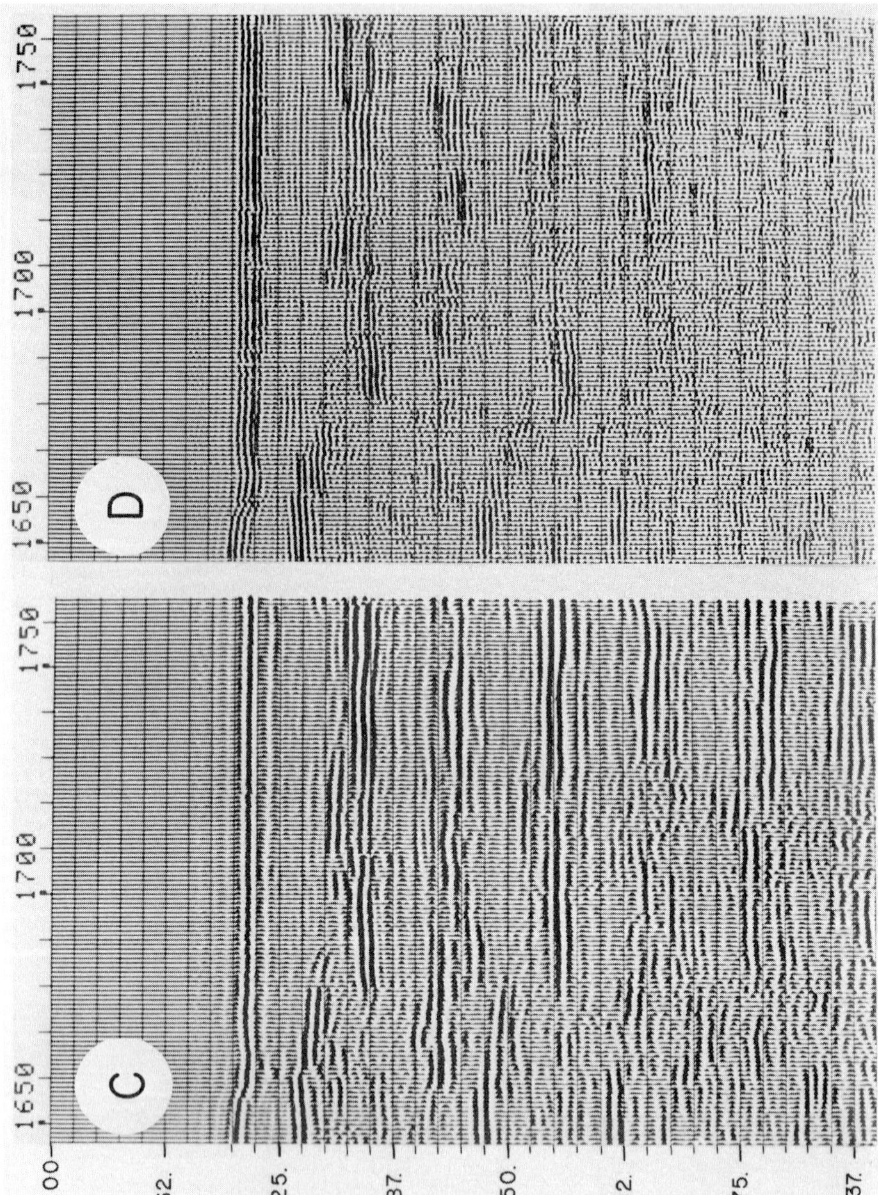

FIGURE 6C. and D. Miniflexichoc 12-fod high resolution seismic reflection section filter anels offshore Venezuela. C: 120 to 240 Hz, and D: 240 to 400 Hz bandpass.

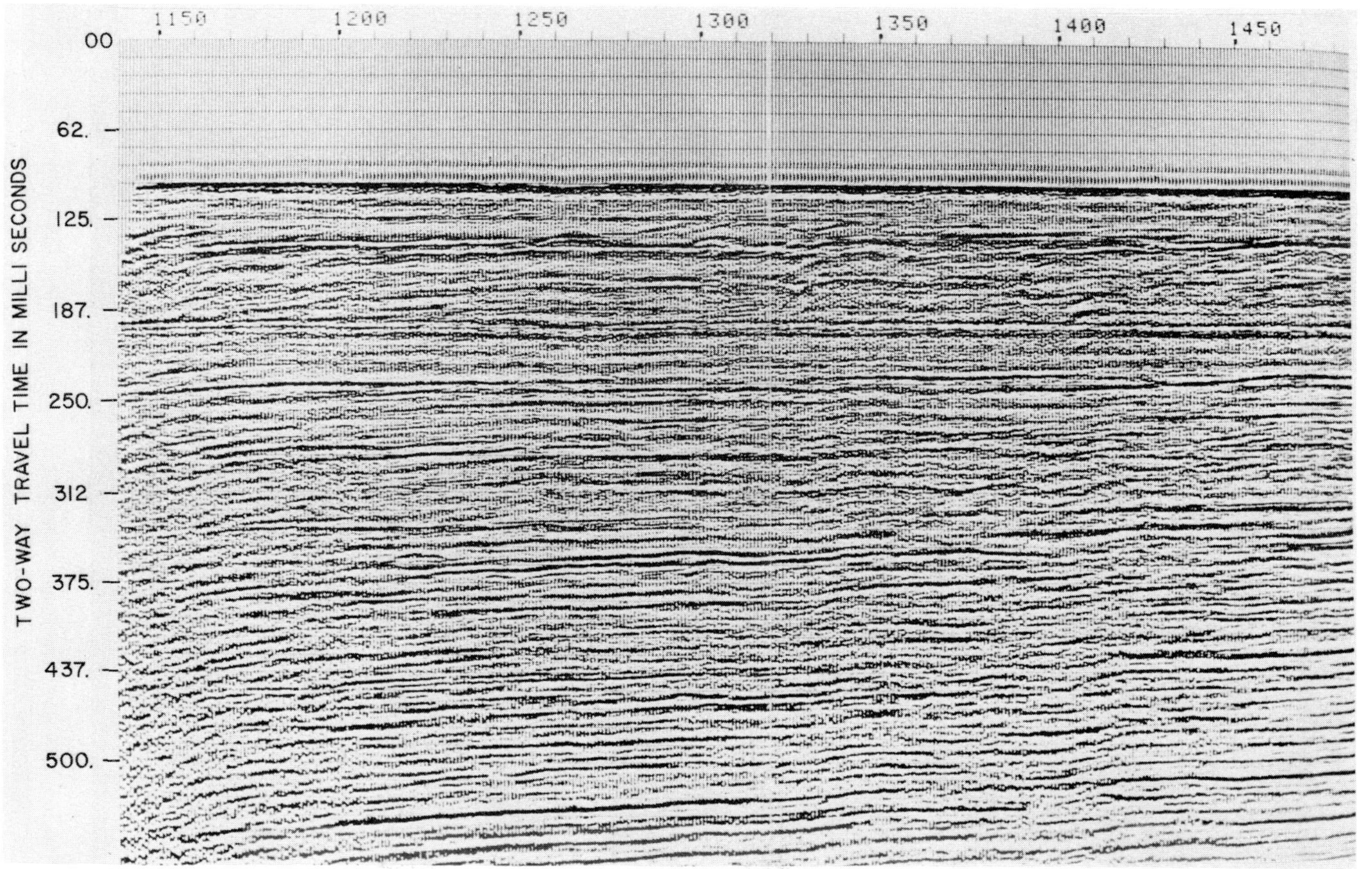

FIGURE 7. Miniflexichoc 12-fold high resolution seismic reflection profile, offshore Venezuela. Deconvolved after stack only.

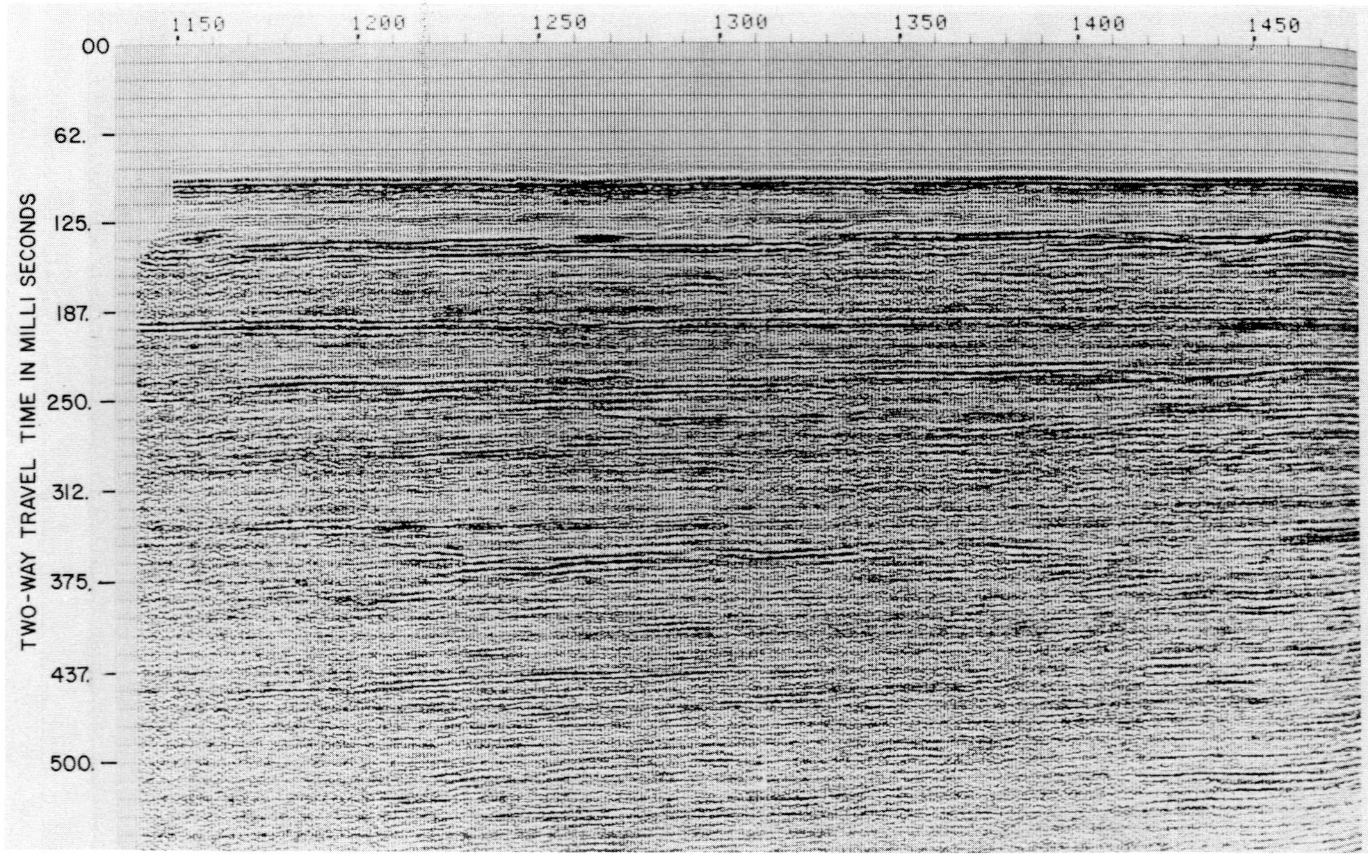

FIGURE 8. Miniflexichoc 12-fold high resolution seismic profile, offshore Venezuela. Deconvolved both before and after stacking, Filter: 64/500 Hz to 200 msec and 56/256 Hz below 200 msec.

REFERENCES

1. **McQuillin, R., Bacon, M., and Barclay, W.,** *An Introduction to Seismic Interpretation,* Gulf Publishing Co., Houston, Tex., 1979.
2. **Dobrin, M. B.,** *Introduction to Geophysical Prospecting,* McGraw-Hill, New York, 1976.
3. **LeTirant, P.,** *Seabed Reconnaissance and Offshore Soil Mechanics for the Installation of Petroleum Structures,* Editions Technip, Paris, 1979.
4. **Sieck, H. C. and Self, G. W.,** Analysis of high resolution data, in *Seismic Stratigraphy — Applications to Hydrocarbon Exploration,* Payton, C. E., Ed., American Association of Petroleum Geologists, Tulsa, 1977, 26.

MAGNETOTELLURIC MEASUREMENTS IN THE GULF OF MEXICO AT 20 M OCEAN DEPTHS

Gustave L. Hoehn and Barry N. Warner

INTRODUCTION

Magnetotellurics (MT) is a geophysical exploration technique employing simultaneous measurements of time-varying electric and magnetic fields to infer the electrical resistivity distribution within the earth beneath the site of the surface field measurements. The MT method, first proposed by Cagniard[1] and Tikonov,[2] differs from other electromagnetic geophysical techniques that require man-made sources of electromagnetic energy to "probe" the earth. The naturally occurring interminable variations in the earth's magnetic field induce eddy currents in the conductive crust of the earth that are detectable as electric field variations at the surface (and within the earth). The low frequency noise-like "signals" have their origin in the sun whose time-varying particulate radiation flux ("solar wind") interacts with the magnetic field of the earth. The energy above 1 Hz in the electromagnetic field variations has an atmospheric source, consisting of resonances in the earth-ionosphere bispherical cavity, excited by thunderstorms.

For magnetotelluric exploration the usual components measured at a single site are three orthogonal magnetic fields (H_x, H_y, and H_z) and two electric field (E_x and E_y) variations as a function of time. The Z-axis reference is generally taken by convention to be directed vertically downward. Several hours of data recording are necessary in order to determine (using Fourier transform computations) the energy content of the fields at various frequencies to as low as 0.1 mHz.

Details on the theory, instrumentation, field procedures, data processing, and interpretation of magnetotelluric surveys on land have been comprehensibly summarized by Vozoff.[3]

The electric and magnetic fields measured at the earth's surface are the resultant fields from an incident and reflected electromagnetic wave, and as a consequence, are a measure of the energy that is refracted into the earth in an essentially vertical direction. The E and H fields decay exponentially in strength as they penetrate into a uniformly conductive earth, the higher frequency components decaying more rapidly than the lower frequencies. The E and H strengths fall to 1/e of their surface values in a characteristic distance called the skin depth δ, where

$$\delta \approx 0.5\sqrt{\rho/f}\ \text{km} \qquad (1)$$

and where ϱ is the earth's resistivity in ohmmeters and f is the frequency in hertz. Thus the skin depth at 1 Hz in ohmmeter earth material is 500 m but only 250 m in seawater of 0.25 ohmmeter. The skin depth in these materials would increase tenfold for a lower frequency of 0.01 Hz.

THEORETICAL VALIDITY

The simple one-dimensional theoretical analysis first introduced by Cagniard[1] is an exact mathematical analogue to the solution of transmission lines in classical electromagnetic theory. The E (electric) field is analogous to voltage on the transmission line, H (magnetic) field analogous to current in the transmission line conductors, resistivity of the earth is related to the propagation constant of the line, and depth in the earth directly equal to distance along the line. In principle, E and H fields could be measured at depth rather than at the earth's surface and the results interpreted in terms of the resistivity of only those earth materials beneath the measurement depth. Thus, magnetotelluric measurements on the ocean bottom are interpreted in terms of earth resistivities beneath the seabed just as though the ocean layers were not present.

Researchers in France[4-6] and Russia[7] have proposed magnetotelluric measurements at sea, and although the Russians have made a limited number of E measurements with electrode arrays in large lakes and inland seas, many workers favor the placement of H sensors on the sea bottom.[8,9] Coil variometers, usually employed as H sensors, are extremely sensitive to motion in the earth's large static magnetic field; mechanical oscillations as small as 10^{-8} radians produce voltages comparable to the typical natural MT signals at 0.1 Hz.

It is possible to evaluate in a precise way the attenuation of the electric and magnetic fields from their surface values to the values they would have on the sea bottom at a depth h. Assume the seawater to have a resistivity of ϱ_w = 0.25 ohmmeters and the resistivity ϱ_B of the bottom to be uniform to great depths. Then

$$\frac{H_{yB}}{H_{ys}} = 1/[\cosh \gamma h + \beta \sinh \gamma h] \tag{2}$$

and

$$\frac{E_{xB}}{E_{xs}} = 1/[\cosh \gamma h + (1/\beta) \sinh \gamma h] \tag{3}$$

where

$$\beta = \sqrt{\rho_B/\rho_w} \quad \text{and} \quad \gamma = \sqrt{\frac{j\mu_o(2\pi f)}{\rho_w}}$$

and μ_o is the permeability of free space. These attenuation relations are plotted in Figure 1, where the abscissa is represented by the normalized variable, $x = \sqrt{f}\, h$, and attenuation is in decibels, −20 dB representing a factor of 10 reduction in field. The solid curve represents attenuation for either H or E when the sea bottom resistivity equals that of the seawater (β = 1). The lower dashed curves represent successively greater attenuation of H fields as the resistivity of the sea bottom is increased from 1 ohmmeter (β = 2) to 100 ohmmeter (β = 20). The upper dashed curves show a lesser attenuation of E fields as the sea bottom resistivity increases. For very deep oceans of 1000 m or more, frequencies higher than about 1 Hz are so severely attenuated that magnetotelluric measurements become ineffective in resolving thin layers of sediments at the ocean bottom.[10,11] From Figure 1 it can be seen that for ocean depths of less than 20 m all frequency components of the E and H fields below 25 Hz would suffer less than 20 dB attenuation, and this has been confirmed in the experiments reported here.

When seawater moves in the earth's static magnetic field, eddy currents are induced in this moving conductor and electromagnetic signals are generated with the same periods (or frequencies) as the motion. Ocean waves with periods from 0.1 to 1000 sec

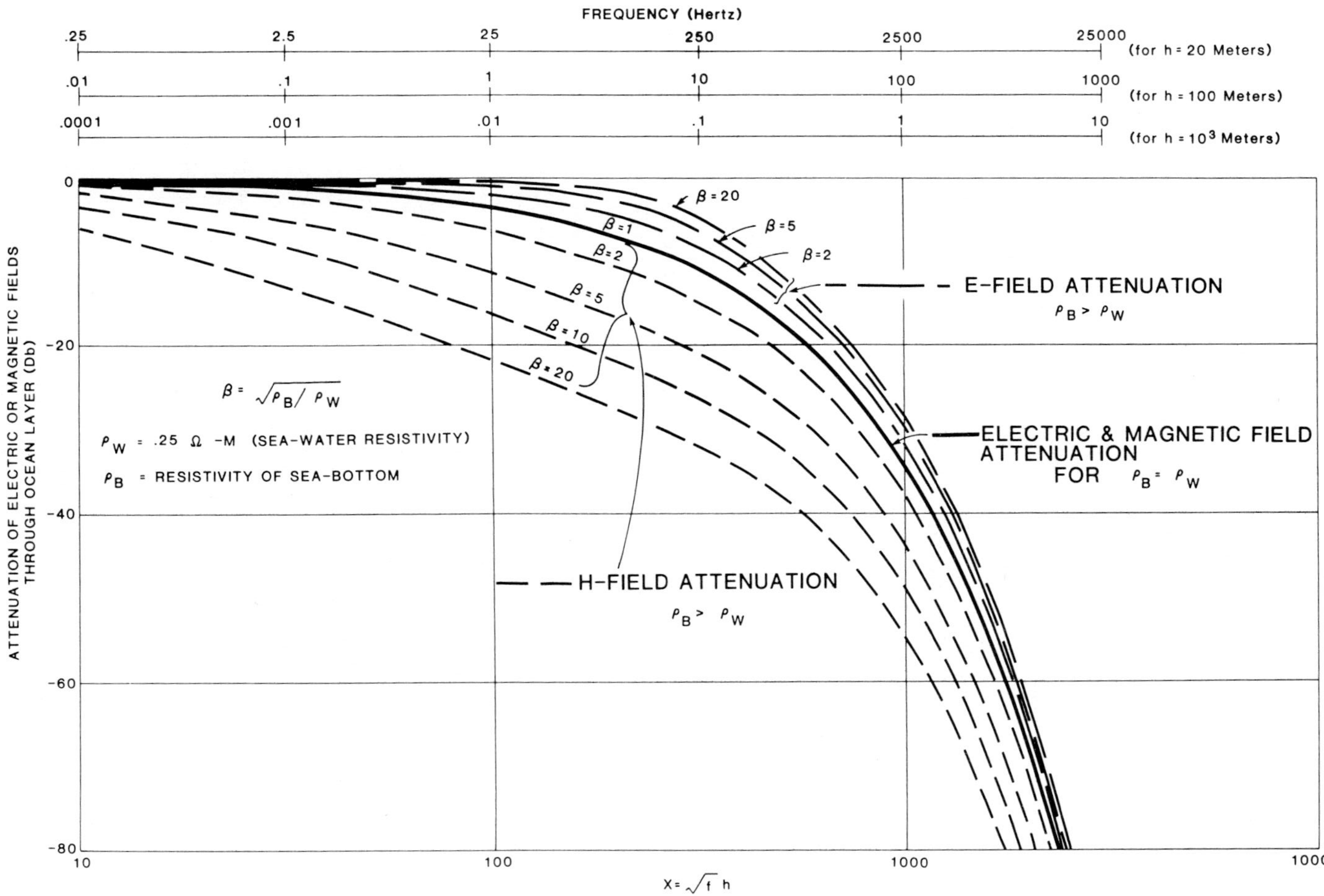

FIGURE 1. Attenuation of electric and magnetic fields in a conductive ocean layer.

are typically referred to as gravity waves. These waves are generated primarily by winds and can travel great distances from their source location, their principal restoring or damping force being the earth's gravity. The maximum for the energy spectra of gravity waves occurs at about 0.1 Hz, or a period of 10 sec. Surface waves can be treated by Airy theory to estimate water motion as a function of depth for a "hard bottom" ocean of finite depth.[12] The wavelength for a gravity wave in very deep water is proportional to the square of the wave period and is directly proportional to the wind velocity driving the wave, when the distance (or reach) over which the wind acts is sufficiently long. These long period waves can travel very great distances from their point of origin, their wave lengths shortening as the effect of shallower ocean bottom becomes manifest. The dispersion relation between period and wavelength is given by the nonlinear equation:

$$L = (gT^2/2\pi)\tanh(2\pi h/L) \tag{4}$$

where g is the acceleration of gravity (32 ft/sec²), T is the wave period in seconds, and L is the wavelength (feet). Figure 2 shows this relationship graphically with solid lines representing various wave lengths, L, as a function of period, T, and ocean depth, h. The dashed straight lines of Figure 2 illustrate the attenuation factor in wave motion,

$$K \triangleq \frac{A_d}{A_o} = \operatorname{sech}(2\pi d/L) \tag{5}$$

where A_o is the amplitude of the water motion at the surface and A_d is the amplitude of water motion at a depth d. At a hard inelastic sea bottom, the oscillatory water motion must be entirely in a horizontal direction. The dashed curves of Figure 2 represent the peak values of this horizontal bottom current velocity in feet/second/foot of surface wave heght. These bottom currents are capable, under certain conditions, of producing movement of magnetic sensors as reported by the French researchers.[13] As their measurement program in the Mediterranean Sea ended, the massive and tall structures that had been used initially for deploying H sensor coils on the ocean bottom were being abandoned in favor of flatter and more streamlined structures.

The theoretical problems of estimating the magnetic fields generated by movement of the conductive seawater near the ocean bottom is difficult and the estimates are highly model dependent. Weaver[14] has considered surface waves, or swells, in a very deep ocean. Klein[5] and Bostick[15] have limited the ocean depth in their analysis. Podney,[16] Beal and Weaver,[17] and Russian researchers[18] have considered internal ocean waves in addition to surface swells, and the Russian model includes an ocean bottom of resistivity differing from that of seawater. Attempts to use these various results to estimate the maximum expected H variations in a 20 m deep ocean with resistivity of 0.25 ohmmeter, a bottom of 1 ohmmeter resistivity, and waves of 10 sec period have yielded a range of 10^{-4} to 0.1 gammas per meter of surface wave height.

EXPERIMENTS IN SHALLOW OCEAN WATER

A number of electrode types have been tested in the laboratory as well as in shallow water areas of Corpus Christi Bay. Solid lead electrodes that had been employed routinely on many magnetotelluric land surveys exhibited impulsive voltage offsets and noise pulses in an unpredictable manner when they were placed in salt water even after several days were allowed for stabilization. Cadmium metal in a saturated gelled elec-

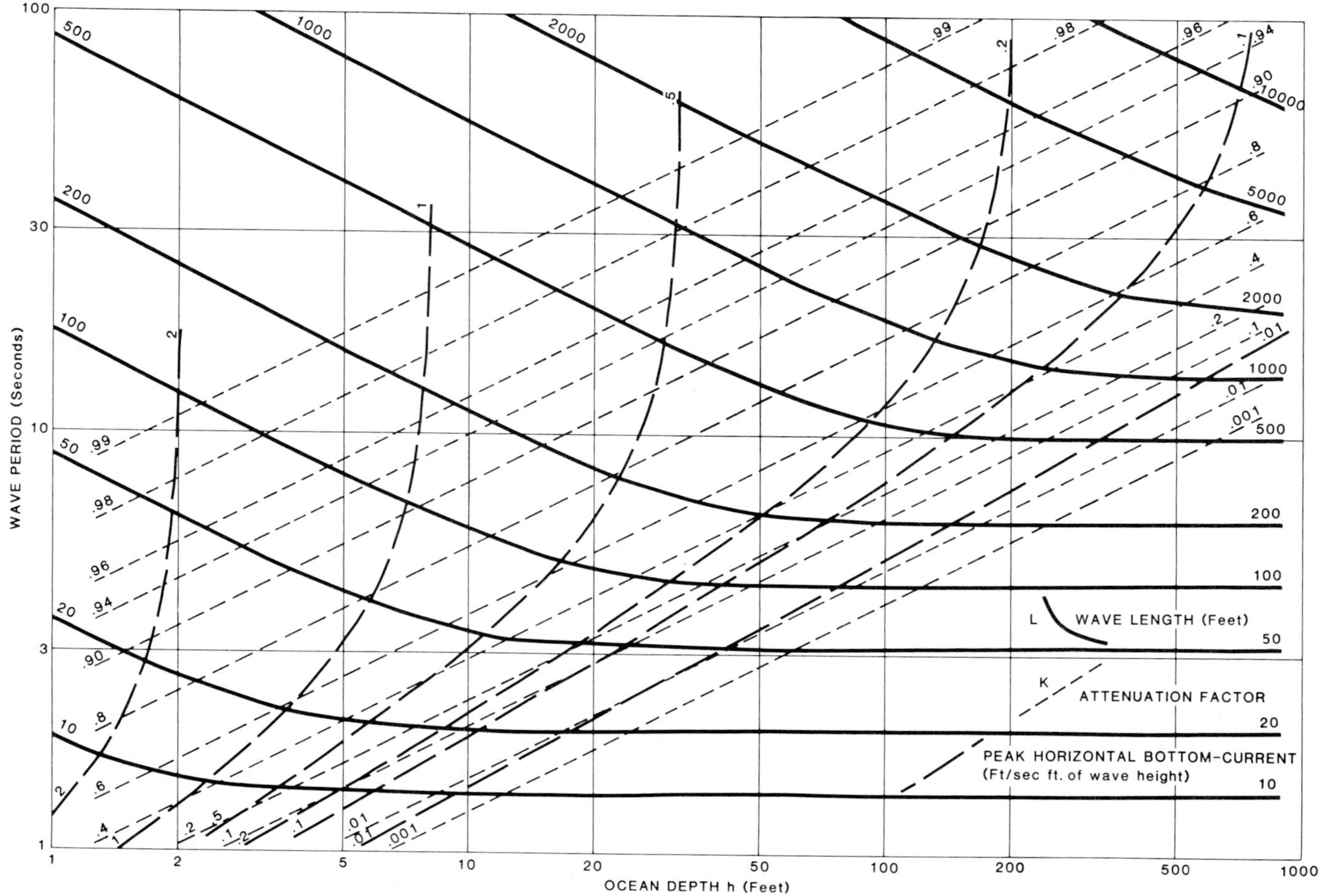

FIGURE 2. Wavelength, attenuation, and bottom currents of surface ocean waves or swells (Airy theory).

trolyte of cadmium chloride proved to be superior to silver chloride or copper sulfate electrode systems, and these, in turn, better than solid metals like copper or brass. Tests involved long-term monitoring of offset voltages and noise. Long-term exposure of the cadmium-cadmium chloride electrodes in a laboratory bell jar of chemically simulated ocean water did not result in a detectable diffusion of cadmium ions into the water, so that environmental safety for ocean use of these electrodes was reasonably assured.

Packaging of the electrodes initially involved capsulation of the porous-frit Buechner funnel with its cadmium chloride electrolyte and cadmium-copper wire electrode within a perforated polyvinyl chloride (PVC) pipe. These short length (PVC) pipes were fitted with pointed ends and wire connections at the opposite ends. These electrodes (with attached wires) were placed by means of poles into the soft bottom of Corpus Christi Bay in water depths of 4 to 5 m. By duplicating on land the ocean electrode array as to spacing and orientation, exact comparison of E_x and E_y fields were made over a wide band of frequencies for a number of days. The onshore and ocean bottom E fields agreed to such an extent that if any spurious electric fields were generated by water motion they were not correlated with water motion or were below the instrument noise. This has been the subsequent experience in moving to deep water electrode placement in the Gulf of Mexico. Occasional wave frequency noise appearing in the ocean bottom E field measurements has thus far been attributable to motion of the electrodes or interconnecting cables and correctable by reanchoring the moving electrode and/or cable.

Before deploying coil variometers on the ocean bottom, coils of a type commonly employed in land surveys (Geotronics Corporation Type MTC-4SS) were repackaged in longer and thicker-walled cases. The cases, with special ocean-type connectors, were designed to withstand the hydrostatic pressure of 50 m ocean depths and redesigned preamplifiers were placed in the cases with the coils. Extensive land tests were made with two of these coils placed in close proximity, one above the other in a cross configuration. H_x and H_y recording comparisons made with nearby standard coil variometer arrays revealed negligible coil interaction, preamplifier instabilities, or sensor sensitivity changes. A massive nonmagnetic flat-lying square frame structure (H-Pod) was designed to support and deploy the crossed H_x-H_y coils to the ocean bottom. An H_z (vertical magnetic field) sensor was not included so that a minimal vertical cross section would be presented by the structure to horizontal water motion.

Figure 3 is a photograph revealing the principle components of the H-Pod with its two coils (H_x and H_y) crossed near the center and clamped above tubular members of the stainless steel structure. A bail is attached to the square tubular frame and is used for hoisting and deployment via a specially constructed multiple conductor marine cable with internal stress members. The hoisting bail folds down to lie along the ocean bottom after deployment and has the necessary connecting cables fastened along its length to the hinge. From near the hinge, cables supply necessary power and receive signals from the upright cylindrical unit containing inclinometer and magnetic compass sensors (used to measure H-Pod orientation) and also from the diagonally placed unit containing modulator electronics to drive the buoy telemetry transmitter with the H_x and H_y preamplifier signals.

While the H-Pod ocean bottom structure was being designed and fabricated and special marine cables were being manufactured, shallow water E and H magnetotelluric measurements in 1 to 2 m water depths were made near the site labeled "SPOIL BANK MT SITE" on the map of Figure 4.

FIGURE 3. H-Pod structure for magnetic sensors in preparation of shipboard deployment. The H_x sensor is located on top of the tubular structure marked with an "X". The H_y sensor is beneath and perpendicular to the H_x sensor. The compass and inclinometers are housed in the vertical tube.

TELEMETRY OF ELECTRIC FIELD MEASUREMENTS FROM OCEAN BOTTOM ELECTRODES

A conventional magnetotelluric instrument truck with multichannel digital data recording equipment was located at a base site near Port Mansfield, indicated in Figure 4. The magnetotelluric H and E data at the basesite were monitored and recorded as a reference with which to compare E and H field data received via VHF radio telemetry from the spoil bank site. Electrodes and coils could be placed by hand and from small boats at surveyed locations near the intracoastal canal in water less than 2 m deep. Cables from these sensors were manually deployed to small spoil bank islands in the tidal flats where they were connected to preamplifiers, which in turn modulated the telemetry system. A directional antenna erected on a 20 ft mast was driven by a 3-W 200 MHz battery-operated transmitter. Standard commercial transmitters and modulators, using the so-called IRIG FM techniques common in the aerospace industry, permitted several analogue E and H preamplifier outputs to be transmitted simultaneously over a single radio frequency link.

Figure 5 shows magnetotelluric data recorded in a frequency band of 0.01 to 0.1 Hz. The correlation of the telemetered data (E_{XT}, E_{YT}, H_{XT}, H_{YT}) with the corresponding components of the base site data was excellent, as shown, for all frequencies below

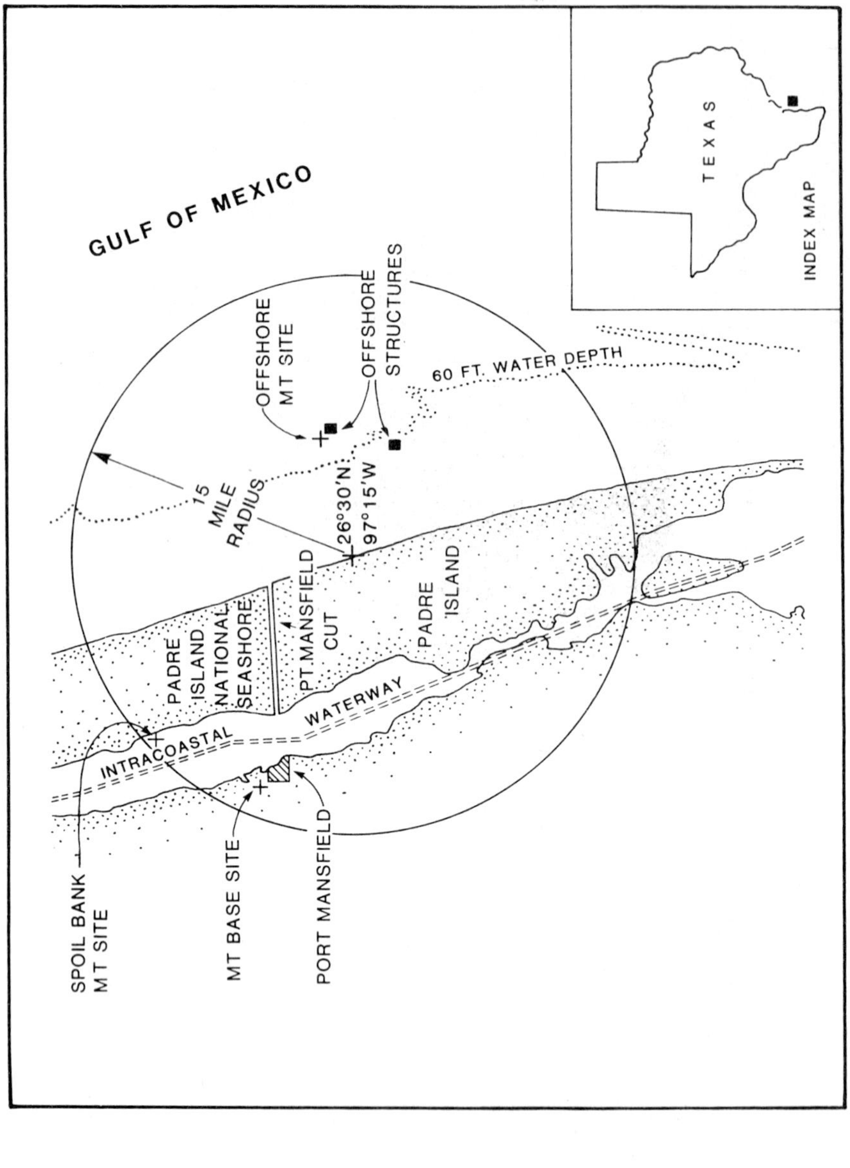

FIGURE 4. Site locations for initial marine magnetotelluric experiments.

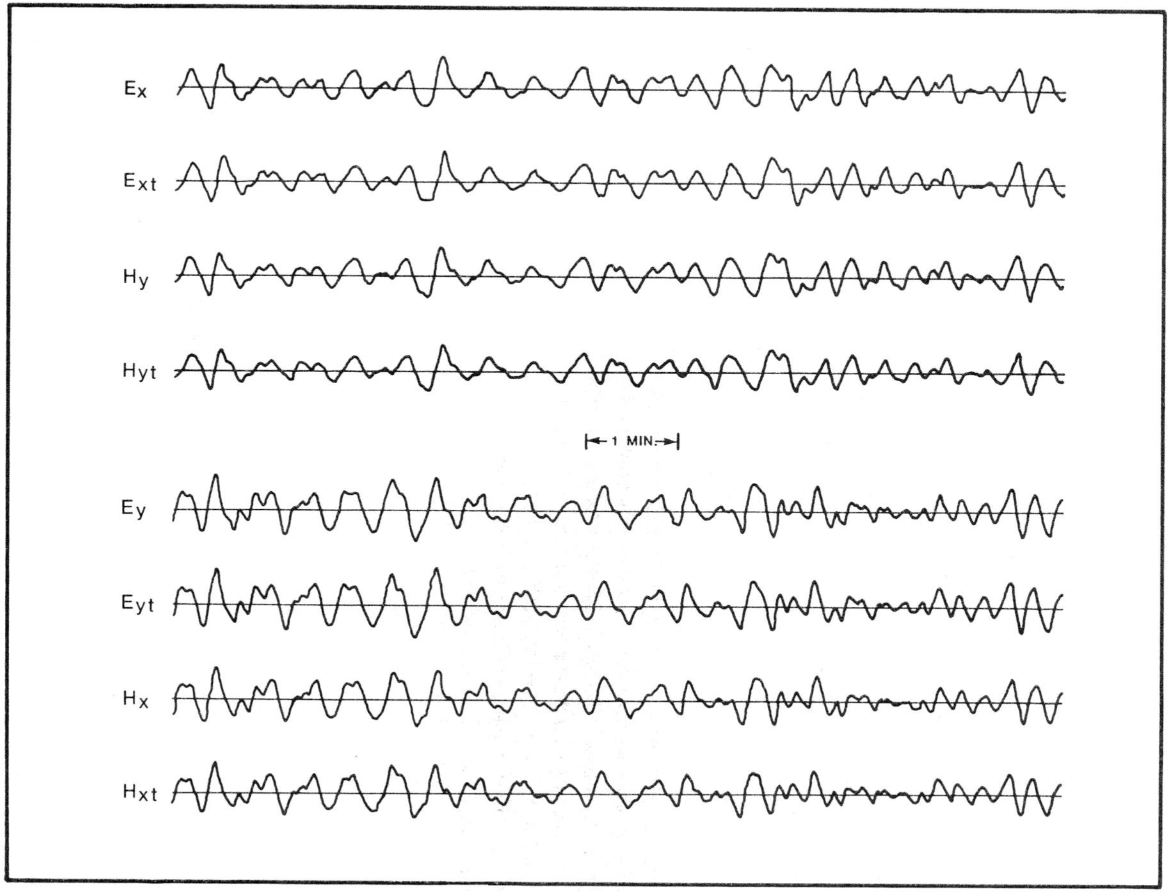

FIGURE 5. Telemetered shallow water magnetotelluric data compared (E_{xt}, E_{yt}, H_{xt}, H_{yt}) with simultaneous base site data (E_x, E_y, H_x, H_y). Data is recorded in the 0.01 to 0.1 Hz band.

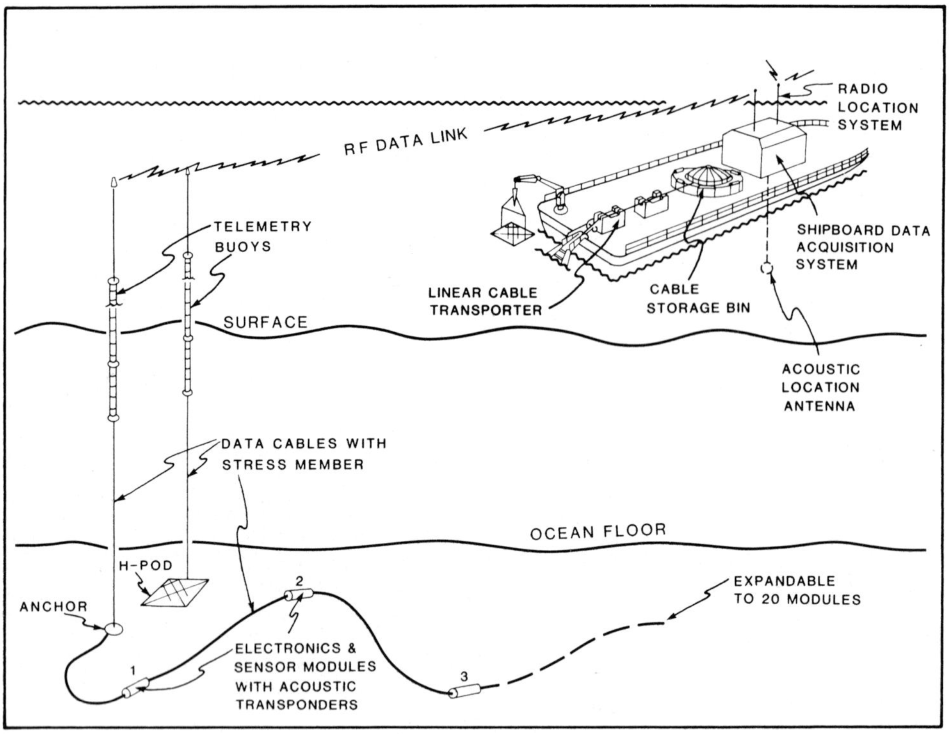

FIGURE 6. Idealized marine magnetotelluric measurements.

about 2 Hz. A narrowband "wavenoise" of 1.4 Hz corrupted the H_{XT} and H_{YT} data and was often twice the amplitude of the magnetotelluric signals. There was obvious visual correlation of the noise with wave-chop on the water surface, but the sea electrodes deployed with a spacing of 100 m did not reveal detectable 1.4 Hz wavenoise. Correlation between corresponding E and H field measurements in the 10 to 100 Hz band was also excellent over the 10 km telemetry distance between base site and spoil bank site.

Successful telemetry of data over distances of up to 30 km was achieved with the 3-W transmitters by utilizing directional transmitting antennas erected on offshore structures with heights as much as 20 m above the ocean. This telemetry link was to be used to test E field recording from electrodes and preamplifiers deployed in 20 m water depths by small boats near the offshore structure on which telemetry had been installed. The preamplifier canister could not be deployed successfully when strong ocean currents prevented further tests with light weight cables employed for E field measurements.

COLLECTION OF SIMULTANEOUS OCEAN BOTTOM ELECTRIC AND MAGNETIC FIELD SENSOR DATA

An idealized schematic illustrating deployment of, and data collection from, ocean bottom E and H sensors is shown in Figure 6. Certain portions of this idealized system were constructed. Spar buoys to house the telemetry transmitters were fabricated from PVC pipe and flange fittings. The lower third of each buoy contained lead-acid gel cells to power the transmitter and its modulator electronics as well as to supply ballast

FIGURE 7. Telemetry buoy deployment.

to the buoy and its omnidirectional antenna. In use the buoys were actually tethered to a separate buoy-anchor system and data cables entered through connectors in the side of the spar buoy. Figure 7 shows one of these telemetry buoys being hoisted by a crane on the fantail of a workboat. Reliable telemetry of up to 5 km was obtained in moderate seas.

Only three ocean bottom electrodes were deployed to measure the E_x and E_y fields rather than a multiple array as envisioned in Figure 6. The electrodes were redesigned, after the difficulties experienced near Port Mansfield, to act as rather massive anchors. Trailer tires weighted with lead shot surrounded plastic hubs containing the electrodes and electrical cable connectors. The electrodes were dropped at premarked locations, surveyed by Raydist N and marked with light anchor-buoy lines. Each marker buoy had a small lightweight radar corner reflector that permitted subsequent checking of the surveyed location by the ship's radar. The electrodes were interconnected to a preamplifier package via heavy flexible cables (welding cable) which tended to lie flat on the ocean bottom and be less subject to motion in ocean currents. The fairly massive preamplifier module acted as a bottom anchor when deployed, and in addition, its output data cable was anchored at several locations before attachment to the telemetry buoy.

Following contruction of the H-Pod a pressure transducer was retrofitted into the preamplifier cannister in order that a third data channel (in addition to H_x and H_y)

FIGURE 8. Recovery of H-Pod (magnetic and pressure sensors).

could be monitored as a crude ocean bottom wave gauge. Figure 8 shows recovery of the H-Pod from a 20-m ocean depth.

Cables were deployed and recovered aboard ship with the use of a large electrically driven cable winch rather than with the system envisioned in Figure 6.

OCEAN BOTTOM MT DATA EVALUATION

Successful magnetotelluric data collection experiments were performed near Dauphin Island, Ala. during August and September of 1976. Water depths at the sites indicated on the map of Figure 9 ranged from 62 to 68 ft.

Aboard ship an eight-channel digital data recorder, that had been used for magnetotelluric surveys on land, was mounted in a small airconditioned instrument hut which could be readily loaded or unloaded onto the deck of a workboat. Telemetry antennas aboard ship could be positioned to receive the VHF telemetry signals from a land-based MT station on the west end of Dauphin Island or from the telemetry buoys.

When the H-Pod was lowered to the ocean bottom, readings of the orientation sensors (compass and inclinometers) were made to determine if the H-sensor coils were approximately in a horizontal plane. This would permit repositioning of the H-Pod if it had landed improperly on a rough bottom. After reasonably horizontal positioning was achieved and orientation of the H-coils measured, the H-Pod cable could be connected to a telemetry buoy.

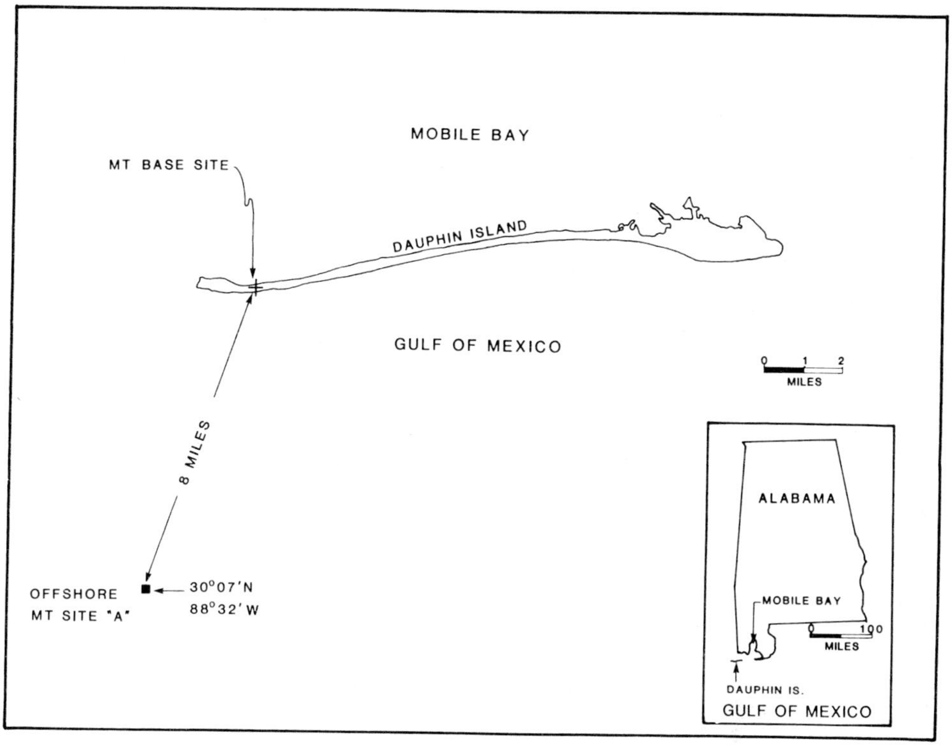

FIGURE 9. Site Locations for Marine Magnetotelluric Measurements.

A small analog computer aboard ship was used to process the H-field signals from the H-Pod coils, using the orientation information recorded during deployment so as to obtain true H_x (north) and H_y (east) signals that could readily be compared via chart recorder monitors to the H_x and H_y signals received from the base station on Dauphin Island. Figure 10 shows chart trackings where H_x, and H_y are the MT signals telemetered from the Dauphin Island base station and H_{xp}, H_{yp} and Pressure are from the H-Pod. This excellent low-frequency data persisted for the entire 30-hr period during which the H-Pod was deployed. When the workboat was within a few hundred meters of the H-Pod large signals were recorded corresponding with the motion of the magnetized hull.

Midband data (0.01 to 0.1 Hz) shown in Figure 11 clearly reveals the effect of ocean waves on the H-Pod signals. A pronounced 0.1 Hz wavetrain, apparent in the pressure recording, is also clearly seen in the H_{xp} and H_{yp} recordings. The peak-to-peak pressure variation at 62 ft depth appears to be about 0.9 in. of water (0.035 lb/in^2) and, from Figure 2, for a wave period of 10 sec and water depth of 62 ft the wavelength should be 390 ft, the attenuation factor K = 0.65, and the peak horizontal bottom currents about 0.25 ft/sec/ft surface wave height. The surface wave or swell should then have a crest-to-trough height of approximately 0.9/0.65 = 1.4 in. with a 390-ft crest-to-crest distance — a very subtle swell that could not be visually observed on the sea surface with the typical chop of several feet that was present during the recording of data.

By subtracting the H_x signal at the base site from the H_{xp} signal measured with the H-Pod sensors a picture of the wave noise introduced into the H measurement can be seen. Such a waveform is illustrated in Figure 12 together with the pressure waveform

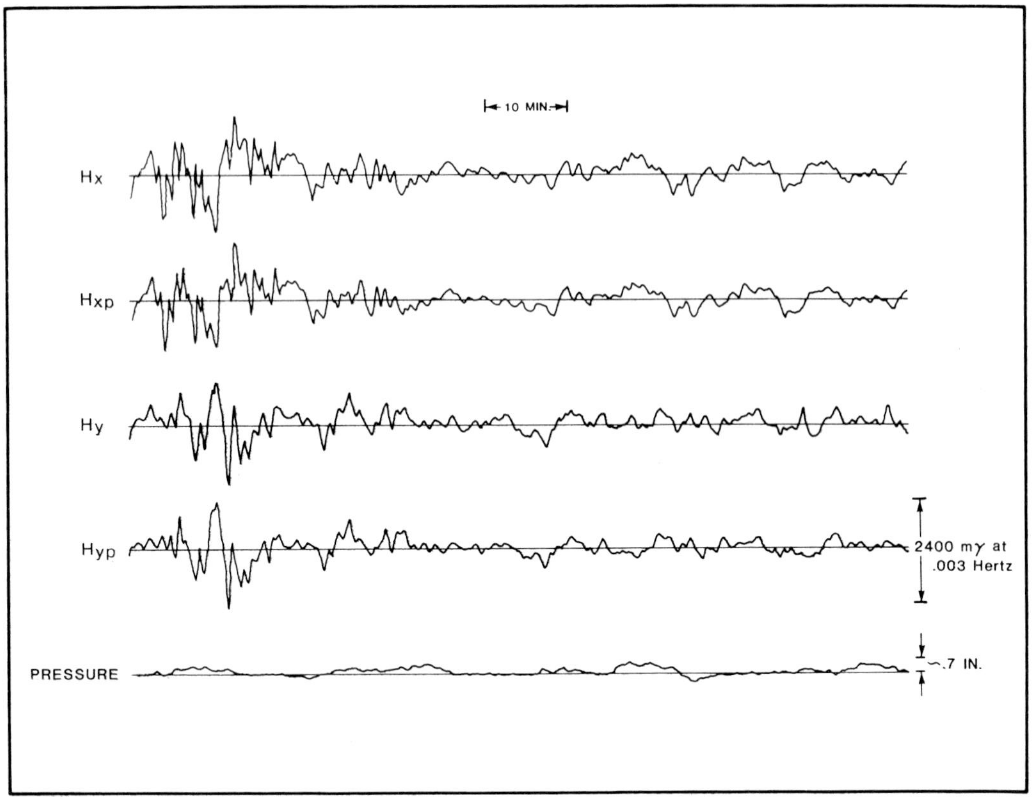

FIGURE 10. Low frequency (0.001 to 0.01 Hz band) ocean bottom H_{xp} and H_{yp} fields compared with base station MT fields.

and a hodagram of the pressure-noise waveforms. The hodagram illustrates a relatively coherent phase relationship between the 0.1 Hz water motion and the "noise" contamination of H_{xp} field measurement. H_{yp} noise did not exhibit nearly as coherent a relationship with water motion.

Digitized data collected in a number of frequency bands was processed and coherences calculated between the field of the land base site (H_x), the field of the H-Pod (H_{xp}) and the pressure variations (P). The dashed curves in Figure 13 show an excellent agreement (coherencies near 1.0) between the H-field measurement on land and on the ocean bottom except over the narrow band of frequencies between 0.1 and 0.3 Hz. The solid curves in Figure 13 show a very strong correlation in the 0.1 to 0.3 Hz band between the ocean bottom field and the pressure measurements. This narrow band noise in the H-field measurement is presumed to be generated by physical motion of the H-Pod rather than electromagnetic signals generated within the seawater, since ocean bottom electrodes have not detected this same frequency noise.

Simultaneous E-field and H-field MT measurements were made at 68 ft water depth using two telemetry buoys near the site A location of Figure 9. Two days of excellent data recording were obtained and the striking coherence between the orthogonal field pairs E_x-H_y and E_y-H_x measured on the ocean bottom are shown in Figure 14. Furthermore, Figure 15 shows an excellent correlation between the E-field recorded both onshore and offshore simultaneously.

Data obtained at Site A on digital magnetic tapes were processed and conventional magnetotelluric tensor impedances and sounding curves were calculated by digital com-

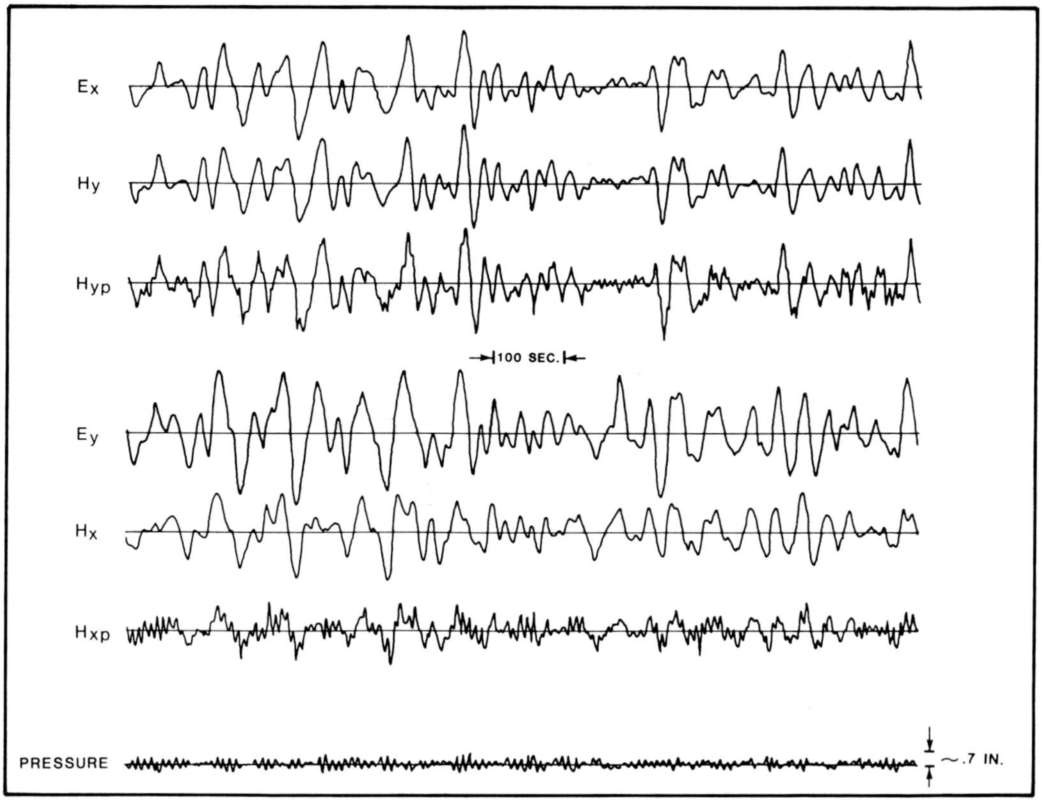

FIGURE 11. Midband (0.01 to 0.1 Hz) ocean bottom H-field data indicating 0.1 Hz water wave noise.

puter using standard techniques.[3,19] Since the computed tensor impedance exhibited a one-dimensional earth property, only a single sounding curve is shown in Figure 16. Some impairment of data quality is apparent in the vicinity of 0.1 Hz because of wavenoise, but interpretation of the data is not seriously handicapped in this example. There are several apparent resistivity values per frequency point due to multiple runs of each recording band. The solid curve (and RMS errors) are based upon a weighted stack for each frequency point.

GEOLOGIC SIGNIFICANCE

The purpose of this study is to document the simultaneous measurements of time-varying electric and magnetic fields in the Gulf of Mexico. It is not within the scope of this study to analyze in detail the subsurface electrical resistivity distribution near Dauphin Island. Nevertheless, it is constructive to compare the magnetotelluric interpretation for one of the offshore sites with the regional geology.

Dauphin Island is one of a chain of barrier islands along the Alabama-Mississippi coast.[21] The islands change shape from year to year with erosion on the eastern ends and deposition taking place on the western ends. The islands are probably of early Pliocene or late Miocene origin.

Mobile, Ala. (see Figure 9), is underlain by terrace and alluvial deposits in the flood plain of the Mobile River.[22] Wells in this area usually encounter 80 to 150 ft of river alluvium before penetrating the underlying Miocene beds. The Miocene deposits ex-

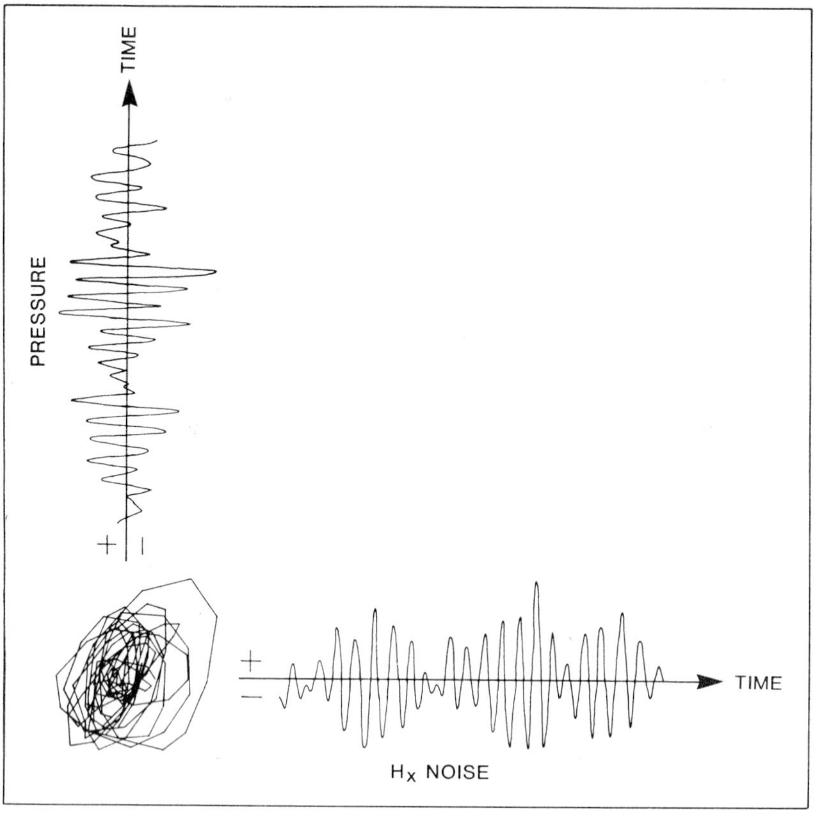

FIGURE 12. Pressure waveform and H_x noise waveform together with hodagram of phase coherence.

tend to a depth of 1300 to 1400 ft. These deposits consist of gray dense clay, sandy clay, sand, and limestone. Beneath the Miocene, within the Cenozoic, are the Wilcox sands and shales down to −2500 ft. Shales of Midway age mark the base of the Cenozoic at −4500 ft.

The upper Cretaceous extends to a depth of −8000 ft with the Austin chalk overlying the shales of the Tuscaloosa formation. A thick Red Bed Facies in the lower Cretaceous extends down to −17,000 ft. There are thick sections (up to 1000 ft) of shales and limestones within these Red Beds.

Limestones predominate in the probable basement Jurassic System. The Louann Salt delineates the base of the Jurassic and overlays the Red Beds of the Eagle Mills formation.

An iterative, least-squares, layered model analysis was made of the offshore magnetotelluric data using an algorithm developed by Laird and Bostick.[23] Figure 17 shows a four-layer model fit relative to the stacked apparent resistivity data (see Figure 16). The four-layer resistivity-depth parameters are shown to the right of the curves. Layer 1 corresponds to the lower resistivity near-surface deposits. Layer 2 is a 1240-ft thick section with a resistivity of 5.4 ohmmeters and is probably related to the Wilcox limestones. Layer 3 is a 6670-ft thick low resistivity section related to the sand-shale sequence above the Washita-Fredericksburg formation. Layer 4 is the high resistivity electrical basement.

Figure 17 also shows the parameters for a five-layer model solution (the five-layer

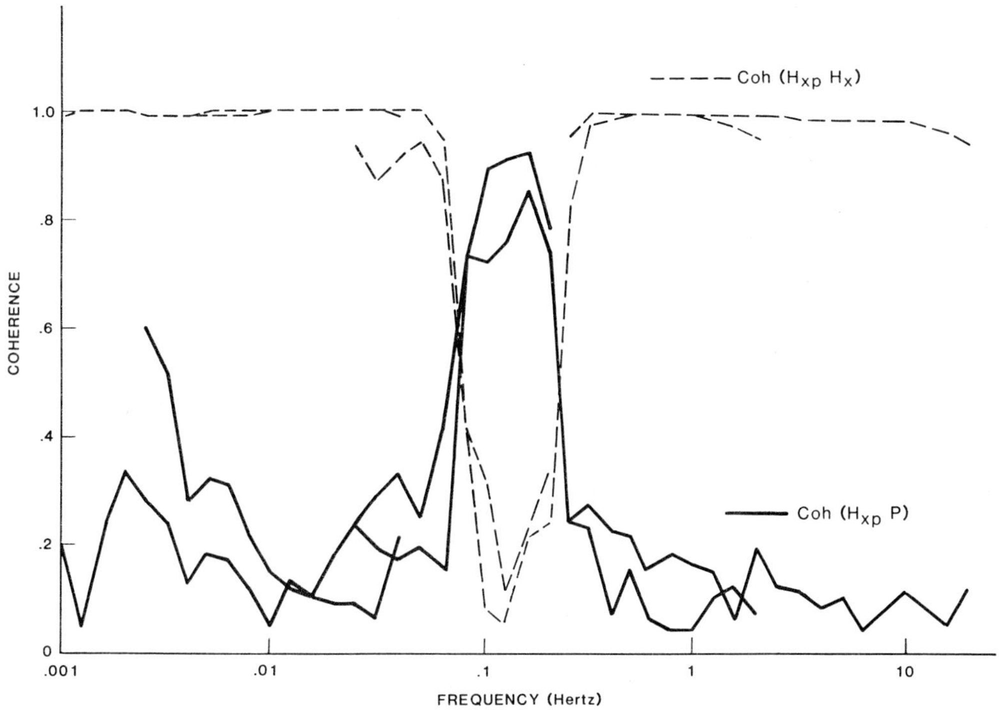

FIGURE 13. Coherence plots of <H_{xp} H_x> and <H_{xp} P>, where H_{xp} is the ocean bottom field, H_x is the land base field, and P is the ocean bottom pressure measurement.

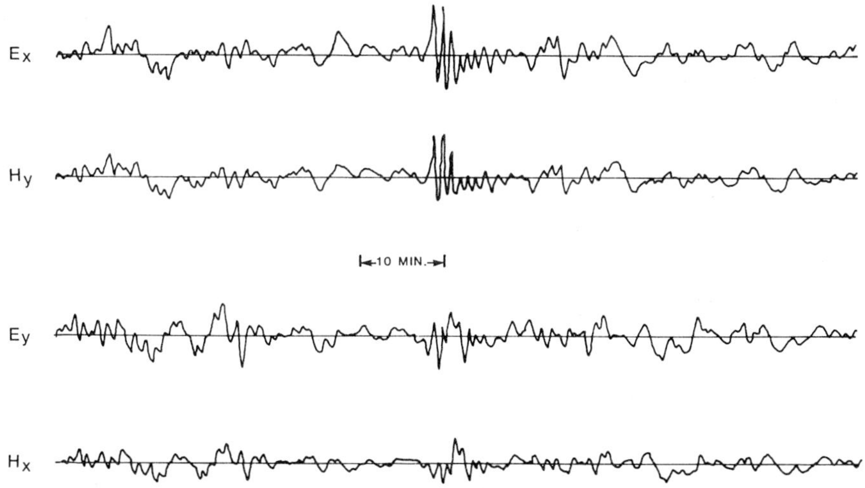

FIGURE 14. Low frequency (.001 to 0.1 Hz) ocean bottom magnetotelluric data, 20-m water depth.

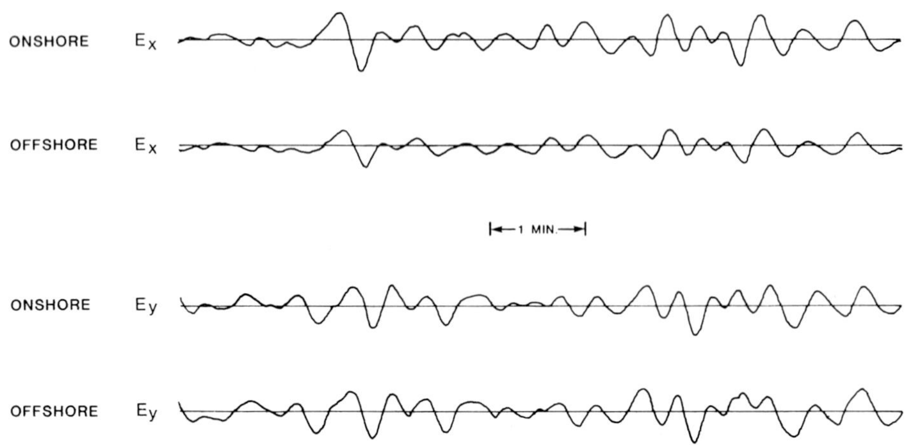

FIGURE 15. Comparison of simultaneously recorded onshore and offshore magnetotelluric data.

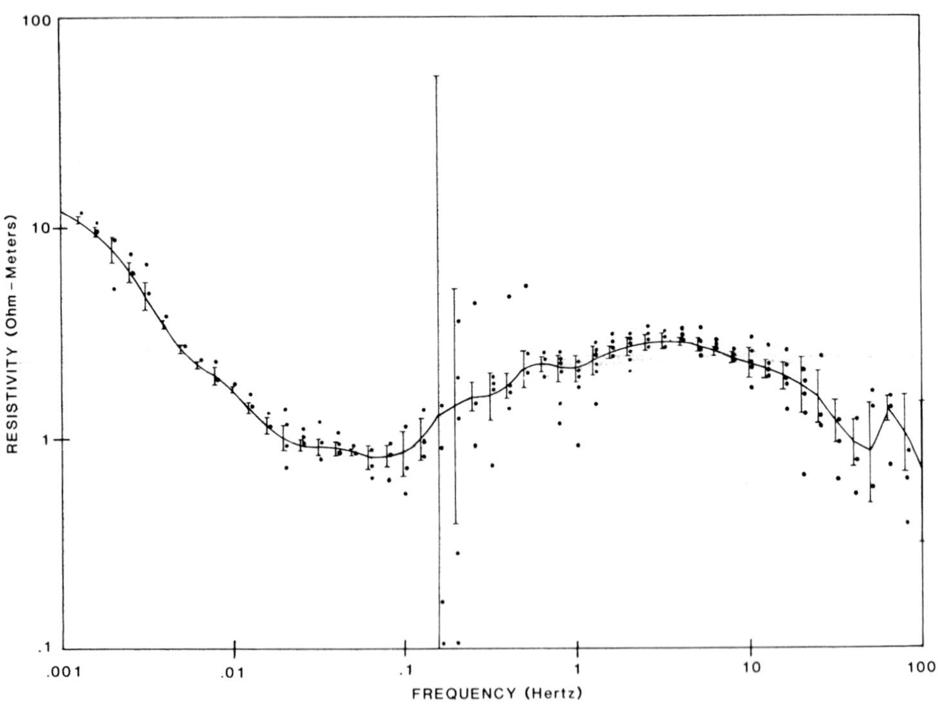

FIGURE 16. Magnetotelluric sounding curve (site A, near Dauphin Island, Ala).

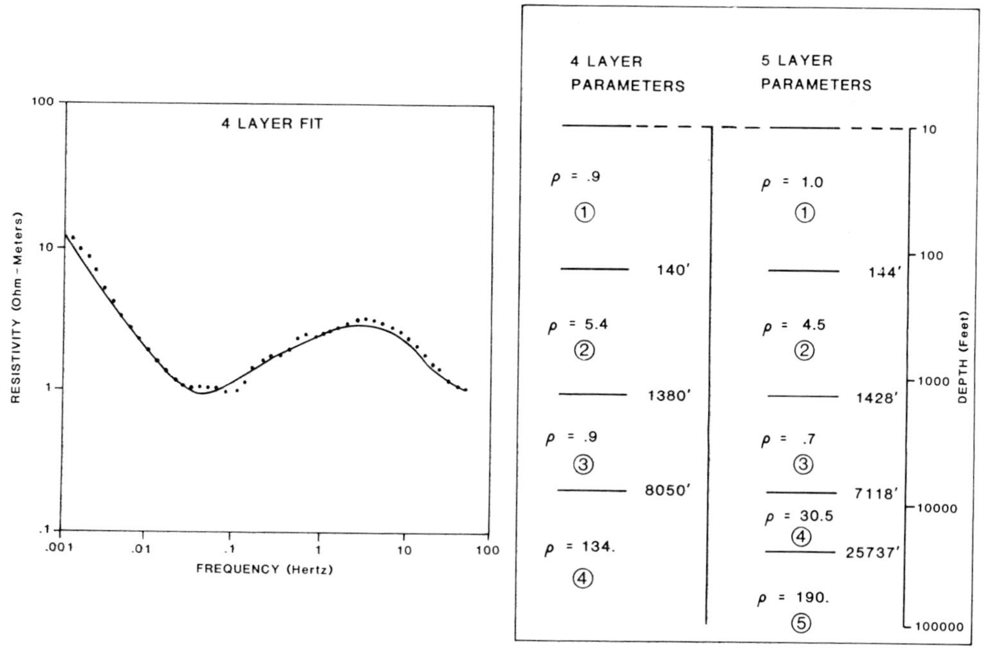

FIGURE 17. Magnetotelluric sounding curve and one-dimensional interpretations for four-layer and five-layer models.

curve virtually overlays the four-layer curve and hence is not shown). In this five-layer model we are attempting to resolve the "geologic basement" whose top is the late Triassic-early Jurassic Louann Salt. Layers 1, 2, and 3 are similar to the four-layer model. However, the five-layer model shows an 18,619-ft thick, 30.5 ohmmeter section overlaying a 190 ohm-meter basement at −25,737 ft. This interface at −25,737 ft may not be significant, however, due to limitations of resolving one resistive layer into two resistive layers.[23] Consequently, the determination of the Triassic-Jurassic basement from the five-layer model should be considered as speculative.

CONCLUSIONS

Ocean bottom magnetotelluric surveying is technically feasible, at least for the water depths of the continental shelf areas. Measurements reported here encountered the anticipated water motion noise in the magnetic fields but negligible noise in the electric field. Mechanical motion of the magnetic sensors appears more serious than electromagnetic signals produced by conductive saltwater motion in generating noise. While these measurements were on ideal "hard bottom" sites, other workers[20] have reported wave-induced physical motion of as much as ½ in. for the "soft bottom" in 20 m water depths near the mouth of the Mississippi River.

ACKNOWLEDGMENTS

The authors wish to thank Mobil Research and Development Corporation for permission to publish this research. Success in these experiments could not have been achieved without the diligent efforts of the technicians, engineers, and scientists contributing to this marine magnetotelluric project.

REFERENCES

1. **Cagniard, L.,** Basic theory of the magnetotelluric method, *Geophysics,* 18, 605, 1953.
2. **Tikhonov, A. N. and Shakhsuvarov, D. N.,** On the use of the impedance of the earth's natural electromagnetic field for the study of its upper levels, *Akad. Nauk. SSSR. Izv. Ser. Geofiz.,* 4, 410, 1956.
3. **Vozoff, K.,** The magnetotelluric method in exploration of sedimentary basins, *Geophysics,* 37, 98, 1972.
4. **Cagniard, L.,** Extension de la prospection magnetotellurique a l'exploration off-shore, *Rev. l'Hydrocarbure,* 104, Suppl. 1967.
5. **Klein, M.,** Influence du Milieu our les Mesures Magnetiques en Magnetotellurique Marine, thesis, Fac. Sci., University of Paris, 1971.
6. **Morat, P.,** Application of measurements of magnetic and electric fields to knowledge of sea swell and currents, Colloq. Trans. 11th Int. ACOBOR Exploit. of Oceans, Centre de Recherches Geophysiques, Loire, France, 2, 1974.
7. **Gol'mshtok, A. Y. and Sochel'nikov, V. V.,** Methods of marine magnetotelluric studies of the sedimentary-layer thickness and the crystalline foundation surface, *Appl. Geophys.,* 62, 93, 1971.
8. **Berdichevskiy, M. N. and Van'yan, L. L.,** Prospects for deep magnetotelluric sounding in bodies of water, *Akad. Nauk. SSSR. Izv. Fizika Zemli,* 11, 51, 1969.
9. **Filloux, J. H.,** An ocean bottom, D-component magnetometer, *Geophysics,* 32, 978, 1967.
10. **Hermance, J. F.,** Resolution of ocean floor magnetotelluric data, *J. Geophys. Res.,* 74, 5527, 1969.
11. **Cox, C. S., Filloux, J. H., and Larsen, J. C.,** *The Sea,* Vol. IV, John Wiley & Sons, New York, 1970, 637.
12. **Kinsman, B.,** *Wind-waves: Their Generation and Propagation on the Ocean Surface,* Prentice-Hall, Englewood Cliffs, N.J., 1965.
13. **Klein, M., Louvet, P., and Morat, P.,** Measurements of electromagnetic effects generated by swell, *Phys. Earth Planetary Inter.,* 10, 49, 1975.
14. **Weaver, J. T.,** Magnetic variations associated with ocean waves and swell, *J. Geophys. Res.,* 70, 1921, 1965.
15. **Bostick, F. X.,** The detection of ULF-ELF emission from moving ships, Final Tech. Rep. ARPA Order No. 2600, Electr. Geophys. Labs., University of Texas, Austin, 1977.
16. **Podney, W.,** Electromagnetic fields generated by ocean waves, *J. Geophys. Res.,* 80, 2977, 1975.
17. **Beal, H. T. and Weaver, J. T.,** Calculations of magnetic variations induced by internal ocean waves, *J. Geophys. Res.,* 75, 6846, 1970.
18. **Sochel'nikov, V. V., Gol'mshtok, A. Y., and Mogilatov, V. S.,** Influence of surface and internal waves on the results of electrical exploration at sea, *Akad. Nauk SSSR. Izv. Fizika Zemli,* 10, 74, 1975.
19. **Word, D. R., Smith, H. W., Bostick, F. X., Jr.,** Crustal investigations by the magnetotelluric tensor impedance method, *The Structure and Physical Properties of the Earth's Crust,* Geophys. Monogr. Ser. Vol. 14, Heacock, J. G., Ed., American Geophysical Union, Washington, D.C., 1971, 145.
20. **Suhayda, J. W., Whelan, T., Coleman, J. M., Booth, J. S., and Garrison, L. E.,** Marine sediment instability of hydrodynamic zones and bottom sediments, Offshore Technol. Conf. Pap. No. OTC 2426, 1976.
21. **Smith, B. L.,** Dauphin Island, Ala., *Fifteenth Field Trip Given by the Mississippi Geological Society,* Sestak, A. A., Ed., Mississippi Geological Society, Jackson, 1960, 27.
22. **Grover, M. E.,** *Geology of the Atlantic and Gulf Coastal Province of North America,* Harper & Row, New York, 1961, chap. 3.
23. **Laird, C. E. and Bostick, F. X., Jr.,** *One-dimensional Magnetotelluric Inversion Techniques,* Austin Electron. Res. Cent., University of Texas, Austin, 1970.

TECTONIC ORIGIN OF THE GULF OF SUEZ, EGYPT, AS DEDUCED FROM GRAVITY DATA

A. I. Bayoumi

INTRODUCTION

The purpose of this chapter is to present the results of a study of the lateral distribution of the evaporite group that comprises the bulk of the "rift" Miocene sediments in the Gulf of Suez shallow water basin (Figure 1), using potential field data together with available seismic and drill hole information.

Critical analysis of such exploratory data reveals that a relationship appears to exist between the abundance of each member of this evaporite group, i.e., salt and anhydrite, and the principal structures of typical rift basins, which are mainly confined to major and minor horsts and grabens (Figures 1 and 2). Typical sections of this group are represented by the Kareem, Belayim, and S. Gharib formations of Helvetian Age (Figure 1). The generalization may be made that anhydrites, with a maximum thickness of 2500 m, are restricted to the uplifted blocks that complicate the eastern and western sides of the major Gulf trough, whereas the salt, with a maximum thickness of 1500 m, is deposited on downfaulted blocks of two major subbasins. This seems to complicate the regional structural setting of the central portion of the Gulf of Suez "graben".

The eastern subbasin is believed to parallel the axis of the major trough that joins the Morga, Gara, Sukna, and Zaafarana local troughs. These troughs include within their boundaries the major oil fields of Morgan, Ramadan, Belayim, and Amal. The western basin also parallels the eastern one and includes the local troughs of Gemsa, Gharib, and Dip. Potential field and seismic methods were used to construct models across the entire width of the Gulf, and these indicate that it is bounded on both the eastern and western sides by two master NW-SE trending faults with basement rocks cropping out on their upthrown blocks (Figure 2). The following generalizations can be made based on these model studies:

1. Areas north of latitude 29° 30'N and south of latitude 27°30'N seem devoid of evaporites, and exhibit a general northward decrease in the thickness of the salt mass within the central portion of the Gulf (Figure 1).
2. The velocity of seismic waves throughout the entire thickness of the Miocene sediment section generally increases with depth. The extrapolated "surface" velocity is 2200 m/sec, and has a vertical gradient of 0.58 m/sec.
3. Trapping of oil in the Gulf area is largely related to its tectonic history and is confined to two major types:
   a. Miocene and postMiocene flexures over preMiocene topography. This results in faulted anticlinal and synclinal structures by dragging the less rigid Miocene strata along the fault planes, i.e., along the flanks of the major uptilted blocks, and
   b. Pre-Miocene tilted structures unconformably overlaid by Miocene and younger beds, i.e., along local structures lying within the inter subbasins located between major uplifts.
4. The structural nature of the Gulf of Suez is quite similar to that of the Red Sea, in spite of the fact that:
   a. The amount of evaporites in the Gulf of Suez is about one half of that present in the Red Sea.
   b. The prerift sections of both are different.

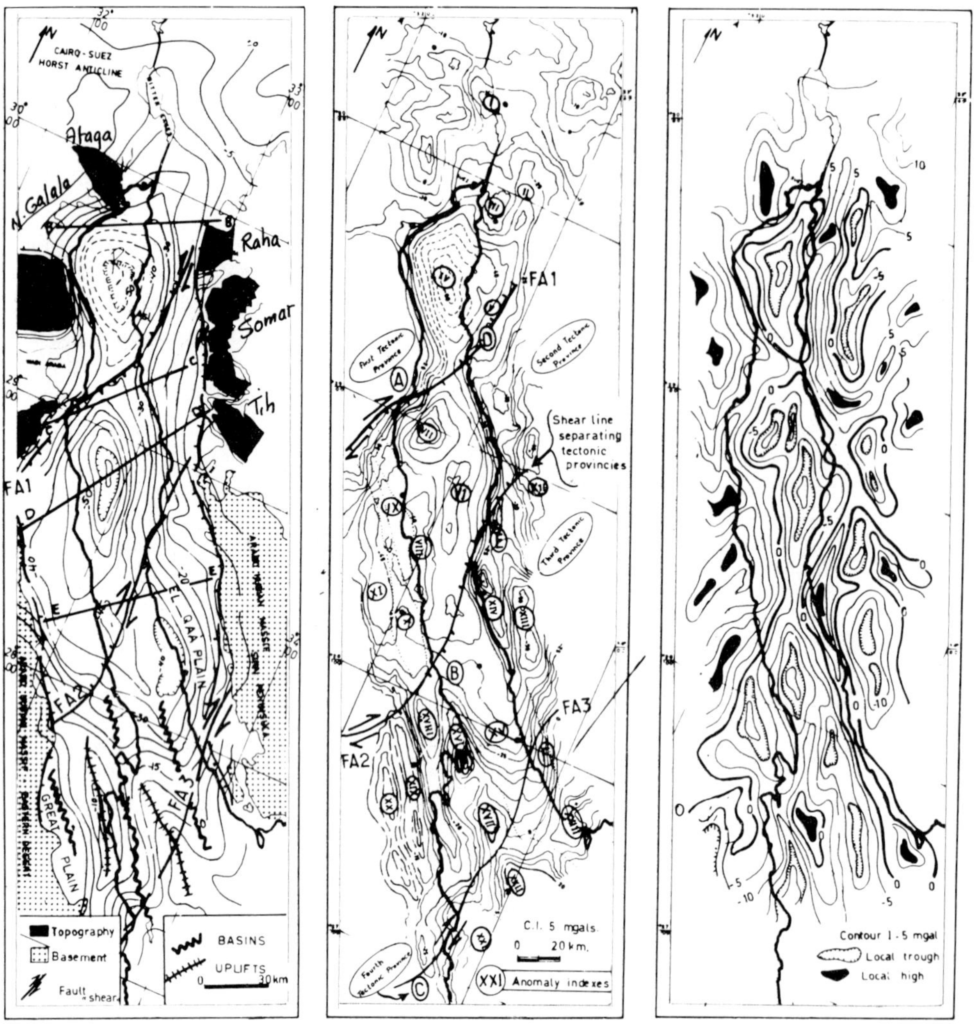

A B C

FIGURE 1A. Major geologic-morphologic features, with regional gravity field in the Gulf of Suez Region. B. Bouguer anomaly map, Gulf of Suez Region. C. Computed residual anomaly map, Gulf of Suez Region.

c. The faulting that controls both the rift and prerift sediments in the Gulf of Suez does not extend to the same depth as that controlling the structure of the Red Sea.

The regional tectonics of the Afro-Eurasian Junction are understood reasonably well.[1,2] However, the structural development of the Gulf of Suez, which is considered one of the most interesting rift basins of this junction, is still a matter of controversy among many investigators.[3-23] This particularly applies to the question of whether such a ''Clysmic'' graben originated entirely by horizontal movements along a major shear zone (wrenching), by displacements of major crustal blocks apart from each other (drifting), or is it simply a graben dominated by normal faulting. In general, this kind of controversy arises from the fact that not enough information is available as to the

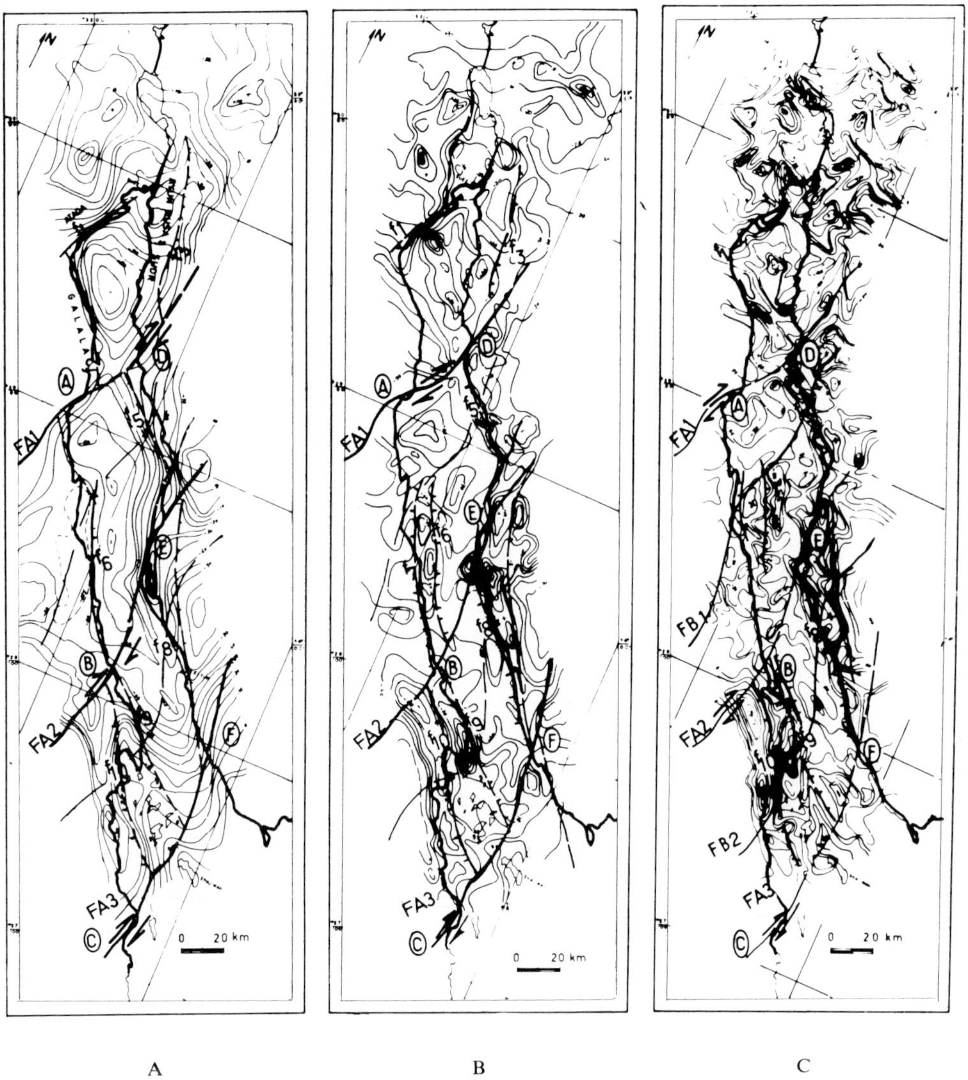

FIGURE 2A. Downward continuation of Bouguer field to a depth of 1.25 km. B. Downward continuation of Bouguer field to a depth of 2.5 km. C. Downward continuation of Bouguer field to a depth of 5 km.

exact nature of the motion along the planes of the different fracture systems[13,16] that have been found to complicate the structure of the Gulf of Suez region as a whole. It is situated principally, as defined in this investigation, between latitudes 27 to 30°00' N and longitudes 32 to 34°00' E, and is bounded on the east and west by the major geologic and geomorphic features of both the Sinai Peninsula (Raha, Somar, Tih, Gulf of Aqaba, and El Qaa Plain) and the eastern desert of Egypt (Galala, Ataqua, El Zeit, Esh Mellaha, and the Great Plain), respectively. It may also be defined by the two master NW-SE trending faults that extend parallel to each other along both sides of the Red Sea. The northern fault is offset about 100 km along the Gulf of Aqaba from the one present along the Red Sea (Figure 1a).

The main objective of this study, however, is to compile all available information on the surface and subsurface geologic conditions of the Gulf of Suez region. This includes data that have been accumulated principally by several investigators[24-29] who critically interpreted, under the supervision of the author, the different anomalous features complicating the normal distribution of the earth's gravitational field in the Gulf of Suez region (Figures 1 and 2). In addition, this compiled information is used in combination with the idea of possible reorientation of stresses[30] resulting from variations in the intensity and direction of the lithostatic pressure at depth. The information from these sources was then combined to propose a reasonable model of the tectonic origin of the Gulf of Suez during its entire geologic history. All other types of exploratory data, such as those obtained from seismic reflection surveys, and from the more than 250 wells drilled in the region and studied by different exploration companies (Standard Oil Co. of Egypt, Anglo-Egyptian Oil Fields, Standard Vacuum Oil Company, South Mediterranean Oil Company, General Petroleum Company of Egypt, Le Companie Orientale de Petrol d'Egypt, Gulf of Suez Petroleum Company, and Pan Am Oil Company) were also used to help resolve any structural ambiguities arising from the compilation process.

MAJOR GEOLOGIC FEATURES OF THE GULF OF SUEZ REGION

The results obtained from this compilation process are given as follows.

1. Almost all of the structural linear elements are strike-slip faults[29,42] of different trend patterns, and appear in places with slight vertical displacements of less than 300 m.
2. Such major fracture systems are, in general, identified as three left-lateral, trending N35°W, N15°E, and N60°E, and three right-lateral, trending N45°W, NNE-SSW, and N85°E, as shown in Figure 3a. They also form a more or less regular fracture system, emphasizing the block faulting nature of the Gulf of Suez basin that has been discussed by some investigators.[10,31,32,35]
3. Some sort of relationship[24,25,35] exists between the lateral distribution of the end members of the Miocene evaporite group (salt and anhydrite), with type sections represented by Kareem, Belayim, S. Gharib, and Zeit formations of Helvetian Age (Figure 4a), and the principle structures of the Gulf of Suez basin, confined mainly to major and minor faulted blocks (Figures 3b and 4b).
4. It was also found that the anhydrites, with a maximum thickness of 2500 m, are restricted to the uplifted blocks, thereby complicating the eastern and western sides of the main trough. However, the salt, with a maximum thickness of 1500 m, is deposited on the downfaulted blocks of two major subbasins.[24,25] This appears to complicate the regional structure of the central and southern portions of the Gulf of Suez main basin.

   The eastern subbasin is expected to parallel the axis of the major trough that joins Morgan, Gara, Sukna, and Zaafarana local troughs and includes the major Morgan, Ramadan, Belayim, and Amal oil fields, whereas the western subbasin also parallels the first and includes Gemsa, Gharib, and Ras Dip local troughs (Figure 4b).
5. Results of investigations also indicate that, among the basement and sedimentary rocks, Precambrian metavolcanics, Oligocene basalt sheets, and Miocene evaporites are the most important to consider when interpreting gravity anomalies in the Gulf of Suez[29,32] region (Figure 5). The sedimentary rocks consist of marine and nonmarine sediments, ranging in age from the Paleozoic era to the Plio-Pleistocene-recent. The post- and preMiocene formations have almost a uniform average density of 2.34 g/cm^3. However, the Miocene section exhibits marked variations in density, ranging between 2.19 g/cm^3 for salt to 2.90 g/cm^3 for an-

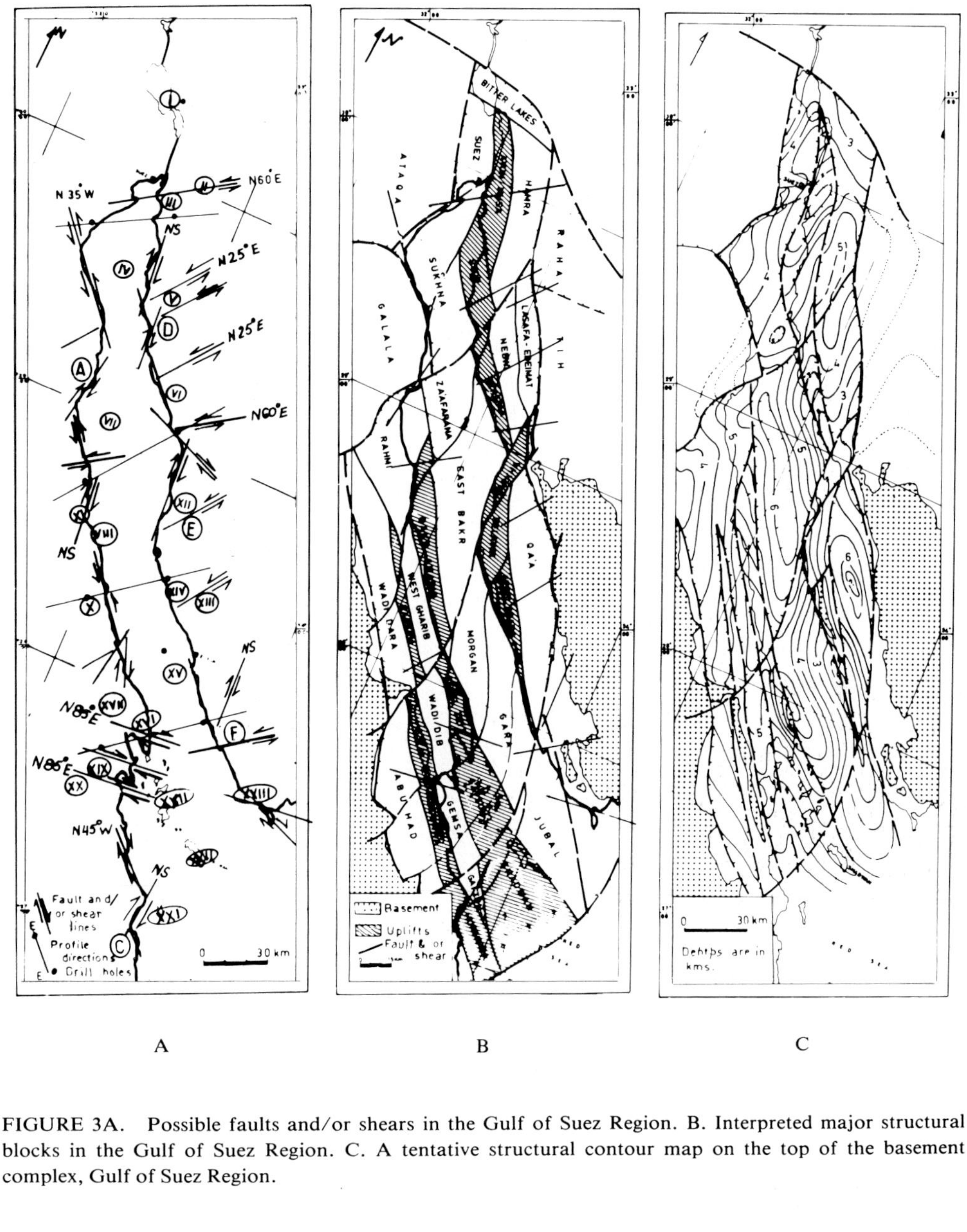

A B C

FIGURE 3A. Possible faults and/or shears in the Gulf of Suez Region. B. Interpreted major structural blocks in the Gulf of Suez Region. C. A tentative structural contour map on the top of the basement complex, Gulf of Suez Region.

hydrites; whereas the average densities of the basalts and the Precambrian metavolcanics were found to be 2.93 g/cm^3 and 2.84 g/cm^3,[24,27,29,36] respectively.

6. The present work also yielded the following information:
   a. The considerable variation in depths to the top of the basement complex across the Gulf of Suez basin varies from 300 m to about 6 km[29,32] and its surface is complicated by major fracture systems (Figure 3c).
   b. The role of sedimentation contemporaneous with faulting in defining the gradual increase of the thickness of a certain rock unit on the successive downdip fault block relative to the next block updip.[35]
   c. The role of rotation in opposite directions of the competent faulted strata lying above and below the incompetent evaporites in defining the marked variations in thickness of the salt and anhydrite observed in some adjacent drill holes in the Gulf of Suez region.[34]

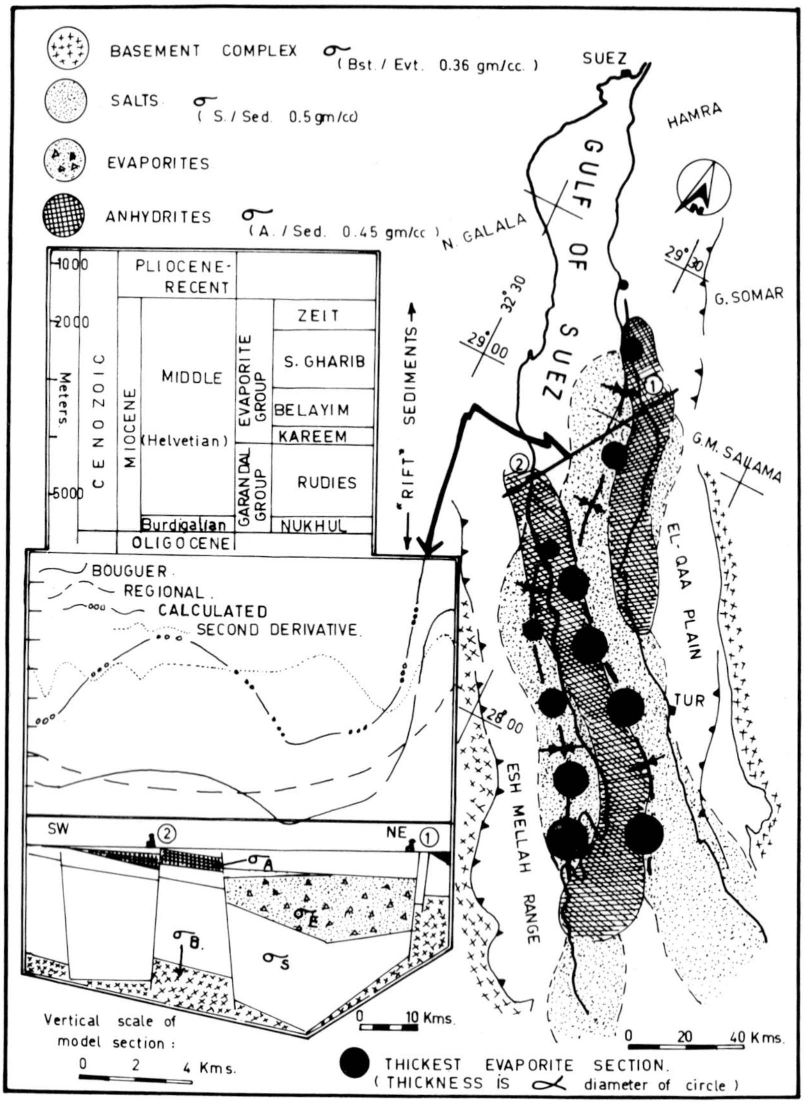

A

FIGURE 4A. Lateral distribution of the salt and anhydrites in the Gulf of Suez Region. B. Major structural features of economic interest, Gulf of Suez Region.

SPECULATION ON THE TECTONIC ORIGIN OF THE GULF OF SUEZ

Based on the abovementioned structural findings, summarized in Figure 6, it is possible to draw some tectonic speculations as to the structural developments of the Gulf of Suez during different geologic ages. These are as follows:

1. During the Precambrian period it seems plausible that under the effect of the tensile stresses that were active at that time, and particularly during the updoming of the Arabo-Nubian shield, several fracture systems having different trend patterns (NW-SE, NE-SW, EW, and NS) and without any significant shearing move-

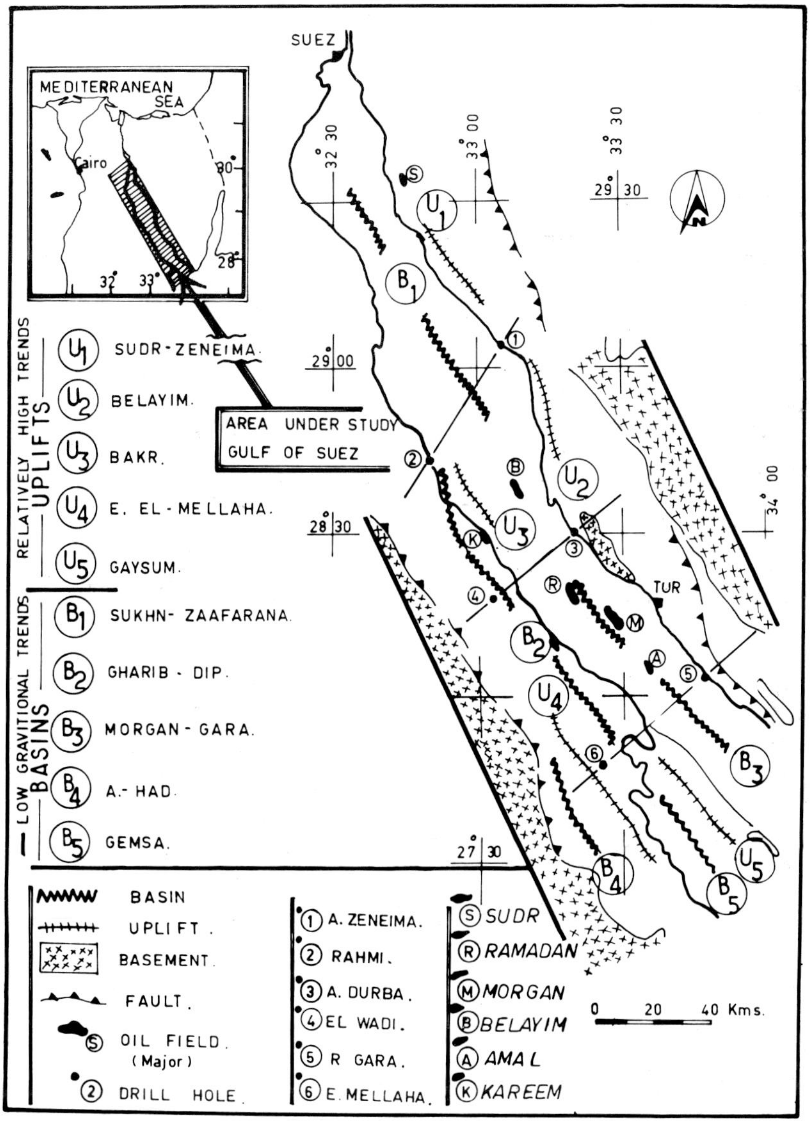

FIGURE 4B.

ments, were developed within the crust of the earth. Evidence for the existence of such old block-fracturing has been cited in the literature and need not be emphasized.

2. By the late Precambrian to early Paleozoic era, specifically in the Carboniferous period however, the lithostatic pressure at some depth somehow must have been modified. It could have resulted from the effect of complicated tectonics arising from the strong epeirogenic movements that prevailed in the region during the Paleozoic era. This could have acted to the extent that the resultant effect appeared as a marked decrease in pressure parallel to the horizontal NE-SW direction, a slight decrease in pressure parallel to the NW-SE direction, and leaving the vertical unchanged. This causes an orientation of the stress ellipsoid in such a way that its greatest principle axis (P) is directed vertically, while both the

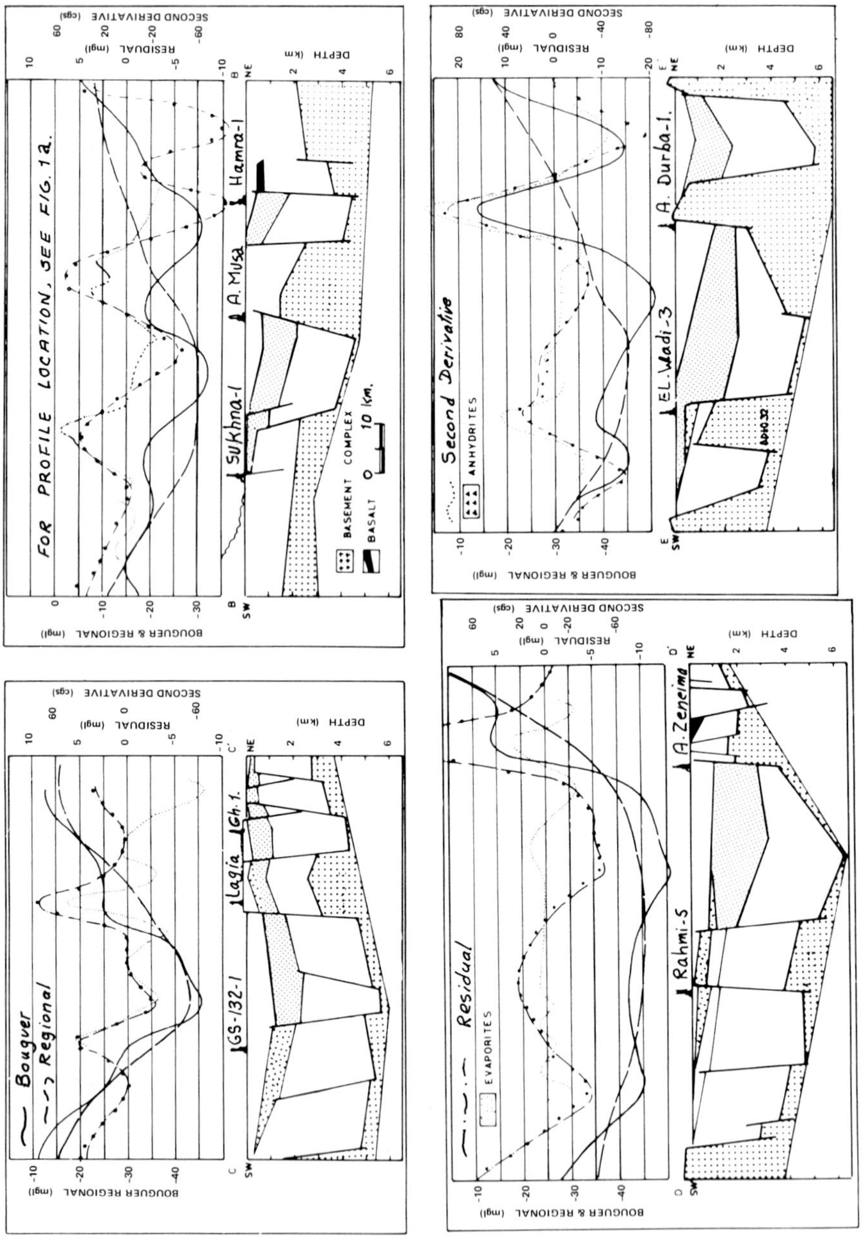

FIGURE 5. Two-dimensional gravity modeling showing the anomaly-causing bodies in the Gulf of Suez Region.

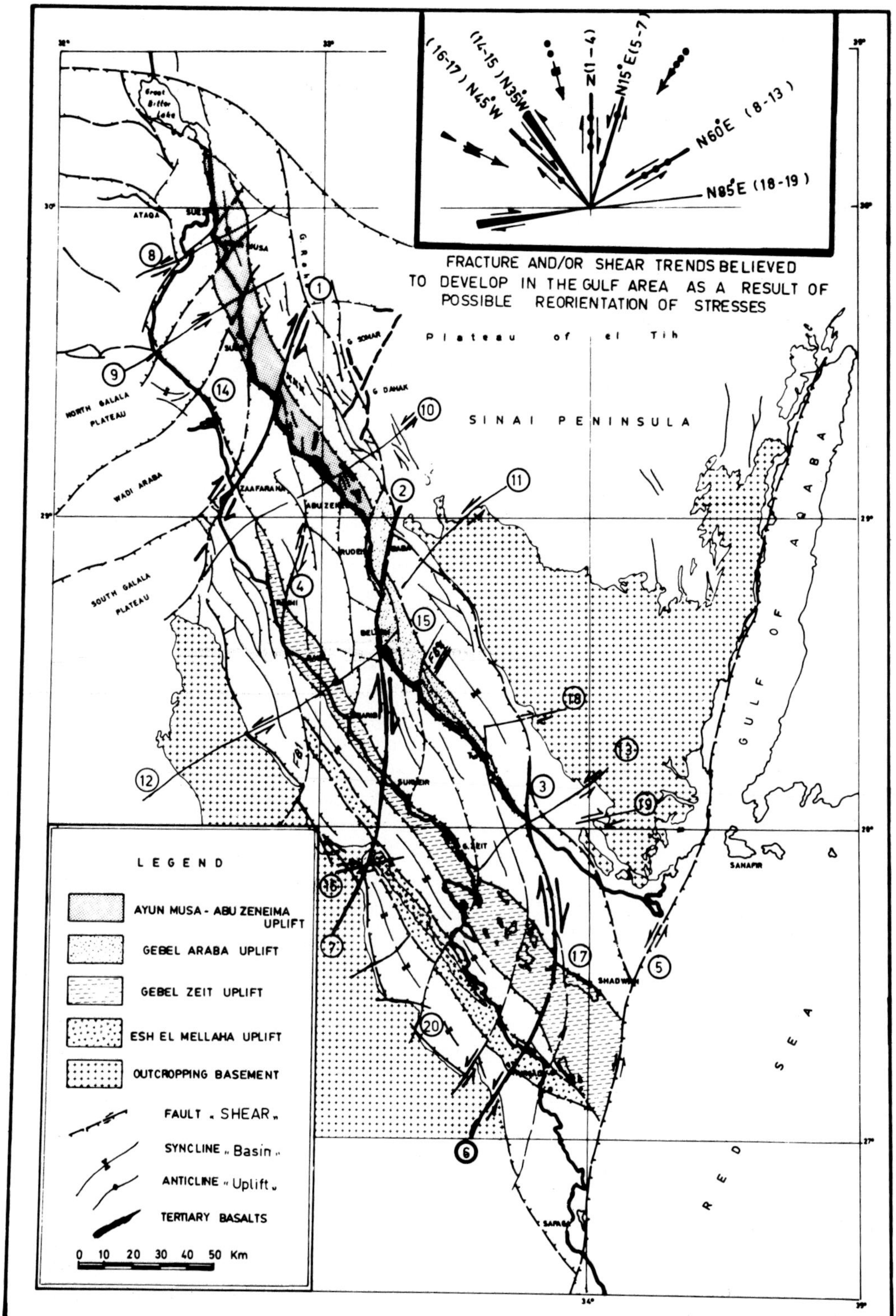

FIGURE 6. The overall structural pattern of the Gulf of Suez, based on all available exploratory data.

intermediate (Q) and least (R) were directed horizontally along NW-SE and NE-SW directions, respectively. Then when the stress difference exceeded the critical limit, probably during the Paleozoic (first phase of movements[29]), displacements did occur along the old NW-SE trending fractures. This, in turn, initiated normal faulting and the forming of the Clysmic or Carboniferous graben, which was subsequently filled (after the marine Paleozoic transgression) by more than 800 m of Carboniferous-Permian sediments.

3. By the early Carboniferous to the early Cretaceous period in the area was subjected to strong lateral compression, together with crustal movements accompanied by another period of marine transgression. This allowed the directed stress system inside the crustal plate of the Arabia-Africa massif to resist deformation and to be again reoriented in such a way as to cause an increase and a decrease in pressure horizontally and along NE-SW and NW-SE directions, respectively. At the same time the vertical component of pressure is assumed to have remained unchanged. This, in turn, forced the stress ellipsoid to be oriented with its (P) axis lying horizontally along N25°-E30° direction and its intermediate axis (Q) lying in a vertical plane; and with the least axis (R) being directed horizontally perpendicular to the direction along which (P) acts. Again, the stress difference exceeded the critical limit, probably during the Mesozoic era. Two sets of vertical shear planes (parallel to Q) were developed; one set striking N60°E and the other striking N-S also the N60°E representing strike-slip faults having left-lateral components of motion; and with the rejuvenated N-S trending fracture constituting their right-lateral conjugates.
4. By the late Mesozoic (early Cretaceous) to the early Cenozoic (early Miocene) era the marked vertical uplifting of the earth's crust, together with the effect of the complicated tectonics that arose from the anticlockwise rotation of Arabia relative to Africa[38,39] that prevailed in the region during this period, affected the area under investigation to the extent that it permitted reorientation of the stresses. The (P) and (R) axes of the stress ellipsoid were directed horizontally along N65°W and N25°E, respectively, and the (Q) axis was directed vertically. A rupture again occurred, probably during the late Cenozoic (post Eocene-Miocene) era, and developed shear planes in the form of left-lateral and right-lateral strike-slip faults, striking N35°W and N85°E, respectively.
5. By the early Miocene epoch the tectonic effects arising from the greatest taphrogeny that had been recorded in the Gulf of Suez region during this period, as well as from couple forces initiated from the clockwise rotation of Africa relative to Arabia[38-40,42] gave way to another reorientation of stresses. This resulted in the development of the Gulf of Aqaba left-lateral fractures trending N15°E and their right-lateral conjugates that strike due N45°W.
6. The Pleistocene era is generally considered as being a period of rejuvenation of possibly all the fracture systems in the Gulf of Suez region.

CONCLUSIONS

The Gulf of Suez region seems to have been subjected to three phases of tectonism associated with five cycles of deformation (Appendix 2).

The first phase seems to have prevailed in the Precambrian to early Mesozoic eras with only three cycles of deformation:

1. The Erythrean or the block faulting cycle that prevailed from the Precambrian to the early Carboniferous period, when the Precambrian basement complex was

deformed by at least four sets of fractures; the most pronounced being the Erythrean (NW-SE) trending ones.

2. The second cycle (Clysmic), prevailing from the early Cambrian to the early Caboniferous period when the Clysmic or Carboniferous graben were initiated along the rejuvenated Precambrian NW-SE trending fractures developed during the first phase of tectonism.
3. The third cycle, known as the Hercynian (early Mediterranean) cycle that prevailed from the late Carboniferous to the early Cretaceous period when a pronounced rejuvenation of the Precambrian E-W trending fracture occurred, but with very little movement along the NW-SE trend. At the same time the initiation of major horizontal movements began along the N-S and N60°E trending fractures; together with the development of the left-lateral N60°E and right-lateral N-S conjugate shears.

The second phase of tectonism seems to have prevailed from the early Mesozoic (early Cretaceous) to the late Cenozoic (Oligocene-early Miocene) era with only one cycle of deformation; namely, the late Mediterranean-late Clysmic cycle. At this time, rejuvenation occurred along the N35°W and E-W fracture zones, with the development of the left-lateral N35°W and the N85°E right-lateral strike-slip conjugates.

The third phase of tectonism seems to have started between the late Cenozoic and the Plio-Pleistocene era with one major cycle of deformation, known as the Aqaba cycle, when the Aqaba left-lateral and the right-lateral N45°W strike-slip conjugates were developed.

The Gulf of Suez cannot be considered in any way as being the conjugate shear of the Gulf of Aqaba, as may authors believe.[13] Instead, the author introduces a simple mechanical model for the structural development of both the Gulf of Suez and the Gulf of Aqaba. This model is presented in Figure 7 and it is believed that it is in good agreement with that introduced by several other authors.[1,3,16,20,23]

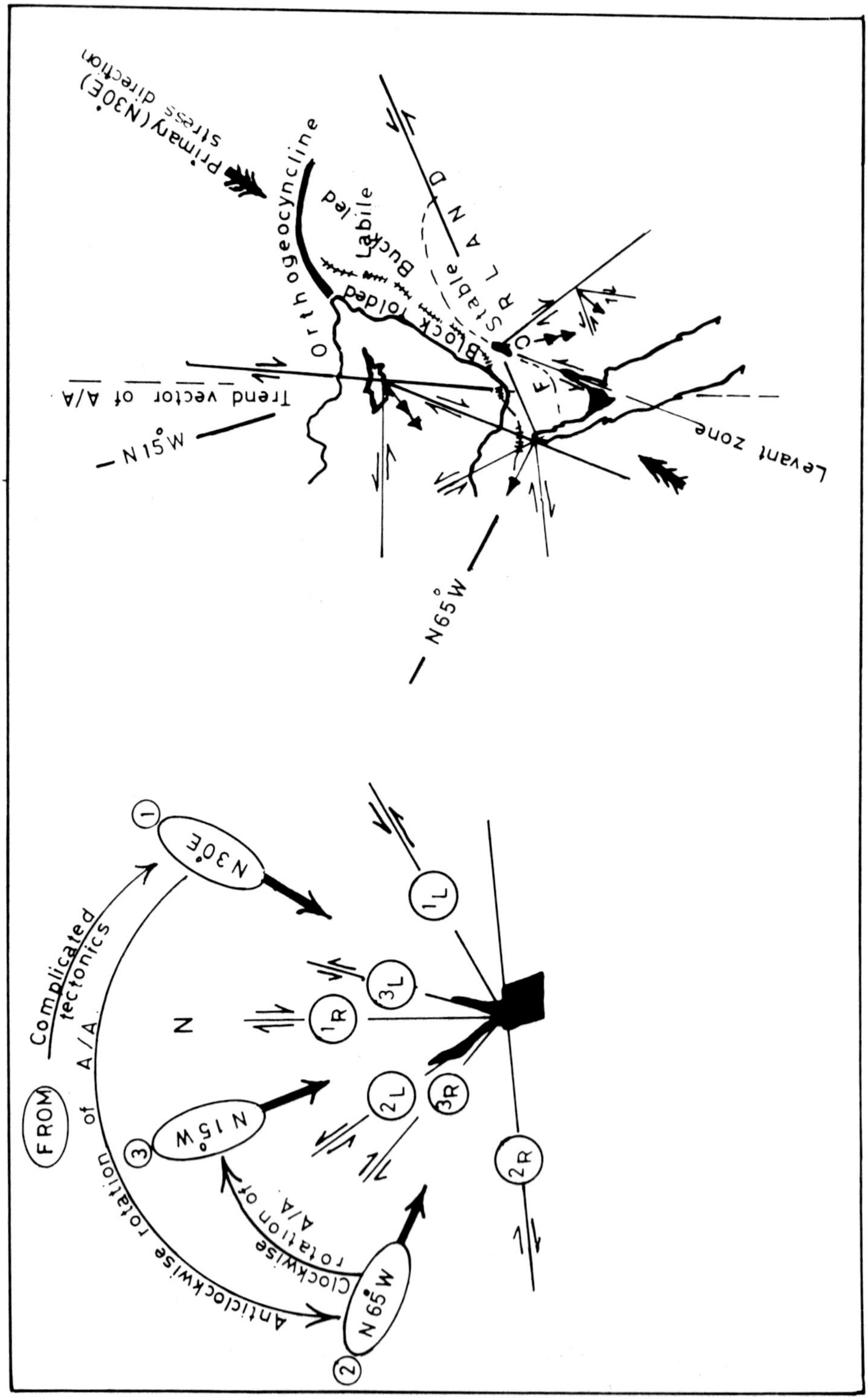

FIGURE 7. Proposed tectonic model of Sinai-Arabia-Africa plates based on detected shear directions.

APPENDIX 1

A COMPOSITE COLUMNAR SECTION IN THE GULF OF SUEZ REGION

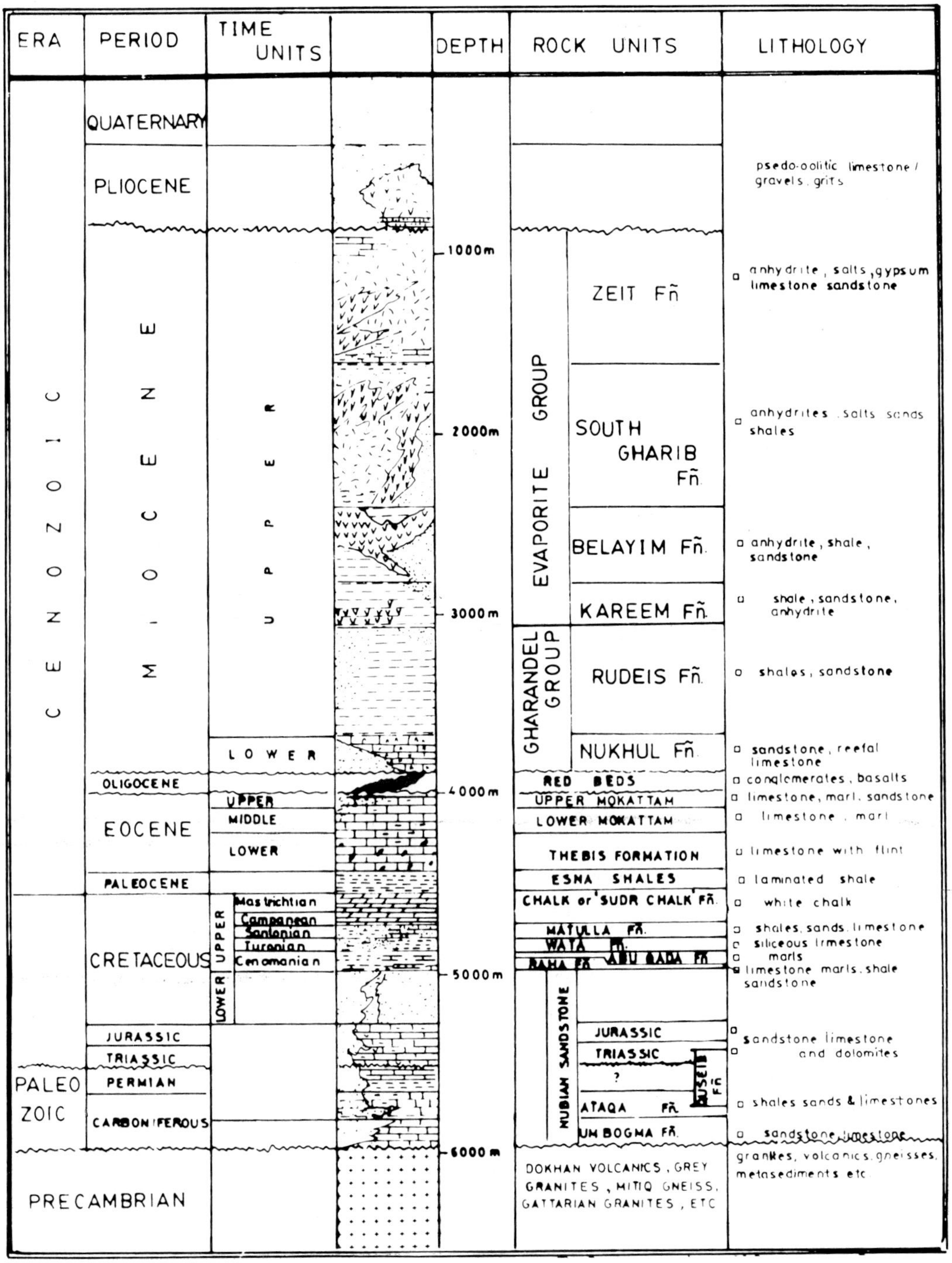

APPENDIX 2

SPECULATIONS ON THE STRUCTURAL DEVELOPMENT OF THE GULF OF SUEZ REGION DURING DIFFERENT GEOLOGIC AGES

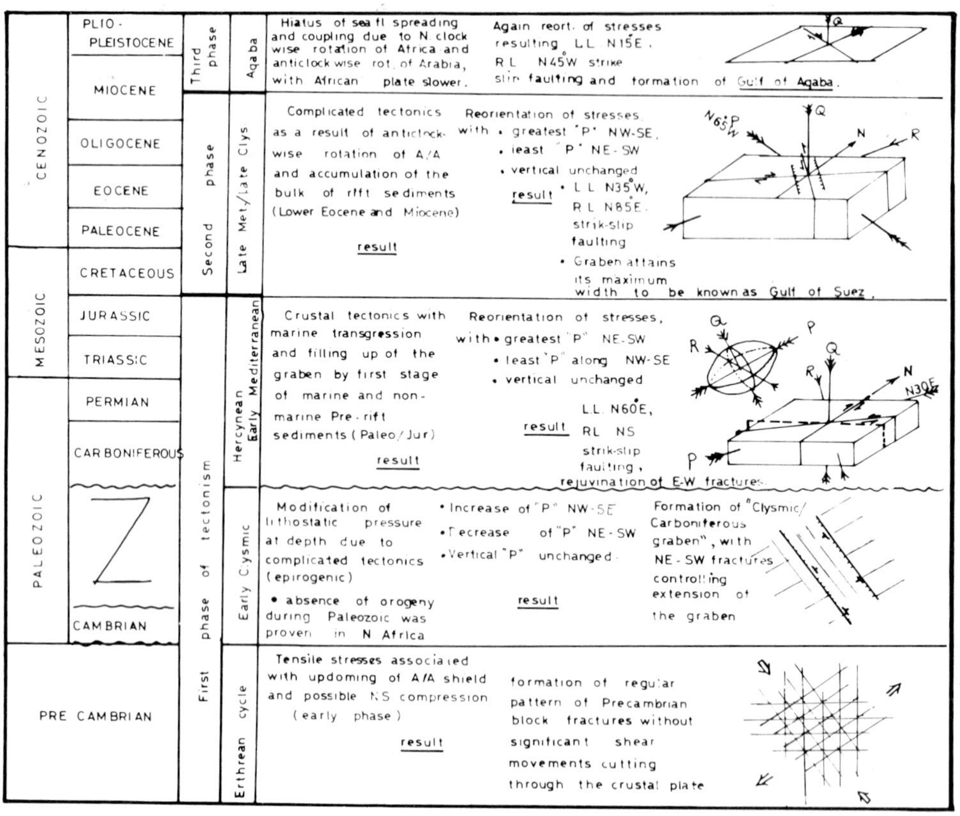

ACKNOWLEDGMENTS

This regional contribution has been derived from the graduate work conducted by F. Abdel Rahman (M.Sc., 1970), A. El Tantawi (M.Sc., 1973), J. Boctor (M.Sc., 1969 and Ph.D., 1978) and N. El Nadi (M.Sc., 1980) under the direct supervision of the author, Dr. A. Bayoumi, Professor of Geophysics, Faculty of Science, Cairo University, Giza, Egypt.

REFERENCES

1. **Ben-Menahem, A., et al.**, Tectonics, seismicity and structure of the Afro-Eurasian Junction — The breaking of an incoherent plate, Phys. of the earth and Planetary Interiors, 12, 1-50, 1976.
2. **Mckenzie, D. P. and Morgan, W. J.**, Evolution of triple junction, *Nature (London)*, 224, 125, 1969.
3. **Freund, R.**, A model of the structural development of Israel and adjacent areas since Cretaceous times, *Geol. Mag.*, 102, 189, 1965.
4. **Cloos, H.**, Hebung-Spaltung-Vulkanismus, *Geol. Rdsch.*, 30, 401, 1939.
5. **Shalem, N.**, The Red Sea and the Erythrean disturbances, C. R. 19th Int. Geol. Congr., Sec. 15, pt. 17, Algiers, 1954, 223.
6. **Swartz, D. H. and Arden, D. D.**, Geologic history of the Red Sea area, *Am. Assoc. Petr. Geol. Bull.*, 44, 1621, 1960.
7. **Knetsch, G.**, Strukturelle Skizze von Aegypten, *Geol. Rdsch.*, 42, 242, 1954.
8. **Shata, A.**, Structural development of the Sinai Peninsula, Egypt, Assoc. Afr. Geol. Survey, Symp. Int. Geo. Congr., Mexico City, 1959, 225.
9. **Vorman, A. J.**, On the Red Sea rift problem, *Bull. Res. Counc. Isr.*, 10G, 321, 1961.
10. **Said, R.**, The Geology of Egypt, Elsevier, New York, 1962.
11. **Picard, L.**, Outline of the tectonics of the earth with special emphasis upon Africa, *Bull. Geol. Dep. Hebrew Univ.*, 2 (No. 3-4), 1, 1939.
12. **Picard, L.**, Thoughts on the grabens system of the Near East, *Geol. Surv. Can. Pap.*, 66(44), 22, 1966.
13. **Youssef, M. I.**, Structural pattern of Egypt and its interpretation, *Am. Assoc. Petrol. Geol.*, 52, 601, 1968.
14. **Whiteman, A. J.**, Formation of the Red Sea Depression, *Geol. Mag.*, 105 (5), 231, 1969.
15. **Abdel-Gawad, M.**, New evidence of transcurrent movements in the Red Sea and petroleum implications, *Am. Assoc. Petrol. Geol.*, 53(7), 1466, 1969.
16. **Abdel-Gawad, M.**, The Gulf of Suez: a brief review of stratigraphy and structure, *Philos. Trans. R. Soc. London, Ser. A*, 267, 41, 1970.
17. **El-Tarabili, A.**, Contribution to the origin of the Red Sea depression, origin of its northern part, 7th. Arab. Petrol. Congr., Pap. 63, Kuwait, 1970.
18. **MacKenzie, D. P. and Davies, D.**, Plate tectonics of the Red Sea and East Africa, *Nature*, 226, 243, 1970.
19. **Le Pichon, X., et al.**, Plate Tectonics, Elsevier, Amsterdam, 1973, 300.
20. **Neev, D.**, Tectonic evolution of the Middle East and the Levantine basin, *Geology*, 3, 683, 1975.
21. **Kenyon, N. H., et al.**, Plan views of active faults and other features of the lower Nile cones, Wormly, Godelming Surv. Inst. Oceanogr. Soc., 1975.
22. **Freund, R., et al.**, The shear along the Dead Sea rift, *Philos. Trans. R. Soc. London, Ser. A*, 267, 1970.
23. **Ben-Manahem, A and Aboodi, E.**, Tectonic pattern in the northern Red Sea region, *J. Geophys. Res.*, 76, 2674, 1971.
24. **El- Nadi, N.**, A geophysical study on the area S.E. Ras Gharib, Gulf of Suez, Egypt, M. Sc. thesis, Fac. Sci., Cairo University, 1980.
25. **El- Tantawi, A.**, On the seismic velocity distribution in the Nile and Gulf of Suez regions for subsurface geological exploration, M. Sc. thesis, Fac. Sci., Cairo University, 1972.
26. **Abdel-Rahman, F.**, A geophysical study on the area between Wadi Dara-Gebel El- Zeit, Gulf of Suez, Egypt for petroleum prospection, M. Sc. thesis, Fac. Sci., Cairo University, 1970.
27. **Boctor, J.**, A geophysical study on Rahmi area, Gulf of Suez, Egypt, M. Sc. thesis, Fac. Sci., Cairo University, 1969.
28. **Shenouda, H.**, A geophysical study on the Cairo-Suez area for the exploration of its subsurface geology, M.Sc. thesis, Fac. Sci., Cairo University, 1970.
29. **Boctor, J.**, A geophysical study on the tectonic framework of the Gulf of Suez region, Ph.D. thesis, Fac. Sci., Cairo University, 1978.
30. **Mckinstry, H. E.**, Shears of the second order, *Am. J. Sci.*, 251, 401, 1953.
31. **Mustafa, A.**, Block faulting in the Gulf of Suez, EGPC 5th Exploration Semin. Bull., Cairo, 1976.
32. **Bayoumi, A. and Boctor, J.**, A contribution to gravity anomalies in the Gulf of Suez, Egypt, 5th Conf. Afr. Geol., Cairo, 1979.
33. **Bayoumi, A. I.**, On the tectonic origin of the Gulf of Suez region, Egypt, as deduced from gravity data, 50th Annu. Meet. SEG, Houston, 1980.
34. **Abdel-Rahman, S. and Bayoumi, A. I.**, Role of rotation in defining thickness variations of incompetent strata in Northern Egypt, submitted for publication.

35. **Bayoumi, A. I.,** A contribution to the Neogene evaporites in the Gulf of Suez basin, Egypt, 51st Annu. Meet. SEG, Los Angeles, 1981.
36. **Schlatter, L. E.,** Geological report on the well Ras Rahmi No. 1 of SOE, GPC Rep., 1951.
37. **Girdler, R. W.,** The relationship of the Red Sea to the East Africa rift system, *Q. J. Soc. London,* 114, 79, 1958.
38. **Girdler, R. W.,** The role of translational and rotational movements in the formation of the Red Sea and the Gulf of Aden, in *The World Rift System,* W.M.C. Symp., Irvine, T-N., Ed.
39. **Girdler, R. W.,** *Drifting and rifting of Africa,* Nature, 217, 1102.
40. **Ewing, M. and Ewing, J.,** Sediment and distribution on the Mid-Ocean ridges with respect to spreading of the sea floor, *Science,* 156, 1580, 1967.
41. **Van der Pleog, P.,** The world's oilfields — the eastern hemisphere, in *The Science of Petroleum,* Vol. 6 (1), Oxford University Press, 1953, 151.
42. **Briden, J. C.,** Recurrent continental drift of Gondwanaland, *Nature,* 215, 1334, 1967.

INDEX

INDEX

A

B

C

D

E

F

G

H

R

T

U

V

W

Z